Nematode-Trapping Fungi

Fungal Diversity Research Series

Volume 23

Series Editor: Kevin D. Hyde

For further volumes:
http://www.springer.com/series/10408

Ke-Qin Zhang • Kevin D. Hyde
Editors

Nematode-Trapping Fungi

Springer

Editors
Ke-Qin Zhang
Laboratory of Conservation and Utilization of Bio-Resources
Yunnan University
Kunming
China

Kevin D. Hyde
Institute of Excellence in Fungal Research, and School of Science
Mae Fah Luang University
Chiang Rai
Thailand

ISSN 1608-6791 ISSN 2212-6236 (electronic)
ISBN 978-94-017-8729-1 ISBN 978-94-017-8730-7 (eBook)
DOI 10.1007/978-94-017-8730-7
Springer Dordrecht Heidelberg New York London

Library of Congress Control Number: 2014935064

Printed on acid-free paper

Springer is part of Springer Science+Business Media (www.springer.com)

Preface

Nematophagous fungi are an important, fascinating group of soil microorganisms that can suppress the populations of plant-parasitic nematodes. They have been studied over several decades and the beautiful micrographs of these fungi trapping nematodes have marveled scientists and students alike. These remarkable fungi are also important potential biocontrol agents of nematodes diseases of plants and animals, they produce a swathe of natural chemicals and are model organisms for gene functional studies. The research team of Zhang Ke-Qin and colleagues has been working on many aspects of nematode-trapping fungi for more than 20 years. These studies have resulted in a wealth of information and it is now considered timely to bring all this information together into a book so that everyone who is interested can read about the amazing fungi.

The book comprises eight chapters, Chapter 1 is an overview of the data on nematode-trapping fungi presented in this book, including their taxonomy, phylogeny and evolution. Four general groups of nematophagous fungi are general accepted and are based on the mechanisms by which they attack nematodes. These consist of (1) nematode-trapping fungi using adhesive or mechanical hyphal traps, (2) endoparasitic fungi using their spores, (3) egg parasitic fungi invading nematode eggs or females with their hyphal tips, and (4) toxin-producing fungi immobilizing nematodes before invasion. The nematode-trapping fungi with adhesive or mechanical hyphal traps are the main focus of this book, while the other types of fungi are discussed but in less detail. Chapter 2 deals with the methodology to study nematode trapping fungi. Although there has been a huge increase in our knowledge of nematophagous fungi, the methodology used to study these organisms are unique and have changed little over the years. These methods are detailed in Chapter 2 and include techniques for obtaining mixed cultures of the fungi from nature, for isolating taxa into pure culture, for observing living material and for making permanent microscope preparations. Chapter 3 deal with history and taxonomy of orbiliaceous nematode-trapping fungi, the filamentous species forming trapping devices to prey nematodes. The taxonomic history of predatory orbiliaceous fungi is reviewed and the system of using trapping devices as the primary morphological criterion for generic delimitation is advocated. Following this taxonomic concept, keys for genera of *Arthrobotrys*, *Drechslerella* and *Dactylellina*, which include all reported species of predatory orbiliaceous fungi are presented. Totally, 54, 14 and 28 species from *Arthrobotrys*, *Drechslerella* and *Dactylellina* respectively, are morphologically described and illustrated. Known asexual-sexual connections (14 pairs) of predatory orbiliaceous fungi are summarized and their taxonomic descriptions and illustrations presented.

Chapter 4 deals with the ecology of nematode–trapping fungi. These fungi have extensively been studied both because of their unique predatory life history and because they are potential biocontrol agents of economically important plant- and animal- parasitic nematodes. Fundamental knowledge of the ecology of these fungi is therefore essential before the value of such biocontrol methods can be assessed. Topics dealt with in Chapter 4 include occurrence and habitats, geographical and seasonal distribution, quantification of abundance, and effects of soil conditions and nematode density on their distribution. Competition between nematodes and nematophagous fungi, the effect of fungistasis and extreme ecological factors such as heavy metals and salinity on these fungi, and their genetic diversity and speciation are included in this chapter.

In Chapter 5 the use of nematode trapping fungi in biological control is discussed. Plant-parasitic nematodes cause severe damage to world agriculture each year. Environmental and health concerns over the use of chemical pesticides has increased the need for alternative measures to control plant-parasitic nematodes. Nematophagous fungi, a natural enemy of nematodes, have received most attention in biological control of plant-parasitic nematodes. This is due to their specific ability to capture and kill nematodes. Nematophagous fungi are divided into four groups according to their mode of action against nematodes, and several fungi such as *Pochonia chlamydosporia* and *Paecilomyces lilacinus* have been developed as commercial biological nematicides. In Chapter 5, important nematode parasitic and antagonistic fungi, and their taxonomy, biology and their mode of action are discussed. Progress in the study of highly virulent fungal strains for nematode biological control, and application of nematode-antagonistic agents are also discussed.

In Chapter 6 we describe the characterization of extracellular enzymes from nematophagous fungi, the expression and regulation of serine protease prC in *Clonostachys rosea*, and the genome and proteomic analyses of the nematode-trapping fungus *Arthrobotrys oligospora*. At present, the detailed molecular pathogenic mechanisms against nematodes by nematophagous fungi have not yet been fully elucidated. However, increasing evidence show that extracellular hydrolytic enzymes including proteases, collagenase, and chitinase may be involved in nematode-cuticle penetration and host-cell digestion. Recently, the crystal structures of proteases (Ver112 and PL646) and chitinase CrChi1 from nematophagous fungi were resolved, which can help us to identify the active site residues and to elucidate the catalytic mechanism of these enzymes involved in infection against hosts. The expression and regulation of protease PrC from *Clonostachys rosea* by different environmental conditions has also been reported. The genome of *Arthrobotrys oligospora* has been sequenced, and a model of nematode trap formation in *A. oligospora* suggested; thus the genome data may serve as a roadmap for further investigations into the interaction between nematode-trapping fungi and their host nematodes, providing broad foundations for research on the biocontrol of pathogenic nematodes.

Nematode-trapping fungi having proven to be a treasure house of novel chemicals. Chapter 7 summarizes more than 200 compounds from fungi that have been shown to possess nematicidal activities. These compounds belong to diverse chemical groups including alkaloid, quinone, isoepoxydon, pyran, furan, peptide, macrolide, terpenoid, fatty acid, diketopiperazine, aphthalene and simple aromatics. They have mainly been isolated from a variety of ascomycetous and basidiomycetous fungal taxa. Their nematicidal activities are described and their potential roles in the biocontrol of nematodes are discussed.

Previous studies based on traditional techniques have revealed much about nematophagous fungi. Modern techniques however, can now help to elucidate the molecular mechanisms underlying infection of nematodes by the nematophagous fungi, and help us to understand virulent factors, the role of proteases, chitinases and small chemical molecules, and the regulation of trap formation. Further studies using modern molecular techniques are needed to better understand these important mechanisms and the overall functioning of nematode trapping fungi and the future studies needed are addressed in Chapter 8.

It has been a privilege to participate in the compilation of this book on nematode-trapping fungi. Although I have barely studied the nematode-trapping fungi through one student, Aung Swe, I believe these fungi to be truly amazing and I have enjoyed editing the writings of the various authors of the world experts at the Yunnan University nematode-trapping fungi laboratory. A huge amount of work on this organisms has come out the laboratories of Zhang Ke-Qin as well as other laboratories and it has been an honour to help put this data together in a modern book on nematode-trapping fungi. I hope the readers will find as much enjoyment in reading the book as I have in editing it.

Chiang Rai
September 2013

Kevin D. Hyde

Abstract

Nematophagous fungi are an important, fascinating group of soil microorganisms that can suppress the populations of plant-parasitic nematodes. They have been studied over several decades and the beautiful micrographs of these fungi trapping nematodes have marvelled scientists and students alike. This book comprises eight chapters providing data on nematode-trapping fungi, particularly those of the Orbiliaceae whose asexual states produce nematode- trapping devices. Chapter 1 provides an overview of nematode-trapping fungi, including their taxonomy, phylogeny and evolution. The nematode-trapping fungi with adhesive or mechanical hyphal traps are the main focus of this book, while the other types of fungi are discussed but in less detail. Chapter 2 deals with the methodology to study nematode trapping fungi. These methods include techniques for obtaining mixed cultures of the fungi from nature, for isolating taxa into pure culture, for observing living material and for making permanent microscope preparations. Chapter 3 deal with history and taxonomy of orbiliaceous nematode-trapping fungi, the filamentous species forming trapping devices to prey nematodes. The taxonomic history of predatory orbiliaceous fungi is reviewed and the system of using trapping devices as the primary morphological criterion for generic delimitation is advocated. Following this taxonomic concept, keys for genera of *Arthrobotrys*, *Drechslerella* and *Dactylellina*, which include all reported species of predatory orbiliaceous fungi are presented. Totally, 54, 14 and 28 species from *Arthrobotrys*, *Drechslerella* and *Dactylellina* respectively, are morphologically described and illustrated. Known asexual-sexual connections (14 pairs) of predatory orbiliaceous fungi are summarized and their taxonomic descriptions and illustrations presented. Chapter 4 deals with the ecology of nematode–trapping fungi. Topics dealt with include occurrence and habitats, geographical and seasonal distribution, quantification of abundance, and effects of soil conditions and nematode density on their distribution. Competition between nematodes and nematophagous fungi, the effect of fungistasis and extreme ecological factors such as heavy metals and salinity on these fungi, and their genetic diversity and speciation are also. In Chapter 5 the use of nematode trapping fungi in biological control is discussed. Nematophagous fungi are divided into four groups according to their mode of action against nematodes, and several fungi such as *Pochonia chlamydosporia* and *Paecilomyces lilacinus* have been developed as commercial biological nematicides. In Chapter 5, important nematode parasitic and antagonistic fungi, and their taxonomy, biology and their mode of action are discussed. Progress in the study of highly virulent fungal strains for nematode biological control, and application of nematode-antagonistic agents are also discussed. In Chapter 6 we describe the characterization of extracellular enzymes from nematophagous fungi, the expression and regulation of serine protease prC in *Clonostachys rosea*, and the genome and proteomic analyses of the nematode-trapping fungus *Arthrobotrys oligospora*. At present, the detailed molecular pathogenic mechanisms against nematodes by nematophagous fungi have not yet been fully elucidated. However, increasing evidence show that extracellular hydrolytic enzymes including proteases, collagenase, and chitinase may be involved in nematode-cuticle penetration and host-cell digestion. The expression and regulation of protease PrC from *Clonostachys rosea* by different environmental conditions has also been reported. The genome of *Arthrobotrys oligospora* has been sequenced, and a model of nematode trap formation in *A. oligospora* suggested; thus the genome data may serve as a roadmap for further investigations into the interaction between nematode-trapping fungi and their host nematodes, providing broad foundations for research on the biocontrol of pathogenic nematodes. Nematode-trapping fungi having proven to be a treasure house of novel chemicals. Chapter 7 summarizes more than 200 compounds from fungi that have been shown to possess nematicidal activities. These compounds belong to diverse chemical groups including alkaloid, quinone, isoepoxydon, pyran, furan, peptide, macrolide, terpenoid, fatty acid, diketopiperazine, aphthalene and simple aromatics. Their nematicidal activities are described and their potential roles in the biocontrol of nematodes are discussed. Previous studies based on traditional techniques have revealed much about nematophagous fungi. Modern techniques however, can now help to elucidate the molecular mechanisms underlying infection of nematodes by the nematophagous fungi, and help us to understand virulent factors, the role of proteases, chitinases and small chemical molecules, and the regulation of trap formation. Further studies using modern molecular techniques are needed to better understand these important mechanisms and the overall functioning of nematode trapping fungi and the future studies needed are addressed in Chapter 8.

Contents

Chapter 1
Nematode-Trapping Fungi

KD Hyde, A Swe and Ke-Qin Zhang

Contents

Abstract This chapter provides an overview of the data on nematode-trapping fungi presented in this book, including their taxonomy, phylogeny and evolution. Four general groups of nematophagous fungi are general accepted and are based on the mechanisms by which they attack nematodes. These consist of (1) nematode-trapping fungi using adhesive or mechanical hyphal traps, (2) endoparasitic fungi using their spores, (3) egg parasitic fungi invading nematode eggs or females with their hyphal tips, and (4) toxin-producing fungi immobilizing nematodes before invasion. The nematode-trapping fungi with adhesive or mechanical

K.-Q. Zhang (✉)
Laboratory for Conservation and Utilization of Bio-Resources, and Key Laboratory of Microbial Diversity in Southwest China, Ministry of Education, Yunnan University, 650091 Kunming, Yunnan, China
e-mail: kqzhang1@ynu.edu.cn

KD Hyde
Institute of Excellence in Fungal Research, Mae Fah Luang University, Chiang Rai, Thailand

School of Science, Mae Fah Luang University, Chiang Rai, Thailand

A Swe
School of Biological Science, University of Hong Kong, Pokfulam, Hong Kong

K.-Q. Zhang, K. D. Hyde (eds.), *Nematode-Trapping Fungi,* Fungal Diversity Research Series 23, DOI 10.1007/978-94-017-8730-7_1, © Mushroom Research Foundation 2014

hyphal traps are the main focus of this book. Fossil nematode-trapping fungi, and the biodiversity, ecology and geographical distribution nematode-trapping fungi, including factors affecting their distribution are also addressed. Terrestrial, freshwater and marine habitats are home to nematode-trapping fungi, but they have rarely been reported from extreme environments. Fungal-nematodes interactions are also discussed.

Keywords Ascomycota • Asexual fungi • Biocontrol • Biodiversity • Fossil fungi • Nematodes

Introduction

Four general groups of nematophagous fungi are categorized based on the mechanisms by which they attack nematodes (Liu et al. 2009); (1) nematode-trapping fungi using adhesive or mechanical hyphal traps, (2) endoparasitic fungi using their spores, (3) egg parasitic fungi invading nematode eggs or females with their hyphal tips, and (4) toxin producing fungi immobilizing nematodes before invasion (Kendrick et al. 2001; Liu et al. 2009). The first nematode-trapping fungus, *Arthrobotrys superba* Corda was described in 1839, but its predatory habit was unknown at that time. The predatory habit was first observed 36 years later by Zopf (1888) and the link to *Orbilia* and its nematode-trapping life style confirmed by Pfister (1994). Biological control of nematodes by nematode-trapping fungi has considerable potential and therefore there have been extensive studies and reviews on their taxonomy, phylogeny, biology, and ecology (Cooke 1963; Kerry 1987; Sayre and Walter 1991; Sikora 1992; Morton et al. 2004; Dong and Zhang 2006). In this chapter, we discuss the taxonomy, phylogeny, diversity, and ecology of nematode-trapping fungi and their potential in use in biocontrol of nematodes. Each of these topics are introduced and then expanded in the subsequent chapters.

Methods to Study Nematode-Trapping Fungi

The methods for studying nematode-trapping fungi are generally distinct from those used to study most other fungi. This is because they do not form aggregated mycelial structures that can easily be seen and the fact that they develop in living nematodes. Techniques, which more or less depart from traditional mycological methods are therefore used to study, isolate and characterize nematophagous fungi. The unique techniques involve the recovery, isolation and maintenance of nematophagous fungi and are detailed in Chapter 2.

Taxonomy and Phylogeny of Nematode-Trapping Fungi

The discovery of predacious activity in *Arthrobotrys oligospora* by Zopf (1888), and link with *Orbilia* by Pfister (1994) have resulted in nematode trapping fungi attracting much interest amongst mycologists. Nematode-trapping fungi are a heterogeneous group of asexual ascomycetes with species previously defined primarily on conidial characteristics such as size, septation, and type of conidiogenous cells (Subramanian 1963) and more recently on their type of trapping devices (Scholler et al. 1999; Li et al. 2005; Yang and Liu 2006; Yang et al. 2007a). This has resulted in considerable changes in classification of these organisms over time. Many generic names have been given to nematode-trapping fungi, but the basis on which species were circumscribed to different genera has been unclear and subjective. We therefore provide a account of the history of nematode-trapping fungi in Chapter 3 to clarify the reasons for the various opinions and name changes.

Arthrobotrys was established by Corda (1839), *A. superba* Corda as the type species. The discovery of predacious activity was first noted for *Arthrobotrys oligospora* by Zopf (1888), and the link with *Orbilia* established by Pfister (1994). *Arthrobotrys* characteristics were considered as hyaline conidiophores which produce conidia asynchronously on short denticles at swollen conidiogenous heads or clusters of pronounced denticles; conidia being subhyaline, obovoidal or clavate and (0–)1(–6)-septate (Pfister 1994). Trapping devices were considered as constricting rings, adhesive nets, hyphae or adhesive knobs (Pfister 1994). This wide generic concept resulted in considerable confusion, although since 1930, nematode-trapping fungal have been mostly described and classified in *Arthrobotrys*, *Dactylaria*, and *Dactylella*, and some other genera.

Morphology-based classification based on conidia was demonstrated to be inadequate in reflecting natural relationships among the nematode-trapping fungi. The first notable phylogeny study on nematode-trapping fungi was that of Rubner (1996). She used molecular data to rationalize the classification of nematode-trapping fungi based on types of trapping devices. Later, phylogenetic studies based on rDNA sequence analysis also found that trapping devices are more informative than other morphological characters in delimiting genera (Liou and Tzean 1997; Pfister 1997; Ahrén et al. 1998; Scholler et al. 1999; Ahrén and Tunlid 2003; Kano et al. 2004; Li et al. 2005; Yang and Liu 2006; Yang et al. 2007a; Liu et al. 2009). For example, Ahrén et al. (1998) revealed nematode-trapping fungi grouping in three lineages based on different types of trapping devices. Scholler et al. (1999) proposed a new genus concept for predatory asexual *Orbiliaceae*. In this classification, phylogenetic significance was given to trapping devices. Scholler et al. (1999) classified nematode-trapping fungi into four genera using 18S and ITS rDNA analysis. Li et al. (2005) re-evaluated the placement of nematode-trapping genera based 28S, 5.8S and β tubulin analysis and the establishment of *Gamsylella* proposed by Scholler et al. (1999) was criticized and not accepted. Because of this confusion we provide a phylogenetic based classification of the nematode-trapping fungi in Chapter 3 where we placed the fungi into three genera; *Arthrobotrys* with 54 species, *Drechslerella* with 14 species and *Dactylellina* with 28 species. In 2012, the International Code of Nomenclature for Algae, Fungi and Plants was revised and no longer allows a single fungal species (or genus) to have two names, with the older name taking priority except in exceptional circumstances (Hawksworth 2012). The above genera are linked with a sexual *Orbilia* morphs and may need to take the *Orbilia* name, or alternatively the above genera may be maintained. Discussions are presently underway by the *Orbilomycetes working group* of the *International Commission on the Taxonomy of Fungi* (http://www.fungaltaxonomy.org/subcommissions) to establish which is the best name (or names) to adopt for this group(s) of fungi.

Li et al. (2005) and Yang et al. (2007a) have studied the evolution of nematode-trapping devices. Phylogenetic analysis of nucleotide sequences demonstrated that early trapping structures evolved along two lines, yielding two distinct trapping mechanisms. One lineage developed into constricting rings and the other into adhesive traps. The adhesive network separated early. The adhesive knob evolved through stalk elongation, with a final development of nonconstricting rings (Yang et al. 2007a). Li et al (2005), on the other hand, showed a conflicting result. This aspect is detailed more in Chapter 3.

Ancient Nematode-Trapping Fungi

The earliest record of a nematode-trapping fungus is that of *Palaeoanellus dimorphus* which lived approximately 100 million years ago in a limnetic-terrestrial microhabitat (Schmidt et al. 2008; Swe et al. 2011). The fossil with unicellular hyphal rings as trapping devices and blastospores which formed from which a yeast stage probably represents an asexual ascomycete (Swe et al. 2011). Because predatory fungi with yeast stages are not known from modern ecosystems, Schmidt et al. (2008) assumed the fungus was not related to any recent nematode-trapping fungi and is probably an extinct lineage. Alternatively, this strange species may yet be found in modern settings.

Biodiversity of Nematode-Trapping Fungi

There have been several publications debating estimates of fungal species (Hammond 1992; Cannon 1997; Huhndorf and Lodge 1997; Hyde 2001; Hyde et al. 2007), however the 1.5 million global species of fungi estimated by Hawksworth (1991) has generally been accepted as a working figure for fungal numbers by many mycologists. Nevertheless, however many fungi exist in nature, there are few (ca 100,000) that have been described. Most of the undescribed fungi are microfungi and they probably occur in poorly investigated areas and less explored niches, substrates, hosts and habitats (Hyde 2001). There has been a relatively large numbers of studies on fungal diversity, however, there are relatively few diversity studies of nematode-trapping fungi (Hao et al. 2005; Mo et al. 2006, 2008; Saxena 2008). In Chapter 4 of this book we therefore provide an account of the biodiversity and ecology of nematode-trapping fungi.

Ecology, Occurrence and Geographical Distribution

Numerous surveys on nematophagous fungi have shown that they are present throughout the world and in all types of climate and habitats (Duddington 1951, 1954; Gray 1987; Sunder and Lysek 1988; Boag and Lopez-Llorca 1989; Saxena and Mukerji 1991; Dackman et al. 1992; Liu et al. 1992; Saxena and Lysek 1993; Persmark and Jansson 1997; Hao et al. 2005). The sexual state of nematode-trapping fungi, *Orbilia*, occurs on decaying wood from terrestrial to freshwater habitats (Pfister 1994; Webster et al. 1998; Liu et al. 2006; Zhang et al. 2007), and the asexual states commonly occur in terrestrial, freshwater and marine habitats (Hao et al. 2004; Li et al. 2006; Swe et al. 2008a, b, 2009), but rarely occur in extreme environments (Onofri and Tosi 1992). Because of their potential in biological control, there have been several studies on nematode-trapping fungi, however most of these have concentrated on agriculture and animal husbandry or forestry (Kerry and Hominick 2002; Ahrén and Tunlid 2003; Jaffee and Strong 2005; Dong and Zhang 2006; Su et al. 2007) or freshwater environments (Maslen 1982; Hao et al. 2005). Numerous fungal-animal associations have been reported from aquatic habitats. The first report of marine predacious fungi was three zoophagous forms discovered in brackish water (Jones 1958). Currently, more than 50 species of predacious hyphomycetes have been recorded from aquatic habitats (Ingold 1944; Peach 1950, 1952, Johnson and Autery 1961; Anastasiou 1964; Hao et al. 2004, 2005). *Arthrobotrys dactyloides* Drechsler was the first species of nematode-trapping fungi to be reported from brackish water (Johnson and Autery 1961), while Swe et al. (2008a, b, 2009) recorded several species from mangroves.

Factors Affecting the Distribution of Nematode-Trapping Fungi

The distribution and occurrence of nematode-trapping species and groups of fungi is associated with specific soil variables in particular pH, moisture, nutrients (N, P, K), heavy metal and nematode density (Gray 1985; Sánchez Moreno et al. 2008; Mo et al. 2008). Gray (1988) revealed that soil nutrients such as N, P and K were all positively correlated with nematode density. Species with stalked knobbed trapping devices (*Dactylellina*) and those species with constricting rings (*Drechslerella*) were isolated more readily from richer soils which contained a greater density of nematodes. However, net-forming species (*Arthrobotrys*) are largely independent of soil fertility, especially low K (Burges and Raw 1967). Interestingly, Mo et al. (2008) found that diversity of nematode-trapping was positively correlated with lead concentration. These soil variables are known to vary with depth, as are the densities of soil bacteria, fungi and nematodes (Mankau and McKenny 1976; McSorley et al. 2006) and a high level of nematode trapping activity have been recorded from the rhizosphere area (Peterson and Katznelson 1964; Persmark and Jansson 1997; McSorley et al. 2006). However, there are large variations depending on plant and soil types (Jansson and Lopez-Llorca 2001). The species of nematode trapping fungi also vary with depth (Gray and Bailey 1985). Peterson and Katznelson (1964) revealed that the greatest diversity occurred in the upper 10–30 cm of soils, and this was a positive correlation between the population density of nematophagous fungi and root-knot nematodes in peanut fields. Gray and Bailey (1985) have examined the vertical distribution of nematophagous fungi in soil cores collected from a deciduous woodland, predators forming constricting rings, adhesive branches and adhesive knobs are restricted to the upper litter and humus layer, while the net-forming predators and endoparasites were isolated at all depths, although they are significantly more abundant in the lower mineral-rich soils. In contrast, predators able to form traps spontaneously are restricted to the organic soils of the hemiedaphic zone which are rich in nematodes. Nematophagous fungi are small enough to be affected by micro-climates within the soil (Gray 1985). Hao et al. (2005) observed that the nematode-trapping were not detected deeper than four meter in a freshwater pond.

The effect of major biotic and abiotic variables such as soil moisture, organic matter, pH, nematode density, soil nutrients (Gray 1987) and submerged water condition on the distribution of nematode-trapping fungi has been extensively studied. The diversity was highest at the depth of 20 cm and no nematode trapping fungi were found at the depth of 4 m (Hao et al. 2005). Heavy metals concentrations affected distribution of nematode trapping fungi and was not negatively affected by Pb concentration (Mo et al. 2006, 2008). Gray (1987) pointed out that the endoparasitic nematophagous fungi are obligate parasites, and unlike the predatory fungi, they appear to be unable to live as saprotrophs in the soil. He also suggested that non-specific method of attraction may rely on a greater density of soil nematodes to ensure infection, as compared with parasites which produce adhesive conidia (Gray 1987).

Fungal–Nematode Interactions

Host Recognition, Adhesion, Host Specificity and Infection Process

Nematophagous fungi are an important group of soil microorganisms that can suppress the populations of plant and animal parasitic nematodes. They can be grouped into three categories according to their mode of infestation: nematode-trapping, endoparasitic, and toxic compound producing (Hertz et al. 2002). The pathogenic mechanisms during the infestation process are diverse. They can be mechanical through the production of specialized capturing devices, or through production of toxins that kill nematodes. During infection, a variety of virulence factors may be involved against nematodes by nematophagous fungi. The infection processes and host range of nematophagous fungi have been studied using various techniques such as light and low temperature electron microscopy and bioassays and have been supported by biochemical, physiological, immunological and molecular techniques (Thorn and Barron 1984; Murray and Wharton 1990; Jansson et al. 2000). The ultrastructure of the nematode-trapping devices has been extensively studied (Heintz and Pramer 1972; Nordbring-Hertz and Stalhammar-Carlemalm 1978; Dijksterhuis et al. 1994). The mode of infection by nematophagous fungi has been reviewed by Yang et al. (2007c). Research on attraction of nematodes to fungi has focused on the host-finding behavior of fungal-feeding nematodes (Bordallo et al. 2002; Wang et al. 2010). Nematophagous fungi are attracted to plant and animal parasitic nematodes and microbivorous nematodes (Balan and Gerber 1972; Jansson and Nordbring-Hertz 1979, 1980). Zuckerman and Jansson (1984) review nematode chemotaxis and possible mechanisms of host/prey recognition. These topics are updated and reviewed in Chapters 6 & 7.

Extracellular Enzymes Involved in Nematode Infestation Process

During the infection process, the cuticle must be penetrated, the nematode is immobilized, and the prey is finally invaded and digested. This sequence of infection seems to be present in most nematophagous fungi, but the molecular mechanisms are not well-understood (Lopez-Llorca and Duncan 1988; Dackman et al. 1989). However, several nematophagous fungi have been reported to produce nematotoxins that immobilize or kill nematodes, and ultrastructural and histochemical studies suggest that the penetration of the nematode cuticle involves the activity of hydrolytic enzymes (Schenck et al. 1980). Much data has recently come to light concerning the enzymes involved in the infection process and this topic is reviewed in detail in Chapter 6.

Biological Control of Nematodes

Nematode-trapping fungi have long been considered promising biological agents for control of plant-parasitic nematodes (Duponnois et al. 2001; Sorbo et al. 2003; Singh et al. 2007) and animal parasitic nematodes (Paraud et al. 2007, Carvalho et al. 2009). This is an important area where basic research is utilized in an applied setting and has important possibilities for the control of nematode disease. For further detailed information on this topic see Chapter 7.

Acknowledgements Aung Swe would like to thank the University of Hong Kong for the award of a postgraduate scholarship to study nematode trapping fungi.

References

Ahrén, D., & Tunlid, A. (2003). Evolution of parasitism in nematode-trapping fungi. *Journal of Nematology, 35,* 194–197.

Ahrén, D., Ursing, B. M., & Tunlid, A. (1998). Phylogeny of nematode-trapping fungi based on 18S rDNA sequences. *FEMS Microbiology Letters, 158,* 179–184.

Anastasiou, C. J. (1964). Some aquatic fungi imperfecti from Hawaii. *Pacific Science, 18,* 202–206.

Balan, J., & Gerber, N. N. (1972). Attraction and killing of the nematode *Panagrellus redivivus* by the predacious fungus *Arthrobotrys dactyloides*. *Nematologica, 18,* 163–173.

Boag, B., & Lopez-Llorca, L. V. (1989). Nematodes and nematophagous fungi associated with cereal fields and permanent pasture in Eastern Scotland. *Crop Research, 29,* 1–10.

Bordallo, J. J., Lopez-Llorca, L. V., Jansson, H. B., Salinas, J., Persmark, L., & Asensio, L. (2002). Colonization of plant roots by egg-parasitic and nematode-trapping fungi. *New Phytologist, 154,* 491–499.

Burges, A., & Raw, F. (1967). *Soil Biology*. London: Academic Press.

Cannon, P. F. (1997). Diversity of the *Phyllachoraceae* with special reference to the tropics. In K. D. Hyde (Ed.), *Biodiversity of tropical microfungi* (pp. 255–278). Hong Kong: Hong Kong University Press.

Carvalho, R. O., Araújo, J. V., Braga, F. R., Ferreira, S. R., Araujo, J. M., Silva, A. R., Frassy, L. N., & Alves, C. D. F. (2009). Biological control of Ancylostomosis in dogs using the nematode-trapping fungus *Monacrosporium thaumasium* in southeastern Brazil. *Veterinary Parasitology, 165,* 179–183.

Cooke, R. C. (1963). Ecological characteristics of nematode-trapping hyphomycetes I. Preliminary studies. *Annals of Applied Biology, 52,* 431–437.

Corda, A. C. I. (1839). Pracht-Flora Europaïscher Schimmelbildungen (i–viii, 55 pp, 2 Tabs.). Leipzig: Gerhard Fleischer.

Dackman, C., Chet, I., & Nordbring-Hertz, B. (1989). Fungal parasitism of the cyst nematode Heterodera schachtii: Infection and enzymatic activity. *FEMS Microbiology Letters, 62,* 201–208.

Dackman, C., Jansson, H. B., & Nordbring-Hertz, B. (1992). Nematophagous fungi and their activities in soil. In G. Stotzky & J. M. Bollag (Eds.) *Soil biochemistry* (pp. 95–103). New York: Marcel Dekker.

Dijksterhuis, J., Veenhuis, M., Harder, W., & Nordbring-Hertz, B. (1994). Nematophagous fungi: Physiological aspects and structure-function relationships. In: *Advances in microbial physiology* (Vol. 36, pp. 111–143) Academic Press.

Dong, L. Q., & Zhang, K. Q. (2006). Microbial control of plant-parasitic nematodes: A five-party interaction. *Plant and Soil, 288,* 31–45.

Duddington, C. L. (1951). The ecology of predaceous fungi. I. Preliminary survey. *Transactions of the British Mycological Society, 34,* 322–331.

Duddington, C. L. (1954). Nematode-destroying fungi in agricultural soils. *Nature, 173,* 500–501.

Duponnois, R., Chotte, J. L., Sall, S., & Cadet, P. (2001). The effects of organic amendments on the interactions between a nematophagous fungus *Arthrobotrys oligospora* and the root-knot nematode *Meloidogyne mayaguensis* parasitizing tomato plants. *Biology and Fertility of Soils, 34,* 1–6.

Gray, N. F. (1985). Ecology of nematophagous fungi: Effect of soil moisture, organic matter, pH, and nematode density in distribution. *Soil Biology and Biochemistry, 17,* 499–507.
Gray, N. F. (1987). Nematophagous fungi with particular reference to their ecology. *Biological Reviews, 62,* 245–304.
Gray, N. F. (1988). Ecology of nematophagous fungi: Effect of the soil nutrients N, P and K, and seven major metals on distribution. *Plant and Soil, 108,* 286–290.
Gray, N. F., & Bailey, F. (1985). Ecology of nematophagous fungi: Vertical distribution in a deciduous woodland. *Plant and Soil, 86,* 217–223.
Hammond, P. M. (1992). Species inventory. In B. Groombridge (Ed.), *Global biodiversity: Status of the Earth's living resources* (pp. 17–39). London: Chapman and Hall.
Hao, Y., Luo, J., & Zhang, K. Q. (2004). A new aquatic nematode-trapping hyphomycete. *Mycotaxon, 89,* 235–239.
Hao, Y., Mo, M., Su, H., & Zhang, K. Q. (2005). Ecology of aquatic nematode-trapping hyphomycetes in southwestern China. *Aquatic Microbial Ecology, 40,* 175–181.
Hawksworth, D. L. (1991). The fungal dimension of biodiversity: Magnitude, significance, and conservation. *Mycological Research, 95,* 641–655.
Hawksworth, D. L. (2012). Managing and coping with names of pleomorphic fungi in a period of transition. *Mycosphere, 3*(2), 143–155.
Heintz, C. E., & Pramer, D. (1972). Ultrastructure of nematode-trapping fungi. *Journal of Bacteriology, 110,* 1163.
Huhndorf, S. M., & Lodge, D. J. (1997). Host specificity among wood-inhabiting pyrenomycetes (Fungi, Ascomycetes) in a wet tropical forest in Puerto Rico. *Tropical Ecology, 38,* 307–315.
Hyde, K. D. (2001). Where are the missing fungi? Does Hong Kong have any answers? *Mycological Research, 105,* 1514–1518.
Hyde, K. D., Bussaban, B., Paulus, B., Crous, P. W., Lee, S., Mckenzie, E. H. C., Photita, W., & Lumyong, S. (2007). Diversity of saprobic microfungi. *Biodiversity and Conservation, 16,* 7–35.
Ingold, C. T. (1944). Some new aquatic hyphomycetes. *Transactions of the British Mycological Society, 27,* 45–46.
Jaffee, B. A., & Strong, D. R. (2005). Strong bottom-up and weak top-down effects in soil: Nematode-parasitized insects and nematode-trapping fungi. *Soil Biology and Biochemistry, 37,* 1011–1021.
Jansson, H. B., & Lopez-Llorca, L. V. (2001). Biology of Nematophagous fungi. In J. K. Misra & B. W. Horn (Eds.) *Trichomycetes and other fungal groups* (pp. 145–172). Plymouth: Science Publishers.
Jansson, H. B., & Nordbring-Hertz, B. (1979). Attraction of nematodes to living mycelium of nematophagous fungi. *Journal of General Microbiology, 112,* 89–93.
Jansson, H. B., & Nordbring-Hertz, B. (1980). Interactions between nematophagous fungi and plant-parasitic nematodes: Attraction, induction of trap formation and capture. *Nematologica, 26,* 383–389.
Jansson, H. B., Persson, C., & Odeslius, R. (2000). Growth and capture activities of nematophagous fungi in soil visualized by low temperature scanning electron microscopy. *Mycologia, 92,* 10–15.
Johnson, T. W., & Autery, C. L. (1961). An *Arthrobotrys* from brackish water. *Mycologia, 53,* 432–433.
Jones, F. R. (1958). Three zoopagales from brackish water. *Nature, 181,* 575–576.
Kano, S., Aimi, T., Masumoto, S., Kitamoto, Y., & Morinaga, T. (2004). Physiology and molecular characteristics of a pine wilt nematode-trapping fungus, *Monacrosporium megalosporum. Current Microbiology, 49,* 158–164.
Kendrick, B. (2001). *The fifth kingdom* (3rd ed). Canada: Mycologue Publications.
Kerry, B. R. (1987). Biological control. In R. H. Brown & B. R. Kerry (Eds.) *Principles and practice of nematode control in crops* (pp. 233–263). New York: Academic Press.
Kerry, B. R., & Hominick, W. B. (2002). Biological control. In D. L. Lee (Ed.), *The biology of Nematodes* (pp. 483–510). London: Taylor and Francis.

Li, Y., Hyde, K. D., Jeewon, R., Lei, C., Vijaykrishna, D., & Zhang, K. Q. (2005). Phylogenetics and evolution of nematode-trapping fungi (*Orbiliales*) estimated from nuclear and protein coding genes. *Mycologia, 97,* 1034–1046.

Li, Y., Jeewon, R., Hyde, K. D., Mo, M. H., & Zhang, K. Q. (2006). Two new species of nematode-trapping fungi: Relationships inferred from morphology, rDNA and protein gene sequence analyses. *Mycological Research, 110,* 790–800.

Liou, G. Y., & Tzean, S. S. (1997). Phylogeny of the genus *Arthrobotrys* and allied nematode-trapping fungi based on rDNA sequences. *Mycologia, 89,* 876–884.

Liu, X., Ding, L., Wu, X., & Shen, C. (1992). Nemotophagous fungi in China. *Mycosystema, 5,* 117–126.

Liu, B., Liu, X. Z., Zhuang, W. Y., & Baral, H. O. (2006). Orbiliaceous fungi from Tibet, China. *Fungal Diversity, 22,* 107–120.

Liu, X., Xiang, M., & Che, Y. (2009). The living strategy of nematophagous fungi. *Mycoscience, 50*(1), 20–25.

Lopez-Llorca, L. V., & Duncan, G. H. (1988). A study of fungal endoparasitism of the cereal cyst nematode (*Heterodera avenae*) by scanning electron microscopy. *Canadian Journal of Microbiology, 34,* 613–619.

Mankau, R., & McKenny, M. V. (1976). Spatial distribution of nematophagous fungi associated with *Meloidogyne incognita* on peach. *Journal of Nematology, 8,* 294–295.

Maslen, N. R. (1982). An unidentified nematode-trapping fungus from a pond on Alexander Island. *British Antarctic Survey Bulletin, 51,* 285–287.

McSorley, R., Wang, K. H., Kokalis-Burelle, N., & Church, G. (2006). Effects of soil type and steam on nematode biological control potential of the rhizosphere community. *Nematropica, 36,* 197–214.

Mo, M. H., Chen, W. M., Su, H. Y., Zhang, K. Q., Duan, C. Q., & He, D. M. (2006). Heavy metal tolerance of nematode-trapping fungi in lead-polluted soils. *Applied Soil Ecology, 31,* 11–19.

Mo, M. H., Chen, W. M., Yang, H. R., & Zhang, K. Q. (2008). Diversity and metal tolerance of nematode-trapping fungi in Pb-polluted soils. *Journal of Microbiology* (*Seoul, Korea*)*, 46,* 16–22.

Morton, C. O., Hirsch, P. R., & Kerry, B. R. (2004). Infection of plant-parasitic nematodes by nematophagous fungi—A review of the application of molecular biology to understand infection processes and to improve biological control. *Nematology, 6,* 161–170.

Murray, D. S., & Wharton, D. A. (1990). Capture and penetration processes of the free-living juveniles of *Trichostrongylus colubriformis* (Nematoda) by the nematophagous fungus, *Arthrobotrys oligospora*. *Parasitology, 101,* 93–100.

Nordbring-Hertz, B., Jansson, H. B. & Tunlid, A. (2002). Nematophagous Fungi: *Encyclopedia of Life Sciences*. Macmillan Publishers Ltd., London.

Nordbring-Hertz, B., & Stalhammar-Carlemalm, M. (1978). Capture of nematodes by *Arthrobotrys oligospora*, an electron microscopic study. *Canadian Journal of Botany, 56,* 1297–1307.

Onofri, S., & Tosi, S. (1992). *Arthrobotrys ferox* sp. nov. a springtail capturing hyphomycete from Continental Antarctica. *Mycotaxon, 44,* 445–451.

Paraud, C., Pors, I., & Chartier, C. (2007). Efficiency of feeding *Duddingtonia flagrans* chlamydospores to control nematode parasites of first-season grazing goats in France. *Veterinary Research Communications, 31,* 305–315.

Peach, M. (1950). Aquatic predacious fungi I. *Transactions of the British Mycological Society, 33,* 148–153.

Peach, M. (1952). Aquatic predacious fungi. II. *Transactions of the British Mycological Society, 35,* 19–23.

Persmark, L., & Jansson, H. B. (1997). Nematophagous fungi in the rhizosphere of agricultural crops. *FEMS Microbiology Ecology, 22,* 303–312.

Peterson, E. A., & Katznelson, H. (1964). Occurrence of nematode-trapping fungi in the rhizosphere. *Nature, 204,* 1111–1112.

Pfister, D. H. (1994). *Orbilia fimicola*, a nematophagous discomycete and its *Arthrobotrys* anamorph. *Mycologia, 86,* 451–453.

Pfister, D. H. (1997). Castor, pollux and life histories of fungi. *Mycologia, 89,* 1–23.
Rubner, A. (1996). Revision of predacious hyphomycetes in the *Dactylella-Monacrosporium* complex. *Studies in Mycology, 39,* 1–134.
Sánchez-Moreno, S., Smukler, S., Ferris, H., O'Geen, A., & Jackson, L. (2008). Nematode diversity, food web condition, and chemical and physical properties in different soil habitats of an organic farm. *Biology and Fertility of Soils, 44*(5), 727–744.
Saxena, G. (2008). Observations on the occurrence of nematophagous fungi in Scotland. *Applied Soil Ecology, 39,* 352–357.
Saxena, G., & Lysek, G. (1993). Observation of nematophagous fungi in natural soils by fluorescence microscopy and their correlation with isolation. *Mycological Research, 97,* 1005–1011.
Saxena, G., & Mukerji, K. G. (1991). Distribution of nematophagous fungi in Varanasi, India. *Nova Hedwigia, 52,* 487–495.
Sayre, R. M., & Walter, D. E. (1991). Factors affecting the efficacy of natural enemies of nematodes. *Annual Review of Phytopathology, 29,* 149–166.
Schenck, S., Chase, T., Rosenzweig, W. D., & Pramer, D. (1980). Collagenase production by nematode-trapping fungi. *Applied and Environmental Microbiology, 40,* 567–570.
Schmidt, A. R., Dörfelt, H., & Perrichot, V. (2008). *Palaeoanellus dimorphus* gen. et sp. nov. (Deuteromycotina): A cretaceous predatory fungus. *American Journal of Botany, 95,* 1328–1334.
Scholler, M., Hagedorn, G., & Rubner, A. (1999). A reevaluation of predatory orbiliaceous fungi. II. A new generic concept. *Sydowia, 51,* 89–113.
Sikora, R. A. (1992). Management of the antagonistic potential in agricultural ecosystems for the biological control of plant parasitic nematodes. *Annual Review of Phytopathology, 30,* 245–270.
Singh, K. P., Jaiswal, R. K., Kumar, N., & Kumar, D. (2007). Nematophagous fungi associated with root galls of rice caused by *Meloidogyne graminicola* and its control by *Arthrobotrys dactyloides* and *Dactylaria brochopaga*. *Journal of Phytopathology, 155,* 193–197.
Sorbo, D. G., Marziano, F., & D'Errico, F. P. (2003). Diffusion and effectiveness of the nematophagous fungus *Hirsutella rhossiliensis* in control of the cyst nematode Heterodera daverti under field conditions. *Journal of Plant Pathology, 85,* 219–221.
Su, H., Hao, Y. E., Mo, M., & Zhang, K. Q. (2007). The ecology of nematode-trapping hyphomycetes in cattle dung from three plateau pastures. *Veterinary Parasitology, 144,* 293–298.
Sunder, A., & Lysek, G. (1988). Quantitative investigations on nematode-trapping hyphomycetes from woodland soils. *FEMS Microbiology Letters, 53,* 285–290.
Swe, A., Jeewon, R., & Hyde, K. D. (2008a). Nematode-trapping fungi from mangrove habitats. *Cryptogamie Mycologie, 29*(4), 333–354.
Swe, A., Jeewon, R., Pointing, S. B., & Hyde, K. D. (2008b). Taxonomy and molecular phylogeny of *Arthrobotrys mangrovispora* sp. nov. a new marine nematode-trapping fungal species. *Botanica Marina, 51*(4), 331–338.
Swe, A., Jeewon, R., Pointing, S. B., & Hyde, K. D. (2009). Diversity and abundance of nematode-trapping fungi from decaying litter in terrestrial, freshwater and mangrove habitats. *Biodiversity and Conservation, 18*(6), 1695–1714.
Swe, A., Li, J., Zhang, K. Q., Pointing, S. B., Jeewon, R., & Hyde, K. D. (2011). Nematode-trapping fungi. *Current Research in Environmental & Applied Mycology, 1*(1), 1–26.
Thorn, R. G., & Barron, G. L. (1984). Carnivorous mushrooms. *Science, 224,* 76–78.
Wang, C. Y., Wang, Z., Fang, Z. M., Zhang, D. L., Gu, L. J., Liu, L., & Sung, C. K. (2010). Attraction of pinewood nematode to endoparasitic nematophagous fungus *Esteya vermicola*. *Current Microbiology, 60,* 387–392.
Webster, J., Henrici, A., & Spooner, B. (1998). *Orbilia fimicoloides* sp. nov., the teleomorph of *Dactylella* cf. *oxyspora*. *Mycological Research, 102,* 99–102.
Yang, J., Tian, B., Liang, L., & Zhang, K. Q. (2007c). Extracellular enzymes and the pathogenesis of nematophagous fungi. *Applied Microbiology and Biotechnology, 75,* 21–31.
Yang, Y., & Lui, X. Z. (2006). A new generic approach to the taxonomy of predatory anamorphic *Orbiliaceae* (Ascomycotina). *Mycotaxon, 97,* 153–161.

Yang, Y., Yang, E., An, Z., & Liu, X. (2007a) Evolution of nematode-trapping cells of predatory fungi of the Orbiliaceae based on evidence from rRNA-encoding DNA and multiprotein sequences. *Proceedings of the National Academy of Sciences, USA, 104,* 83–79.
Zhang, Y., Yu, Z. F., Baral, H. O., Qiao, M., & Zhang, K. Q. (2007). *Pseudorbilia* gen. nov. (Orbiliaceae) from Yunnan, China. *Fungal Diversity, 26,* 305–312.
Zopf, W. (1888). Zur Kenntnis der Infektionskrankheiten niederer Thiere und Pflanzen. Nova Acadamy of Caes. *Leop. German. Nat. Cur, 52,* 314–376.
Zuckerman, B. M., & Jansson, H. B. (1984). Nematode chemotaxis and possible mechanisms of host/prey recognition. *Annual Review of Phytopathology, 22,* 95–113.

Chapter 2
Methodology for Studying Nematophagous Fungi

Juan Li, KD Hyde and Ke-Qin Zhang

Contents

K.-Q. Zhang (✉) · J. Li
Laboratory for Conservation and Utilization of Bio-Resources, and Key Laboratory of Microbial Diversity in Southwest China, Ministry of Education, Yunnan University, 650091 Kunming, Yunnan, China
e-mail: kqzhang1@ynu.edu.cn

K. Hyde
Institute of Excellence for Fungal Research, Mae Fah Luang University, Chiang Rai, Thailand

School of Science, Mae Fah Luang University, Chiang Rai, Thailand

K.-Q. Zhang, K. D. Hyde (eds.), *Nematode-Trapping Fungi,* Fungal Diversity Research Series 23, DOI 10.1007/978-94-017-8730-7_2, © Mushroom Research Foundation 2014

Abstract There has been a huge increase in our knowledge of nematophagous fungi, however, the methodology used to study these organisms are unique and have changed little over the years. Nematophagous fungi are easy to find and cultivate, but their peculiar mode of life makes it necessary to use certain techniques to bring them under observation. These methods are detailed in this chapter and include techniques for obtaining mixed cultures of the fungi from nature, for isolating taxa into pure culture, for observing living material and for making permanent microscope preparations.

Keywords Baermann funnel • Differential centrifugation • Field collection; Soil sprinkling • Soil dilution • Slide preparations

Introduction

Nematophagous fungi are classified as either nematode-trapping fungi, which catch nematodes by producing sophisticated mycelial trapping devices, or endoparasitic fungi, which infect nematodes by adhesive conidia or modified conidia, which germinate within the host after ingestion. The protocols for studying nematophagous fungi differ to those used to study most other fungi, because they do not form aggregated mycelial structures visible to the unaided eye. Therefore, certain techniques differ from orthodox mycological methods are used to isolate and characterize nematophagous fungi. The techniques related to the recovery, isolation and maintenance of nematophagous fungi involve a number of unusual techniques which are unique to the study of this group.

Habitats of Nematophagous Fungi

Ecological surveys of nematophagous fungi suggest that this group has extensive worldwide distribution, in all climates and habitats examined (Gray 1987). Therefore, it can be relatively easy to isolate nematophagous fungi, particularly from soils and organic matter.

Gray (1984b) investigated ten different terrestrial habitats for nematophagous fungi in Ireland and found that although nematophagous fungi were abundant in all the habitats examined, agricultural pasture, coniferous leaf litter and coastal vegetation had the greatest number of samples with nematophagous fungi. It was also established that nematophagous fungi in soil are associated with specific soil variables, including soil pH, depth, moisture, nitrogen, phosphorus and potassium amount, and the densities of soil bacteria, fungi and nematodes (Gray 1985, 1988). Studies on the vertical distribution of nematophagous fungi in deciduous woodland with various biotic and abiotic soil factors indicated that this group is restricted to the upper litter and humus layers (Gray and Bailey 1985). Mitsui (1985) also found similar results indicating that most of nematophagous fungi occur in the top 10–30 cm of soil.

Persmark and Jansson (1997) investigated the nematophagous fungi in rhizosphere soils of barley, pea and white mustard, and found that the densities of nematophagous fungi slightly increased in the rhizosphere as compared to root-free soil. They also found that the most common species of nematophagous fungi in these soils was *Arthrobotrys oligospora*. However, *Arthrobotrys musiformis*, *A. robusta* and *Dactylella lobata* are also significantly more abundant than most other species (Jaffee et al. 1996; Persmark and Jansson 1997; Jaffee and Strong 2005).

The best source for isolation of nematophagous fungi is rotting vegetable matter, preferably on or within soil (Duddington 1955). Nematophagous fungi are abundant in forest leaf-mould, and even more so in rotting bark and twigs and in decomposing leaf and stem litter from the forest floor or from a garden compost heaps (Duddington 1955). Therefore, it is preferable to collect soil samples from these habitats whenever possible. A high nematode population density is however, no indicator of the possible biodiversity, and so should not be used as the sole criterion for site selection (Gray 1984b).

Field Collection

The size and quantity of samples is dependent on the subsequent isolation procedure adopted in the laboratory. Sampling methodology and subsequent treatment before specialized isolation are important factors affecting the success of recovering the fungi. The commonest and most efficient sampling system is to use cork borers with different specifications (Shepherd 1955; Fowler 1970; Farrell et al. 2006). Shepherd (1955) used a 20 mm diameter cork borer to take 225 mm long cores of soil, while Fowler (1970) used a 50 mm cork borer, removing 75 mm long cores. A 75 mm diameter bulb planter was used to remove large soil cores, some 125–150 mm long, along with its associated plant material.

Little or no change is likely to have occurred in the centre of the core if the samples are processed within 24 h of collection. After sample collecting, the cores are immediately placed in polythene bags and double-sealed to prevent evaporation and deterioration. Samples in polythene bags can be stored at room temperature for 21 days, or at 4 °C for nine weeks, with no reduction in the isolation efficiency (Fowler 1970). Samples are however, preferably processed on the same or subsequent days of collection to minimize any changes in biological or soil variables. Shepherd (1955) sterilized the cork borer, used to take the soil samples, in 70 % alcohol in between sampling to avoid cross-contamination. This was not possible using the larger bulb-planter, but as cores were eventually split longitudinally and subsamples taken from the centre of each core, it was considered adequate to clean the bulb-planter between sampling with fresh moist tissue.

It is not known whether nematophagous fungi are evenly distributed throughout the soil or whether they are restricted to particular zones with the nematodes. It is better to take subsamples from the nematode layer whenever possible. Where no distinct nematode zones exists, then the material is gently mixed with a pestle and mortar and subsamples taken randomly from the homogenous sample.

Table 2.1 Media used for isolating nematodes

Media	Recipe	Methods	Reference
PDA	200 g peeled potato, 20 g agar; 20 g glucose; 1,000 ml water	Boil potatoes for 30 min, filter through cheesecloth	
CMA	20 g cornmeal; 20 g agar; 1,000 ml water	Boil cook cornmeal for 30 min, filter through cheesecloth	
Water agar (WA)	20 g agar; 1,000 ml water		
Oat meal medium	20 g oatmeal; 20 g agar; 1,000 ml water	Boil oatmeal for 30 min, filter through cheesecloth	
Maize meal agar	20 g maize meal (or crushed maize grains), 20 g agar; 1,000 ml water	The maize and water are warmed to about 70 °C, for 1 h	Duddington (1955)
Difco CMA	8.5 g Difco cornmeal;10 g agar, 1,000 ml water		Rubner (1996)
Rabbit-dung agar[a]	Rabbit pellets; 20 g agar; 1,000 ml water	Soak rabbit pellets in tap water for 2 or 3 days, and pour off the supernatant fluid, filter and dilute with tap water until a pale straw colour	Duddington (1955)
Selective media	1 % water agar or 1 % tryptone glucose agar; 1,000 mg/l triton, 50 mg/l streptomycin sulfate, 50 mg/l rose Bengal		Lopez-Llorca and Duncan (1986)

[a] This media encourages the multiplication of nematodes, but does not support good vegetative growth even of predacious fungi, which makes it useful for the study of internal parasites of nematodes or for material where, as sometimes happens, mould contamination is severe. A disadvantage is that the agar often becomes rather soft, that the nematodes are plentiful and the surface may be badly disturbed

Culture Medium

There are several culture media (Table 2.1) that have been used for incubation and isolation of nematophagous fungi. However, as a general rule, low nutrient media should be used for isolation and high nutrient media should be used for incubation. Once isolated the most commonly used media for incubation of nematophagous fungi are PDA and CMA. However, it should be noted that for nematophagous fungi, if more nutrients are present then they will generally produce less conidiophores.

Baits for Nematophagous Fungi

The addition of nematodes as baits in isolation media can effectively stimulate the growth and trap formation of nematophagous fungi, facilitating their accurate qualitative and quantitative assessment (Wyborn et al. 1969). Several nematode species have been used as the target organism in tests involving nematophagous fungi (*e.g.*, Jansson 1994; Zuckerman et al. 1994; Stirling et al. 2005) or nematicidal compounds (*e.g.*, Lambert et al. 1995; Sharma et al. 2003; Schwarz et al. 2004). These studies have helped elucidate pathogenicity factors and antagonistic agents of plant-parasitic nematodes (Dong et al. 2007).

The most common nematode used in studies of nematophagous fungi is a free-living soil nematode *Panagrellus redivivus* which is a well known commercially produced nematode used as a food source for fish larvae (Gray 1983) and can be bought online (http://www.atcc.org/). *Panagrellus redivivus* does not lay eggs, but the juveniles hatch internally. They have a short life cycle which has four larval stages before becoming adults. The first larval stage is intrauterine, but the remaining stages are free-living (Stock and Nadler 2006). This nematode is easy to rear in large quantities in culture and has a high fecundity. Several methods have been designed to culture *P. redivivus*, however, oatmeal is the most commonly used medium [10 g oat meal in 23 ml tap water, sterilized by autoclaving at 121 °C for 30 min] (Gray 1983; Dong et al. 2007). After 7–10 days of incubation at room temperature, *P. redivivus* can be separated from the oatmeal by using a Baermann-funnel which allows active nematodes to pass through a filter (Schindler 1961; Coolen 1979). Aliquots (0.01 ml) of the nematode suspension are placed on slides and gently heated to immobilize the nematodes which are then counted at 60× magnification to estimate the population density.

Rhabditis terricola has also extensively been used as a target organism to stimulate trap formation in nematode-trapping fungi (Barron 1977). The population density of these nematodes, however, never reaches the levels as can be achieved with *Panagrellus redivivus*. Barron (1977) developed an improved method for culturing this species which was to centrally inoculate a water agar plate with 1 ml of the nematode suspension, and then sprinkled the area with 0.5 g of autoclaved Lipton's green pea dried soup (probably any green pea or other dried soup can be tested). Plates are inoculated at either 20 or 30 °C to obtain large populations of nematodes within a few days (Barron 1977).

Caenorhabditis elegans has extensively been used as a model organism in biological research (Brenner 1974), and is also extensively used to an invertebrate infection model to screen for virulence factors involving nematophagous fungi (Jansson 1994; Zuckerman et al. 1994; Stirling et al. 2005; Vaitkevicius et al. 2006). In general, juveniles and adults are used. Dong et al. (2007) has designed a simple, but reproducible, cost-effective and efficient way to obtain quantities of eggs in bulk.

This method is follows:

- Cultivate *Caenorhabditis elegans* at room temperature on oat meal medium (10 g oat meal in 23 ml tap water, sterilized by autoclaving at 121 °C for 30 min) and seeded with *Escherichia coli* OP50 in an Erlenmeyer flask. Nematode cultures are transferred to fresh medium on a fortnightly basis.
- Nematode inoculum is prepared by scraping those individuals wriggling on the interface of the medium and the sides of the flask and suspending them in deionised water (DW). A 400 µl suspension containing *ca* 5,000 worms is inoculated onto the Luria-Bertani (LB) medium (tryptone 10 g, yeast extract 5 g, NaCl 5 g,

tap water to 1,000 ml, pH adjusted to 7.0 with 5 M NaOH, sterilized by autoclaving at 121 °C for 30 min) in a Petri-dish (6 cm diam.) and spread evenly by repeated rocking.

- Following inoculation, the culture is left at room temperature. Cultures are observed microscopically daily.
- From the third day (72 h post inoculation), a large quantity of eggs could be seen on the surface, whilst the juveniles and adults wriggle through the superficial softened medium.
- The number of eggs laid, peaks on days 3–4. After day 4 the eggs on the culture begin to hatch. For the removal of non-egg developmental stages from the medium, the medium is repeatedly flooded with the M9 buffer ($Na_2HPO_4 \cdot 12H_2O$ 6 g, KH_2PO_4 3 g, NaCl 0.5 g, NH_4Cl 1 g, DW to 1,000 ml, sterilized by autoclaving at 121 °C for 30 min) for three cycles of 1 min, until all the juveniles or adults are suspended and removed.
- These worms are then transferred to fresh LB medium where they continued to lay eggs. The eggs from the original flooded medium can be scraped from the surface and suspended in the M9 buffer. The eggs are surface sterilized by suspending in freshly prepared sodium hypochlorite (0.5 % available chlorine) for 5 min and then rinsed with sterile DW for five cycles. A total of *ca* (1.20–1.40) $\times 10^5$ eggs could be obtained from one Petri-dish culture, on day 3.
- To test the hatch rate of these harvested eggs, a 400 μl suspension of prepared eggs is inoculated onto the LB medium. The cultures are placed at room temperature and microscopically inspected on a daily basis.
- Five days later, juveniles from 80–90 % of the eggs hatch. The methods reported herein were conducted three times with four replicate plates per trial. The results presented are from the final trial. The method provides a reproducible, cost-effective and efficient way to obtain quantities of eggs in bulk. This method could also be used to obtain synchronized juvenile stages. The surface-bound eggs are in various stages of development. However, juveniles of the same age could be obtained by serially flooding the culture whilst leaving the unhatched eggs intact.

Wyborn et al. (1969) found that 1,000 nematodes per plate was below the threshold for stimulation of the fungal species studied, whereas at the 10,000 nematodes per plate the fungi attained epidemic proportions and swamped the nematode population. A level of 5,000 nematodes per plate was demonstrated to be the most effective level, and this is recommended. Larger inocula of nematodes have been found to cause excessive damage to the surface of the agar, making subsequent microscopic observation difficult (Gray 1984c). Therefore, it is more effective to centrally inoculate plates if the soil samples are either very wet or clay-based, otherwise this material was will be spread thinly over the surface of the medium by the nematodes, leaving no clear areas of agar for direct microscopic observation (Gray 1984c).

Although trap formation of nematode-trapping fungi can be stimulated by the addition of living nematodes (Cooke 1962), a number of other substances, such as nemin, sodium cyanide, sodium azide, colchicine, cytochalasin B, abscisic acid, nitric oxide, yeast extract, and valine are also capable of stimulating trap formation (Higgins and Pramer 1967; Chen et al. 2001; Wang and Higgins 2005; Herrera-Medina et al. 2007; Xu et al. 2011). Nemin, which is a metabolic product of nematodes rapidly induces trap formation in culture without the problem of nematode contamination (Pramer and Stoll 1959; Feder et al. 1963; Nordbring-Hertz 1972; Barron 1977).

Treatment of Samples

Several techniques have been devised to isolate nematophagous fungi, such as the soil sprinkling method, the Baermann funnel technique and the differential centrifugation technique (Duddington 1955; Cooke 1961; Barron 1969). These methods are efficient and time saving and give an overview of the spectrum of nematophagous fungi infecting living nematodes in soils. Quantitative examinations are also possible. The soil sprinkling technique and the soil dilution method are good for quantifying nematode-trapping fungi (Dackman et al. 1987; Persmark and Jansson 1997), while the differential centrifugation technique are good for quantifying endoparasitic fungi. The soil sprinkling technique and the soil dilution method employ nematodes as baits, enabling microscopic observation of the fungi. Limitations are that they do not allow determination of whether the fungi are present in soil as spores or hyphae, and that they are time-consuming and labour-intensive (Persson et al. 2000). The Baermann funnel technique and differential centrifugation technique are mainly used to isolate endoparasitic species.

Soil Sprinkling Technique

The soil sprinkling technique was originally devised by Drechsler (1941), and developed by Duddington (1955), and then further by Eren and Pramer (1965). Duddington (1955) first isolated nematode-trapping fungi by plating out small amounts of soil onto nutrient agar. However, the slow growing nematode-trapping fungi were overgrown by the faster growing saprobic fungi that rapidly colonized plates. Eren and Pramer (1965) changed the nutrient agar to 2 % water agar, with adding nematodes to be selective for nematode-trapping fungi. The low nutrient status of this agar reduces saprobic fungal growth, and by providing a separate nutrient source in the form of nematodes, the method becomes highly selective for nematode-trapping fungi. In a later experiment, CMA nutrient adjusted to 15 %, was found to be a more effective isolation media (Fowler 1970).

A soil dilution technique combined with the sprinkling plate method gave a higher probability of isolating nematode-trapping fungi rather than the sprinkling plate method mentioned above, when similar soils were compared (Dackman et al. 1987). The soil dilution method is described below.

The soil sprinkling technique is widely used for the recovery of nematode-trapping fungi (Persson et al. 2000; Park et al. 2002; Farrell et al. 2006). The limitations of the soil sprinkling method are that they do not allow determination of whether the fungi are present in soil as spores or hyphae, and that they are time-consuming and labour-intensive (Persson et al. 2000).

Soil Dilution Method

A similar technique to isolate nematode-trapping fungi is the soil dilution method. This method were first proposed by Dackman et al. (1987) who found that the soil dilution technique can give approximately 10 times as many fungal species as the sprinkle plate method, when similar soils are compared. At present, the soil dilution method which was a development of the earlier sprinkling technique have also been used to isolate nematode-trapping fungi (Persmark and Jansson 1997).

The soil dilution method is performed in the following manner:

- 15 g soil is shaken in 15 ml 0.01 % (w/v) sodium hexametaphosphate in a 100-ml Erlenmeyer flask for 10 min.
- Heavy material is allowed to settle before the soil suspension is decanted into a graduated cylinder and the volume adjusted to 15 ml with water.
- A dilution series is prepared giving 1, 0.1, 0.01, and 0.001 g soil ml^{-1}.
- From each dilution a sample of 0.5 ml is spread on a water agar plate (1.5 % w/v) in such a way that a star is formed (Barron 1977), and the plates are left with the lid half open to allow excess water to evaporate.
- Five replicates are made from each dilution.
- On the following day, the plates are baited with approximately 500 nematodes per plate.
- The plates are incubated in the dark at approximately 22 °C and examined for growth of predatory fungi after 2 and 6 weeks.
- More frequent examination were found unnecessary. Plates showing growth of predatory fungi, manifested in the formation of trapping organs, are recorded as positive. The proportions of positive plates in each dilution is used to calculate the most probable numbers from an appropriate MPN table.

Baermann Funnel Technique

Living nematodes in soils will normally be colonized by fungal parasites. Therefore, nematophagous fungi can be isolated by extracting nematodes from soil and plating them onto low nutrient agar (Capstick et al. 1957; Giuma and Cooke 1972). Although many methods have been described for separating nematodes from the soil (Whitehead and Hemming 1965; Southey 1986), the Baermann funnel technique (Staniland 1954; Giuma and Cooke 1972) has been most extensively used, especially for extracting nematodes infected by nematophagous fungi.

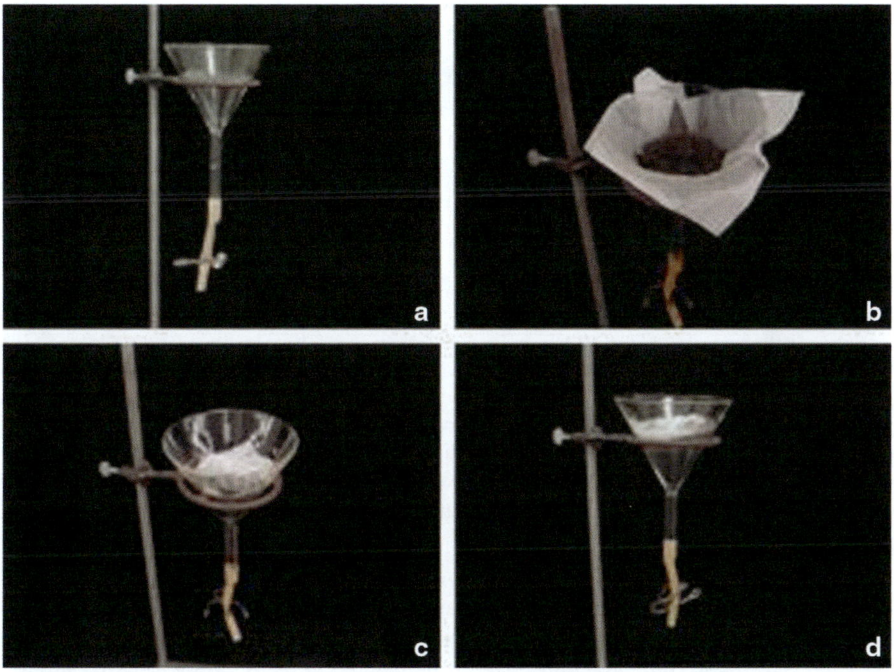

Fig. 2.1 Procedure for isolating living nematodes in soil by Baermann funnel technique

- The standard Baermann funnel apparatus is shown in Fig. 2.1a.
- Wrap approximately 30 g of fresh soil sample in a double layer of soft tissue (Fig. 2.1b).
- Placed this in the standard Baermann funnel apparatus with distilled water (Fig. 2.1c, d), the water level can be readjusted until it is in contact with the soil.
- The nematodes wriggle through the tissue and pass down through the funnel into a small collecting tube.
- After 24 h the collecting tube is removed and the nematode suspension concentrated by centrifugation at 1,000 rpm/min for 3 min.
- Supernatants are discarded and the nematode mass is resuspended in 3 ml of distilled water and incubated at room temperature (18–22 °C).
- Microscopic examination of suspensions should be carried out immediately after centrifugation and this may show that very few nematodes have obvious symptoms of fungal attack.
- After 4 days or more of incubation in distilled water, place samples into glass cavity blocks and re-examine. The nematodes which become sluggish or immobile may contain fungi.
- Remove the nematodes to a sterile 10 ml centrifuge tube containing 5 ml of sterile distilled water by means of a fine Pasteur pipette and concentrated as before.
- Discard the supernatant and resuspend the nematodes in a further 5 ml of sterile distilled water and centrifuge again.

- Transfer the wash nematodes to glass cavity blocks in a small amount of distilled water.
- Transfer those nematodes with emerging hyphae to 2% cornmeal agar plates, and leave the remaining nematodes containing thalli in the cavity blocks
- Examine at daily intervals in order to observe the development of the parasites. Fertile hyphae usually grow rapidly from the nematodes and can be subsequently identified.
- When conidia form, transfer them aseptically to fresh media, where they should usually germinate. In this way pure cultures of a number of species of nematophagous fungi can be obtained (Giuma and Cooke 1972).

Although some active nematodes may be caught on the sides of the funnel or may die from a lack of oxygen in the funnel system, this technique allows reasonably rapid isolation of endozoic parasites without the necessity for their maintenance on nematode populations or the use of antibiotics in the agar medium employed. A comparative investigation has shown the Baermann funnel technique to be significantly more effective in isolating endoparasitic nematophagous fungi than the soil sprinkling method (Gray 1984a).

However, only nematodes in early stages of infection seem mobile enough to do this. The chance of parasites with very rapid rates of development within their hosts being isolated using this method might therefore be small (Giuma and Cooke 1972), and the Baermann funnel technique is influenced by the indigenous nematode population density (Gray 1984a).

Differential Centrifugation Technique (Barron 1969)

Differential centrifugation is a common procedure in microbiology and cytology used to separate certain organelles from whole cells for further analysis of specific parts of cells (Wilchek and Bayer 1990). This technique has also been modified to investigate nematophagous fungi. The principle of differential centrifugation technique is to spin down the large predaceous type conidia at low speeds leaving the smaller conidia of the endoparasitic fungi in the supernatant. Therefore, this method can be used to separate the heavier spores of nematode-trapping fungi from the lighter spores of endoparasites. The supernatant is then centrifuged at a higher speed and the pellet containing the spores spread on plain agar baited with nematodes which subsequently become infected (Barron 1969).

The procedure for this technique is as follows:

- Approximately 200 cm^3 of the soil sample is added to 250 cm^3 of water and blended for 30–60 s.
- The resultant mixture is passed through a series of screens down to a No. 35 screen to remove any coarse material.
- The soil suspension is poured into 50-ml centrifuge tubes and spun down for 2 min at 1,000–2,000 rpm to remove heavy soil particles and large spores.

- The supernatant is decanted and retained to be centrifuged again at 4,500 rpm for 1 h.
- The supernatant is then discarded and the residue saved.
- This is broken up and mixed with a few drops of sterile water using a glass rod and then spread onto a water agar plate.

The residue is spread over half of the plate and then seeded with a few drops of concentrated nematode solution containing 2,000–3,000 nematodes. The centrifugation method results in lower efficiency and diversity of nematophagous fungi than the Baermann funnel technique. However, this technique can isolate conidia as well as the Baermann funnel, making this isolation technique especially useful for studies in specificity of infection and on the conidia (Gray 1984a).

Observation and Identification

Observation

The identification of nematophagous fungi is normally established by direct observation. Nematophagous fungi are observed in culture and conidiophores and trap structure characteristics are determined. Conidiophores are often observed on top of crumbs of soil or on organic debris, where they are only visible at lower magnifications. Emergence times for nematophagous fungi vary considerably, and in order to recover the majority of species in a particular sample, the soil plates should be scanned frequently. Fowler (1970) compared 32 samples of garden soil at weekly intervals for 14 weeks, finding that most of nematophagous fungi present had been recorded within six weeks. Therefore Barron (1978) advised that plates should be examined daily for up to ten days to ensure that all the possible nematophagous species were observed. Presently, plates with small amounts of soil on nutrient agar are examined as frequently as possible, and at least at weekly intervals, for up to ten weeks. In this way even the most slowly developing species are recovered. Less frequent observations will result in fewer species being recorded.

Identification of the fungi by direct observation at $\times 400$ magnification using an objective lens with a long working distance is highly successful and significantly reduces the number of mounted preparations. When fungi are immature, the bottom of the Petri-dish can be marked using a indelible marker pen so that the specimen can be relocated and examined at a later date. A thin section of the agar (clear water agar) is better as it allows the researcher to focus on infected nematodes. In general, plates should be scanned at $\times 100$ magnification for clear observation of nematophagous fungi. Lower magnifications such as $\times 40$ or $\times 60$ magnification should only be used by experienced workers. Subsequent identification is carried out directly at $\times 100$ magnification using an objective with a long working distance. The entire surface of each plate is scanned in this way using a template fixed to a movable stage.

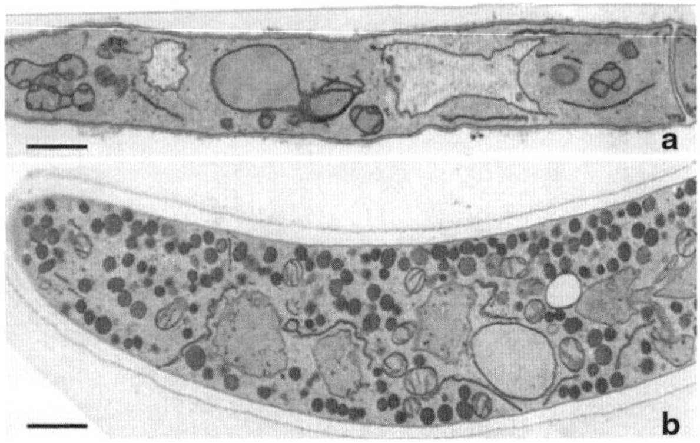

Fig. 2.2 Transmission electron micrograph (TEM) of vegetative hypha (upper panel) and a trap cell. Bar, 1 mm. Note dense bodies only in the trap cell. Figure reproduced from Nordbring-Hertz (1984) with kind permission from Cambridge University Press

Electron Microcopy

Electron microscopy is a method which uses a beam of electrons to illuminate the specimen and produce a magnified image. This method has a greater resolving power than a light-powered optical microscope, because electrons have wavelengths about 100,000 times shorter than visible light (photons), and can achieve better than 50 pm resolution (Erni et al. 2009) and magnifications of up to about 10,000,000x, whereas ordinary, non-confocal light microscopes are limited by diffraction to about 200 nm resolution and useful magnifications below 2,000x. There are two forms of electron microscopy which are commonly used to provide surface information: Secondary Electron Microscopy (SEM), which provides a direct image of the topographical nature of the surface from all the emitted secondary electrons, and Scanning Auger Microscopy (SAM), which provides compositional maps of a surface by forming an image from the Auger electrons emitted by a particular element. Both of these techniques employ focusing of the probe beam (a beam of high energy electrons, typically 10–50 keV in energy) to obtain spatial localization.

Electron microscopy can provide valuable morphological information, as well as some remarkable views of nematophagous fungi and have been extensively used to observe species (Nordbring-Hertz 1972; Barron 1979; Saikawa and Morikawa 1985; Tunlid et al. 1991; Bordallo et al. 2002). Many workers, such as Heintz and Pramer (1972), Nordbring-Hertz (1972) and Janssona (Janssona 1982) have provided useful details of the techniques of transmission and scanning electron microscopy of nematophagous fungi. For example, Lysek and Nordbring-Hertz (1983) used SEM to obtain a clear figure of an adhesive network trap of *Arthrobotrys oligospora*, while Nordbring-Hertz (1984) used TEM to reveal a common feature observed in all traps of trap-forming fungi (Fig. 2.2). The presence of numerous cytosolic organelles (dense bodies) are formed directly on initiation of the trap. Normal vegetative hyphae invariably lack dense bodies (Fig. 2.2). The dense bodies develop from specialized regions of the endoplasmic reticulum and exhibit catalase and amino acid oxidase activity and thus are peroxisomal in nature. They are supposed to be involved in penetration and digestion of the nematode.

Fluorescence Microscopy

With sophisticated equipment and an increasing number of fluorochromes, fluorescence techniques have become an effective tool in microscopy to directly observe soil dwelling fungi *in situ* (Jensen and Lysek 1991; Saxena and Lysek 1993; Jensen 1994). This has been made possible by the use of specific fluorochromes such as fluorescein-di-acetate (FDA) and calcofluor white. Fluorescein-di-acetate (FDA) is a substance which does not give any fluorescence by itself; but it can be taken up by actively metabolizing cells or hyphae, and hydrolyzed enzymatically to yield free fluorescein, a true fluorochrome. Thus, only living and actively metabolizing microorganisms become visible when stained with FDA (Correa et al. 1986; Soderstrom and Erland 1986; Jensen 1994). Calcofluor white (correct name Calcofluor white M2R new) is an optical brightener with a high affinity to β-1, 4-glucans such cellulose or chitin. It is thus specific for hyphal walls, and does not differentiate between living and dead tissue. Calcofluor stains many types of fungal structures, for example spores or fruiting body initials (Von Sengbusch et al. 1983).

Investigation of fungi in the soil by fluorescence microscopy is now possible. Staining of key taxonomic features by fluorochromes, for example developing conidiophores and spores of *A. oligospora*, make it an ideal tool for the rapid identification of species of fungi that grow actively in soil, and which would be extremely difficult to observe by conventional microscopy. The technique can also be used to study fungal growth in soils over time, which exhibit periodic effects and even allows some considerations concerning their ecology (Jensen et al. 1997; 1998). It also has been used to study the colonization of soils contaminated with urban pollutants and to study the occurrence of fungi in the ventilation systems of buildings (Neumeister et al. 1996). Here, we give examples of using fluorescence microscopy to examine the trap formation of nematophagous fungi. Figure 2.3 shows the constricting rings of *Arthrobotrys dactyloides* which was stained with calcofluor white before (a) and after contraction (b) (Jensen et al. 1998). These inflation acts very rapidly, triggered when a nematode touches the inside of the ring. The three cells enlarge their volume such that the ring is completely filled. The nematode is quickly ensnared and cannot escape. Figure 2.3 also shows a recently captured nematode, but due to the selectiveness of calcofluor white for glucans, it is stained poorly and hence is only visible as a shadow (Jensen et al. 1998).

Figure 2.4 shows an example of double staining of *A. oligospora* with FDA and calcofluor white (Jensen et al. 1998). If the filters are chosen adequately, the stains highlight different cellular components: FDA (Fig. 2.4a) stains actively metabolizing areas such as capture organs, while calcofluor white makes the entire complex visible (Fig. 2.4b; Jensen et al. 1998).

When using fluorescence microscopy to investigate nematophagous fungi in soil we recommend the use of sterilized (autoclaved) compost, to avoid any effect of soil

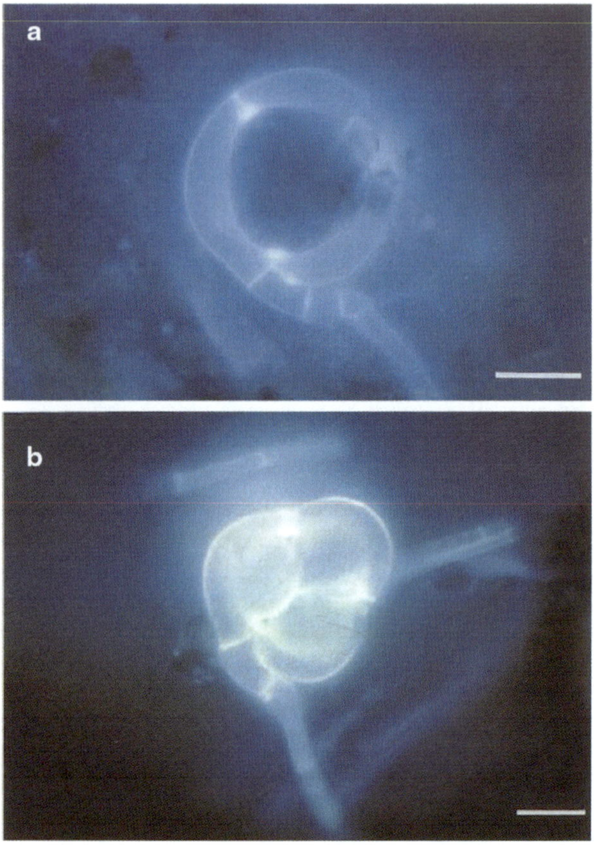

Fig. 2.3 Constricting rings of *Arthrobotrys dactyloides* before (**a**) and after contraction (**b**). In (**b**) the captured nematode captured is just visible. Staining with calcofluor white. Bar = 10 μm. (Figure reproduced from Jensen et al. (1998) with kind permission from Elsevier)

fungistasis and inoculation of the fungi from pre-cultures on malt extract agar plates. Incubation should be done in Petri-dishes at room temperature. In order to preserve the exact localization of hyphae, traps, or other specific structures and surrounding soil particles, the soil samples need to be treated very carefully. Soil samples are cut vertically from the surface to the bottom of the Petri-dish with cover slips to obtain a number of soil slides to an average depth of 2.5 mm. Each soil slide is placed on one cover slip. Staining and manipulation should be done with much care to avoid distortion of the fungal structures. Observation is best through a cover slip using the hanging drop technique (Jensen 1994; Jensen and Lysek 1995). When staining by fluorochromes, if stain by FDA, a stock solution should be prepared from 500 mg FDA dissolved in 100 ml acetone and stored at −18 °C. A working solution should be prepared by diluting 0.1 ml of the stock in 5 ml 60 mM phosphate buffer pH 6.88. Since this degrades very rapidly, the working solution should be freshly prepared for every observation. It is mixed into the soil and microscopic observation started immediately; images are visible 1 min after preparation. Due to the rapid photofading of fluorescein, staining only lasts 20 min. If stain by Calcofluor white, a stock solution

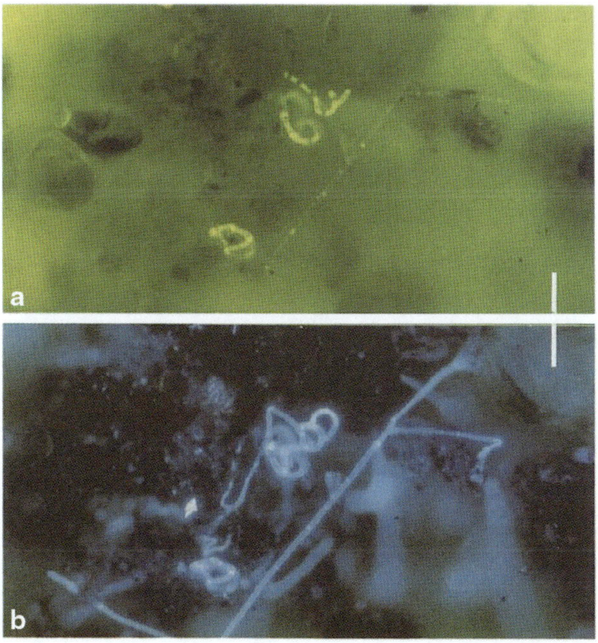

Fig. 2.4 Hyphae with sticky networks of *A. oligospora* between soil particles stained with FDA (**a**) and calcofluor white (**b**). Due to the triggering light and filtering the FDA (**a**) and calcofluor white (**b**) give different images of the same structure. Bar = 100 μm. (Figure reproduced from Jensen et al. (1998) with kind permission from Elsevier).

should be prepared from 50 mg calcofluor dissolved in 5 ml buffer (Tris pH 9.0). The working solution consists of 0.1 ml of stock solution in 9.9 ml buffer. Both solutions are stable and should be stored in the dark at room temperature. For microscopic observations, the working solution should be simply mixed with the soil to be investigated immediately prior to observation. The fluorescence is very stable; for example, mounts kept in a wet chamber could be studied for at least 24 h after staining.

Isolation of Pure Culture Species

In order to identify and characterize the morphology of nematophagous fungi, it is necessary to obtain pure cultures of species (Gray 1983). The method used to obtain pure subcultures of nematophagous fungi is to transfer a small piece of agar which contains numerous fungal conidia to a new fresh plate with nutrition medium. This method has the disadvantage of picking up surface contamination from the agar, which causes unwanted bacterial and mould growth on the plates. Heintz (1978) used two techniques for transferring nematophagous fungi from pure cultures onto fresh plates, a standard method uses plug inocula and macerated inocula.

Standard method

- 3 mm diameter disks of agar are cut from the periphery of an actively growing culture on nutrient agar using a sterile cork borer
- The disks are transferred to new plates.

Macerated Inocula Method (Heintz 1978)

- A 50–60 mm diameter colony of the fungus is removed from an agar plate with a sterile scalpel, removing as little agar as possible and placed in 100 ml of sterile distilled water.
- The colony is macerated until a homogenous suspension is produced.
- 0.5 ml aliquots are then spread uniformly over the entire surface of each agar plate.

The second technique uses macerated inocula, which allows a more uniform growth to be produced on the new plate. Heintz (1978) noted that this method resulted in the development of a typical fungal thallus within three days.

Numerous methods have been used to transfer conidia to obtain pure cultures of nematophagous species. Shepherd (1955) used a sterile platinum wire while Barron (1977) used a sterile dissecting needle to remove the conidia. However, Gray (1984b) used a single haired sable paint brush to pick off individual conidia. He believed that the single hair is extremely fine, short and flexible, and can be easily directed to particular conidia under 60 × or 100 × magnification. A popular method to transfer conidia is to use a sterile toothpick (Hao et al. 2005; Zhang et al. 2005; Li et al. 2006). The single hair is extremely fine, short and flexible, and can be easily directed to particular conidia under 60 × or 100 × magnification. The collected conidia are streaked onto nutrient agar.

In addition, pure cultures of some endoparasites can also be obtained by transferring nematodes, which are infected but with the fungus not yet sporulating, onto nutrient agar supplemented with aureomycin (30 g/l) to reduce bacterial contamination. Endoparasites are more difficult to transfer and grow in culture. The simplest method is to remove the infected nematode with a needle and wash it several times in saline to remove any contamination. This is best done by using a series of cavity slides and transferring the nematode from slide to slide. The infected nematode is then placed on a fresh water agar plate and inoculated with nematodes.

Identification

Identification of nematophagous fungi depends on the morphology and morphometry of conidia (shape, septa and size) and conidiophore branching, and modifications at the apex (Drechsler 1937; Cooke and Godfrey 1964), but also the trapping structure, and presence or absence of chlamydospores. Five kinds of trapping devices have been recognized, including adhesive networks, adhesive hyphae, constricting rings, adhesive knobs and nonconstricting rings (Fig. 2.5). The adhesive hypha is a short erect branch of a few swollen cells produced on a hypha with an adhesive layer covering its surfaces (Figs. 2.5, 1–4). The adhesive knob is a globose or subglobose

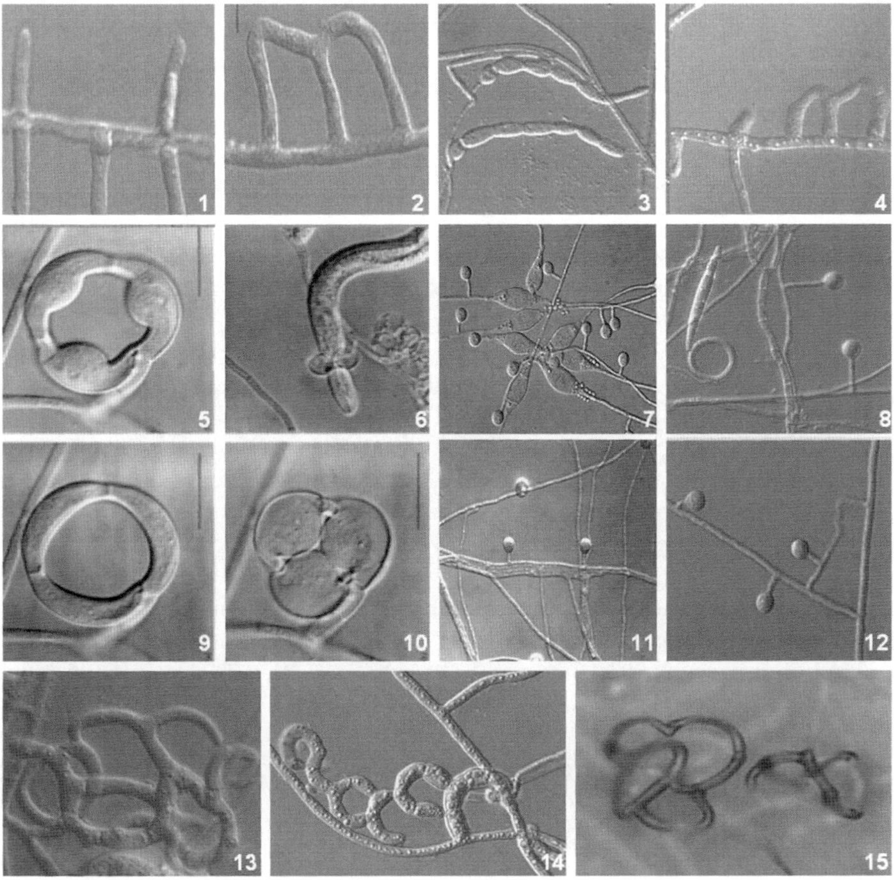

Fig. 2.5 Morphology on traps of nematode-trapping fungi. *1–4*. Adhesive hypha. *5–6*, *9–10*. Constricting rings. *7*, *11–12*. Adhesive knobs. *8*. Adhesive knobs and nonconstricting rings. *13–15*. Adhesive networks

cell that is either sessile on the hypha or with an erect stalk (Figs. 2.5, 7–8, 11–12). Nonconstricting rings always occur alongside adhesive knobs, and are produced when erect lateral branches from vegetative hypha thicken and curve to form a generally three-celled ring that then fuses to the supporting stalk (Figs. 2.5, 8). The constricting ring contains three ring cells. When a nematode enters a constricting ring, the three ring cells are triggered to swell rapidly within 1–2 s (Figs. 2.5, 5–6, 9–10). The adhesive network is formed by an erect lateral branch growing from a vegetative hypha, curving to fuse with the parent hypha and developing more loops exterior to the original loop or on the parent hypha (Figs. 2.5, 13–15; Barron 1979; Yang et al. 2007).

Use of Molecular Techniques in Identifying Nematophagous Fungi

Rapidly developing molecular techniques such as the polymerase chain reaction (PCR) have revolutionized identification and classification of fungi (White et al. 1990; Foster et al. 1993). Application of such techniques for identification of nematophagous fungi can be important tools for rapid and accurate identification. For example, Hu et al. (2006) has identified *Arthrobotrys multisecundaria* (YMF1.01821) as a new nematode-trapping fungus based on the production of unicellular secondary conidia from both distal and basal ends. However, the extracellular serine protease genes, ITS and β-tubulin fragments which were cloned from *Arthrobotrys multisecundaria* show a high degree of similarity to those from *Monacrosporium microscaphoides*. Therefore, although these two fungi exhibited obvious differences in secondary conidia morphology, Li et al. (2008) has considered *Arthrobotrys multisecundaria* to be a spontaneous mutant of *Monacrosporium microscaphoides.* This example suggests that there may be several mistakes regarding the identification of nematode-trapping fungi when based only on the morphological observations.

Another example indicating the importance of molecular technology concerns the taxonomic revisions of nematode-trapping fungi. With molecular technology, the systemic classification of nematode-trapping fungi has been redefined because there is multitude inconsistencies in the evolutionary relationships as compared to the established taxonomy, which was traditionally based on morphology of conidia and conidiophores. Since the pioneering work by Drechsler (Drechsler 1937), nematode-trapping fungi have been classified in a number of genera based on morphology of conidia and conidiophores. Based on the phylogenetic analyses of the 5.8 S rDNA, ITS1 and ITS2 of 28 nematode-trapping fungi, Liou and Tzean (1997) found that the 28 nematode-trapping fungi can be separated into four clades, each with a unique trapping device, and were first to propose that trapping organs may reflect evolutionary relationships, and appear more significant than conidiogenous cells and conidia for genus and species delimitation. Pfister (1997) also found that trapping devices are more informative than other morphological characters in delimiting genera. Subsequently, based on the small subunit (SSU) ribosomal DNA (18 S rDNA) from 15 species of nematode-trapping fungi, Ahren et al. (1998) found that nematode-tapping fungi clustered into three lineages: species with constricting rings, species with various adhesive structures (net, hyphae, knobs and nonconstricting rings) and species have no trapping devices. Based on results obtained from morphological and molecular characters, Hagedorn and Scholler (1999) and Scholler et al. (1999) classified nematode-trapping fungi into four genera: *Dactylellina* characterized by stalked adhesive knobs including species characterized by nonconstricting rings and stalked adhesive knobs; *Gamsylella* characterized by adhesive branches and unstalked knobs; *Arthrobotrys* characterized by adhesive networks; and *Drechslerella* characterized by constricting rings. Based on sequence analyses of 28 S and 5.8 S rDNA and β-tubulin, Li et al. (2005) proposed an emended generic concept for nematode-trapping fungi. *Arthrobotrys* is characterized by adhesive networks, *Dactylellina* by adhesive knobs, and *Drechslerella* by constricting-rings. Phylogenetic placement of taxa characterized by stalked adhesive knobs and nonconstricting rings also is confirmed in *Dactylellina*. Species that produce unstalked adhesive knobs that grow out to form loops were transferred from *Gamsylella* to *Dactylellina*, and those that produce unstalked adhesive knobs that grow out to form networks were transferred from *Gamsylella* to *Arthrobotrys*. *Gamsylella* as currently circumscribed cannot be treated as a valid genus.

The use of molecular techniques is widespread and has helped to establish relationships between different organisms. However, these techniques only came into use in the previous 15 years and thus much can be learnt from these techniques. They can be used not only for the investigation of phylogenetic, taxonomic, and diagnostic problems in nematophagous fungi, but also provide the basis for gaining deeper insights in their biological research. However, we should not use molecular data in isolation to other data such as morphological data because researchers may infer wrongly perceived ideas from incorrectly generated phylogenetic trees due to the unwise selection of ingroups and outgroups.

In 2012, the International Code of Nomenclature for Algae, Fungi and Plants was revised and no longer allows a single fungal species (or genus) to have two names, with the older name taking priority except in exceptional circumstances (Hawksworth 2012). The asexual genera of nematode-trapping fungi are linked with a sexual *Orbilia* morphs and may need to take the *Orbilia* name, or alternatively their names may be maintained. Discussions are presently underway by the *Orbilomycetes working group* of the *International Commission on the Taxonomy of Fungi* (http://www.fungaltaxonomy.org/subcommissions) to establish which is the best name (or names) to adopt for this group(s) of fungi.

Preservation and Permanent Slides

Storage

In general, nematophagous fungi lose their viability if maintained for long periods in pure culture. The storage methods most suited to each fungal species or genera needs to be determined (Ryan et al. 2000). The maintenance of predatory activity of the fungal isolates is one of the basic prerequisites to ensure predatory activity. Several methods have therefore, been devised for preserving fungal cultures for prolonged periods and thus retaining the original characteristics of the isolates. Methods include preservation in mineral oil, cryopreservation, cold storage at 4 °C and preservation of species in silica gel (Duddington 1955). However, all these methods can more or less reduce the predatory activity and sporulation capacity.

Barron (1977) developed a soil storage method which can preserve nematophagous fungi for much longer periods without loss of viability. In this method 30-ml screw top vials are half filled with moist greenhouse soil and sterilized. Fresh nematodes are added to a plate containing infected nematodes, and softly mixed together. The resulting suspension, including bacteria and other contaminants, is then used to inoculate the vials. Approximately 0.5 ml of the suspension is added to each vial, which is allowed to stand at room temperature for several days to ensure infection of the healthy nematodes and is then stored at 5 °C. When required the vial is shaken to loosen the soil, and the material plated out as normal using the soil sprinkling technique.

Slides

In order to identify the nematode-trapping fungi that grow on the culture medium, microscopic slides must be prepared. Fowler (1970) initially used a dissecting microscope to locate fungi by examining the plates at 60× magnification. Alternatively where immediate identification was not possible subcultures were prepared. Shepherd (1955) used high-power microscopy to identify fungi directly by adding a drop of water to the agar, lowering a cover slip over the specimen, and examining it under oil immersion. Wood (1973) examined his isolates in temporary water mounts or in lactophenol acid fuschin, while live nematodes were fixed in 5 % formalin containing acid fuschin. It is important to retain the hyphae *in situ* for subsequent examination and so permanent preparations were made using the method originally described by Dixon and Duddington (1951).

Lactophenol cotton blue (LPCB) wet mounts are the most widely used method for staining and observing fungi (Leck 1999). A thin section of the agar containing the fungus is removed using a scalpel and transferred onto the slide. The agar thickness is reduced by gently warming the slide, and the specimen is stained using cotton blue in lactophenol. A piece of clear adhesive tape can be gently touched on the colony and then placed on a drop of stain on the slide to observed intact specimens with string of spores or other structures that break up when making agar mounts.

It is difficult to make good permanent preparations of nematophagous fungi and these are less satisfactory to study than water mounts of living material. Several techniques for making permanent mounts of nematophagous and other fungi have been described, including lactophenol mounts with nail varnish sealant (Hawkswell 2001), Kohlmeyer's permanent mount method (Volkmann-Kohlmeyer and Kohlmeyer 1996) and Venetian turpentine method (Duddington and Dixon 1951, 1953).

Semi-permanents Mounts in Lactophenol Sealed with Nail Varnish

When examining spores of *Russulaceae* in Melzer's reagent under the microscope and subsequently mounting them in lactophenol, Hawkswell (2001) found that the iodine staining faded, rendering details of the spore ornamentation indistinct.

However, by adding a quantity of iodine to lactophenol, he found that the staining intensity was enhanced and it did not fade and the details of spore ornamentation remained clear (Hawkswell 2001). The routines for staining and mounting are as follows:

- Take a small piece of fungal tissue or infected nematodes, place it on a microscope slide and add a drop of iodine.
- Place a cover glass over the specimen, cover with a strip of filter paper and exert firm pressure with a finger to compress the sample.
- Pour a few drops of distilled water over the slide, gently remover the cover slip and carefully pour off the surplus water.
- Give the specimen a second rinse with distilled water, pour off the surplus water and blot dry with tissue.
- Add a drop of iodine-lactophenol and place a cover slip over the specimen, carefully avoiding air bubbles.
- Remover the surplus iodine-lactophenol with tissue and seal the cover glass with nail varnish.

All specimens mounted in lactophenol must be sealed with nail varnish or a similar sealant if it is intended to keep the slide on a permanent basis. Permanent slides prepared by this method will be good for at least 18 months Hawkswell (2001) and the specimens retain the stain and ornamentations details clearly.

Double Cover-Glass Method

The "double cover-glass method" was introduced by Diehl (1929) who described the mounting of the specimen between two cover glasses of different sizes. It was revised by Kohlmeyer and Kohlmeyer (1972) and also adapted for fungal cultures by Connell and Padgett (1988).

- A large square cover glass (25 × 25 mm, No. 1) is attached on a clean slide (76 × 26 mm) by means of two droplets of distilled water (Figs. 2.6, 1).
- A larger drop of distilled water is placed in the centre of the cover glass, to which the specimen is added in the form of whole mounts, squash mounts or sections (Figs. 2.6, 2).
- The object is then covered with a smaller cover glass (18 × 18 mm, No. 1, Figs. 2.6, 3) allowing immediate investigation of the living material with bright-field, phase contrast or Nomarski interference contrast. The use of immersion oil is also possible, although easier in the later stages, since special care must be taken when wiping off the oil.
- After the necessary observations, measurements, drawings and/or photographs have been made, a droplet of concentrated glycerin is added to the water at one or two sides of the small cover slip (Figs. 2.6, 4).

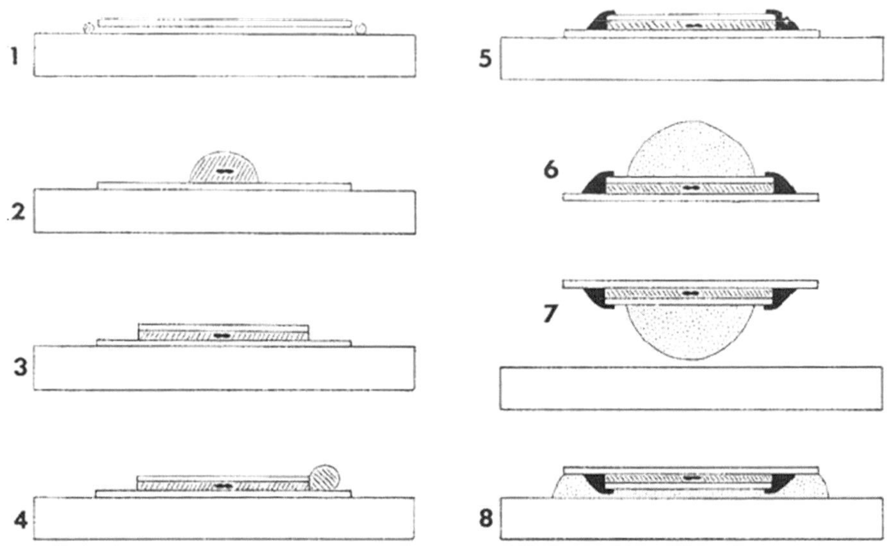

Fig. 2.6 Double cover glass method for permanent microscopic mounts. (From Kohlmeyer and Kohlmeyer 1972)

- The slide is stored horizontally in a dustproof container for a few days to allow the water to evaporate. Excessive glycerin is removed from the edges of the larger cover glass easily done with filter paper—or if needed, more glycerin is added.
- The mount is then sealed with a thin ring of clear fingernail polish; it is best to repeat this step after an hour to be sure the glycerin is perfectly sealed in (Figs. 2.6, 5).
- When the nail polish is fully dried, the large cover glass is removed from the slide with the aid of a needle and a drop of mounting medium is placed on the small cover glass (Figs. 2.6, 6).
- Then the preparation is turned over (Figs. 2.6, 7) and placed on the slide. The drop of mounting medium flattens out (more effectively if one puts a small weight e.g., old AA battery on top for a few minutes), surrounding the edges of the small cover glass (Figs. 2.6, 8), there by permanently sealing in the small cover glass and nail polish.
- Any excess medium is squeezed out at the sides and can be taken off with a needle. The preparation is now stored horizontally until the medium is hardened, but it can be used after a day, should further inspection at the microscope be necessary.
- The sealing procedures are best done under a hood to avoid breathing the toxic fumes of the media. Nearly as important as the sealing technique is the labelling.

For this method, paper labels are not always permanent especially if written in pencil, neither with pencil written on frosted glass. Kohlmeyer and Kohlmeyer (1972) recommend to used slides with one frosted end, and the labelling being done with waterproof ink. Kohlmeyer and Kohlmeyer have slides as old as 30 years that are in perfect condition; the fungi being permanently sealed, there is no drying out or cracking and the writing is as clear as on the first day. Again, this method is more time-consuming than just using fingernail polish, but the results are so much more satisfying for anybody who needs to prepare voucher or holotype specimens, and it should always be used for preserving type material.

Erythrosin-Glycerine and Venetian Turpentine Methods

Duddington and Dixon (1951) have introduced methods for making rapid preparations for nematophagous fungi from agar culture. They believed that formalin-acetic acid fixative is excellent for preservation of material. For endoparasitic nematophagous fungi, the Erythrosin-glycerine method is most useful. The procedure of this method is as follows:

- A small piece of agar containing the specimen is fixed in the following mixture: commercial formalin, 10 ml; glacial acetic acid, 5 ml; distilled water, 85 ml for not less than 24 h.
- The specimen is washed in water until no smell of fixative is detectable.
- The material is placed in a saturated aqueous solution of erythrosin for 24 h.
- The specimen is transfer to 10 % glycerine and allowed to concentrate until most of the water has evaporated.
- Addition of 0.5 % of glacial acetic acid is added to improve staining.
- The dehydration is completed in a desiccator, and the material can then be mounted in glycerine or in glycerine jelly.
- Duddington (1955), however, believed that nematode-trapping fungi can be mounted satisfactorily using the Venetian turpentine method. The procedures of this method are as follows:
- Fix agar slices in the formalin-acetic acid for not less than 24 h.
- Washed with water until no smell of acetic acid.
- Stain in haematoxylin for about 5 min.
- ‘blue’ in tap water or ammonia vapour.
- Place in 10 % glycerine.
- When the concentration of the glycerine is saturated, wash in industrial methylated spirit until all glycerine is removed.
- Dehydrate in absolute alcohol.
- Place in a 10 % solution of turpentine in absolute alcohol.
- When the dish containing the material is left in the desiccator. Cleared with xylol and mounted in Canada balsam.

Please note that the dehydration and clearing must be carried out very gradually. For the preservation of material, the formalin-acetic acid fixative is excellent. The method causes no noticeable distortion, and the fungi can be left in the gelatin indefinitely without harm.

Conclusions

Although many standard experimental techniques, as presented above, have been developed for studying nematophagous fungi, they may not be suitable for all studies. Future studies should develop and adopt more efficient and better methods and researchers are encouraged to improve or design novel techniques to facilitate the investigation of nematophagous fungi.

References

Ahrén, D., Ursing, B. M., & Tunlid, A. (1998). Phylogeny of nematode-trapping fungi based on 18 S rDNA sequences. *FEMS Microbiology Letters, 158,* 179–184.
Barron, G. L. (1969). Isolation and maintenance of endoparasitic nematophagous hyphomycetes. *Canadian Journal of Botany, 47,* 1899–1902.
Barron, G. L. (1977). *The nematode-destroying fungi*. Canadian Biological Publications Ltd.
Barron, G. L. (1978). Nematophagous fungi: Endoparasites of *Rhabditis terricola. Microbial Ecology, 4,* 157–163.
Barron, G. L. (1979). Observations on predatory fungi. *Canadian Journal of Botany, 57,* 187–193.
Bordallo, J., Lopez-Llorca, L., Jansson, H. B., Salinas, J., Persmark, L., & Asensio, L. (2002). Colonization of plant roots by egg-parasitic and nematode-trapping fungi. *New Phytologist, 154,* 491–499.
Brenner, S. (1974). The genetics of *Caenorhabditis elegans*. *Genetics, 77,* 71–94.
Capstick, C., Twinn, D., & Waid, J. (1957). Predation of natural populations of free-living nematodes by fungi 1). *Nematologica, 2,* 193–201.
Chen, T. H., Hsu, C. S., Tsai, P. J., Ho, Y. F., & Lin, N. S. (2001). Heterotrimeric G-protein and signal transduction in the nematode-trapping fungus *Arthrobotrys dactyloides*. *Planta, 212,* 858–863.
Connell, S. L., & Padgett, D. E. (1988). An improved technique for making permanent slide cultures of fungi. *Mycopathologia, 101,* 165–166.
Cooke, R. C. (1961). Agar disk method for the direct observation of nematode-trapping fungi in the soil. *Nature, 191,* 1411–1142.
Cooke, R. C. (1962). The ecology of nematode-trapping fungi in the soil. *Annals of Applied Biology, 50,* 507–513.
Cooke, R. C., & Godfrey, B. (1964). A key to the nematode-destroying fungi. *Transactions of the British Mycological Society, 47,* 61–74
Coolen, W. (1979). Methods for the extraction of *Meloidogyne* spp. and other nematodes from roots and soil. In F. Lamberti & C. E. Taylor (Eds.), *Root-knot nematodes* (*Meloidogyne species*). *Systematics, biology and control* (pp. 317–329). New York: Academic Press.
Correa, B., Purchio, A., Paula, C., & Gambale, W. (1986). Evaluation of a fluorescent method (fluorescein diacetate and ethidium bromide solution) in the study of the viability Cryptococcus neoformans strains. *Mycopathologia, 96,* 91–96.
Dackman, C., Olsson, S., Jansson, H. B., Lundgren, B., & Nordbring-Hertz, B. (1987). Quantification of predatory and endoparasitic nematophagous fungi in soil. *Microbial Ecology, 13,* 89–93.
Diehl, W. W. (1929). An improved method for sealing microscopic mounts. *Science, 69,* 276–277.
Dixon, S. M., & Duddington, C. L. (1951). Permanent preparation of fungi growing on agar. *Nature, 168,* 38–39.
Dong, L. Q., Mo, M. H., Yang, J. K., & Zhang, K. Q. (2007). A method for obtaining quantities of *Caenorhabditis elegans* eggs. *Nematology, 9,* 743–744.

Drechsler, C. (1937). Some hyphomycetes that prey on free-living terricolous nematodes. *Mycologia, 29,* 447–552.
Drechsler, C. (1941). Predacious fungi. *Biological Reviews of the Cambridge Philosophical Society, 16,* 265–290.
Duddington, C. (1955). Notes on the technique of handling predacious fungi. *Transactions of the British Mycological Society, 38,* 97–103
Duddington, C., & Dixon, S. M. (1951). Permanent preparations of fungi growing on agar. *Nature, 168,* 38–39.
Duddington, C., & Dixon, S. M. (1953). Some methods for making permanent preparations of micro-fungi. *The Journal of the Quekett Microscopical Club, 3,* 305
Eren, J., & Pramer, D. (1965). The most probable number of nematode-trapping fungi in soil. *Soil Science, 99,* 285.
Erni, R., Rossell, M. D., Kisielowski, C., & Dahmen, U. (2009). Atomic-resolution imaging with a sub-50-pm electron probe. *Physical Review Letters, 102,* 96101.
Farrell, F., Jaffee, B., & Strong, D. (2006). The nematode-trapping fungus *Arthrobotrys oligospora* in soilof the Bodega marine reserve: Distribution and dependenceon nematode-parasitized moth larvae. *Soil Biology and Biochemistry, 38,* 1422–1429.
Feder, W. A., Everard, C., & Wootton, L. (1963). Sensitivity of several species of the nematophagous fungus *Dactylella* to a morphogenic substance derived from free-living nematodes. *Nematologica, 9,* 49–54.
Foster, L. M., Kozak, K. R., Loftus, M. G., Stevens, J. J., & Ross, I. K. (1993). The polymerase chain reaction and its application to filamentous fungi. *Mycological Research, 97,* 769–781.
Fowler, M. (1970). New Zealand predaceous fungi. *New Zealand Journal of Botany, 8,* 283–302.
Giuma, A., & Cooke, R. (1972). Some endozoic fungi parasitic on soil nematodes. *Transactions of the British Mycological Society, 59,* 213–218
Gray, N. (1983). Ecology of nematophagous fungi: *Panagrellus redivivus* as the target organism. *Plant and Soil, 73,* 293–297.
Gray, N. (1984a). Ecology of nematophagous fungi: Comparison of the soil sprinkling method with the Baermann funnel technique in the isolation of endoparasites. *Soil Biology and Biochemistry, 16,* 81–83.
Gray, N. (1984b). Ecology of nematophagous fungi: Methods of collection, isolation and maintenance of predatory and endoparasitic fungi. *Mycopathologia, 86,* 143–153.
Gray, N. (1984c). The effect of fungal parasitism and predation on the population dynamics of nematodes in the activated sludge process. *Annals of Applied Biology, 104,* 143–149.
Gray, N. (1985). Ecology of nematophagous fungi: Effect of soil moisture, organic matter, pH and nematode density on distribution. *Soil Biology and Biochemistry, 17,* 499–507.
Gray, N. (1987). Nematophagous fungi with particular reference to their ecology. *Biological Reviews, 62,* 245–304.
Gray, N. (1988). Ecology of nematophagous fungi: Effect of the soil nutrients N, P and K, and seven major metals on distribution. *Plant and Soil, 108,* 286–290.
Gray, N., & Bailey, F. (1985). Ecology of nematophagous fungi: Vertical distribution in a deciduous woodland. *Plant and Soil, 86,* 217–223.
Hagedorn, G., & Scholler, M. (1999). A reevaluation of predatory orbiliaceous fungi. I. Phylogenetic analysis using rDNA sequence data. *Sydowia, 51,* 27–48.
Hao, Y., Mo, M., Su, H., & Zhang, K. Q. (2005). Ecology of aquatic nematode-trapping hyphomycetes in southwestern China. *Aquatic Microbial Ecology, 40,* 175–181.
Hawkswell, A. (2001). Iodine-lactophenol as a mycological mounting medium. *Field Mycology, 2,* 12.
Hawksworth, D. L. (2012). Managing and coping with names of pleomorphic fungi in a period of transition. *IMA Fungus, 3,* 15–24.
Heintz, C. E. (1978). Assessing the predacity of nematode-trapping fungi *in vitro*. *Mycologia, 70,* 1086–1100
Heintz, C., & Pramer, D. (1972). Ultrastructure of nematode-trapping fungi. *Journal of Bacteriology, 110,* 1163–1170.

Herrera-Medina, M. J., Steinkellner, S., Vierheilig, H., Ocampo Bote, J. A., & Garrido, J. M. G. (2007). Abscisic acid determines arbuscule development and functionality in the tomato arbuscular mycorrhiza. *New Phytologist, 175,* 554–564.

Higgins, M., & Pramer, D. (1967). Fungal morphogenesis: Ring formation and closure by *Arthrobotrys dactyloides. Science, 155,* 345–346.

Hu, W., Li, Y., Mo, M., & Zhang, K. Q. (2006). A new nematode-trapping hyphomycete of *Arthrobotrys. Mycotaxon, 95,* 181–184.

Jaffee, B., & Strong, D. (2005). Strong bottom-up and weak top-down effects in soil: Nematode-parasitized insects and nematode-trapping fungi. *Soil Biology and Biochemistry, 37,* 1011–1021.

Jaffee, B., Strong, D., & Muldoon, A. (1996). Nematode-trapping fungi of a natural shrubland: Tests for food chain involvement. *Mycologia, 88,* 554–564.

Jansson, H. B. (1994). Adhesion of conidia of Drechmeria coniospora to Caenorhabditis elegans wild type and mutants. *Journal of Nematology, 26,* 430–435.

Janssona, H. B. (1982). Attraction of nematodes to endoparasitic nematophagous fungi. *Transactions of the British Mycological Society, 79,* 25–29.

Jensen, C. (1994). *Fluoreszenzmikroskopische Möglichkeiten zur in situ-Beobachtung nematophager Pilze im Boden. Dissertationes Botanicae, 217.* Berlin: Cramer.

Jensen, C., & Lysek, G. (1991). Direct observation of trapping activities of nematode-destroying fungi in the soil using fluorescence microscopy. *FEMS Microbiology Letters, 85,* 207–210.

Jensen, C., & Lysek, G. (1995). Fluorescence microscopy of fungi in native soil-improvement by additional substances. *Microscopy and Analysis, 49,* 7–9.

Jensen, C. H., Neumeister, H., & Lysek, G. (1997). Nematophagous fungi-study by fluorescence microscopy and edx-technique of the periodicity of trap formation in soil. *Biological Rhythm Research, 28,* 365–373.

Jensen, C., Neumeister-Kemp, H., & Lysek, G. (1998). Fluorescence microscopy for the observation of nematophagous fungi inside soil. *Mycologist, 12,* 107–111.

Kohlmeyer, J., & Kohlmeyer, E. (1972). Permanent microscopic mounts. *Mycologia, 64,* 666–669.

Lambert, M., Kremer, S., & Anke, H. (1995). Antimicrobial, phytotoxic, nematicidal, cytotoxic, and mutagenic activities of 1-hydroxypyrene, the initial metabolite in pyrene metabolism by the basidiomycete Crinipellis stipitaria. *Bulletin of Environmental Contamination and Toxicology, 55,* 251–257.

Leck, A. (1999). Preparation of lactophenol cotton blue slide mounts. *Community Eye Health, 12,* 24.

Li, Y., Hyde, K. D., Jeewon, R., Cai, L., Vijaykrishna, D., & Zhang, K. Q. (2005). Phylogenetics and evolution of nematode-trapping fungi (*Orbiliales*) estimated from nuclear and protein coding genes. *Mycologia, 97,* 1034–1046.

Li, Y., Jeewon, R., Hyde, K. D., Mo, M. H., & Zhang, K. Q. (2006). Two new species of nematode-trapping fungi: Relationships inferred from morphology, rDNA and protein gene sequence analyses. *Mycological Research, 110,* 790–800.

Li, J., Yang, J., Liang, L., & Zhang, K. Q. (2008). Taxonomic revision of the nematode-trapping fungus *Arthrobotrys multisecundaria. Journal of Microbiology, 46,* 513–518.

Liou, G. Y., & Tzean, S. S. (1997). Phylogeny of the genus *Arthrobotrys* and allied nematode-trapping fungi based on rDNA sequences. *Mycologia, 89,* 876–884

Lopez-Llorca, L., & Duncan, J. (1986). New media for the estimation of fungal infection in eggs of the cereal cyst nematode. *Nematologica (Netherlands), 32,* 486–490.

Lysek, G., & Nordbring-Hertz, B. (1983). Die Biologie nematodenfangender Pilze. Forum Mikrobiologie. *Darmstadt, 6,* 201–208.

Mitsui, Y. (1985). Distribution and ecology of nematode-trapping fungi in Japan. *JARQ, 18,* 182–193.

Neumeister, H., Moritz, M., Schleibinger, H., & Martiny, H. (1996). Investigation of allergic potential induced by fungi on air filters of HVAC systems. Proceedings of Indoor Air '96, Nagoya, Japan.

Nordbring-Hertz, B. (1972). Scanning electron microscopy of the nematode-trapping organs in *Arthrobotrys oligospora. Physiologia Plantarum, 26,* 279–284.

Nordbring-Hertz, B. (1984). Carbohydrate interactions in nematode-trapping fungi. In DH Jennings & ADM Rayner (Eds.), *The ecology and physiology of the fungal mycelium* (pp. 419–432). Cambridge: Cambridge University Press.
Park, J., Gams, W., Scholler, M., Ghisalberti, E., & Sivasithamparam, K. (2002). Orbiliaceous nematode-trapping fungi and related species in Western Australia and their biological activities. *Australasian Mycologist, 21,* 45–52.
Persmark, L., & Jansson, H. B. (1997). Nematophagous fungi in the rhizosphere of agricultural crops. *FEMS Microbiology Ecology, 22,* 303–312.
Persson, C., Olsson, S., & Jansson, H. B. (2000). Growth of *Arthrobotrys superba* from a birch wood resource base into soil determined by radioactive tracing. *FEMS Microbiology Ecology, 31,* 47–51.
Pfister, D. H., (1997). Castor, Pollux and life histories of fungi. *Mycologia. 89*(1), 1–23
Pramer, D., & Stoll, N. R. (1959). Nemin: A morphogenic substance causing trap formation by predaceous fungi. *Science, 129,* 966–967.
Rubner, A. (1996). Revision of predacious hyphomycetes in the *Dactylella-Monacrosporium* complex. *Studies in Mycology, 39,* 1–134.
Ryan, M., Smith, D., & Jeffries, P. (2000). A decision-based key to determine the most appropriate protocol for the preservation of fungi. *World Journal of Microbiology and Biotechnology, 16,* 183–186.
Saikawa, M., & Morikawa, C. (1985). Electron microscopy on a nematode-trapping fungus, *Acaulopage pectospora. Canadian Journal of Botany, 63,* 1386–1390.
Saxena, G., & Lysek, G. (1993). Observation of nematophagous fungi in natural soils by fluorescence microscopy and their correlation with isolation. *Mycological Research, 97,* 1005–1011.
Schindler, A. (1961). A simple substitute for a Baermann funnel. *Plant Disease Reporter, 45,* 747–748.
Scholler, M., Hagedorn, G., & Rubner, A. (1999). A reevaluation of predatory orbiliaceous fungi. II. A new generic concept. *Sydowia, 51,* 89–113.
Schwarz, M., Köpcke, B., Weber, R. W. S., Sterner, O., & Anke, H. (2004). 3-Hydroxypropionic acid as a nematicidal principle in endophytic fungi. *Phytochemistry, 65,* 2239–2245.
Sharma, V., Walia, S., Kumar, J., Nair, M. G., & Parmar, B. S. (2003). An efficient method for the purification and characterization of nematicidal azadirachtins A, B, and H, using MPLC and ESIMS. *Journal of Agricultural and Food Chemistry, 51,* 3966–3972.
Shepherd, A. M. (1955). Some observations on the distribution and biology of fungi predaceous on nematodes. Ph. D. Thesis, University of London.
Soderstrom, B., & Erland, S. (1986). Isolation of fluorescein diacetate stained hyphae from soil by micromanipulation. *Transactions of the British Mycological Society, 86,* 465–468.
Southey, J. F. (1986). *Laboratory methods for work with plant and soil nematodes*. HMSO Books.
Staniland, L. (1954). A modification of the Baermann funnel technique for the collection of nematodes from plant material. *Journal of Helminthology, 28,* 115–118.
Stirling, G., Wilson, E., Stirling, A., Pankhurst, C., Moody, P., Bell, M., & Halpin, N. (2005). Amendments of sugarcane trash induce suppressiveness to plant-parasitic nematodes in a sugarcane soil. *Australasian Plant Pathology, 34,* 203–211.
Stock, S. P., & Nadler, S. A. (2006). Morphological and molecular characterisation of *Panagrellus* spp. (Cephalobina: Panagrolaimidae): Taxonomic status and phylogenetic relationships. *Nematology, 8,* 921–938.
Tunlid, A., Johansson, T., & Nordbring-Hertz, B. (1991). Surface polymers of the nematode-trapping fungus *Arthrobotrys oligospora. Journal of General Microbiology, 137,* 1231–1240.
Vaitkevicius, K., Lindmark, B., Ou, G., Song, T., Toma, C., Iwanaga, M., Zhu, J., Andersson, A., Hammarström, M. L., Wai, S. N., Tuck, S. (2006). A Vibrio cholerae protease needed for killing of *Caenorhabditis elegans* has a role in protection from natural predator grazing. *Proceedings of the National Academy of Sciences, 103,* 9280–9285.
Volkmann-Kohlmeyer, B., & Kohlmeyer, J. (1996). How to prepare truly permanent microscope slides. *Mycologist, 10,* 107–108.

Von Sengbusch, P., Hechler, J., & Müller, U. (1983). Molecular architecture of fungal cell walls. An approach by use of fluorescent markers. *European Journal of Cell Biology, 30,* 305–312.

Wang, J., & Higgins, V. J. (2005). Nitric oxide has a regulatory effect in the germination of conidia of *Colletotrichum coccodes*. *Fungal Genetics and Biology, 42,* 284–292.

White, T. J., Bruns, T., Lee, S., & Taylor, J. (1990). Amplification and direct sequencing of fungal ribosomal RNA genes for phylogenetics. In: *PCR Protocols: a guide to methods and applications*. (Innis, M. A., Gelfand, D. H., Sninsky, J. J., White, T. J., eds). Academic Press, New York, USA: 315–322.

Whitehead, A., & Hemming, J. (1965). A comparison of some quantitative methods of extracting small vermiform nematodes from soil. *Annals of Applied Biology, 55,* 25–38.

Wilchek, M., & Bayer, E. (1990). *Methods in enzymology*. San Diego: Academic Press Inc.

Wood, F. (1973). Nematode-trapping fungi from a tussock grassland soil in New Zealand. *New Zealand Journal of Botany, 11,* 231–240.

Wyborn, C., Priest, D., & Duddington, C. (1969). Selective technique for the determination of nematophagous fungi in soils. *Soil Biology and Biochemistry, 1,* 101–102.

Xu, L. L., Lai, Y. L., Wang, L., & Liu, X. Z. (2011). Effects of abscisic acid and nitric oxide on trap formation and trapping of nematodes by the fungus *Drechslerella stenobrocha* AS6.1 Fungal Biology, *2*(114), 97–101.

Yang, Y., Yang, E., An, Z., & Liu, X. (2007) Evolution of nematode-trapping cells of predatory fungi of the Orbiliaceae based on evidence from rRNA-encoding DNA and multiprotein sequences. *Proceedings of the National Academy of Sciences, USA, 104,* 83–79.

Zhang, J., Mo, M., Deng, J., Liu, X. F., Bi, T. J., & Zhang, K. Q. (2005). *Dactylella zhongdianensis sp. nov*, a new predacious antagonist of nematodes. *Mycotaxon, 92,* 289–294.

Zuckerman, B., Matheny, M., & Acosta, N. (1994). Control of plant-parasitic nematodes by a nematicidal strain of *Aspergillus niger*. *Journal of Chemical Ecology, 20,* 33–43.

Chapter 3
Taxonomy of Nematode-Trapping Fungi from *Orbiliaceae*, Ascomycota

ZeFen Yu, MingHe Mo, Ying Zhang and Ke-Qin Zhang

Contents

Abstract Nematode-trapping fungi of *Orbiliaceae* include those filamentous species forming trapping devices to prey on juveniles of nematodes. In this chapter, the taxonomic history of predatory orbiliaceous fungi is reviewed and the system of using trapping devices as the primary morphological criterion for generic delimitation is advocated. Following this taxonomic concept, keys for genera of *Arthrobotrys*, *Drechslerella* and *Dactylellina*, which include all reported species of predatory orbiliaceous fungi are presented. Totally, 54, 14 and 28 species from *Arthrobotrys*, *Drechslerella* and *Dactylellina* respectively, are morphologically described and illustrated. Known asexual-sexual connections (14) of predatory orbiliaceous fungi are summarized and their taxonomic descriptions and illustrations presented.

Keywords Nematode-trapping fungi · Taxonomy · Asexual-sexual connections

K.-Q. Zhang (✉) · Z.-F. Yu · M.-H. Mo · Y. Zhang
Laboratory for Conservation and Utilization of Bio-Resources, and Key Laboratory of Microbial Diversity in Southwest China, Ministry of Education, Yunnan University, 650091 Kunming, Yunnan, China
e-mail: kqzhang1@ynu.edu.cn

K.-Q. Zhang, K. D. Hyde (eds.), *Nematode-Trapping Fungi,* Fungal Diversity Research Series 23, DOI 10.1007/978-94-017-8730-7_3, © Mushroom Research Foundation 2014

Introduction

In various natural environments, especially in soil, there are a wide and diverse range of microorganisms that function as antagonists of nematodes and include fungi, bacteria, viruses and rickettsias (Li et al. 2000). Nematophagous fungi or nematode-destroying fungi are antagonists of nematodes, and comprise a great variety of fungi belonging to widely divergent orders and families. Based on nematicidal mechanisms, these antagonists are assigned to the groups, endoparasitic fungi, opportunistic fungi, toxic fungi or nematode-trapping fungi (Li et al. 2000). Endoparasitic fungi are regarded as obligate parasites infecting nematodes via special spores and mycelia development takes place from spores within the nematode and only the reproductive-conidiophores are externalized (Li et al. 2000). Toxic fungi refer to those producing metabolites toxic to nematodes. Opportunistic fungi can colonize nematode reproductive structures and have the ability to seriously affect their reproductive capabilities (Siddiqui and Mahmood 1996).

Nematode-trapping fungi capture nematodes by trapping devices produced from the vegetative mycelia. The trap types reported include adhesive networks, adhesive knobs, constricting rings, non-constricting rings, adhesive branches, undifferentiated or unmodified adhesive hyphae, stephanocysts, spiny balls and acanthocytes (Zhang et al. 2001; Tzean and Liou 1993; Luo et al. 2004, 2007). The first five types of trap are known only to be produced by the predacious asexual morphs of *Orbiliaceae* of Ascomycota. Adhesive knobs are however, also formed by *Zoophagus* species of Zygomycetes (Li et al. 2000) and *Nematoctonus* species of basidiomycetes (Barron 1977).

Orbiliaceous members represent the majority of nematode-trapping fungi, which include 96 species, and are currently assigned in the asexual genera *Arthrobotrys* (53 species), *Dactylellina* (28 species) and *Drechslerella* (14 species). Traps of adhesive hyphae are restricted to the genera *Stylopage* and *Cystopage* of Zygomycetes (Drechsler 1941), while stephanocysts is restricted to the genus *Hyphoderma* of basidiomycetes (Tzean and Liou 1993). Spiny balls and acanthocytes are formed respectively by the macrobasidiomycetes *Coprinus comatus* (Luo et al. 2004, 2007) and *Stropharia rugosoannulata* (Luo et al. 2006). As mentioned above, the orbiliaceous nematode-trapping are the most complex and taxonomically unsettled.

In this chapter we focus on the largest group of nematode-trapping fungi in the Ascomycota. We discuss their historical classification and the taxonomic system currently accepted. All species of orbiliaceous nematode-trapping fungi are described and illustrated. Keys to the genera *Arthrobotrys*, *Dactylellina* and *Drechslerella* are provided in order to facilitate the readers to compare the species features. Furthermore, asexual-sexual state connections of the orbiliaceous nematode-trapping fungi are discussed in species descriptions and illustrations.

Systematics of Nematode-Trapping Fungi

Asexual nematode-trapping hyphomycetous fungi belong to a predatory fungal lineage. Distinctive trapping devices such as adhesive hyphae, adhesive knobs, adhesive networks, constricting rings, and non-constricting rings develop from extensions of the mycelia (Barron 1977). The trapping devices enable capture and cause subsequent mortality of nematodes, other small animals, and protozoans and the nematode-trapping fungi then obtain nutrients from these prey (Barron 1977). Since nematode-trapping fungi play an important role as antagonists of plant-parasitic and animal-parasitic nematodes, there is a great interest in using these fungi as biological control agents. Information concerning the systematic position of nematode-trapping fungi is therefore essential for understanding and exploiting them in biocontrol.

Asexual nematode-trapping hyphomycetous fungi are placed within the family *Orbiliaceae* (*Orbiliales*, *Orbiliomycetes*), based on morphological and molecular studies (Pfister 1997; Eriksson et al. 2003; Yu et al. 2011). The family *Orbiliaceae* was not known to be nematophagous until Pfister (1994) reported that a collection of *Orbilia fimicola* Jeng & Krug produced an *Arthrobotrys* asexual state named *A. superba* Corda. Subsequently, more and more connections between *Orbiliaceae* and predacious hyphomycetes have been established (Liu et al. 2005a; Mo et al. 2005a; Pfister 1995; Yu et al. 2006; Qiao et al. 2012; Li et al. 2009; Yu et al. 2009a). Connections between *Orbiliaceae* and other non-predacious hyphomycetous genera have also been reported (Liu et al. 2005b; Yu et al. 2009b, 2007a, 2007b, 2011; Mo et al. 2005b; Yang and Liu 2005). At least nine asexual genera have been linked to sexual *Orbiliaceae* species (Pfister 1997; Yu et al. 2011).

The family *Orbiliaceae* was shown to be monophyletic in the phylogenetic analyses of 18S and ITS rDNA sequences (Pfister 1997). Morphologically, the unique nature of the ascospores of this family, especially the inclusion of an elaborated spore body makes it distinguishable from other discomycetes. According to phylogenetic analysis of SSU rDNA sequences of 21 taxa, the Orbiliomycetes clustered near the base of the tree near the Pezizomycotina. A signature of six base pairs was consistently found in every studied species of Orbiliomycetes, which is a further evidence in the isolated position of the class (Eriksson et al. 2003). However, a range of conidial morphology is found in nematode-predacious and non-predacious asexual morphs in some phylogenetic clades. For example, in earlier classification systems, conidiophore characters were used to distinguish asexual genera. Within the series *Auricolores* in the sexual genus *Orbilia*, asexual morphs with a candelabrelloid or more simple type of conidiogenesis actually cluster, often separately, from those with an arthrobotryoid type. However, the arthrobotryoid type occurs scattered in several clades, therefore, this character does not permit a split of series *Auricolores* into two groups. Conidiophore types however, appear to characterize small groups within that series (Baral, pers. comm.).

Nematode-trapping fungi were formerly classified into a number of genera based on the morphology of conidia (shape, septa and size) and conidiophores (branching, modifications of the apex), such as *Arthrobotrys*, *Dactylellina*, *Drechslerella*, *Monacrosporium*, *Dactylella*, *Gamsylella*, *Didymozoophaga*, *Trichothecium*, *Dactylaria*, *Candelabrella*, *Genicularia*, *Duddingtonia*, *Nematophagus*, *Monacrosporiella* and *Woroninula*. This has led to a situation where species with diverse types of trapping devices have been assigned to one genus, while others with similar trapping devices can be found in different genera (Glockling and Dick 1994; Liu and Zhang 1994, 2003; Zhang et al. 1996a). The genus *Arthrobotrys* was first established for the type species, *A. superba* (Corda 1839), which is characterized by the formation of two-celled conidia on denticles in a whorled arrangement at the apex of the nodes of the simple, erect, septate conidiophore. Schenck et al. (1977) expanded the genus to include species with aseptate and multi-celled conidia. Scholler et al. (1999) subsequently limited *Arthrobotrys* to only those species with adhesive networks.

Subramanian (1963) erected *Drechslerella* for *D. acrochaeta* (Drechsler) Subram. based on the filiform apical appendage on its conidia. Although Liu and Zhang (1994) synonymized *Drechslerella* with *Monacrosporium*, detailed molecular analyses (Hagedorn and Scholler 1999) convinced Scholler et al. (1999) to propose new generic concepts for orbiliaceous nematode-trapping fungi based on mode of trapping device. We accept these concepts and recognize *Drechslerella* as characterized by forming three-celled constricting ring traps.

The genus *Dactylellina* was described by Morelet (1968) with *Dactylellina leptospora* (Drechsler) M. Morelet as type species. Species included in this genus were characterized by elongate, fusoid conidia, while microconidia and microconidiophores rarely formed. *Dactylellina* was recognized by Scholler et al. (1999) for species characterized by stalked adhesive knobs and included those species producing non-constricting rings. The knob-forming species of the nematode-trapping fungi however, have diverse configurations such as sessile knobs, stalked knobs, a combination of stalked knobs with non-constricting rings, or stalked knobs with proliferated knobs.

Earlier phylogenetic analyses based on ribosomal RNA (rRNA) gene sequences suggested that the trapping devices could rationalize the classification of nematode-trapping fungi (Rubner 1996). Indeed, lineages including species with stalked adhesive knobs, adhesive networks and constricting rings consistently separate from each other in molecular phylogenetic analysis. Recently, phylogenetic analyses based on ribosomal RNA (rRNA) and protein-coding gene sequences suggested that the trapping devices could serve as robust indicators of generic delimitation for these asexual fungi (Yang and Liu 2006; Ahrèn et al. 1998; Li et al. 2005; Scholler et al. 1999), different trapping devices-forming species were separated into different genetic lineages and new generic concepts of the taxonomy of predatory sexual *Orbiliaceae* were proposed.

Ahrén et al. (1998) found that nematode-tapping fungi clustered into three lineages: species with constricting rings, species with various adhesive structures (net, hyphae, knobs and non-constricting rings) and species have no trapping device. Based on results obtained from morphological and molecular characters, Hagedorn and Scholler (1999) and Scholler et al. (1999) classified nematode-trapping fungi into four genera: *Dactylellina* characterized by stalked adhesive knobs including species characterised by non-constricting rings and stalked adhesive knobs; *Gamsylella* characterized by adhesive branches and unstalked knobs; *Arthrobotrys* characterized by adhesive networks and *Drechslerella* characterized by constricting rings. Furthermore, the systematic classification of nematode-trapping fungi was redefined based on phylogenies inferred from sequence analyses of 28S rDNA, 5.8S rDNA and ß-tubulin genes and an emended generic concept of nematode-trapping fungi is provided (Li et al. 2005). *Arthrobotrys* is characterized by adhesive networks, *Dactylellina* by adhesive knobs, and *Drechslerella* by constricting-rings. Phylogenetic placement of taxa characterized by stalked adhesive knobs and non-constricting rings is also confirmed in *Dactylellina*. Based on careful morphological observation of culture, Li et al. (2005) found that none of these taxa only formed unstalked adhesive knobs, which was just one temporary structure formed at the first stage in the fungus development. These unstalked knobs could then grow out to form branches, which may further change in different species, they proposed that unstalked adhesive knobs should not be treated as a unique type of trapping-device and this character should not be given taxonomic importance, and the genus *Gamsylella* is invalid. Species which produce unstalked adhesive knobs that grow out to form loops were transferred from *Gamsylella* to *Dactylellina*, and those which produce unstalked adhesive knobs that grow out to form networks were transferred from *Gamsylella* to *Arthrobotrys*.

Based on the combined phylogenetic analysis of rDNA and protein-coding genes and morphological studies, a hypothesis for the evolution of trapping-devices was also presented (Li et al. 2005). Predatory and non-predatory fungi appear to have been derived from non-predatory members of *Orbilia*. The adhesive knob is considered to be the ancestral type of trapping-device from which constricting rings and networks were derived via two pathways. In the first pathway adhesive knobs retained their adhesive material forming simple two-dimension networks, eventually forming complex three-dimension networks. In the second pathway adhesive knobs lost their adhesive materials, with their ends meeting to form non-constricting rings and they in turn formed constricting rings with three inflated-cells. Furthermore, using phylogenetic analysis of nucleotide sequences of three protein-coding genes (RNA polymerase II subunit gene, *rpb2*; elongation factor 1-α gene, *ef1*-α; and ß tubulin gene, *bt*) and ITS rDNA region, Yang et al. (2007) described the evolution of nematode-trapping cells of predatory fungi, they thought the initial trapping structure evolved along two lineages yielding two distinct trapping mechanisms: one developed into constricting rings and the other developed into adhesive traps. Among adhesive trapping devices, the adhesive network separated from the others early and evolved at a steady and gentle speed. The adhesive knob evolved through stalk elongation, with a final development of non constricting rings. The derived adhesive traps are indicated as a highly differentiated stage.

The non-predacious genus *Dactylella* was established by Grove (1884) on the basis of one species, *D. minuta* Grove. The genus is characterized by erect, simple, hyaline conidiophores with conidia produced singly at the apex. Conidia are ellipsoidal, fusoid or cylindrical, one-celled at first and later having 2 to many septa. This genus has been emended several times and both non-predacious and predacious fungi have been included (Subramanian 1963; Schenck et al. 1977; Zhang et al. 1994). *Dactylella* includes diverse taxa in morphology and behavior. There are considerable differences in conidiophore length and conidial size among species. Some species are saprotrophic, while others are oospore or nematode egg parasites (Zhang et al. 1994). Rubner (1996) revised the generic concept of *Dactylella* and excluded the nematode-trapping species. However, the classification has not commonly been accepted and several predacious species had been described under *Dactylella* (Liu and Zhang 2003). Chen et al. (2007a, b, c) emended the asexual state genus *Dactylella* complex based on ITS phylogeny, all the tested strains cluster into three monophylogenetic clades corresponding to three genera, e.g. *Dactylella*, *Vermispora* and a distinct group comprising species with very short conidiophores. The circumscription of *Dactylella* and *Vermispora* were emended and a new genus *Brachyphoris* characterized by very short conidiophores that are scarcely longer than conidia was established, which sexual states corresponding to the genus *Hyalorbilia* within *Orbiliaceae*, and other species separated into genera *Dactylella* and *Vermispora*, respectively.

Although trapping devices have proven as vital in generic delimitation of nematode-trapping fungi, species in the same genus are mainly delimited from each other by morphology of conidia and conidiophores. With more and more connections between nematode-trapping fungi and *Orbiliaceae* being established, phylogenetic analysis including existing asexual species indicated that conidial morphology showed only little correlation while the sexual state characters showed a high level of correlation. For instance, *Arthrobotrys* belong to Series *Auricolores*, which is defined by the ability to form adhesive networks and the asexual *Trinacrium*-type belong to species of *Hyalorbilia*, *Orbilia* subgenus *Hemiorbilia* and subgenus *Orbilia* (Baral, pers. comm.). Further studies are needed to set up a natural classification system including both asexual and sexual nematode-trapping fungi. In fact this is now a requirements of the International Code of Nomenclature for Algae, Fungi and Plants as two names can no longer be used for asexual and sexual morphs of the same fungus.

Key to Genera of Nematode-Trapping Fungi

1 Trapping-device a constricting ring, which consists of three inflated cells with a short, strong stalk .... ***Drechslerella***
2 Trapping-device not a constricting ring, but various adhesive trapping devices 3
3 Trapping device unstalked adhesive knobs which develops into an adhesive network or adhesive networks only .... ***Arthrobotrys***
4 Trapping device stalked adhesive knobs, some with non-constricting rings, or unstalked adhesive knobs which grow out to form adhesive branches and loops .... ***Dactylellina***

Arthrobotrys Corda, Pracht-Fl. Eur. Schimmelbild.: 43 (1839)
Type species: *Arthrobotrys superba* Corda, Pracht-Fl. Eur. Schimmelbild.: 43 (1839)
Synonyms
= *Monacrosporium* Oudem., Ned. Kruidk. Arch., Ser.2, 4: 250 (1885)
Lectotype species: *Monacrosporium elegans* Oudem., Ned. Kruidk. Arch., Ser.2, 4: 250 (1885), designated by Clements and Shear (1931)
= *Didymozoophaga* Soprunov & Galiulina, Mikrobiologiya 20: 493 (1951) (invalid; Art.36; illegit. Art. 52 ICBN)
Type species: *Didymozoophaga superba* (Corda) Soprunov & Galiulina, Mikrobiologiya 20: 494 (1951) (invalid; Art. 36 ICBN).
≡ *Arthrobotrys superba* Corda, Pracht–Fl. Eur. Schimmelbildung.: 43 (1839)
= *Candelabrella Rifai* & R.C. Cooke, Trans. Br. mycol. Soc. 49 (1): 160 (1966)
Type species: *Candelabrella javanica* Rifai & R.C. Cooke, Trans. Br. Mycol. Soc. 49: 160 (1966).
= *Duddingtonia* R.C. Cooke (1969), Trans. Br. mycol. Soc. 53 (2): 315 (1969)
Type species: *Duddingtonia flagrans* (Dudd.) R.C. Cooke, Trans. Br. Mycol. Soc. 53: 315 (1969)
≡ *Trichothecium flagrans* Dudd., Trans. Br. Mycol. Soc. 32: 287 (1949)
= *Genicularia Rifai* & R.C. Cooke, Trans. Br. Mycol. Soc. 49: 153 (1966) [non Genicularia Rouss. ex Desv. 1808, non de Bary 1858] (nom. illeg.;. Art. 53 ICBN; replaced by Geniculifera)
Type species: *Genicularia cystosporia* (Dudd.) Rifai & R.C. Cooke, Trans. Br. Mycol. Soc. 49: 154 (1966)
≡ *Trichothecium cystosporium* Dudd., Trans. Br. Mycol. Soc. 34: 600 (1951a)
= *Nematophagus* Mekht., Mikol. Fitopatol. 9 (3): 250 (1975)
Type species: *Nematophagus azerbaidzhanicus* Mekht., Mikol. Fitopatol. 9: 250 (1975)
= *Monacrosporiella* Subram., Kavaka 5: 94 (1978) [1977]
Type species: *Monacrosporiella megalospora* (Drechsler) Subram., Kavaka 7: 94 (1977).
≡ *Dactylella megalospora* Drechsler, Mycologia 46: 769 (1954)
= *Woroninula* Mekht., Khishchnye Nematofagovye Griby—Gifomitsety (Baku): 109 (1979)
Type species: *Woroninula polycephala* (Drechsler) Mekht., Khishchnye nematofagovye Griby—Gifomitsety: 110 (1979)
≡ *Dactylaria polycephala* Drechsler, Mycologia 29: 530 (1937)

Characteristics: Mycelium fast growing. Hyphae septate, branching, hyaline. Conidiophores simple or branched, apex mainly with short denticles or with geniculate, candelabrelloid, or percurrent proliferations, rarely simple without modifications. Conidia either singly or in clusters at the apex of the conidiophore. Conidia holoblastic, hyaline, non-septate to multi-septate, mainly ellipsoidal, spindle-shaped or ovoid, rarely clavate, cylindrical, pyriform, or turbinate. Microconidia and microconidiophores frequently formed. Chlamydospores, when present, intercalary or rarely terminal, singly or in chains, thick-walled, sphaerical to ovoid, yellow pigmented. Sexual state when know, belonging to *Orbilia*. Trapping nematodes by means of adhesive networks or modified devices, with good saprotrophic capabilities. Formation of traps induced by nematodes in most species.

Key to Species of *Arthrobotrys*

1. Conidia produced singly at the apices of conidiophores or on subapical, lateral branches 2
1. Conidia not produced singly, often in groups 16
2. Conidia clavate, with 2–9 septa, mostly 3–7-septate, without equal-sized cells .... ***A. shizishana***
2. Conidia other type 3
3. Conidia obovoid, obconical, broad turbinate or subglobose, ratio of length to width less than 2 4
3. Conidia fusiform, ratio of length to width more than 2 8
4. Conidia broadly turbinate to napiform, with 1–2 septa, distal cell largest .... ***A. janus***
4. Central or distal cell of conidia largest 5
5. Conidia with 0–2 septa, or 0–3-septate 6
5. Conidia with 1–3 septa, or 2–3-septate 7
6. Conidia ellipsoid to obovoid to broadly turbinate, with 0–2 septa, 17.5–30 (23.2) × 12.5–20 (14.8) .... ***A. indica***
6. Conidia globose to obovoid, with 0–3 septa, 20–44 (32) × 17–25 (20.4) μm .... ***A. sphaeroides***
7. Conidia globose to turbinate, with 2–3 septa, 7–47.5 (32.2) × 17.5–27.5 (22) μm .... ***A. rutgeriense***
7. Conidia subglobose, with 1–3 septa, 23.5–30 (27.6) × 17–25 (20) μm .... ***A. sinensis***
8. Conidia with 4–12 septa, conidia longer .... ***A. multiformis***
8. Conidia with 2–5 septa 9
9. Conidial less than 20 μm wide 10
9. Conidial wider than 20 μm 11
10. Conidia fusiform to ellipsoid, with 2–3 septa, 32.5–47.5 (41) × 12.5–17.5 (15.5) μm .... ***A. fusiformis***
10. Conidia fusiform to ellipsoid, with 0–4 septa, mostly 3-septate .... ***A. salina***
11. Conidia mostly 4-septate 12
11. Conidia mostly with 2 or 3–4 septa 13
12. Conidia broadly fusiform, with 2–4 septa, mostly 4-septate, with larger size .... ***A. megalospora***
12. Conidia fusiform, sometimes with 5 septa, 50–65 × 20–25 μm .... ***A. reticulata***
13. Conidia turbinate, obovoid, fusiform, mostly 2–3-septate, with narrower and longer basal cell .... ***A. cookedickinson***
13. Conidia fusiform to obpyriform 14
14. Conidia 2–5-septate, mostly 3–4-septate, 40–90 (54) × 15–27.5 (18) μm .... ***A. longiphora***
14. Conidia mostly 2–3-septate 15
15. Conidia mostly 3-septate, 40–52–65 × 17–20–23 μm .... ***A. oudemansii***
15. Conidia 1–4-septate, mostly-2 septate, obpyriform to fusiform .... ***A. huaxiensis***
16. Conidiophores often merged into a bundle .... ***A. dendroides***
16. Conidiophores not merged into bundles 17

17. Conidia 0–1-septate 18
17. Conidia 1- or more than 1-septate 23
18. Conidia lacking septa 19
18. Conidia 0–1-septate 20
19. Conidia obovoid to ellipsoid, 15–31 (23.6) × 10–20 (15.9) μm ... ***A. amerospora***
19. Conidia elongate-ellipsoid, 11–16.8 × 5–6.6 μm .... ***A. nonseptata***
20. Conidia produced from longer sterigmata, conidia occasionally 1-septate, 17.5–32.5 (22.57) × 2.75–7.5 (5.5) μm .... ***A. yunnanensis***
20. Conidia producing from swollen denticles of apex 21
21. Conidia broadly ovoid to ellipsoid, 14.8–21.5 × 10.1–16.3 μm .... ***A. latispora***
21. Conidia other type 22
22. Conidia ellipsoid, occasionally 1-septate, 12–28 × 10–15 μm ..... ***A. botryospora***
22. Conidia cylindrical to clavate 23
23.Conidiophores with apical as well as intercalary clusters of conidia, conidia cylindrical, occasionally clavate, minority 1-septate at centre .... ***A. anomala***
23. Conidiophores with apical clusters of conidia only, conidia clavate, septum near the base .... ***A. pseudoclavata***
24. Conidia always 1-septate 25
24. Conidia sometimes more than 1-septate 39
25. Conidiophores with longer sterigmata at the apex, of the candelabrelloid type 26
25.Conidiogenous loci denticles, or shorter sterigmata 27
26. Conidia obovoid to clavate, narrowed at septum .... ***A. javanica***
26. Conidia elongate-obovoid or ovoid, not narrowed at septum, with narrower and slender basal cell, 33.5–57 × 11–15.5 μm .... ***A. shahriari***
27. Conidiophores geniculation because of repeated elongation, each conidiogenous loci with 1–3 conidia 28
27. Conidiophores erect, not geniculation 30
28.Conidiogenous loci without obvious denticles, conidia narrow pyriform, 24–40 × 12.5–18.8 μm .... ***A. paucispora***
28. Conidiogenous loci with obvious denticles 29
29. Conidia broadly pyriform, 25–35 × 18–24 μm, capturing nematodes by adhesive networks .... ***A. cystosporia***
29. Conidia obpyriform, 24–32.5 × 12.5–20 μm, capturing nematodes by simple hyphae .... ***A. perpasta***
30. Conidiogenous loci without obvious denticles, capturing nematodes by simple hyphae .... ***A. flagrans***
30. Conidiogenous loci with typical denticles or sterigmata 31
31. Conidiophores with several groups denticles 32
31. Conidiophores with apical or occasionally subapical denticles or sterigmata 34
32. Conidia subellipsoid, not narrowed at septum, 7.5–27.5 (5.8) × 5–10.5 (6.6) μm .... ***A. superba***
32. Conidia obovoid to pyriform 33
33. Conidia obovoid, not narrowed at septum, 28.5–32 (30) × 18–20.5 (20) μm .... ***A. obovata***

33. Conidia pyriform to obovoid, narrowed at septum, 17–35 (23) × 8.5–16 (12.1) μm .... *A. oligospora*
34. Conidiophores unbranched 35
34. Conidiophores branched or occasionally branched 36
35. Conidia obconical, produced from denticles, 15–37.5 (8.4) × 7.5–14.5 (11.8) μm .... *A. conoides*
35. Conidia ellipsoid, produced from sterigmata, 20–47.5 (30.9) × 7–12.5 (10.3) .... *A. musiformis*
36. More than 10 conidia arranged in a close capitate head 37
36. 1–6 conidia arranged in a loose capitate head 38
37. Conidia pyriform, occasionally ovoid, 26–40 × 12–23 μm .... *A. apscheronika*
37. Conidia ellipsoid to ovoid, 10–20 (17.5) × 5–8 (6.2) μm .... *A. cladodes*
38. Conidia elongate-ovoid, produced from denticles, 22.5–32 × 11–22.5 μm .... *A. chazarica*
38. Conidia obovoid, produced from sterigmata, 20–27.5 (24.4) × 7.5–12.5 (10.8) μm .... *A. robusta*
39. Conidia with more septa, mostly 2–5-septate, capturing nematodes by simple two-dimensional networks .... *A. dianchiensis*
39. Conidia less than 4-septate, capturing nematodes by three dimensional networks 40
40. Conidia sometimes non-septate 41
40. Conidia at least 1-septate 42
41. Conidia multiform, broadly turbinate to narrow fusiform, ellipsoid, clavate, obovoid .... *A. mangrovispora*
41. Conidia fusiform, largest cell obvious .... *A. microscaphoides*
42. Conidia fusiform or other type 43
42. Conidia other type, never fusiform 48
43. Conidia fusiform to other type 44
43. Conidia only fusiform 45
44. Conidia fusiform, obovoid or broadly conical, 3-septate, 37–55 (49) × 17.5–35 (28) μm .... *A. eudermata*
44.Conidia pyriform or fusiform, 1–3-septate, mostly 1-septate .... *A. guizhouensis*
45. Conidia mostly 3-septate 46
45. Conidia mostly 4 or 3–4-septate 47
46. Conidia of two types, macroconidia arranged loosely, 3–4-septate, 30–50 (45.1) × 8–16.5 (12.2) μm .... *A. polycephala*
46. Conidia arranged closely, 1–4-septate, 30–60 (36.2) × 15–30 (20.2) μm .... *A. thaumasia*
47. Conidia sometimes curved, mostly 4-septate, 40–70 × 9–14 μm .... *A. gampsospora*
47. Conidia mostly 3–4-septate, 46–70 (62.3) × 21–29 (24.7) .... *A. psychrophila*
48. Sterigmata or denticles formed from flank of conidiophores 49
48. Conidiophores with apical and at least subapical clusters of conidia 50
49. Conidia pyriform to broadly clavate,1–3 septa, mostly 2-septate, 25–40 (17.5) × 7.5–19 μm .... *A. clavispora*

49. Conidia elongate-ovoid to ellipsoid, 1–3-septate, mostly 1–2 septa, 28.5–56 × 11.5–22.5 μm .... ***A. tabrizica***
50. Conidia mostly 1-septate or with 1–2 septa 51
50. Conidia mostly 2-septate or with 2–3 septa 52
51. Conidia obovoid to ellipsoid, mostly 1–2-septate, 18–36 (28.1) × 12–20 (15.3) μm .... ***A. azerbaijanica***
51. Conidia pyriform, 1–2-septate, mostly with 1 septum .... ***A. oviformis***
52. Conidia broadly pyriform to subcylindrical, mostly 2–3-septate, occasionally 1-septate, 20–35 × 10–25 μm .... ***A. pyriformis***
52. Conidia ellipsoid, obovoid or cymbiform 52
53. Conidia of two types, macroconidia ellipsoid or cymbiform, 1–3-septate, mostly 2-septate .... ***A. scaphoides***
53. Conidia ellipsoid to obovoid or cymbiform, 2–3-septate .... ***A. vermicola***

Accepted Species of *Arthrobotrys*

1. *Arthrobotrys amerospora* S. Schenck, W.B. Kendr. & Pramer, Can. J. Bot. 55 (8): 979 (1977) [1976]

Characteristics: Mycelium hyaline, septate, branched. Conidiophores simple, erect, septate, unbranched, 75–250 μm long, producing 2–10 conidia singly from retrogressive conidiogenous loci on broad, conspicuous denticles at and near the apex. Conidia holoblastic, hyaline, obovoid, one-celled, 15–31 (23.6) × 10–20 (15.9) μm with a small truncate protuberance at the base. Germinating from the base, and often subsequently also from the apex. Chlamydospores yellowish, smooth-walled, sphaerical to elongate-ellipsoidal, 18–50 × 18–23 μm, usually intercalary, single or in chains. Capturing nematodes on adhesive hyphal loops.

Distribution: Canada, China (Guansi), India, USA.

Notes: The description is based on the protologue. Among *Arthrobotrys*, there are five species with single celled conidia, i.e. *A. amerospora* (Schenck et al. 1977), *A. anomala* (Barron and Davidson 1972), *A. botryospora* (Barron 1979), *A. yunnanensis* (Mo et al. 2005a) and *A. nonseptata* (Yu et al. 2009b). However, only conidia of *A. amerospora* and *A. nonseptata* are consistently non-septate, and conidia of the other three species are occasionally 1-septate. The obovoid conidia of *A. amerospora* distinguishes it from the elongate-ellipsoidal conidia of *A. nonseptata*. (Fig. 3.1)

Arthrobotrys anomala G.L. Barron & J.G.N. Davidson, Can. J. Bot. 50: 1773 (1972)

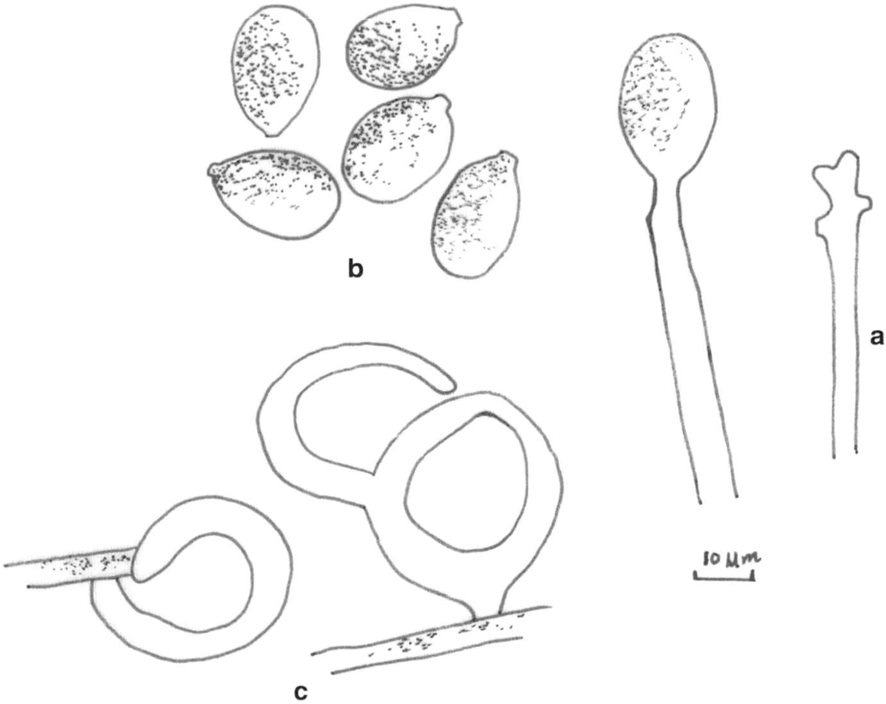

Fig. 3.1 *Arthrobotrys amerospora*. **a** conidiophore; **b** conidia; **c** adhesive network. Bar = 10 µm

Characteristics: Conidiophores short, erect or suberect, hyaline, indeterminate, 20–80 × 4–6 µm; conidia borne in an apical cluster or series of clusters, sometimes borne irregularly along the spore-bearing apex; conidia cylindric to long ellipsoid, occasionally clavate, hyaline, non-septate when attached, frequently forming a septum before germination, produced sympodially on pronounced denticles, 13–22 × 3–7 µm. Nematodes trapped by means of short, erect adhesive branches or more commonly by complex three dimensional adhesive networks.

Distribution: Canada.

Notes: The description is based on the protologue. *Arthrobotrys anomala* is characterized by cylindric to elongate ellipsoidal, 0 (–1)-septate conidia, which distinguish it from other species of *Arthrobotrys*. (Fig. 3.2)

Arthrobotrys apscheronica Mekht., Nov. Sist. Niz. Rast., 10: 174 (1973)

Characteristics: Conidiophores erect or curved, sometimes branched, occasionally repeated propagated, conidiophore apex expanded irregularly, bearing 12 (–16) conidia on denticles of apex; conidia pyriform, seldom ovoid, 1-septate at the centre

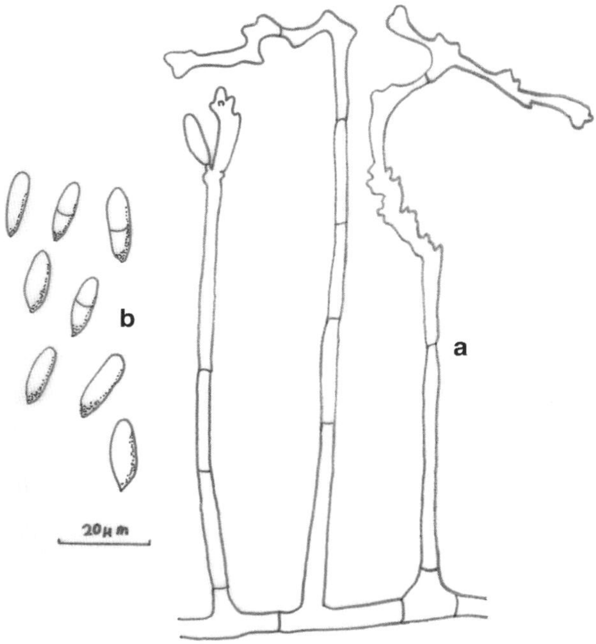

Fig. 3.2 *Arthrobotrys anomala*. **a** conidiophores; **b** conidia. Bar = 20 μm

of conidia, constricted at the septum. Nematodes trapped by three dimensional adhesive networks.

Distribution: Russia.

Notes: The description is based on the protologue. This species is similar to *A. conoides* in conidial shape, but it differs in having longer conidia (Fig. 3.3).

Arthrobotrys azerbaijanica (Mekht.) Oorschot [as 'azerbaidzhanica'], Stud. Mycol. 26: 70 (1985)
≡ *Nematophagus azerbaijanicus* Mekht. [as 'azerbaidzhanicus'], Mikol. Fitopatol. 9: 250 (1975)

Characteristics: Conidiophores erect, branched or unbranched, repeatedly propagating at least once, apex expanded, bearing 10–12 conidia on short and slender denticles; conidia obovoid to ellipsoidal, 1–4-septate, constricted at septum, sometimes 1–2 cells expanded, 18–36 (28.1) × 12–20 (15.3) μm; Chlamydospores present. Nematodes trapped by three dimensional adhesive networks.

Distribution: Russia.

Notes: The description is based on the protologue. This species resembles *A. pyriformis*, but has wider conidia (conidia of *A. pyriformis* 17–38 × 6.5–11.5 μm, mostly 2–3-septate). (Fig. 3.4)

Arthrobotrys botryospora G. L. Barron, Can. J. Bot. 57: 1371 (1979)

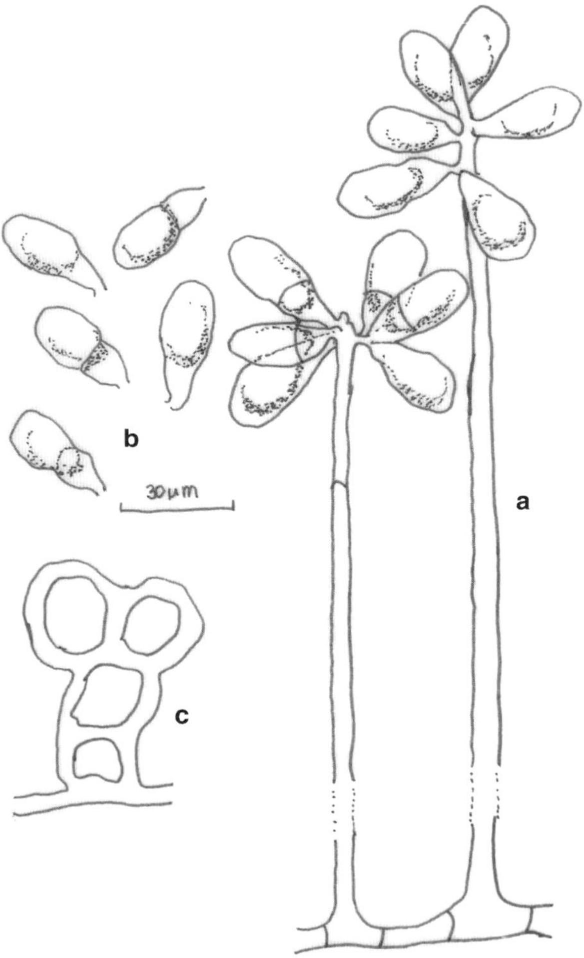

Fig. 3.3 *Arthrobotrys apscheronika.* **a** conidiophore; **b** conidia; **c** adhesive network. Bar = 30 µm

Characteristics: Colonies on CMA whitish. Mycelium spreading, vegetative hyphae hyaline, septate branched. Conidiophores hyaline, erect, septate, 250–350 µm long, 4.5–6.5 µm wide at the base, gradually tapering upwards to a width of 3–5 µm at the apex, simple or sparingly branched in strikingly loose capitate arrangement. Conidia 12–28 × 10–15 µm, hyaline, ellipsoid, non-septate or occasionally 1-septate (8 %) just below the centre of the spore and having a narrowly truncate protuberance at the base. Capturing nematodes by means of three dimensional adhesive networks.

Distribution: Canada (Ontario), China (Anhui, Beijing, Guizhou, Hebei)

Material examined: C9–1, C21–1, isolated from forest soil in Cangshan Mountain, Dali in 1999 by Lu Cao; 01–3, isolated from forest soil in Shigu, Lijiang in 1999 by Lu Cao.

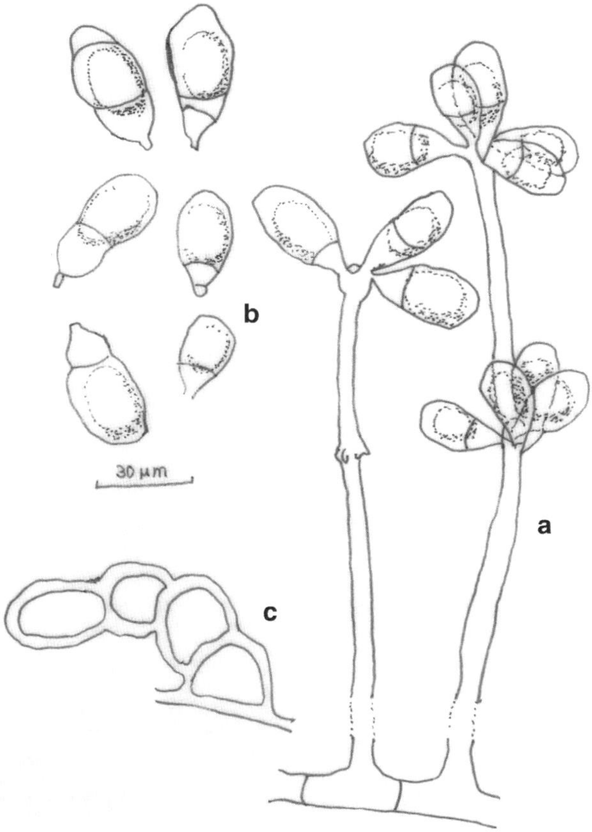

Fig. 3.4 *Arthrobotrys azerbaijanica*. **a** conidiophore; **b** conidia; **c** adhesive network. Bar = 30 µm

Notes: The only species of *Arthrobotrys* described with non-septate conidia are *A. anomala* and *A. amerospora*. *Arthrobotrys anomala* differs from *A. botryospora* in its cylindrical conidia. In *A. amerospora*, the conidia are larger than those of *A. botryospora*. The former also produce abundant chlamydospores, while chlamydospores are not produced in *A. botryospora*. In *A. amersospora* the conidia form in a solitary terminal cluster, produced from a retrogressive conidiogenous locus which remains relatively unswollen. In *A. botryospora*, successive production of conidia produces a swollen apex, with the main axis branching or elongating sympodially to produce a succession of spore clusters. (Fig. 3.5)

Arthrobotrys chazarica Mekht., Mycol. Res. 102: 683 (1998)

Characteristics: The fungus forms a delicate, fluffy, pale grey colony in pure culture on malt agar with pigments diffusing into the agar. Sporulation is abundant. Conidiophores simple, erect, branched. Conidia obovate to elongate-obovate,

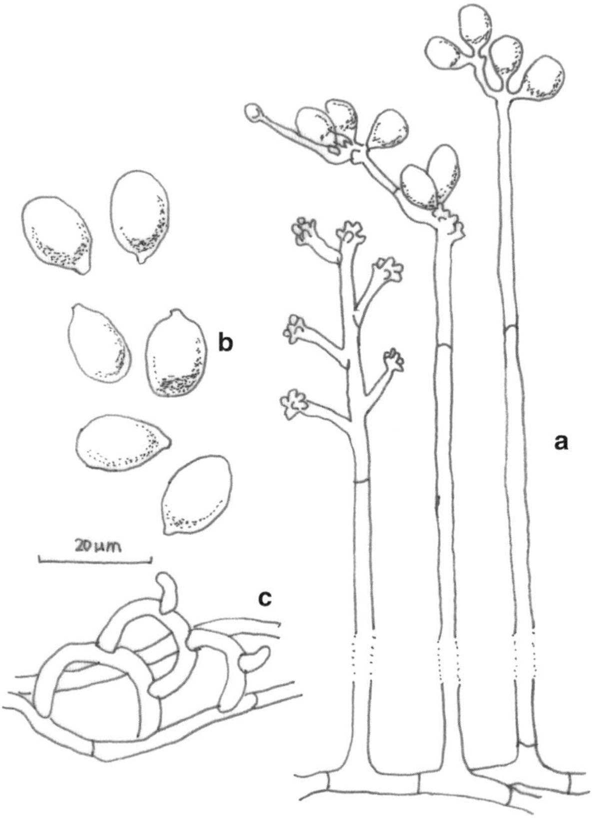

Fig. 3.5 *Arthrobotrys botryospora.* **a** conidiophore; **b** conidia; **c** adhesive network. Bar = 20 μm

1-septate, 22.5–32 × 11–22.5 μm, growing on very closely arranged, thin denticles. Capturing nematodes by means of three dimensional adhesive networks.

Distribution: Azerbaijan (Baku).

Notes: The description is based on the protologue. This species is similar to *A. robusta* in being 1-septate at the centre of the conidia, but conidia of *A. robusta* are mainly oblong-pyriform, and conidia are smaller. (Fig. 3.6)

Arthrobotrys cladodes Drechsler, Mycologia 29: 463 (1937)

≡ *Trichothecium cladodes* (Drechsler) Soprunov, Predacious fungi—Hyphomycetes and their use in the control of pathogenic nematodes: 113 (1958)

Characteristics: Colonies on CMA whitish, slow growing. Mycelium spreading, scanty; vegetative hyphae hyaline, septate, branched. Conidiophores hyaline, erect, septate, more or less branched, 68–236 μm long, 2.5–5 μm wide at the base, tapering gradually upwards to a width of 2–3 μm below the irregularly expanded, globose or somewhat coralloid apex, whereon are borne 10–20 conidia usually in a dense

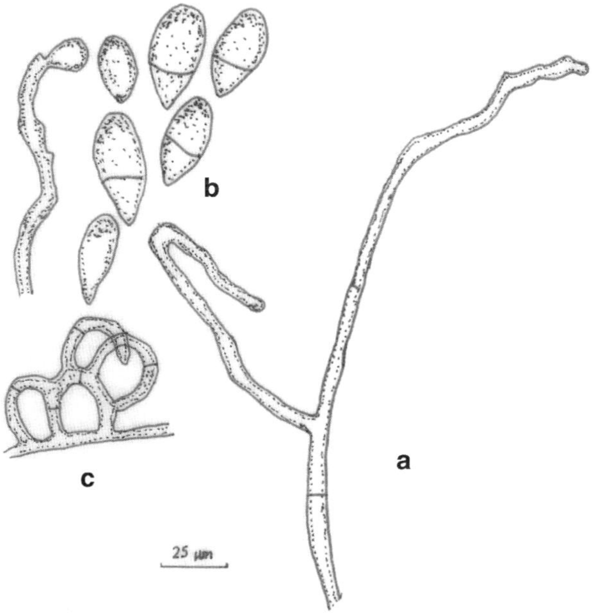

Fig. 3.6 *Arthrobotrys chazarica*. **a** conidiophore; **b** conidia; **c** adhesive network. Bar = 25 µm

capitate arrangement. Conidia hyaline, ellipsoid or elongate obovoid, mostly 10–20 (17.5) × 5–8 (6.2) µm, 1-septate near the centre of the spore. Chlamydospores not formed. Capturing nematodes by means of three dimensional adhesive networks.

Distribution: China (Beijing, Guizhou, Hebei, Taiwan, Xizang, Yunnan), Cuba (Prov. Ciudad de La Habana), UK, USA (Wisconsin)

Material examined: YMF1.00038, isolated from forest soil in Deqin, Yunnan, in 2002 by Jing Zhang; ①–33, isolated from field soil in Guiyang, Guizhou, in May 2002 by Ke-Qin Zhang; XZA–3, isolated from field soil in Zhouxian, Xizang in August 2002, by Minghe Mo. Permanent slide: H1–33.

Notes: This species was first isolated from leaf mold of deciduous woods in Virginia and Maryland, which is similar to *A. superba*, *A. oligospora* and *A. conoides* in conidia shape, but branch mode of conidiophores differ. The apex of the *A. cladodes* conidiophore is irregularly expanded, on which 10–20 conidia form in a densely capitate arrangement. (Fig. 3.7)

Arthrobotrys clavispora (R.C. Cooke) S. Schenck, W.B. Kendr. & Pramer, Can. J. Bot. 55: 982 (1977)
≡ *Dactylaria clavispora* R.C. Cooke, Trans. Br. Mycol. Soc. 47: 307 (1964)
= *Genicularia clavispora* (R.C. Cooke) Rifai, Reinwardtia 7 (4): 367 (1968)
= *Geniculifera clavispora* (R.C. Cooke) Rifai, Mycotaxon 2 (2): 216 (1975)
= *Nematophagus clavisporus* (R.C. Cooke) Mekht., Khishchnye Nematofagovye Griby—Gifomitsety (Baku): 107 (1979)

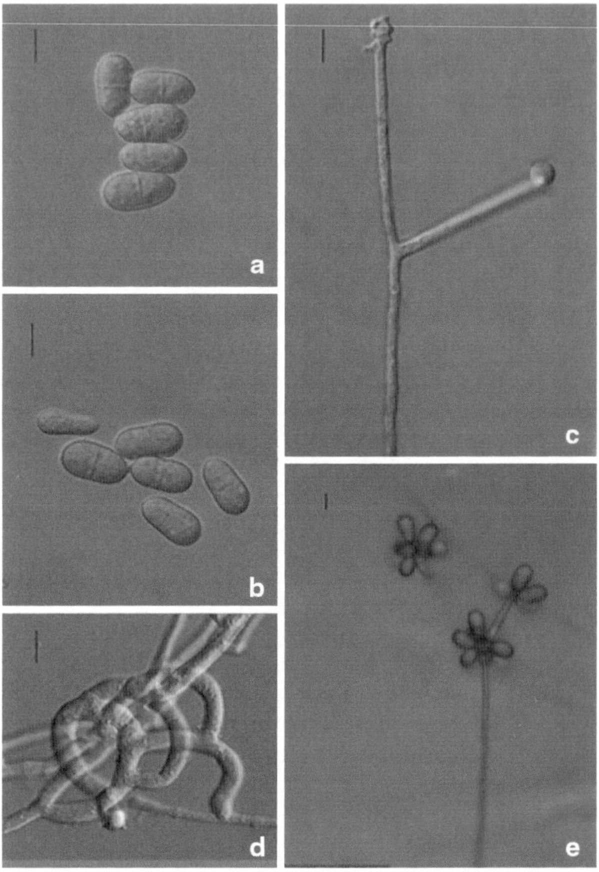

Fig. 3.7 *Arthrobotrys cladodes*. **a**, **b** conidia; **d** adhesive network; **c**, **e** conidiophore. Bar = 10 μm

Characteristics: Conidiophores erect, unbranched, septate, 150–350 μm high, at the apex of each bearing 3–11 widely spaced conidia, pleurogenous on small sterigmatal branches. A single, terminal conidium is formed at the conidiophore apex, which continues to grow for a short distance so that the conidium is displaced laterally. A second conidium is then formed at the new apex followed by growth of the conidiophore displacing this conidium. This process is repeated giving rise to the arrangement of conidia typical of this species. Conidia pyriform to broadly clavate, 25–40 (17.5) × 7.5–19 μm, rounded distally and tapering proximally to a narrow, truncate base. They are 1–2-septate, about twice as many having one septum as two. In both forms of conidium the distal cell is usually the larger. Capturing nematodes by means of three dimensional adhesive networks.

Distribution: France.

Notes: The description based on the protologue. This species is characterized by widely spaced conidia borne pleurogenous on small sterigmatal branches. (Fig. 3.8)

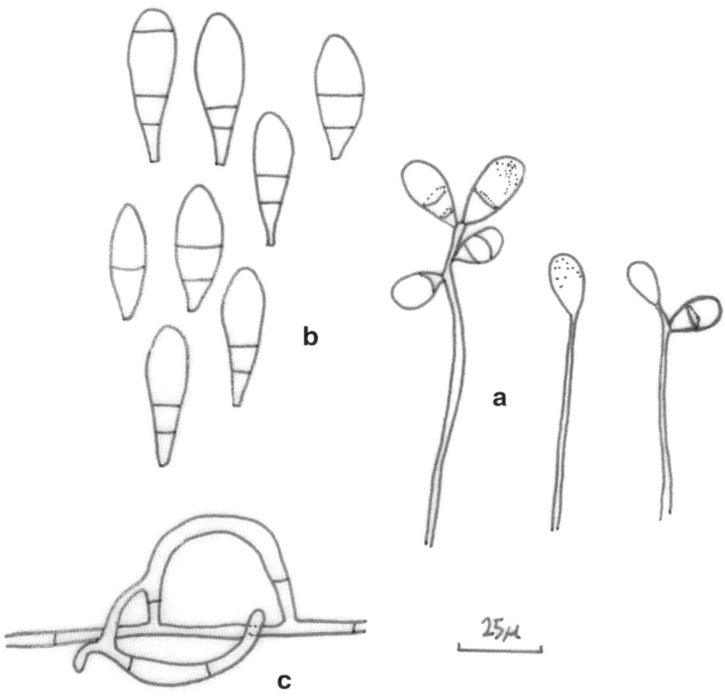

Fig. 3.8 *Arthrobotrys clavispora*. **a** conidiophores with conidia; **b** conidia; **c** adhesive network. Bar = 20 μm

Arthrobotrys conoides Drechsler, Mycologia 29 (4): 476 (1937)
= *Arthrobotrys pravicovii* (Soprunov) Sidorova, Gorlenko & Nalepina [as 'pravicovi'], Bot. Zh. SSSR 49: 1598 (1964)
= *Arthrobotrys pravicovii* (Soprunov) Mekht. [as 'pravicovi'], Dokl. Akad. Nauk Azerb. SSR 20 (6): 71 (1964)
= *Arthrobotrys tortor* Jarow., Acta Mycologica, Warszawa 4: 241 (1968)
= *Trichothecium pravicovii* Soprunov [as 'Pravicovi'], Predacious fungi—Hyphomycetes and their use in the control of pathogenic nematodes: 117 (1958)

Characteristics: Colonies on CMA whitish then turned to yellowish; mycelium spreading, vegetative hyphae hyaline, septate, branched, except for occasional storage filaments that are densely filled with protoplasm and up to 12 μm wide, measuring mostly 2–8 μm in diameter. Conidiophores hyaline, erect, 5–10-septate, usually not branched, mostly 4–7.5 μm wide at the base, tapering gradually upwards to a width of 2.5–5 μm in attaining a height of 120–420 μm before bearing on a globose or more irregularly expanded apex as many as 30 conidia in dense capitate arrangement; subsequently often, following repeated elongation, giving rise successively to additional clusters of conidia. Conidia hyaline, obconical, 15–37.5 (28.4) × 7.5–14.5 (11.8) μm, somewhat flattened at the base, broadly rounded at the apex, 1-septate, usually perceptibly constricted at the at the septum. Chlamydospores yellowish, globose or prolate ellipsoidal, 17–24 μm in diameter, or sometimes narrower, oblong-cylindrical, 30–50 μm long and approximately 15 μm wide. Capturing nematodes by means of three dimensional adhesive networks.

Distribution: China (Guizhou, Hainan, Hubei, Tianjing, Xizang, Yunnan), France, Germany (Berlin), Netherlands, Russian, South Africa (North West Province), Spain (Chavio verde), USA.

Material examined: 8705, isolated from dunghill in Huaxi, Guizhou in 1988 by Ke-Qin Zhang; GZHS–3, isolated from forest soil in Huishui, Guizhou in 1996 by Ke-Qin Zhang; J9–1, J11–2, J5–3, isolated from soil in Jianshui, Yunnan in 1999 by Yanju Bi; TJ–W1Z03, isolated from soil in Wanglanzhuang, Tianjing in 2000 by Wenpeng Li; XZA–4, isolated from pasture soil in Xizang in August 2000 by Minghe Mo; DL1–1, isolated from forest soil in Dali, Yunnan in September 2002 by Jing Zhang; YMF1.00009, YMF1.00541, YMF1.00866, J49–1, isolated from field soil in Huaxi, Guizhou in 1996 by Ke-Qin Zhang; YMF1.00009, YMF1.00541, isolated from forest soil in Baoshan, Yunnan in June 2002 by Jing Zhang; YMF1.00551, YMF1.00866, isolated from forest soil in Lijiang, Yunnan in October 2002 by Jing Zhang. Permanent slide: J49–1.

Notes: *A. conoides* has an extensive worldwide distribution, and was first isolated from delaying leaves from a greenhouse (Drechsler 1937). It resembles *A. oligospora* and *A. superba* in most morphological characters, with the exception of conidial shape and the position of septum. In *A. conoides*, conidia are elongate obconical, 1-septate in the lower third and constricted at the septum. In *A. oligospora*, conidia are obovoid, 1-septate near the base of the spore, and constricted at the septum. In *A. superba*, conidia are ellipsoidal, and 1-septate at the centre of the conidia.

The characters of examined strains match well with the original description (Drechsler 1937). Van Oorschot (1985) rechecked *A. tortor* deposited in CBS and ATCC, and found that *A. tortor* cannot be distinguished from *A. conoides*, and treated *A. tortor* as the synonym of *A. conoides*. (Fig. 3.9)

Arthrobotrys cystosporia (Dudd.) Sidorova, Gorlenko & Nalepina, Bot. Zh. SSR 49: 1598 (1964)
≡ *Trichothecium cystosporium* Dudd. Trans. Br. Mycol. Soc. 34: 600 (1952)
= *Arthrobotrys cystosporia* (Dudd) Mekht.[as 'cystosporium'], Dokl. Akad, Nauk Azerb. SSR 20 (6): 70 (1964)
= *Genicularia cystosporia* (Dudd.) Rifai, Mycotaxon 2 (2): 215 (1975)

Characteristics: The mycelium consist of straight, septate hyphae, 3–6 μm wide, sparingly branched. Conidiophores hyaline, erect or geniculate due to repeated propagated, usually not branched, attaining a height of 100–400 μm. The loose group of spores having the appearance of a panicle thus formed at each conidiogenous loci. The conidia hyaline, broadly pyriform, 25–35 × 18–24 μm, 1-septate near the base, proximal cell much larger. No chlamydospores were observed. Capturing nematodes by means of three dimensional adhesive networks.

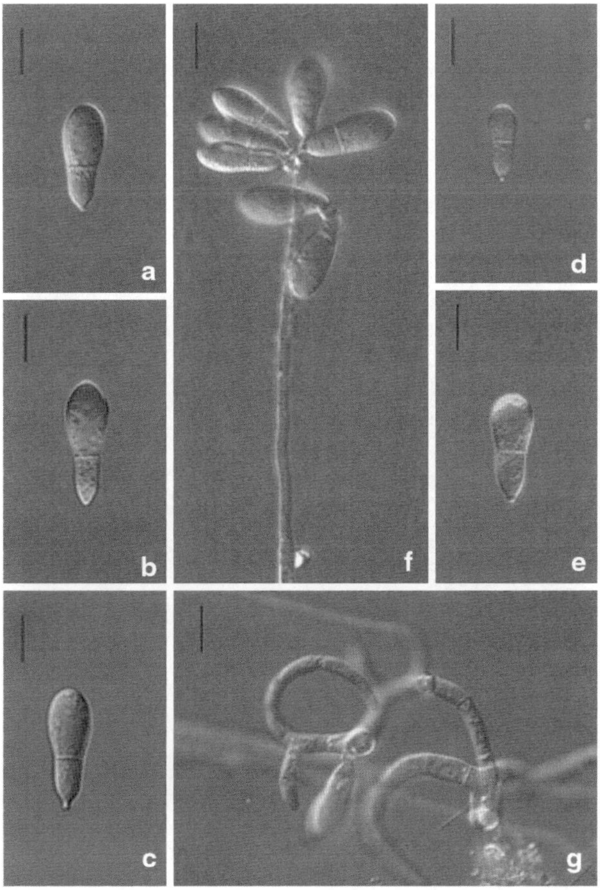

Fig. 3.9 *Arthrobotrys conoides*. Drechsler **a–e** conidia; **f** conidiophore; **g** adhesive network. Bars = 10 μm; Strain number: YMF1.00009

Distribution: Japan, UK.

Notes: The description based on the protologue. Conidia of this species are arranged loosely, not grown from typical denticles, but from sterigmata. (Fig. 3.10)

Arthrobotrys cookedickinson (Cooke & Dickinson) Z.F. Yu, **comb. nov**.
≡ *Monacrosporium cystosporum* Cooke & Dickinson, Trans. Br. Mycol. Soc. 48: 623 (1965)
MB 804791

Characteristics: Vegetative hyphae hyaline, septate, branched. Conidiophores hyaline, erect, simple, 8–11-septate, 60–420 μm long, 5–7.5 μm wide at the base, gradually tapering upwards to a width of 3–5 μm at the apex, bearing a single conidium. Conidia hyaline, broadly clavate or broadly turbinate to obovoid, 30–52.5 (42) × 15–22.5 (17.6) μm, 1–3-septate, mostly 2–3-septate with an almost globose terminal cell, distally rounded and tapered to a long tail formed by the two basal

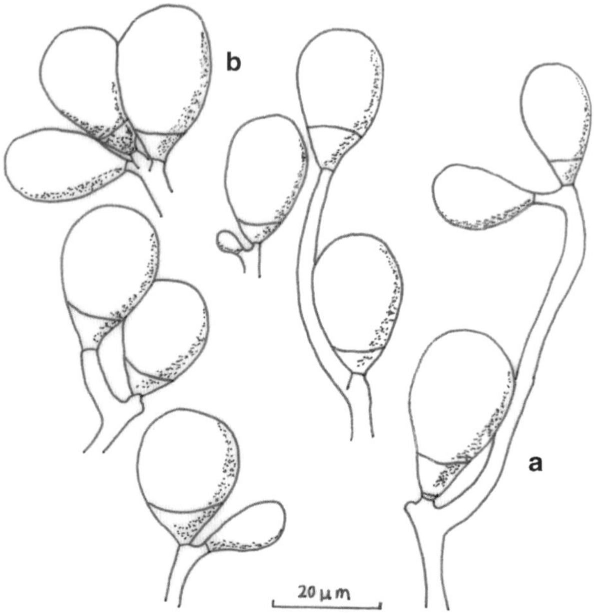

Fig. 3.10 *Arthrobotrys cystosporia*. **a** conidiophore; **b** conidia. Bar = 20 µm

cells, the proximal one being narrowly truncate. Chlamydospores present. Capturing nematodes by means of three dimensional adhesive networks.

Distribution: China (Anhui, Guizhou, Sichuan, Xizang, Yunnan), Germany (former FRG)

Material examined: ②–1–1, isolated from field soil in Huaxi, Guizhou in 1996 by Ke-Qin Zhang; XZM–2, isolated from soil in Xizang in August 2000 by Minghe Mo; DH4–2, isolated from soil in Ruili, Yunnan in October 2002 by Jing Zhang; YMF1.00024, isolated from forest soil in Jiuzhaigou, Sichuan in 2003 by Xuefeng Liu. Permanent slide: ②–1–1

Notes: This species is transferred to *Arthrobotrys* from *Monacrosporium* and renamed as *A. cookedickinson* to avoid confusing with *A. cystosporia* (Dudd.) Mekht. (1964). Cooke & Dickinson indicated *A. cookedickinson* was closely related to *A. eudermata*, but conidia of the latter were described as being much larger. Based on our observation, *A. cookedickinson* resembles *A. sphaeroides* in conidial shape, but *A. cookedickinson* is characterised by tail-like basal cell, which is absent in *A. sphaeroides*. (Fig. 3.11)

Arthrobotrys dendroides Kuthub. Muid & J. Webster Trans. Br. mycol. Soc. 84 (3): 564 (1985)

Characteristics: Colonies on CMA whitish, effuse, with scattered erect hyaline synnemata. Mycelium superficial, hyaline, septate, branched, 2.5–4 µm wide. Conidiophores simple, septate, erect, solitary or synnematous; mainly synnematous

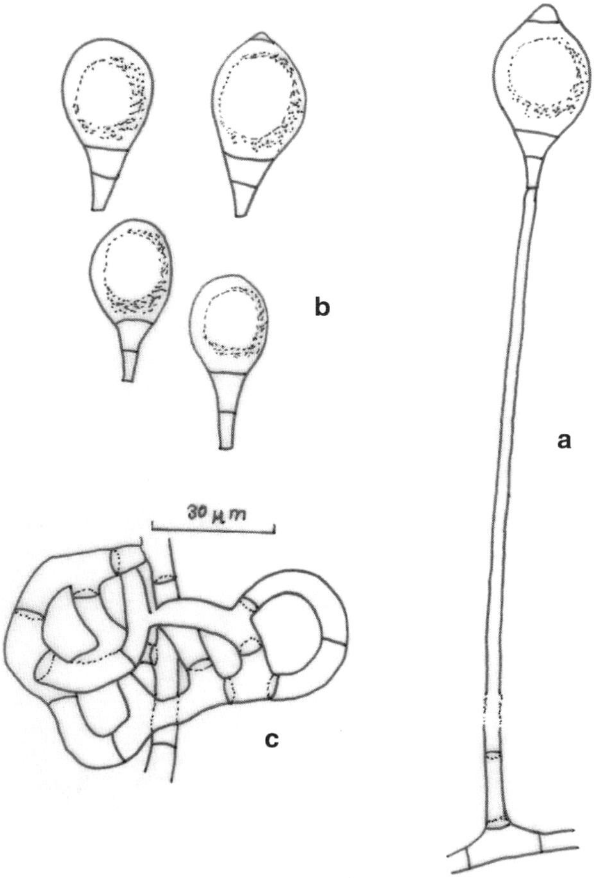

Fig. 3.11 *Arthrobotrys cookedickinson*. **a** conidiophore; **b** conidia; **c** adhesive network. Bar = 30 μm strain number: YMF1.00024

on natural substrata, synnemata 0.5–1.2 mm wide, 1 cm tall; conidiophores sometimes elongating sympodially to form additional heads of conidia at higher levels; when synnematous individual threads slender, septate, smooth, straight except near the apex, splaying out to become flexuous near apex, individual conidiophores up to 800 μm long and 2–3 μm wide. Conidiogenous cells polyblastic, integrated, terminal, raduliform. Conidia 10–20 (14.6) × 2.5–5 (4) μm, hyaline, ovate to oblong, 1-septate. Capturing nematodes by means of three dimensional adhesive networks.

Distribution: China (Yunnan), Malaysia (Kuala Lumpur), USA

Material examined: YMF1.00010, isolated from forest soil in Shuangbai, Yunnan in February 2001 by Xuefeng Liu. Permanent slide: Sb02–4.

Notes: *A. dendroides* is different from other species of *Arthrobotrys* because synnematous conidiophores even in cultures. (Fig. 3.12)

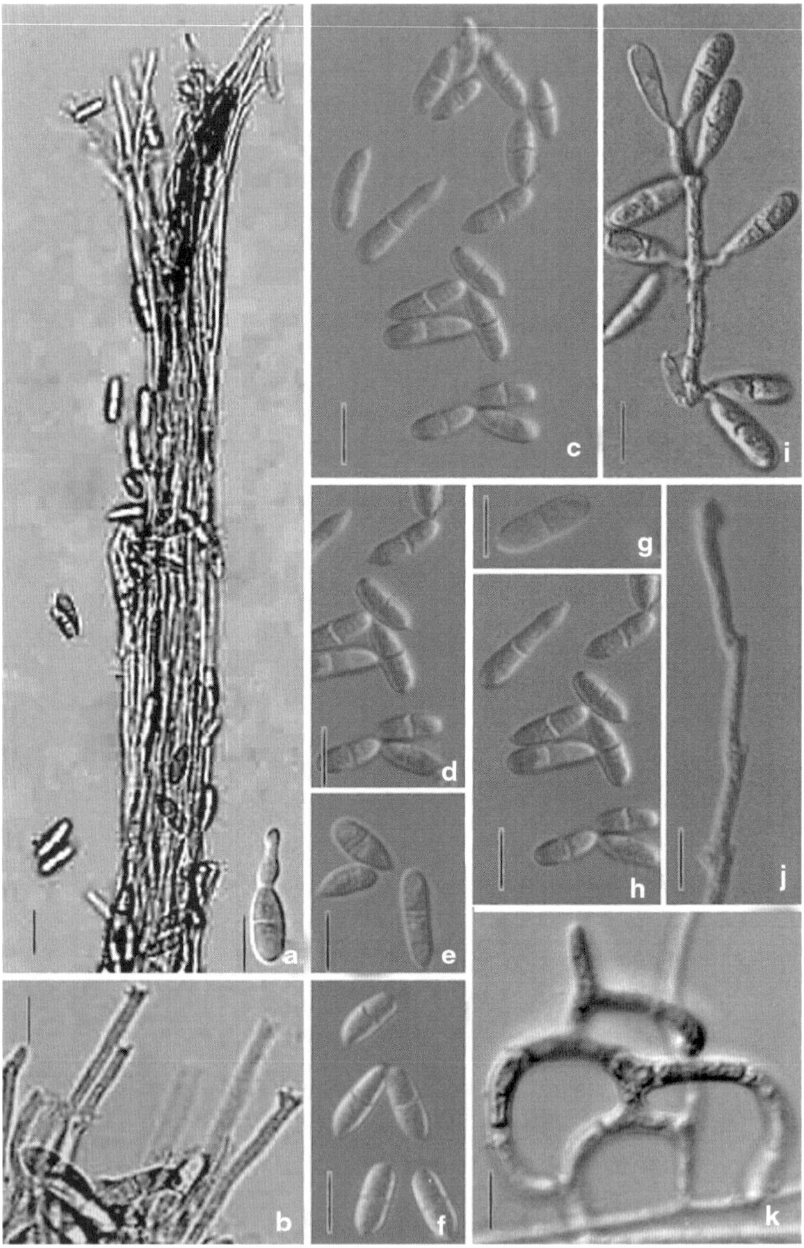

Fig. 3.12 *Arthrobotrys dendroides*. **a**, **b** conidiophore cluster; **c**–**h** conidia; **i**–**j** conidiophore; **k** adhesive network. Bars = 10 μm; Strain number: YMF1.00010

Arthrobotrys dianchiensis (Y. Hao & K.Q. Zhang) Z. F. Yu, **comb. nov.**
≡ *Dactylella dianchiensis* Y. Hao & K.Q. Zhang, Mycotaxon 82: 236 (2004)
MB 804792

Characteristics: Colony on PDA initially whitish and turning orange white after 10 days of incubation, rapidly growing and extending 7 cm in diameter at 28 °C, 5 cm at 25 °C within 4 days. At the same time the reverse side of the media becomes faintly yellow to reddish orange. Colonies on CMA whitish, rapidly growing and extending to a diameter of 6 cm at 25 °C within 6 days. Mycelium hyaline, scanty, hyphae septate, branched, 2.5–5 μm wide. Conidiophores erect, simple or branched, 2–4-septate, 245–425 μm long, 5–7.5 μm wide at the base, gradually tapering upwards to a width of 2.5–5 μm at the apex, initially a width of 2.5–5 μm at the apex, later often producing a few short branches or spurs near the apex, and bearing 2–3 conidia in sympodial arrangement. Conidia hyaline, spindle-shaped or clavate, narrowly round at the distal end, truncate at the base, 37.5–100 (70) × 10–17.5 (14.3) μm, 1–7-septate, mainly 2–5-septate. The proportion of conidia with 1, 2, 3, 4, 5, 6 and 7 septa is 10, 17, 14, 24, 22, 9 and 4 %, respectively. Some conidia had small tubercles at the both ends and could germinate from these tubercles. Conidia could produce secondary conidiophores and secondary conidia. The secondary conidia are spindle-shaped, 23.9 × 5 μm and 1-septate. Capturing nematodes by means of adhesive two dimensional networks.

Distribution: China (Yunnan)

Material examined: YMF1.00471, isolated from fresh water soil in Dianchi, Yunnan in 2002 by Yu'e Hao.

Notes: The description is based on the protologue. *A. dianchiensis* is similar to *A. multiformis* in primary conidial shape, but it can be distinguished from *A. multiformis* by the size and septation of primary conidia. (Fig. 3.13)

Arthrobotrys eudermata (Drechsler) M. Scholler, Hagedorn & A. Rubner, Sydowia 51 (1): 102 (1999)
≡ *Dactylella eudermata* Drechsler, Mycologia 42 (1): 40 (1950)
= *Dactylella eudermata* (Drechsler) Seifert & W.B. Kendr., in Seifert, Kendrick & Murase, Univ. Waterloo Biol. Ser. 27: 30 (1983)
= *Genicularia eudermata* (Drechsler) Rifai, Reintiwarda 7 (4): 367 (1968)
= *Geniculifera eudermata* (Drechsler) Rifai, Mycotaxon 2 (2): 216 (1975)
= *Golovinia eudermata* (Drechsler) Mekht., Dokl. Akad. Nauk Azerb. SSR 27 (2): 73 (1971)
= *Monacrosporium eudermatum* (Drechsler) Subram., J. Indian bot. Soc. 42: 293 (1964)[1963]

Characteristics: Colonies on CMA whitish, rapidly growing. Mycelium spreading, vegetative hyphae hyaline, septate, branched, 1.8–7.5 μm wide. Conidiophores hyaline, 2–8-septate, erect, simple, 400–680 μm long, 5–8 μm wide at the base, gradually tapering upwards to a width of 2–3 μm at the apex, bearing a single

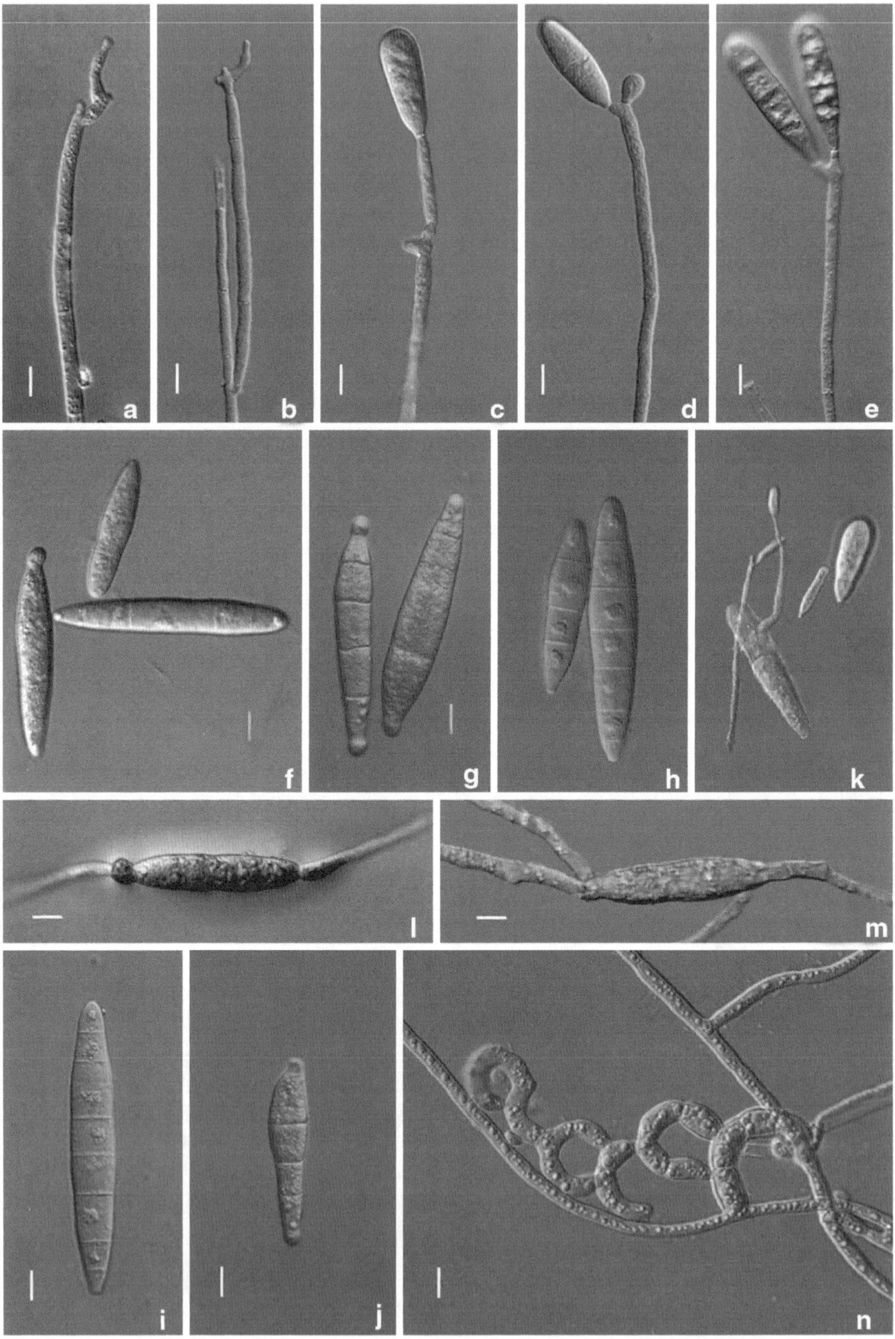

Fig. 3.13 *Arthrobotrys dianchiensis*. **a–e** conidiophore; **f–h**, **i–j** conidia; **k** conidiophore growing from conidia; **l–m** germinating conidia; **n** adhesive network. Bars = 10 μm; Strain number: YMF1.00571

conidium or 2–4 additional conidia. Conidia globose to broadly turbinate or broadly spindle-shaped, rounded at the distal end and truncate at the base, the central cell the largest, 37–55 (49) × 17.5–35 (28) μm, 1–3-septate; microconidia ellipsoid, aseptate, rounded at the distal end and truncate at the base, 10–15 × 5–6 μm. Chlamydospores present on aged cultures, globose, forming in chain or clusters, smooth-walled, yellow. Capturing nematodes by means of three dimensional adhesive networks.

Distribution: China (Anhui, Beijing, Fujian, Guangxi, Hebei, Huibei, Huinan, Shangdong, Xizang, Yunnan, Zhejiang), Burkina Faso, Germany (Berlin-Dahlem), Spain (La Palma, Tenerife)

Material examined: J42–1, J38–2, isolated from soil in Jianshui, Yunnan in 1999 by Yanju Bi; XZM–4, isolated from forest soil in Xizang in August 2000 by Minghe M; YMF1.00120, YMF1.00542, YMF1.00545, isolated from soil in Lijiang, Yunnan in September 2002 by Jing Zhang. Permanent slide: LJ3–4.

Notes: *A. eudermata* resembles *A. psychrophila* and *A. thaumasia* in conidial shape, but differs in conidiophores and width of conidia, respectively. Conidia of *A. eudermata* are considerably wider than in *A. thaumasia*. Moreover, in *A. eudermata*, conidiophores are simple, bearing a single conidium at the apex, or 2–4 additional conidia, while in *A. thaumasia*, conidiophores are usually branched near the apex, bearing 3–15, sometimes up to 25 conidia on blunt sterigmata in loose capitate arrangement. The difference between *A. eudermata* and *A. psychrophila* is that conidia of *A. psychrophila* are longer than in *A. eudermata*. (Fig. 3.14)

Arthrobotrys flagrans (Dudd.) Sidorova, Gorlenko & Nalepina, Bot. Zh. SSSR: 1598 (1964)
≡ *Trichothecium flagrans* Dudd. Trans. Br. mycol. Soc. 32 (3–4): 287 (1950)
= *Arthrobotrys flagrans* (Dudd.) Mekht., Dokl. Akad. Nauk Azerb. SSR 20 (6): 70 (1964)
= *Duddingtonia flagrans* (Dudd.) R.C. Cooke, Trans. Br. Mycol. Soc. 53 (2): 316 (1969)

Characteristics: Colonies effused, scanty, with few aerial hyphae. Conidiophores erect, straight, unbranched, attaining 150 μm long, bearing 2–6 conidia at the apex. Conidia obconical to ellipsoidal, 1-septate, the proximal cell broadly truncate, 25–50 × 10–15 μm. Intercalary chlamydospores present abundantly. Capturing nematodes by means of adhesive three dimensional hyphae.

Distribution: Germany, Soviet, UK.

Notes: The description is based on the protologue. This species has not typical denticles or sterigmata which exist among other species of *Arthrobotrys*, but it has typical adhesive three dimensional hyphae. (Fig. 3.15)

Arthrobotrys fusiformis (R.C. Cooke & C.H. Dickinson) M. Scholler, Hagedorn & A. Rubner, Sydowia 51 (1): 102 (1999)
≡ *Monacrosporium fusiforme* R.C. Cooke & C.H. Dickinson [as '*fusiformis*'], Trans. Br. mycol. Soc. 46: 628 (1965)
= *Golovinia fusiformis* (R.C. Cooke & C.H. Dickinson) Mekht., Khishchnye Nematofagovye Griby—Gifomitsety (Baku): 160 (1979)

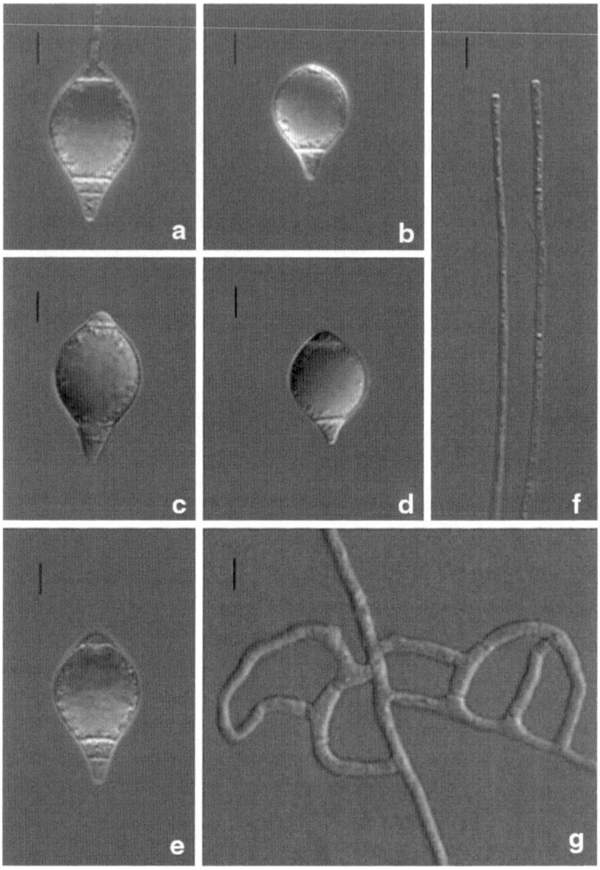

Fig. 3.14 *Arthrobotrys eudermata*. **a–e** conidia; **f** conidiophore; **g** adhesive network. Bars = 10 μm; Strain number: YMF1.00545

Characteristics: Vegetative hyphae hyaline, septate, branched. Conidiophores hyaline, erect, single, 3–7-septate, 250–390 μm long, 5 μm wide at the base, gradually tapering upwards to a width of 2.5–3 μm at the apex, bearing a single conidium, sometimes a second conidium formed on an about 15 μm branch near the conidiophore apex. Conidia hyaline, fusiform-ellipsoidal, 32.5–47.5 (41) × 12.5–17.5 (15.5) μm, 2–3-septate, rounded at the distal end and truncate at the base. Chlamydospores not observed. Capturing nematodes by means of three dimensional adhesive networks.

Distribution: China (Anhui, Hebei, Guizhou, Tianjing, Yunnan, Zhejiang), USA

Material examined: YMF1.00034, isolated from field soil in Balitai, Tianjingin 2001 by Wenpeng Li. Permanent slide: LT–11–03–3

Notes: According to Cooke & Dickinson (1965), *A. fusiformis* is similar to *A. psychrophila* and *A. salina*, but the dimensions, the shape and the septation of conidia distinguish it from these species. (Fig. 3.16)

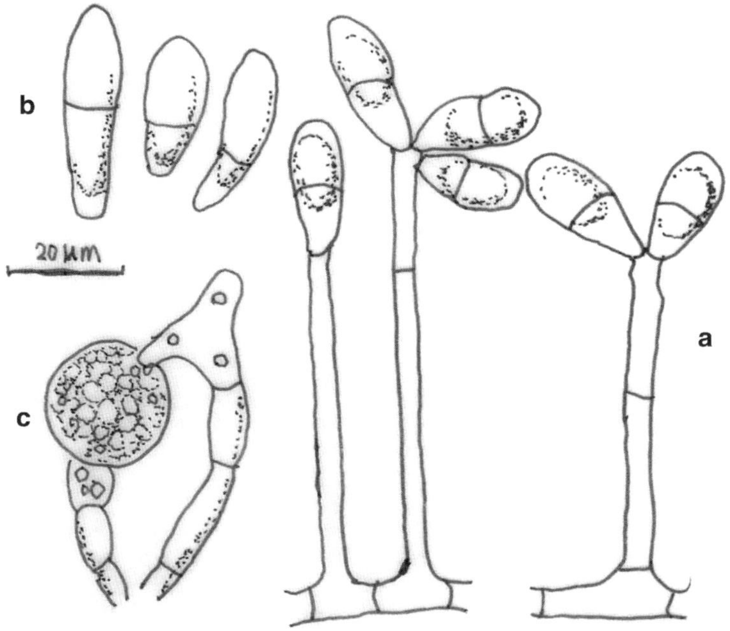

Fig. 3.15 *Arthrobotrys flagrans*. **a** conidiophore; **b** conidia; **c** chlamydospores. Bars = 20 µm

Arthrobotrys gampsospora (Drechsler) S. Schenck, W.B. Kendr. & Pramer, Can. J. Bot. 55 (8): 982 (1977)
= *Dactylaria gampsospora* (Drechsler) de Hoog & Oorschot, Stud. Mycol.26: 110 (1985)
≡ *Dactylaria gampsospora* Drechsler, Sydowia 15 (1–6): 9 (1962)
= *Monacrosporium gampsosporum* (Drechsler) Xing Z. Liu & K.Q. Zhang, Mycol. Res. 98 (8): 865 (1994)
= *Woroninula gampsospora* (Drechsler) Mekht., Khishchnye Nematofagovye Griby- Gifomitsety (Baku): 112 (1979)

Characteristics: Mycelium colourless, branched, septate. Conidiophores erect, mostly 150–626 µm high, 5–7 µm wide at the base, tapering to 2–3 µm at the apex, bearing several conidia in a loose head on tooth–like projecting often 3–5 µm long. Conidia hyaline, spindle-shaped, curved, 1–4-septate, mostly 4-septate, 25–76 × 7–16 µm. Microconidiophores often 1–3-septate, 75–125 µm high, 4–5 µm wide at the base, tapering to a width of 1.2–2 µm long at the apex, furnished distally with several tooth-like projections 1.5–2 µm long on which bearing microconidia. Microconidia colourless, elongate obovoid, tapering towards the base, broadly rounded at the distal end, 10–17 × 4–6 µm. Chlamydospores present, 8–21 × 6–17 µm Capturing nematodes by means of three dimensional adhesive networks.

Distribution: Canada, USA.

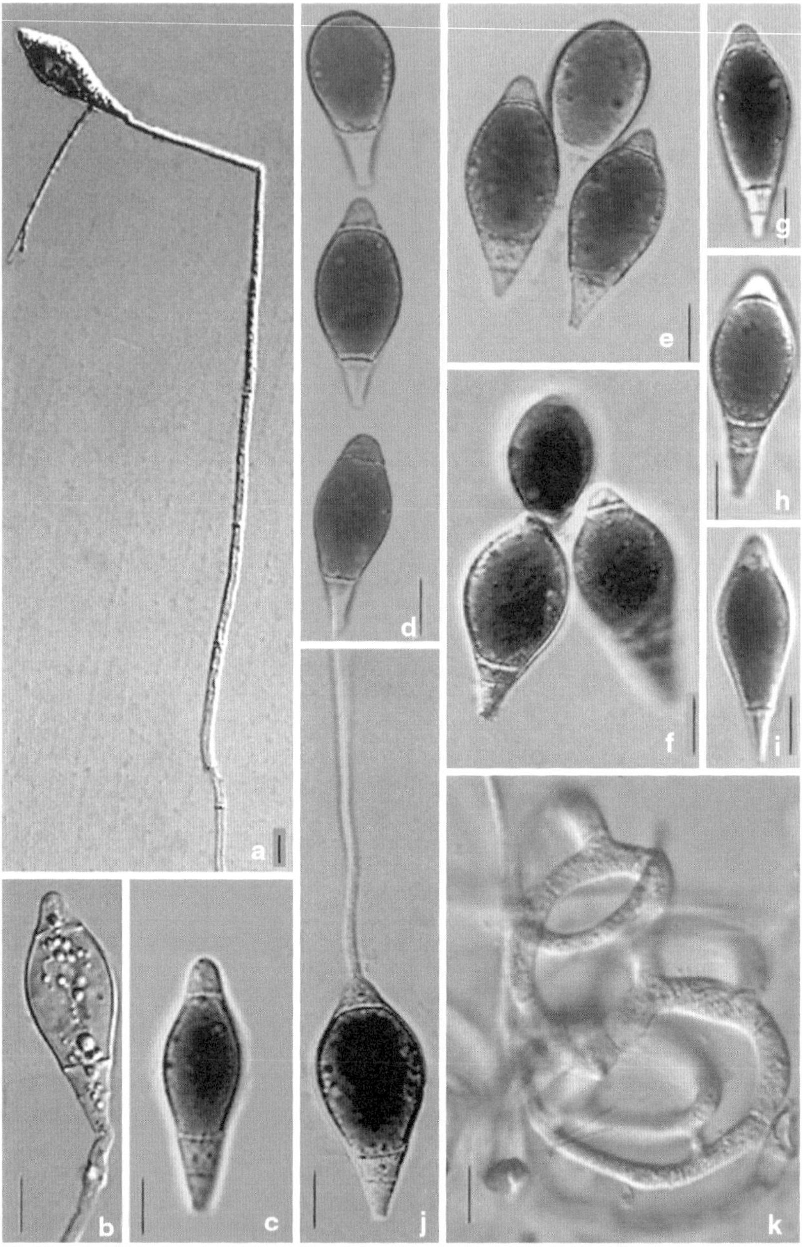

Fig. 3.16 *Arthrobotrys fusiformis*. **a–b** conidiophore; **c–j** conidia; **k** adhesive network. Bars = 10 μm; Strain number: YMF1.00034

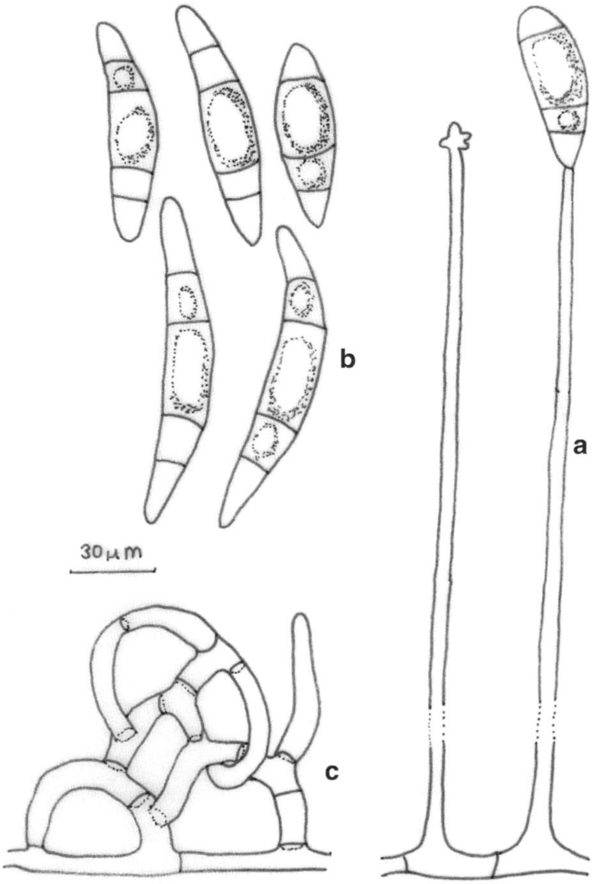

Fig. 3.17 *Arthrobotrys gampsospora*. **a** conidiophore; **b** conidia; **c** adhesive network. Bar = 20 µm

Notes: The description is based on the protologue. This species is different from other species of *Arthrobotrys* in its curved and more septate conidia. (Fig. 3.17)

Arthrobotrys guizhouensis K.Q. Zhang, Acta Mycol. Sin. 13: 101 (1994)

Characteristics: Colonies on CMA initially whitish and turned to reddish, rapidly growing. Mycelium spreading, dense, vegetative hyphae hyaline, septate, branched, mostly 3–9 µm wide. Conidiophores hyaline, erect, branched, 5–10-septate, 120–370 µm long, 5–7.5 µm wide at the base, tapering upwards to a width of 4–5 µm before bearing on irregularly expanded apex whereon form conidia; subsequently often, following repeated elongation, giving rise successively to additional clusters of conidia. Two type of conidia appear, macroconidia fusoid to ellipsoidal, 34–42.5 × 14–17.5 µm, 1–3-septate, proportion of conidia with 1, 2, and 3 is 95,

4 and 1%, respectively. Microconidia pyriform, 1-septate, 19–30 × 8.4–14 μm. Chlamydospores yellowish, globose, 7.5–12.5 μm in diameter. Capturing nematodes by means of three dimensional adhesive networks.

Distribution: China (Guizhou, Yunnan).

Material examined: 8736, isolated from soil in Guiyang, Guizhou in September 1987 by Ke-Qin Zhang; GZFJ-48, isolated from forest soil in Fanjing Mountain, Guizhou in 1996 by Ke-Qin Zhang;YMF1.00014, isolated from field soil in Wenshan, Yunnan in April 2002 by Zhiwei Zhao; LJ1–9-2, isolated from soil in Lijiang, Yunnan in September by Jun Zhang. Permanent slide: P4. 50–1.

Notes: *A. guizhouensis* morphologically resembles *A. oviformis* and *A. vermicola*. However, *A. guizhouensis* differs from *A. oviformis* in irregularly expanded conidiophores at the apex, and larger conidia. It also differs from *A. vermicola* in branched and irregularly expanded conidiophores at the apex, and mostly 1-septate conidia. In *A. vermicola*, conidiophores are unbranched and not expanded at the apex, conidia are mostly 2–3-septate. (Fig. 3.18)

Arthrobotrys huaxiensis (K.Q. Zhang, Xing Z. Liu & L. Cao) Z.F. Yu, **comb. nov.**
≡ *Monacrosporium guizhouense* K.Q. Zhang, Xing Z. Liu & L. Cao, Mycol. Res. 100: 275, 1996
MB 804790

Characteristics: Colonies on CMA pink. Hyphae hyaline, thin-walled, septate, frequently branched, 1.5–5 μm wide. Conidiophores hyaline, septate, single, simple, erect, 172–350 μm long, 3.5–6 μm wide at the base, gradually tapering upwards to a width of 2–2.5 μm at the apex, bearing a single conidium. Conidia hyaline, obpyriform to fusiform, 1–4-septate, mostly 2-septate, 30.5–71.5 (52.7) × 18.5–28.5 (23.9) μm. Capturing nematodes by means of three dimensional adhesive networks.

Distribution: China (Guizhou)

Material examined: GAU1001, isolated from soil in Huaxi, Guizhou in October 1993 by Ke-Qin Zhang.

Notes: This species traps nematodes by means of three dimensional adhesive networks, which was transferred from *Monacrosporium* to *Arthrobotrys* based on the present classification and renamed as *A. huaxiensis* Z.F. Yu to avoid confusing with *A. guizhouensis* K.Q. Zhang (1994). *A. huaxiensis* resembles *A. cookedickinson* and *A. fusiformis* in conidial septation, but differs in its fusiform to obpyriform conidia, conidial size and also in its unbranched conidiophores. (Fig. 3.19)

Arthrobotrys indica (Chowdhry & Bahl) M. Scholler, Hagedorn & A. Rubner, Sydowia 51 (1): 102 (1999)
≡ *Monacrosporialla indica* Chowdhry & Bahl [as 'indicum'], Curr. Science 51: 895 (1982)

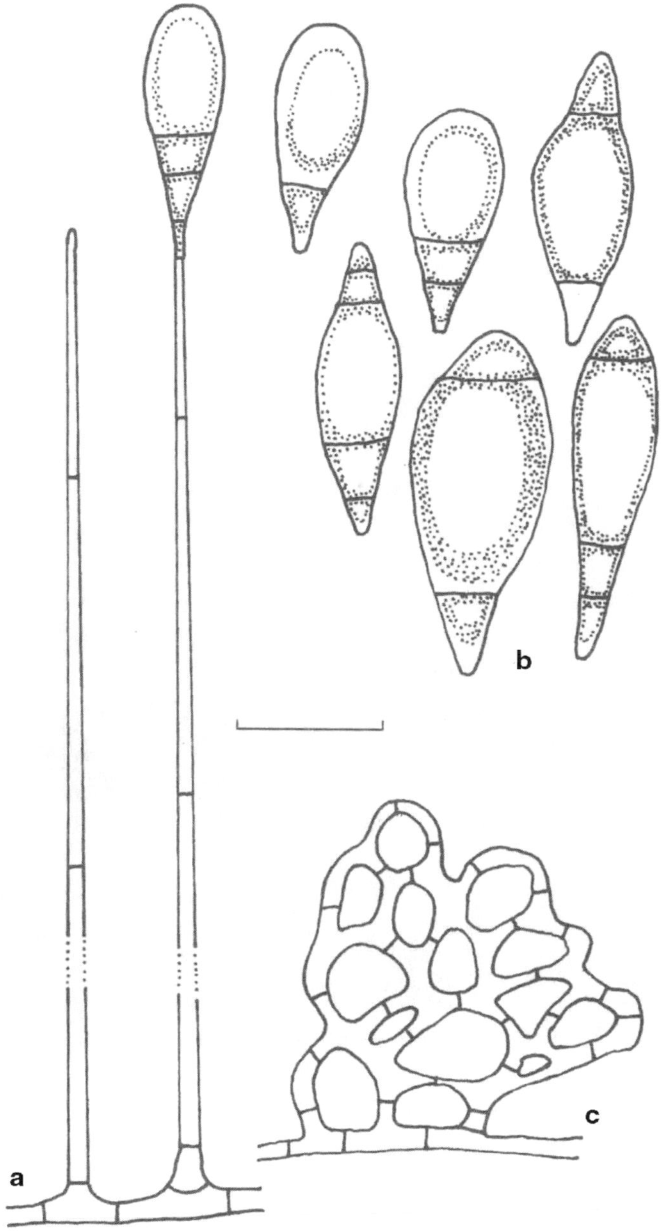

Fig. 3.18 *Arthrobotrys guizhouensis*. **a** conidiophore; **b** conidia; **c** adhesive network. Bars = 30 μm

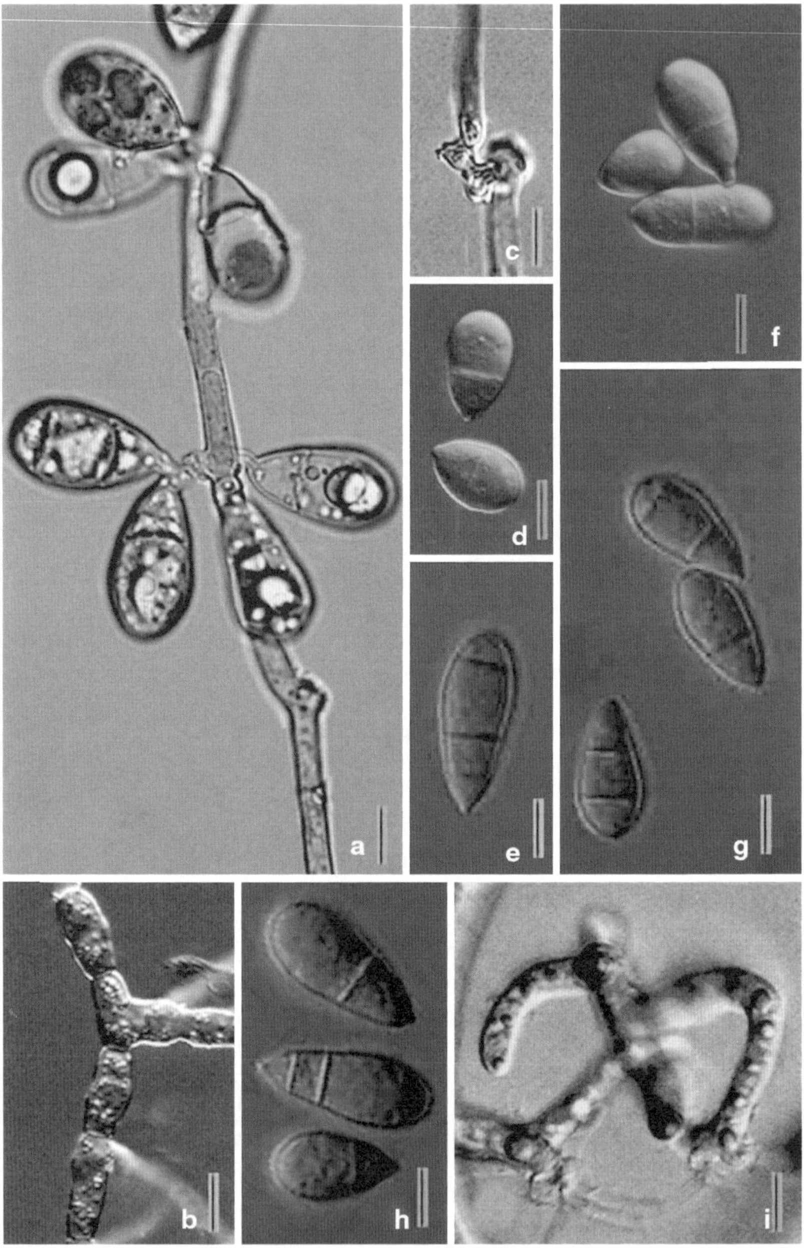

Fig. 3.19 *Arthrobotrys huaxiensis*. **a** conidiophore; **b** chlamydospore; **d–h** conidia; **i** adhesive network. Bar = 10 μm; Strain number: YMF1.00014

Characteristic: Vegetative hyphae hyaline, septate, branched, 2–3 μm wide. Conidiophores hyaline, erect, simple or occasionally branched, 8–20-septate, 150–350 μm long, 3–6 μm wide at the base, gradually tapering upwards to a width of 1–2 μm at the apex. Conidia hyaline, elliptic or broadly top–shaped, 0–2-septate, rounded at the distal end and truncate at the base, a navel at the base, 17.5–30 (23.2) × 12.5–20 (14.8) μm. Chlamydospores present. Capturing nematodes by means of three dimensional adhesive networks.

Distribution: China (Guizhou, Xizang), India (Delhi)

Material examined: Sjz3.11.1, isolated from field soil in Huaxi, Guizhou in 1996 by Ke-Qin Zhang; XZM–7, isolated from soil in Xizang in August 2000 by Minghe Mo.

Notes: *A. indica* resembles *A. eudermata* and *A. cookedickinson* in conidial shape, but the dimension of conidia in *A. eudermata* is larger than in *A. indica*. *A. indica* produces navels at the base of conidia while this character is absent in *A. cookedickinson*. (Fig. 3.20)

Arthrobotrys janus (S.D. Li & Xing Z. Liu) Z.F. Yu, **comb. nov.**
≡ *Monacrosporium janus* S.D. Li & X.Z. Liu, Mycol. Res. 107: 890 (2003)
MB 804791

Characteristics: Colonies growing rapidly on PDA, attaining 8 cm diameter within 7 days at 25 °C, producing orange reddish pigment resulting in medium colored. Aerial mycelium fluffy, whitish when young and becoming light yellowish when aged. Hyphae hyaline, septate branched. Conidiophores hyaline, simple, slender, erect, 185–250 μm long, 5–6.5 μm wide at the base, tapering to 1.5–2.5 μm wide at the apex, bearing a single conidium. Conidia broadly turbinate to napiform, 1–2-septate but mostly (>65 %) 1-septate at basal part, 15–26 (22.5) × 17.5–37.5 (28.5) μm. The distal cell are sphaerical and much bigger than others, proximal cell obconic, tapering to a short truncate base, central cell disciform if any. Chlamydospores sphaerical to ellipsoid, intercalary, 7.5–25 (15.5) × 4–11.5 (7.5) μm. Capturing nematodes by means of three dimensional adhesive networks.

Distribution: China (Shandong)

Material examined: YMF1.01312, isolated from field soil in Mixian, Shandong in September 2000 by Shidong Li.

Notes: *A. janus* closely resembles, but differs from, *A. indica*, *A. sphaeroides* and *A. sinensis* in conidial morphology. According to the original description of these species, the conidia of *A. indica* are 22–30 × 14–20 μm, and mostly 2-septate with distinct basal hila (Chowdhry and Bahl, 1982); those of *A. sphaeroides* are 28–42 × 19–29 μm, usually 2-septate with one basal and one terminal septum (46 %) or 1-septate with only one basal septum (29 %) (Castaner 1968); and those of *A. sinensis* are 25–30.5 × 15–18 μm with 37 % 3-septate, 42 % 2-septate, and 21 % 1-septate (Liu and Zhang 1994). While conidia of *A. janus* are 20–25 × 24.5–32.5 μm

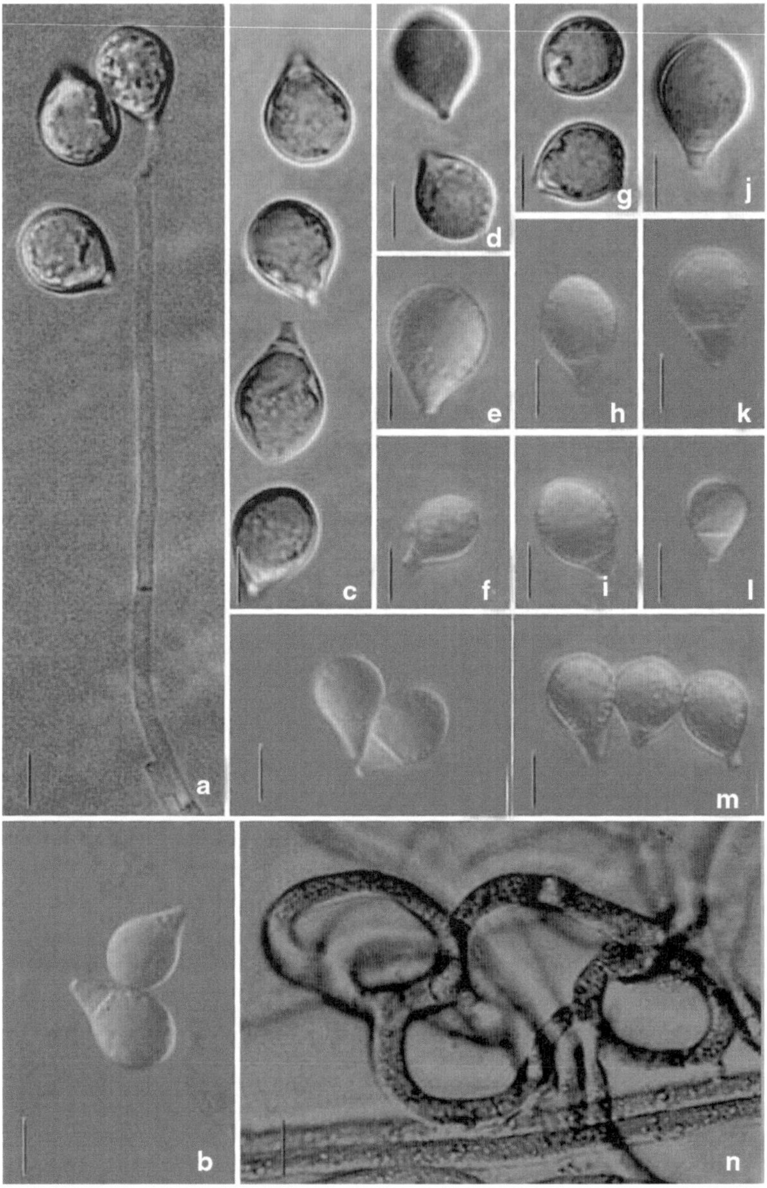

Fig. 3.20 *Arthrobotrys indica*. **a** conidiophore; **b–m** conidia; **n** adhesive network. Bars = 10 μm; Strain number: XZM-7

with 1–2 transverse septa, mostly (>65 %) are 1-septate in the basal part. When compared morphologically, *A. janus* can be easily distinguished from *A. sinensis* by its conidial shape and septation. This fungus also similar in conidial size, shape and septation to *A. cystosporia* (conidia 24–35.5 × 18–25 μm), *A. obovata*

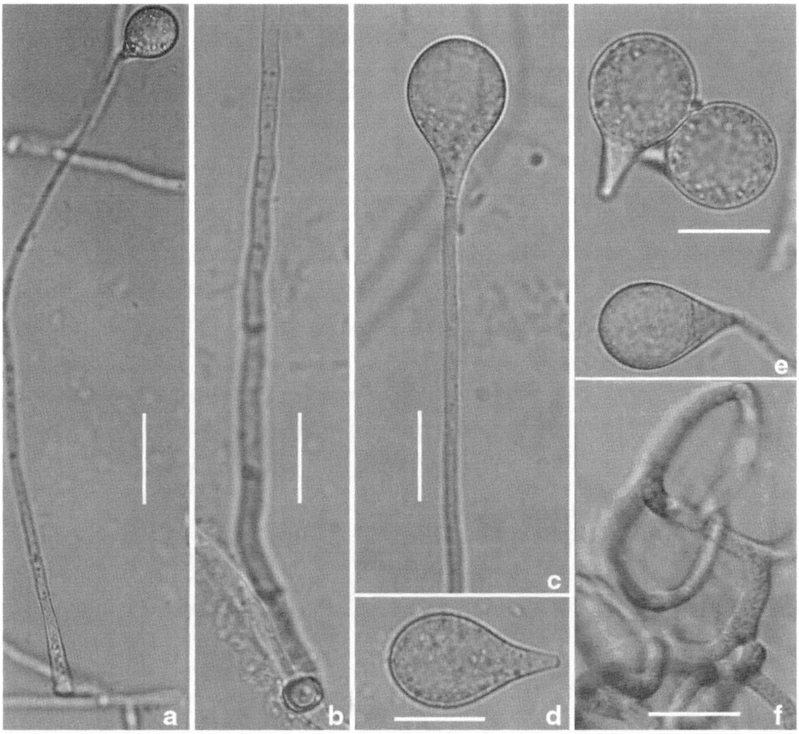

Fig. 3.21 *Arthrobotrys janus*. **a–c**. conidiophore; **d–e** conidia; **f** adhesive network. Bars: **a** = 35 μm, **b–f** = 20 μm; Strain number: YMF1.01312

(28.5–32 × 18–20.5 μm) and *A. perpasta* (24–32.5 × 12.5–20 μm), but differ in that the conidiophores of all the three latter species are geniculately branched (Duddington 1951a, Rifai and Cooke 1966, Zhang et al. 1996a). (Fig. 3.21)

Arthrobotrys javanica (Rifai & R.C. Cooke) Jarow., Acta Mycologica, Warszawa 6 (2): 373 (1970)
≡ *Candelabrella javanica* Rifai & Cooke, Trans. Br. mycol. Soc. 49 (1): 162 (1966)

Characteristics: Colonies on CMA almost hyaline; mycelium spreading, vegetative hyphae hyaline, septate, branched, smooth–walled, 3–7 μm wide. Conidiophores erect, straight, subulate, hyaline, smooth, 5–10-septate, 220–400 μm long, about 5–7.5 μm wide at the base, gradually tapering to 2.5–5 μm at the apex, enlarged slightly as ramification took place and bearing a lax, short, candelabrum-like apical branching system. Conidia arose singly as blown-out ends of the conidiophore and of the subsequently developed growing points, 20–37.5 (27.9) × 7.5–10 (8.8) μm, smooth-walled and hyaline, narrowly obovoid or clavate, 1 septum, slightly constricted at the septum, rounded distally with the obconical basal cell tapering to a truncate base. Chlamydospores yellowish, subglobose, 7.5–15 μm in diameter. Capturing nematodes by means of three dimensional adhesive networks.

Distribution: China (Guizhou, Xizang, Yunnan), France (Limousin), Indonesia (Java)

Material examined: Dian5 (4), isolated from fresh water soil in Dianchi Lake, Yunnan in 1996 by Ke-Qin Zhang et al. YMF1.00015, YMF1.00548, isolated from field soil in Lincang, Yunnan in April 1999 by Lu Cao; MW17–2, isolated from forest soil in Xishuangbanna, Yunnan in 1999 by Yanju Bi; XZA–5, isolated from field soil in Xizang in August 2000 by Minghe Mo; DH6–2, isolated from soil in Ruili, Yunnan in October 2002 by Jing Zhang. Permanent slide: LL4–1.

Notes: Rifai and Cooke (1966) established a new genus *Candelabrella* Rifai and Cooke (1966) based on long, distinctive, subcylindrical conidial pegs and the candelabrum-like branching of the conidiophore. However, this genus was not accepted by most scholars, so this species was transferred into *Arthrobotrys* (Jarowaja, 1970). *A. javanica* resembles *A. musiformis* in the candelabrum-like branching of the conidiophores, but differs in conidia shape. In *A. javanica*, conidia are 20–37.5 (27.9) × 7.5–10 (8.8) μm, narrowly obovoid or clavate, while in *A. musiformis*, conidia ellipsoid, mostly slightly curved, 20–47.5 (30.9) × 7–12.5 (10.3) μm. (Fig. 3.22)

Arthrobotrys latispora Hong Y. Su & X.Y. Yang, in Su, Liu, Li, Cao, Chen & Yang, Mycotaxon 117: 32 (2011)

Characteristics: Colonies grew rapidly on CMA medium, attaining 6 cm diameter in 8 days at 25 °C. Mycelium spreading, vegetative hyphae hyaline, septate and branched, mostly 2–4 μm wide. Conidiophores colourless, erect, simple, septate frequently, 60–120 μm high, 2–5 μm wide at the base and 1.8–4 μm at the apex, producing 4–12 conidia singly from conidiogenous loci on conspiculus, partly superposed nodes at and near the apex. Conidia colourless, broadly ovoid-oval, broadly rounded at the apex, rounded truncate at the narrowed base, 1-septate or aseptate at the centre, 14.8–21.5 × 10.1–16.3 (average18.3 × 13.5) μm (living state). Approximately 41 % of the conidia were non-septate and 59 % 1-septate. Chlamydospores sphaerical to ellipsoidal. Nematodes are trapped by three dimensional adhesive networks.

Distribution: China (Yunnan)

Material examined: YMF1.03168, isolated from soil in Yongping, Yunnan in October 2009 by Hongyan Su.

Notes: *A. latispora* is characterized by its broad, ovoid or slightly oval, partly aseptate conidia, which are borne on nodes. Based on conidial shape, *A. latispora* most closely resembles *A. amerospora*. The conidia of both species are ovoid, but differ in conidial size and septation. (Fig. 3.23)

Arthrobotrys longiphora (Xing Z. Liu & B.S. Lu) M. Scholler, Hagedorn & A. Rubner, Sydowia 51 (1): 102 (1999)
≡ *Monacrosporium longiphorum* Xing Z. Liu & B.S. Lu, Mycosystema 6: 65 (1993)

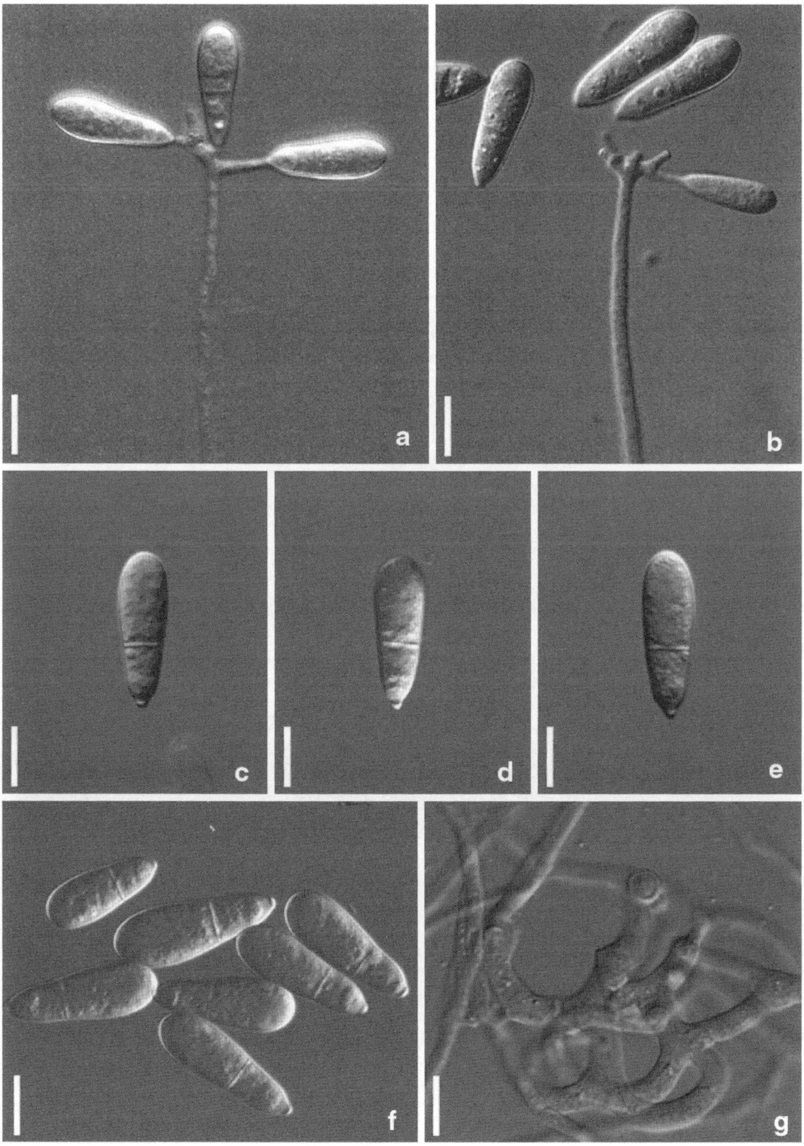

Fig. 3.22 *Arthrobotrys javanica*. **a**–**b** conidiophore; **c**–**f** conidia; **g** adhesive network. Bars = 10 μm; Strain number: YMF1.00015

Characteristics: Colonies on CMA white, growing moderately. Mycelium effuse, vegetative hyphae smooth-walled, hyaline, septate. Conidiophores hyaline, erect, septate, simple or sparsely branched, up to 200–550 μm long, 5 μm wide at the base, gradually tapering upwards to a width of 2–3 μm at the apex, bearing a single conidium. Conidia hyaline, fusiform, 2–5-septate, mainly 3–4-septate, conidia

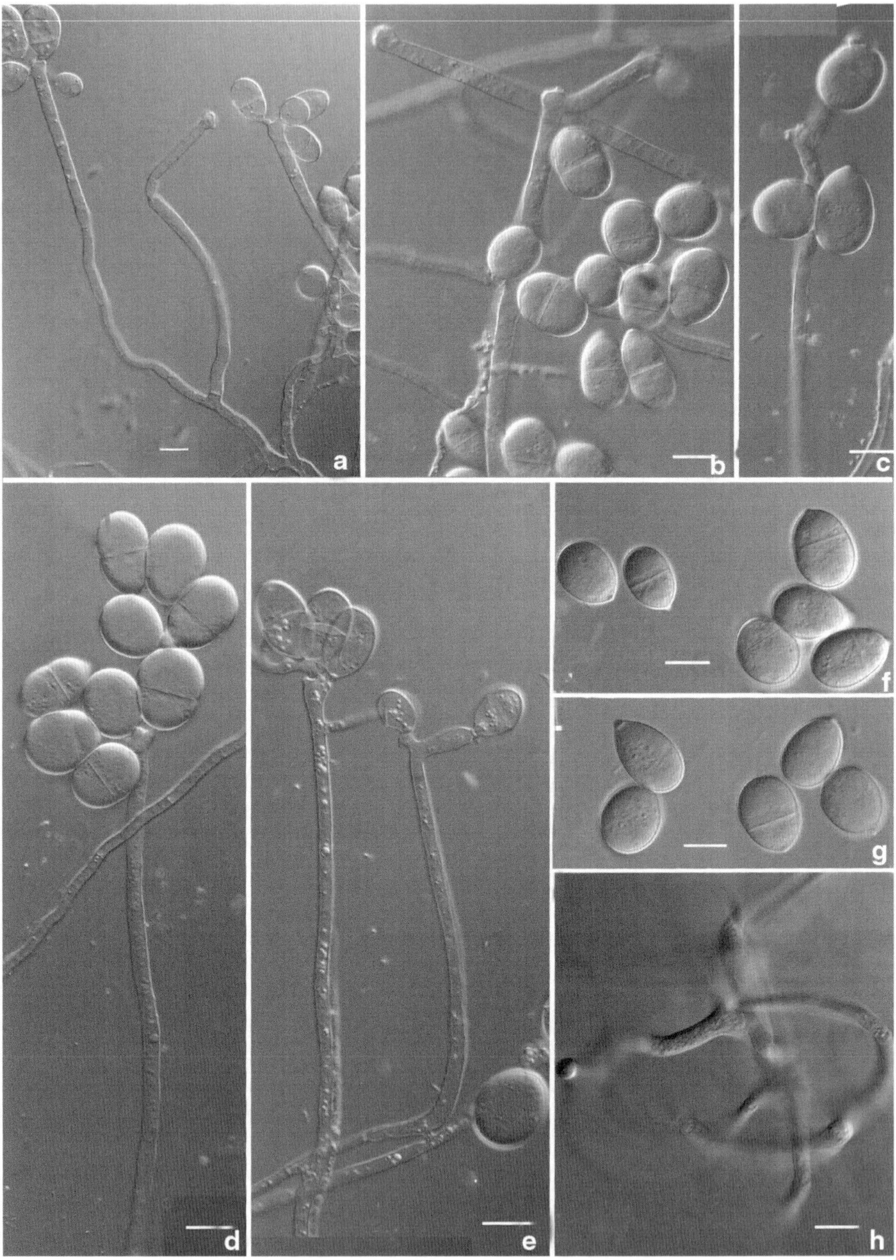

Fig. 3.23 *A. latispora*. **a–e** conidiophore; **f–g** conidia; **h** adhesive network. Bars = 10 μm; strain number: YMF1.00538

holoblastic, 40–90 (54) × 15–27.5 (18) µm. Chlamydospores not observed. Capturing nematodes by means of three dimensional adhesive networks.

Distribution: China (Hebei, Sichuan, Xigjiang, Yunnan)

Material examined: YMF1.00538, isolated from soil in Baoshan, Yunnan in October 2002 by Jing Zhang. Permanent slide: BS5–19

Notes: This taxon most closely resembles *A. psychrophila* in having 3–4-septate conidia and in conidial shape, but the conidia size especially width of the original description of *A. longiphora* (38–65 × 13–22 µm) is less than that of *A. psychrophila* (46–71 × 21–29 µm). The taxon also resembles *A. reticulata* and *A. megalospora* in spore shape, but differs in that the taxon has 3–4-septate conidia, while *A. reticulata* and *A. megalospora* have mainly 4-septate conidia. (Fig. 3.24)

Arthrobotrys mangrovispora Swe, Jeewon, Pointing & K.D. Hyde, Bot. Mar. 51 (4): 332 (2008)

Characteristics: Mycelium spreading, scanty, irregularly projecting, hyaline, septate, branched. Conidiophores 70–330 µm high, 1.5–5 µm wide at the base, hyaline, erect, septate, bearing first a single conidium at the apex, later additional conidia from the lateral branches or longer geniculate branches below the apex, commonly producing 1–6 conidia; each long denticle or branch bearing a single conidium. Conidia 25– (38.9)–50 × 12– (17.3)–24 µm, colourless, variable in shape; broadly turbinate to elongate-fusoid, ellipsoidal or fusiform-ellipsoidal, clavate, broadly obovoid, elongate–obovoid, 1–3-septate; proportions of conidia with 0, 1, 2 or 3 septa are 6, 23.5, 50.0 and 20.5 %, respectively. Chlamydospores present. Capturing nematodes by means of three dimensional adhesive networks.

Distribution: China (Hong Kong)

Notes: The description is based on the protologue. *A. mangrovispora* closely resembles, but differs from, *A. eudermata*, *A. sphaeroides*, *A. microscaphoides* and *A. scaphoides* in conidial morphology. According to the original descriptions of these species, the conidia of *A. eudermata* are 31–40 × 19–28.5 µm, and mostly 3-septate (Chowdhry and Bahl 1982). Those of *A. sphaeroides* are 28–42 × 19–29 µm, usually 2-septate with one basal and one terminal septum or 1-septate with only one basal septum (Castaner 1968), and those of *A. scaphoides* are 26–83 × 12–17 µm with 1–3 septa (Peach 1952). *A. microscaphoides* has conidia that are 25–50 × 12–24 µm with 1–3 transverse septa and most (53 %) are 2-septate. Of the above mentioned species, *A. mangrovispora* most closely resembles *A. scaphoides*, but can be distinguished by the following differences: *A. scaphoides* forms broadly fusiform conidia with 3 septa: non-septate microconidia and chlamydospores are absent, while *A. mangrovispora* possesses broadly turbinate to elongate-fusiform conidia and produces chlamydospores. (Fig. 3.25)

Fig. 3.24 *Arthrobotrys longiphora*. **a–b** conidiophore; **c–e** conidia; **f** adhesive network. Bars = 10 μm. strain number: YMF1.00538

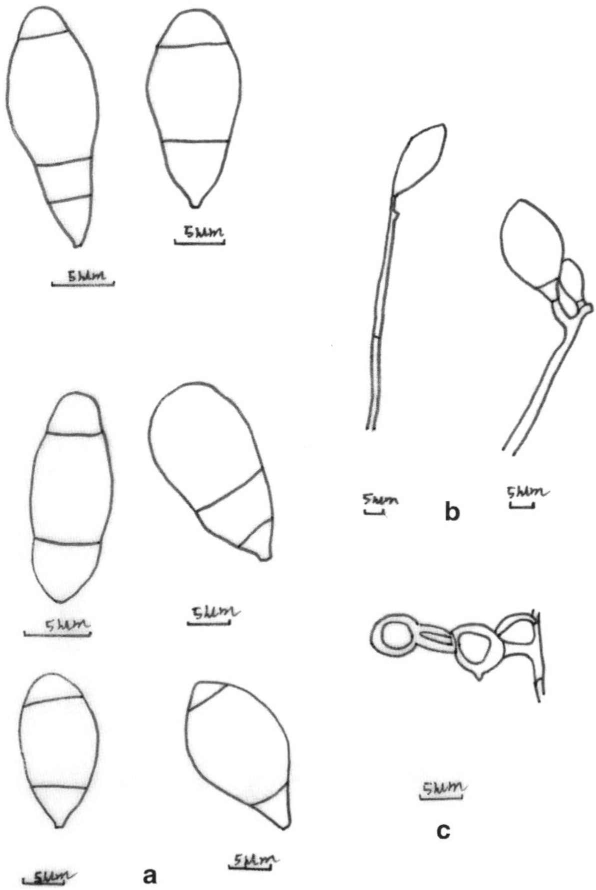

Fig. 3.25 *Arthrobotrys mangrovispora*. **a** conidia; **b** conidiophore; **c** adhesive network. Bars = 5 μm

Arthrobotrys megalospora (Drechsler) M. Scholler, Hagedorn & A. Rubner, Sydowia 51 (1): 103 (1999)
≡ *Dactylella megalospora* Drechsler, Mycologia 46: 769 (1954)
= *Golovinia megalospora* (Drechsler)Mekht., Khishchnye Nematofagovye Griby—Gifomitsety (Baku): 155 (1979)
= *Monacrosporium megalospora* (Drechsler) Subram., Kavaka 5: 94 (1978)[1977]
= *Monacrosporium megalospora* (Drechsler) Subram., J. Indian bot. Soc.42: 293 (1964) [1963]

Characteristics: Mycelium spreading; vegetative hyphae colourless, branched, septate at moderate intervals, mostly 2–6 μm wide. Conidiophores colourless, erect, often containing 1–7 cross walls, commonly 6–8 μm wide at the base, tapering gradually upwards to a distal width of 2–3 μm, 350–450 μm high, unbranched and bearing a single conidium at the apex. Sometimes 1–5 conidia formed on 10–40 μm long branch near the conidiophore apex. Conidia colourless, broadly fusoid or elongate-ellipsoidal or obovoid, mostly 40–75 long and 18–35 wide, commonly divided

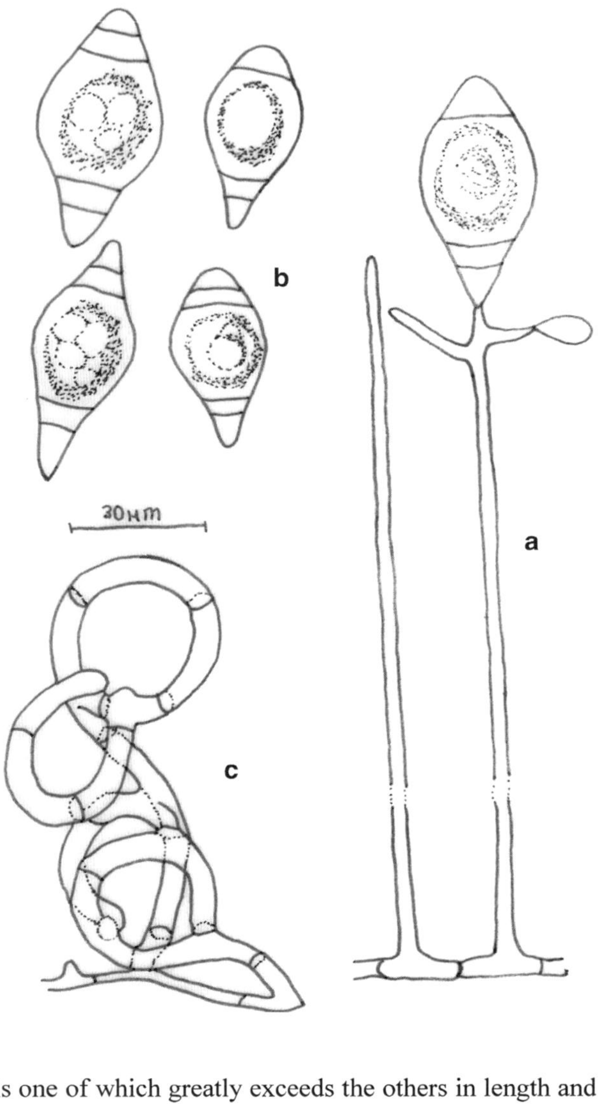

Fig. 3.26 *Arthrobotrys megalospora*. **a** conidiophore; **b** conidia; **c** adhesive network. Bars = 30 μm

by cross-walls into 3–6 cells one of which greatly exceeds the others in length and width; the large cell somewhat variable in position, but occurring most often in median position in the especially distinctive large, broadly fusoid, quadriseptate conidia, frequently 55–75 μm long, and 23–35 μm wide, that are produced singly on nematode-infested substratum. Capturing nematodes by means of three dimensional adhesive networks.

Distribution: USA (Florida)

Notes: The description is based on the protologue. This species resembles *A. psychrophyla* in conidiophore and conidial shape, but conidia of the latter mainly 3–4-septate, conidia of the former are 2–5-septate. (Fig. 3.26)

Arthrobotrys microscaphoides (Xing Z. Liu & B.S. Lu) M. Scholler, Hagedorn & A. Rubner, Sydowia 51 (1): 103 (1999)
≡ *Monacrosporium microscaphoides* Xing Z. Liu & B.S. Lu, Mycosystema 6: 68 (1993)

Characteristics: Colonies on CMA hyaline. Mycelium effuse, vegetative hyphae smooth-walled, hyaline, septate, and branched. Conidiophores hyaline, erect, septate, simple or with several branches normally at the apex, 230–460 µm long, 3–5 µm wide at the base, gradually tapering upwards to a width of 2–2.5 µm at the apex, bearing a single conidium or occasionally several conidia. Conidia hyaline, cymbiform, 0–3-septate, mostly 2-septate, 22.5–45 (27.2) × 10–20 (13.9) µm. The proportion of conidia with 0, 1, 2, and 3 septa is 4, 27, 55, 7 %, respectively. Chlamydospores present in aged cultures, in chain. Capturing nematodes by means of three dimensional adhesive networks.

Distribution: China (Beijing, Hainan, Henan, Sichuan, Yunnan)

Material examined: YMF1.00028, isolated from soil in Lincang, Yunnan in 1999 by Lu Cao; JLCB, isolated from forest soil in Changbai Mountain, Jilin in 1999 by Ke-Qin Zhang; YMF1.00546, isolated from soil in Yinjiang, Yunnan in October 2002 by Jing Zhang. Permanent slide: L32–6

Notes: This fungus most resembles *A. scaphoides* in its fusiform or boat-like shaped spores, but the conidia of *A. microscaphoides* is smaller than those of *A. scaphoides*. Although the conidia of *A. scaphoides* are 23–39 × 12–17 µm in Peach's (1952) description covered the range of *A. microscaphoides* 23–39 × 8–15.5 µm in original description (Liu and Lu 1993), but the average length and width of conidia of *A. scaphoides* is much longer than that of *A. microscaphoides* according to our measurement of the conidia of the figure in Peach's paper. (Fig. 3.27)

Arthrobotrys multiformis (Dowsett J. Reid & Kalkat) M. Scholler, Hagedorn & A. Rubner, Sydowia 51 (1): 103 (1999)
≡ *Dactylella multiformis* Dowsett J. Reid & Kalkat, Mycologia 76: 563 (1984)
= *Monacrosporium multiformis* (Dowsett J. Reid & Kalkat) A. Rubner, Stud.Mycol.39 84 (1996)

Characteristics: Colonies on CMA whitish, rapidly growing and extending to a diameter of 7–8 cm at 25 °C within 7 days. Mycelium spreading, scanty, vegetative hyphae hyaline, septate, branched. Conidiophores hyaline, simple, erect, septate, 80–220 µm long, bearing a single conidium at the apex. Conidia elongate-fusiform, 4–12-septate, 47–198 × 7–20 µm, solitary. Conidia may raise to secondary conidiophores up to 50–120 µm long, on apices of which bearing clavate to cylindric-clavate, 0–1 septum, secondary conidia, 15–35 × 2.5–5 µm. Chlamydospores present in older cultures. Capturing nematodes by three dimension adhesive networks.

Distribution: Canada (Manitoba), China (Hebei, Guizhou, Yunnan), Japan

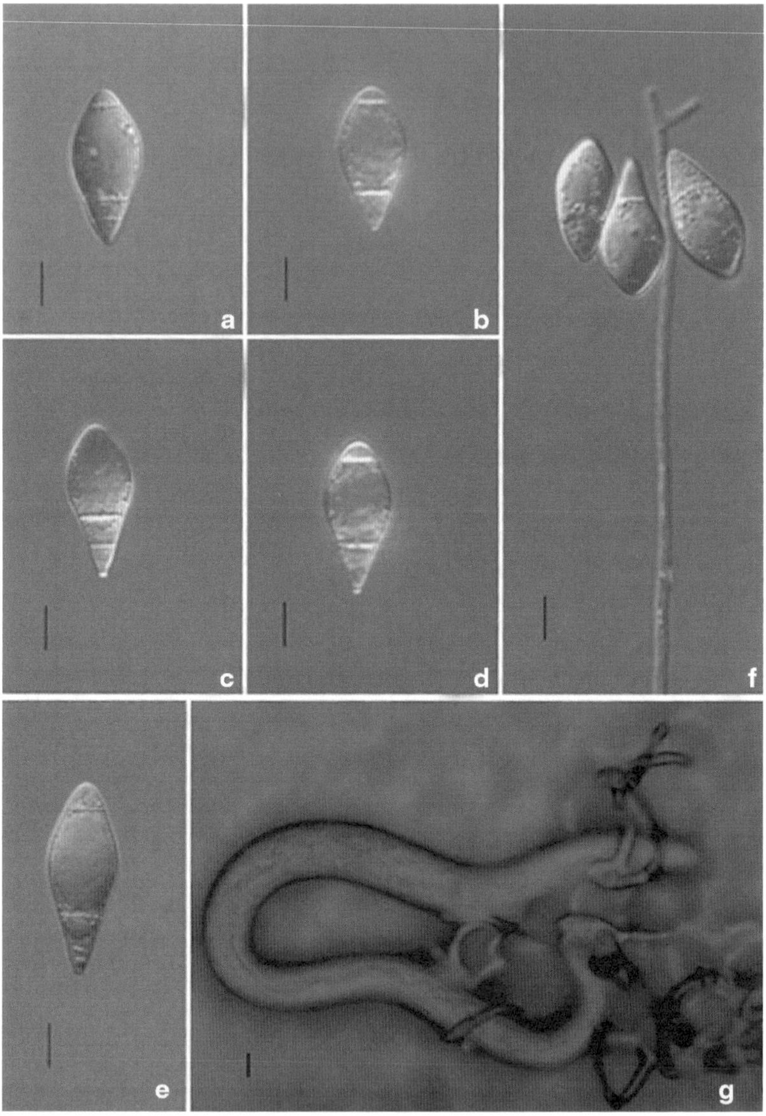

Fig. 3.27 *Arthrobotrys microscaphoides*. **a–e** conidia; **f** conidiophore; **g** nematode trapped by adhesive network. Bars = 10 μm; Strain number: YMF1.00546

Material examined: CBS773.84 (*A. multiformis*), IFO32554 (*A. iridis*), 89019–1 (*A. iridis*), AN–1101 (*A. iridis*), provided by Dr. Xingzhong Liu. YMF 1.01477, isolated from soil in Xundian, Yunnan in April 2004 by Ye'e Hao. Permanent slide: YMF1.01477.

Notes: *A. multiformis* resembles *Dactylellina leptospora* in its rather delicate, spreading mycelium and terminally produced, single conidia which are mostly straight, elongate-fusiform to cylindrical, and with many septa; both taxa produce secondary conidia. However, *Da. leptospora* traps nematodes by non-constricting rings and adhesive knobs, while *A. multiformis* traps by means of adhesive three-dimensional networks. Conidia of *Da. leptospora* are 40–105 × 4–5.8 μm, with 5–15 septa; those of *A. multiformis*, 35–90 × 4–7.5 μm with 4–12 septa. Secondary conidia of *Da. leptospora* are somewhat clavate 20–50 × 4–5.8 μm, 3–8-septate; those of *A. multiformis* are cylindric to clavate, 20–25 × 5 μm, 0–1-septate, and borne singly or on short sympodially proliferated denticles. *A. multiformis*, *Dactylella ramiformis* and *Dactylella iridis* share most morphological characteristics based on our observed strain, so we treat them as conspecific under the name *A. multiformis*. (Fig. 3.28)

Arthrobotrys musiformis Drechsler, Mycologia 29 (4): 481 (1937)
= *Candelabrella musiformis* (Drechsler), Rifai & R.C. Cooke, Trans. Br. mycol. Soc. 49 (1): 163 (1966)
= *Dactylella musiformis* (Drechsler) Matsush., Microfungi of the Solomon Islands and Papua-New-Guinea (Osaka): 22 (1971)

Characteristics: Colonies on CMA whitish, rapidly growing. Mycelium spreading, vegetative hyphae, septate, branched, mostly 1–10 μm wide. Conidiophores hyaline, septate, erect, not branched blow, 104–640 μm long, 5–7.5 μm wide at the base, gradually tapering upwards to a width of 2.5–5 μm near the apex, where are borne on divergent, slightly tapering, simple or branched sterigmata, mostly 1–2.5 μm wide, 3–35 μm long, usually 5–15 conidia in loose capitate arrangement. Conidia hyaline, ellipsoid, straight or slightly curved, broadly rounded at the wider distal end, tapering noticeably towards the slightly protruded base, 20–47.5 (30.9) × 7–12.5 (10.3) μm, 1-septate close to the base of the spore. Chlamydospores yellow, globose or less frequently ellipsoidal, mostly 14–22 μm in diameter. Capturing nematodes by means of three dimensional adhesive networks.

Distribution: Brazil (Sao Paulo), China (Anhui, Beijing, Guangxi, Guizhou, Hubei, Shangdong, Sichuan, Taiwan, Xizang, Yunnan), Ecuador (Tungurahua), France (Limousin), Netherlands (Den Haag, Meerdink–bos near Winterswijk), Nigeria (Ibadan), UK (Scotland), USA (Maryland, Virginia)

Material examined: Z327, Z377, isolated from soil in Guiyang, Guizhou in 1992 by Ke-Qin Zhang; H101, H093, isolated from field soil in Kunming, Yunnan in 1991 by Ke-Qin Zhang; GZFJ–15, AHHS–22, isolated from soil in Fanjing Mountain, Guizhou and Huangshan Mountain, Anhui in 1996 by Ke-Qin Zhang; DL2–2, isolated from forest soil in Dali, Yunnan in September 2002 by Jing Zhang; XM43–1, isolated from forest soil in Xiangmi, Yunnan in 1999 by Yanju Bi; XZA–6, isolated from humus in Xizang in August 2000 by Minghe Mo; YMF1.00581, YMF1.00122, YMF1.00575, isolated from forest soil in Ruili and Lijiang, Yunnan in October 2002 by Jing Zhang. Permanent slide: 187.

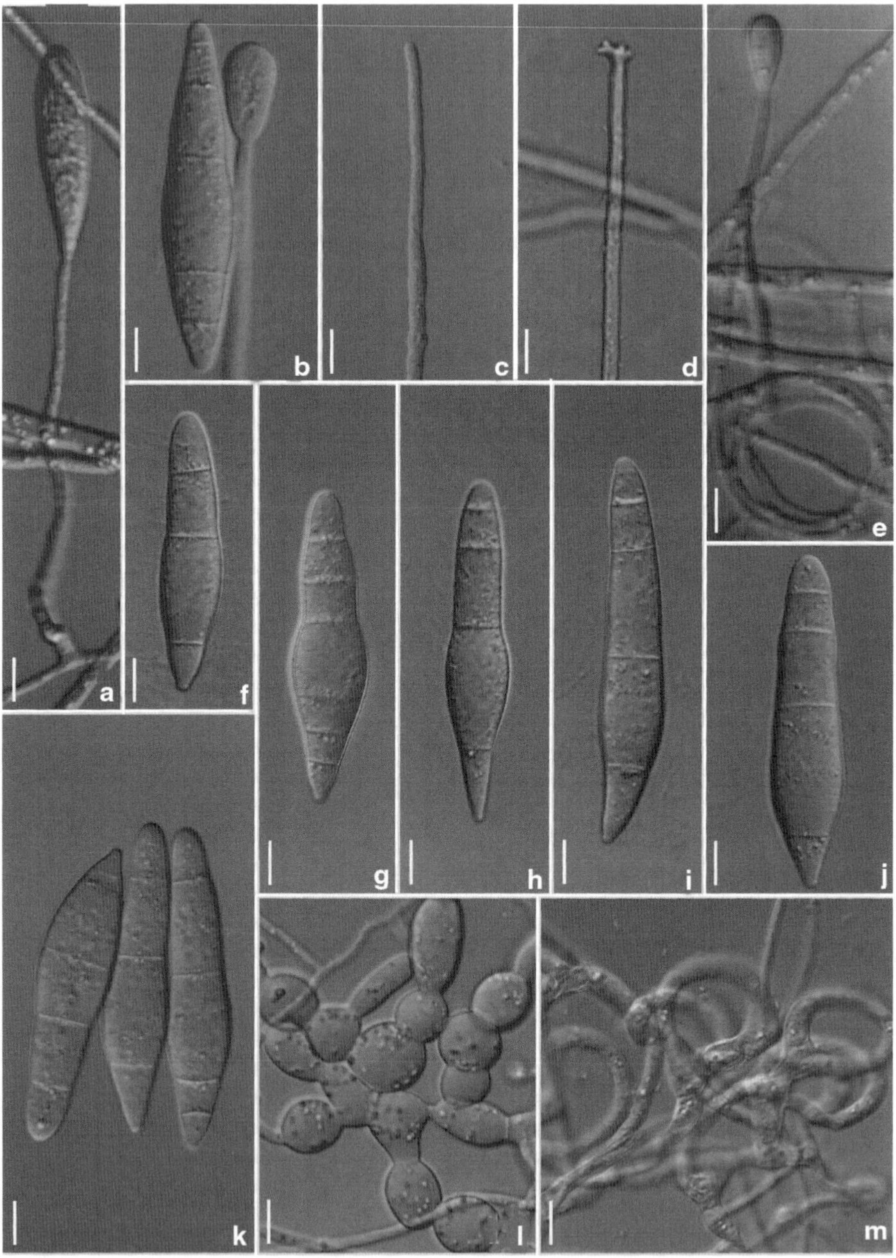

Fig. 3.28 *Arthrobotrys multiformis*. **a–e** conidiophore; **f–k** conidia; **l** chlamydospore; **m** adhesive network. Bars = 10 μm; Strain number:YMF1.01477

Notes: *A. musiformis* resembles *Drechslerella dactyloides* in having ellipsoid, straight or slightly curved conidia, but differs in trapping-devices. The former captures nematodes by three dimensional adhesive networks, while the later captures nematodes by constricting rings. *A. musiformis* also resembles *A. javanica* in candelabrum-like branching of the conidiophore and trapping-devices, but differs in conidia. In *A. musiformis,* conidia are ellipsoid, mostly slightly curved, 20–47.5 (30.9) × 7–12.5 (10.3) μm, while in *A. javanica*, conidia are 20–37.5 (27.9) × 7.5–10 (8.8) μm, narrowly obovoid or clavate.

Strain H093 can produce microconidiophores and microconidia from mycelium or from conidia. Microconidiophores 40–50 μm long, microconidia oviform, single-cell, 12–14 × 4.8 μm. These characters were not described in previous references, but other characters match *A. musiformis*. (Fig. 3.29)

Arthrobotrys nonseptata Z.F. Yu, S.F. Li & K.Q. Zhang, Mycotaxon 109: 249 (2009)

Characteristics: Colonies slow-growing on CMA medium, attaining less than 35 mm diameter in 10 days at 25 °C. Vegetative hyphae hyaline, septate, 3.5–4 μm wide, aerial mycelium sparse, hyaline, septate, branched, 2.5–4 μm wide. Conidiophores erect, septate, unbranched, 40–120 μm high, 2–4 μm wide in the lower part, 1.5–2 μm wide at the apex, producing 3–10 conidia from retrogressive conidiogenous loci on conspicuous denticles at and near the apex. Conidia hyaline, aseptate, 11–16.8 × 5–6.6 μm, elongate ellipsoid, constricted at the base by forming a small truncate protuberance. Nematodes are captured by means of three dimensional adhesive networks.

Distribution: China (Yunnan)

Material examined: YMF1.01852, Yimen, Yunnan, Yimen, in Aug. 2006 by Z.F. Yu.

Notes: There are five known species of *Arthrobotrys* with aseptate or occasionally 1-septate conidia. Conidia of *A. amerospora* (Schenck et al. 1977) are consistently non-septate, those of *A. anomala* (Barron and Davidson 1972), *A. botryospora* (Barron 1979) and *A. yunnanensis* (Mo et al. 2005) are occasionally 1-septate. *A. nonseptata* differs from other four species in conidial shape and size. Among four species, *A. nonseptata* most resembles *A. yunnanensis* in regard to conidial shape. Both species were anamorph state of *Orbilia* sp., and conidia vary in shape, but part conidia of both species are elongate, ellipsoid. In addition, the denticles of *A. yunnanensis* are longer than those of *A. nonseptata*. (Fig. 3.30)

Arthrobotrys obovata K.Q. Zhang & Xing Z. Liu, in Zhang, Liu, Cao & Gao, Mycol. Res. 100 (5): 529 (1996)

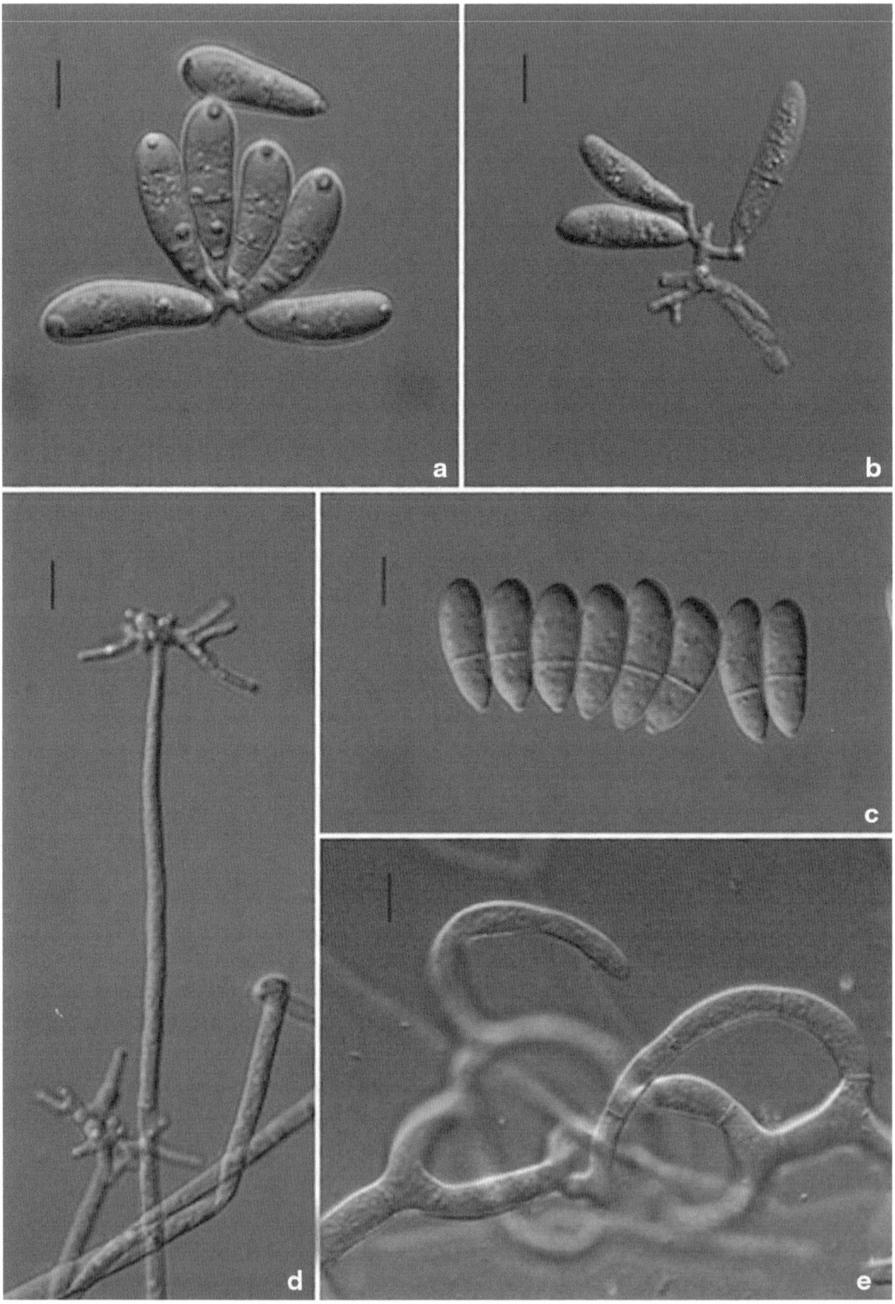

Fig. 3.29 *Arthrobotrys musiformis*. **a–c** conidia; **d** conidiophore; **e** adhesive network. Bars = 10 μm; Strain number: YMF1.00575

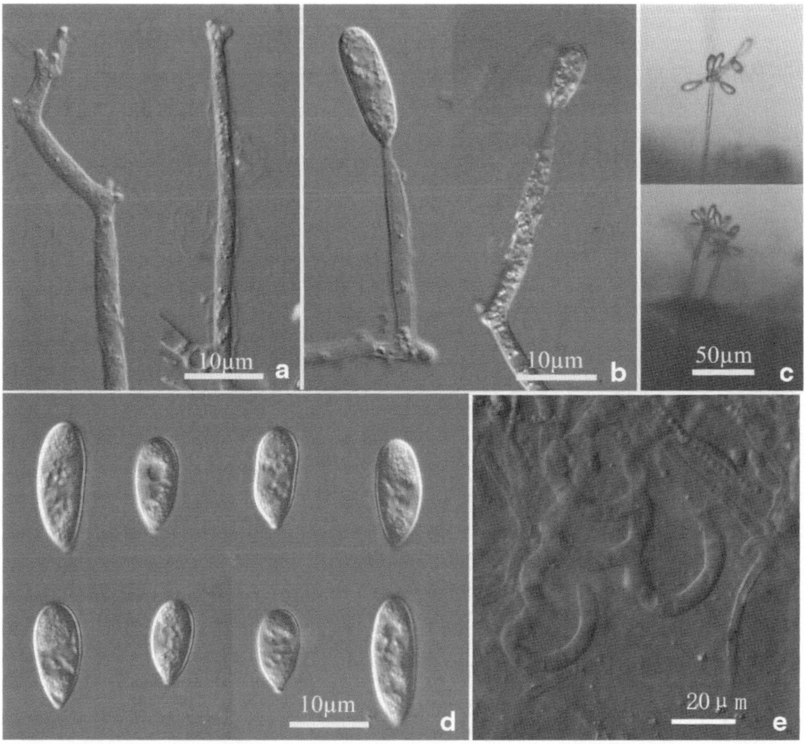

Fig. 3.30 *Arthrobotrys nonseptata.* **a–c** conidiophore; **d** conidia; **e** adhesive network. Bars: **a–b**, **d**=10 µm, **c**=50 µm, **e**=20 µm

Characteristics: Colonies on CMA initially whitish and turned to tinged pink after about one month, rapidly growing, attaining a diameter of 9 cm within 5 days at 25 °C. Mycelium hyaline, vegetative hyphae septate, branched. Conidiophores single or branched, erect, septate, 200–400 µm long, 5–7.5 µm wide at the base, gradually tapering to 2.5–3 µm wide below the first node, branches up to 8–38 µm long, conidiogenous loci with denticle nodes; conidiophores usually continue development by repeated elongation and successive production of 2–9 additional denticle nodes, often containing 5–15 conidia in loose capitate arrangement. Conidia hyaline, obovoid, 1-septum in about one third from the basal end, broadly rounded at the apex, obconical at the base, not constricted at the septum, 28.5–32 (30) × 18–20.5 (20) µm. The basal cells are very small, usually 5–6.5 µm long, and the distal cells are much larger, usually 20.5–25.5 µm long. Chlamydospores absent. Capturing nematodes by means of three dimensional adhesive networks.

Distribution: China (Guizhou, Yunnan)

Material examined: CBS618.95. YMF1.00011, isolated from forest soil in Fanjing Mountain, Guizhou in June 1994 by Ke-Qin Zhang. Permanent slide: FJ92.

Notes: This fungus closely resembles *A. cystosporia* and *A. perpasta* in conidial shape and size but differs in conidiogenous loci and conidial arrangement. In *A. cystosporia* and *A. perpasta*, conidiogenous loci are simple denticles, and conidia are widely spaced arrangement, while in *A. obovata*, conidiogenous loci are denticle nodes and conidia are loose capitate arrangement. (Fig. 3.31)

Arthrobotrys oligospora Fresen., Beitr. Mykol. 1: 18 (1850)
= *Arthrobotrys superba* var. oligospora (Fresen.) Coem.
= *Didymozoophaga oligospora* (Fresen.) Soprunov & Galiulina, Mikrobiol., Reading 20: 493 (1951)

Characteristics: Colonies on CMA whitish or reddish, rapidly growing. Mycelium spreading, dense, vegetative hyphae hyaline, septate, branched. Conidiophores single, erect or curved slightly, septate, 110–440 μm long, 5–10 μm wide at the base, gradually tapering upwards to a width of 3.5–5 μm below the first node; conidiophores usually continue development by repeated elongation and successive production of 1–6 additional denticle nodes, where on forming several conidia. Conidia hyaline, pyriform or obovoid, 1-septate, constricting at the septum, broadly rounded at the apex, 17–35 (23) × 8.5–16 (12) μm. The basal cell is smaller than the distal cell. Chlamydospores yellow, globose, thick-walled, coarse, 15–24 μm in diameter. Capturing nematodes by means of adhesive three dimensional networks.

Distribution: cosmopolitan

Material examined: Z254, isolated from root rubble in Zunyi, Guizhou in 1992 by Ke-Qin Zhang et al.; H014, H058, isolated from forest soil in Guiyang, Guizhou in 1992 by Ke-Qin Zhang et al. ZE102, isolated from Oxford, UK in 1991 by Ke-Qin Zhang; GZDF–3, isolated from field soil in Dafang, Guizhou in 1996 by Ke-Qin Zhang; BJXS–5, isolated from forest soil in Fragrant Mountain, Beijing in 1996 by Ke-Qin Zhang; J8–1, isolated from soil in Jianshui, Yunnan in 1999 by Yanju Bi; XZA–7, isolated from pasture soil in Xizang in August 2000 by Minghe Mo; DL1–3, isolated from soil in Dali, Yunnan in September 2002 by Jing Zhang; YMF1.00012, YMF1.00110, YMF1.00544, YMF1.00051, YMF1.00056, isolated from soil in Dali and Deqin, Yunnan in September 2002 by Jing Zhang. Permanent slide: SJt3.10.8.

Notes: *A. oligospora* resembles *A. conoides* and *A. superba* in most morphological characters except conidia shape and the position of septum. In *A. oligospora*, conidia are obovoid, 1-septate near the base of the spore, and constricted at the septa. In *A. conoides*, conidia elongate obconical, 1-septate about one third from the basal end, and constricted at the septum. In *A. superba*, conidia are elliptical, 1-septate at the centre of the spore, and not constricted at the septum. (Fig. 3.32)

Arthrobotrys oudemansii M. Scholler Hagedorn & A. Rubner, Sydowia 52 (1): 60 (2000)
= *Arthrobotrys elegans* (Oudem.) M. Scholler, Hagedorn & A. Rubner, Sydowia 51 (1): 102 (1999)
= *Golovinia elegans* (Oudem.) Mekht., Khishchnye Nematofagovye Griby—Gifomitsety (Baku): 142 (1979)
= *Monacrosporium elegans* Oudem., Aanwinsten Fl. Mycol. Nederl. 4: 48 (1884)

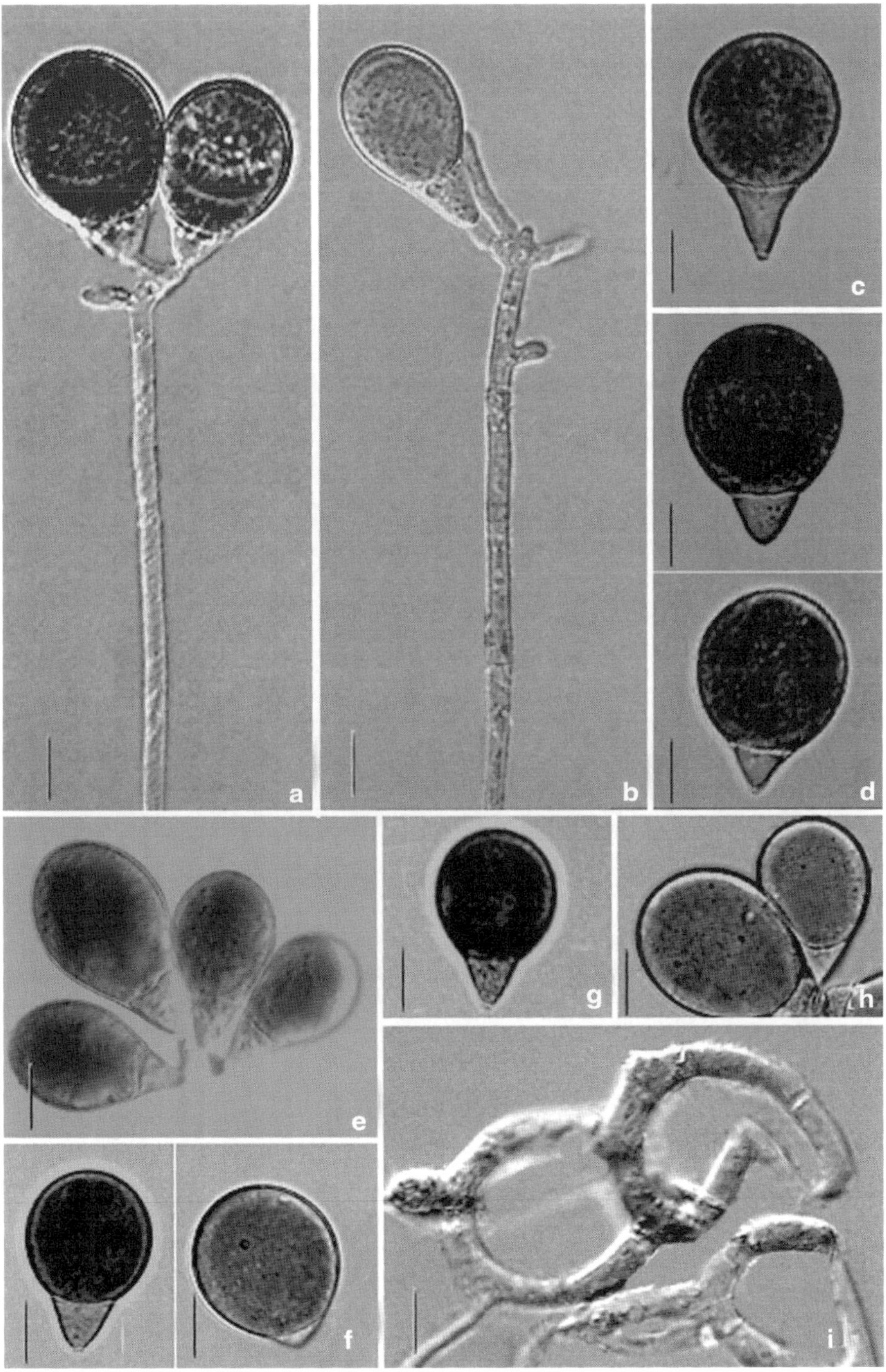

Fig. 3.31 *Arthrobotrys obovata*. **a–b** conidiophore; **c–h** conidia; **i** adhesive network. Bars = 10 μm; Strain number: YMF1.00011

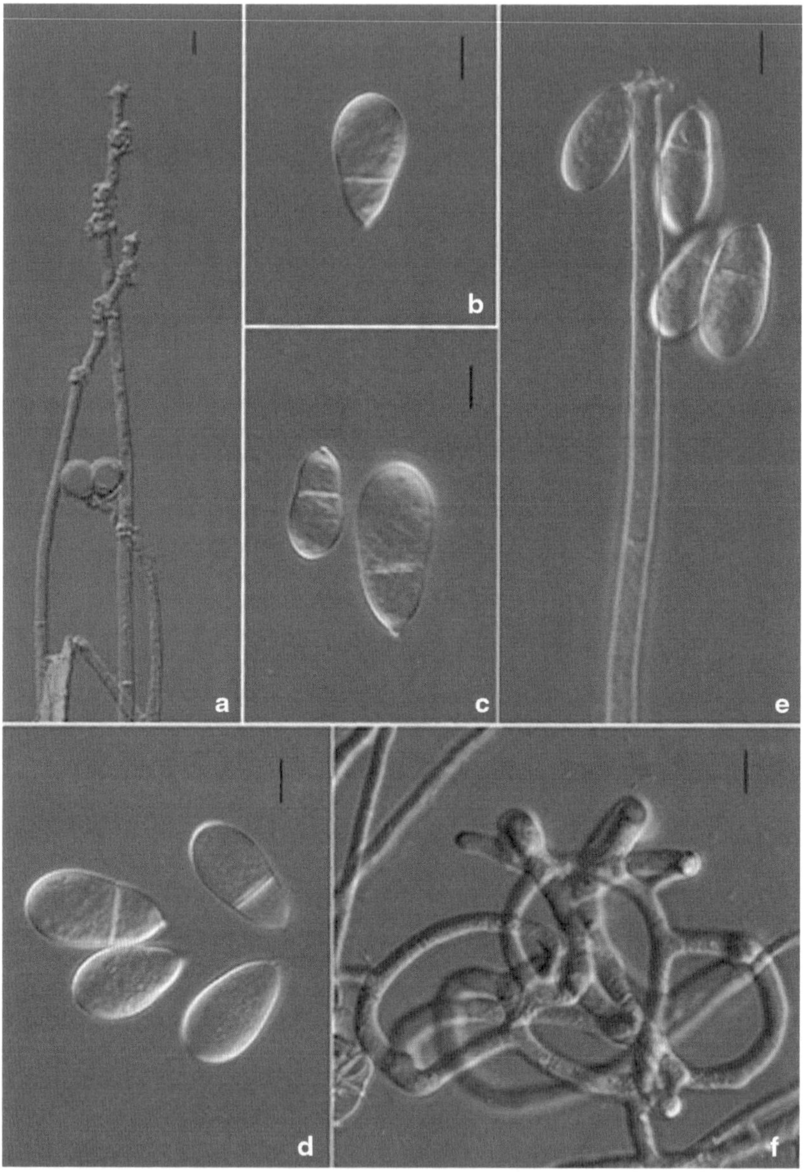

Fig. 3.32 *Arthrobotrys oligospora*. **a, e** conidiophore; **b–d** conidia; **f** adhesive network. Bars = 10 μm; Strain number: YMF1.00544

Characteristics: Conidiophores 150–265–400 μm high, bearing a single conidium on the apex or some additional conidia on small perpendicular and sometimes longer, geniculate branches. Conidia top-shaped, 40–52–65 × 17–20–23 μm, commonly three-septate, the inequidistantly distributed septa demarcating a small apical cell, a

Fig. 3.33 *Arthrobotrys oudemansii*. **a** conidia; **b** microconidia; **c** conidiophores. Bars: **a** = 30 μm, **b** = 10 μm, **c** = 50 μm

strongly ventricose and large central cell, a parabasal and a basal cell. Microconidia obovoid, protracted at the base, 13–16.5 × 5–6 μm, aseptate. Chlamydospores not observed. Trapping nematodes by means of three dimensional networks.

Material examined: CBS 299.94, from soil of a laurel forest, Las Mercedes, Tenerife, coll./isol, G.Lysek,1989. CBS300.94, from moss cushion, Berlin–Dahlem, coll./isol.S.Behnke, 1992, by Rubner (1996).

Distribution: Germany, Netherlands, Sweden.

Notes: This species is the lectotype species of *Monacrosporium*. The size of conidia of all other isolations are within that of type strain, only some strains have branches. (Fig. 3.33)

Arthrobotrys oviformis Soprunov Khishchnye Nematofagovye Griby—Gifomitsety (Baku): 136 (1958)

= *Nematophagus oviformis* (Soprunov) Mekht., Khishchnye Nematofagovye Griby—Gifomitsety (Baku): 106 (1979)

Characteristics: Colonies on CMA whitish or yellowish. Mycelium hyaline, vegetative hyphae branched, septate. Conidiophores hyaline, 6–9-septate, branched, 200–600 μm long, 2.5–5 μm wide, conidiogenous loci with denticle nodes;

conidiophores usually continue development by repeated elongation and successive production of several additional denticle nodes, where on forming 20–39 conidia. Conidia pyriform, broadly rounded at the apex, 1–2-septate, mainly 1-septate at the centre of the spore, slightly constricted at the septum, 7.5–22.5 (15.8) × 5–10 (6.6) μm. Capturing nematodes by means of three dimensional adhesive networks.

Distribution: Burkina Faso (Houet), China (Anhui, Beijing, Guangxi, Guizhou, Hebei, Heilongjiang, Hubei, Neimeng, Xizang, Yunnan), Russian.

Material examined: Z12, isolated from soil in Yushan, Guangxi in 1987 by Ke-Qin Zhang; 8739, isolated from field soil in Huaxi, Guiyang in 1988 by Ke-Qin Zhang; BJHD–3, isolated from soil in Haidian, Beijing, in 1996 by Ke-Qin Zhang; XbL15, isolated from forest soil in Kunming, Yunnan in April 1999 by Lu Cao; XZA–8, isolated from pasture soil in Xizhang in August 2000 by Minghe Mo; YMF1.00121, DL1–3, isolated from forest soil in Dali, Yunnan in September 2002 by Jing Zhang. Permanent slide: XbL15–3.

Notes: *A. oviformis* resembles *A. oligospora* in almost all morphological characters except branched conidiophores. Basically the conidia of *A. oviformis* are not easy to distinquish from those of *A. oligospora*. However, *A. oviformis* has 2-septate conidia, in *A. oligospora* conidia are consistently 1-septate. (Fig. 3.34)

Arthrobotrys paucispora (R.C. Cooke) Jarow., Acta Mycologica, Warszawa 6 (2): 381 (1970)
≡ *Genicularia paucispora* R.C. Cooke, in Rifai & R.C. Cooke, Trans. Br. Mycol. Soc. 49: 157 (1966)
= *Genicularia paucispora* (R.C. Cooke) Rifai, Mycotaxon 2 (2): 215 (1975)

Characteristics: Conidiophores simple, erect, septate, straight at first but sometimes becoming geniculate through repeated subapical proliferation, 300 μm long and 4 μm wide, bearing widely spaced conidia. Conidia narrowly obpyriform, smooth, thin-walled, 1-septate, 24–40 × 12.5–18.8 μm, with a distinct constriction at the septum, distally rounded and tapered below to an obconical truncate base, In pure culture on cornmeal agar they were smaller, 25–33.8 × 12.5–16.3 μm and were sometimes curved or asymmetrical. capturing nematodes by means of three dimensional adhesive networks.

Distribution: UK.

Notes: The description is based on the protologue. This species has not typical denticles, conidia of which borne from small sterigmata of conidiogenous loci. (Fig. 3.35)

Arthrobotrys perpasta (R.C. Cooke) Jarow., Acta Mycologica, Warszawa 6 (2): 381 (1970)
≡ *Genicularia perpasta* R.C. Cooke, in Rifai & R.C. Cooke, Trans. Br. Mycol. Soc. 49: 156 (1966)
= *Genicularia perpasta* (R.C. Cooke) Rifai, Mycotaxon 2 (2): 216 (1975)

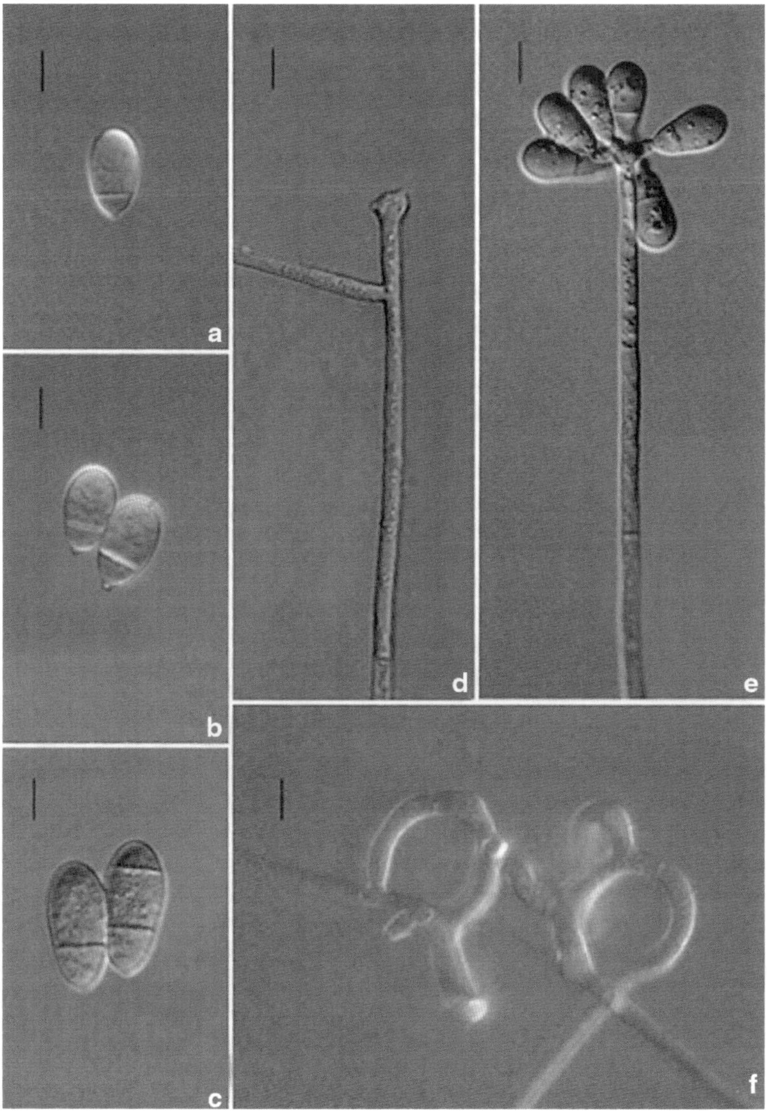

Fig. 3.34 *Arthrobotrys oviformis*. **a–c** conidia; **d–e** conidiophore; **f** adhesive network. Bars = 10 μm; Strain number: YMF1.00121

Characteristics: Conidiophores erect, straight, septate, hyaline and typically unbranched, up to 400 μm long, later became geniculate or sinuous, bearing widely spaced conidia. Conidia plump, obpyriform, 24–32.5 × 12.5–20 μm, smooth and thin-walled, hyaline or subhyaline, 1-septate, sometimes constricted at the septum. In pure culture on cornmeal agar conidia were smaller, 21.5–30 × 12.5–19 μm. Capturing nematodes by simple, branched adhesive hyphae.

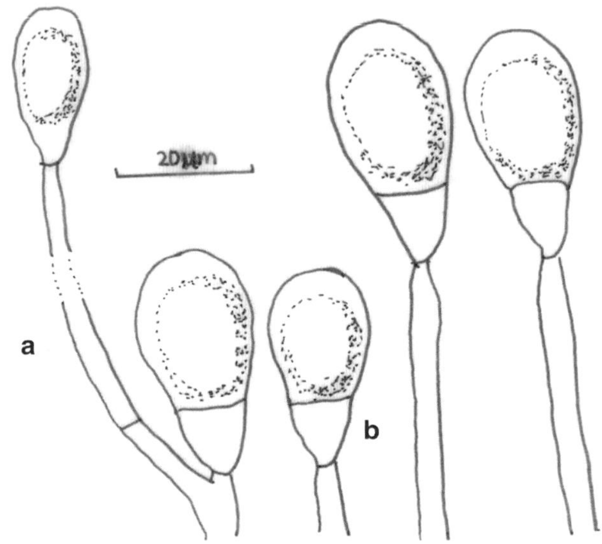

Fig. 3.35 *Arthrobotrys paucispora* (R.C. Cooke) Jarow **a** geniculate conidiophore with conidia; **b** conidiophores and terminal conidium. Bar = 20 µm

Distribution: UK.

Notes: The description is based on the protologue. This species resembles *A. cystosporia* in conidiogenous structure and conidial shape, but differs from the latter by trapping device and the size of conidia. (Fig. 3.36)

Arthrobotrys polycephala (Drechsler) Rifai, Reinwardtia 7 (4): 371 (1968)
≡ *Dactylaria polycephala* Drechsler, Mycologia 29: 530 (1937)
= *Woroninula polycephala* (Drechsler) Mekht, Khishchnye Nematofagovye Griby—Gifomitsety (Baku): 110 (1979)

Characteristics: Colonies on CMA whitish, rapidly growing and extending to a diameter of 10 cm at 25 °C within one week. Mycelium spreading, scanty, vegetative hyphae hyaline, septate, branched, mostly 1.8–4.5 µm wide. Conidiophores hyaline, septate, at first erect, later often more or less procumbent, 90–270 µm long, 4–5 µm wide at the base, gradually tapering upwards to a width of 2.5–3 µm below the first denticle node producing 1–5 conidia, then following repeated elongation often giving rise successively to additional conidial clusters. Initially forming ellipsoidal-shaped conidia, hyaline, rounded at the distal end, somewhat acute at the base,1–3-septate, 12.5–25 (20) × 3–5 (4.5) µm; then forming fusoid-shaped conidia, 2–4-septate, occasionally 2-septate, often 4-septate, most often 3-septate, sometimes slightly constricted at the septa, 30–50 (45.1) × 8–16.5 (12.2) µm; proportion of conidia with 2, 3 and 4 septa is 3, 79 and 18, respectively. Capturing nematodes by means of three dimensional adhesive networks.

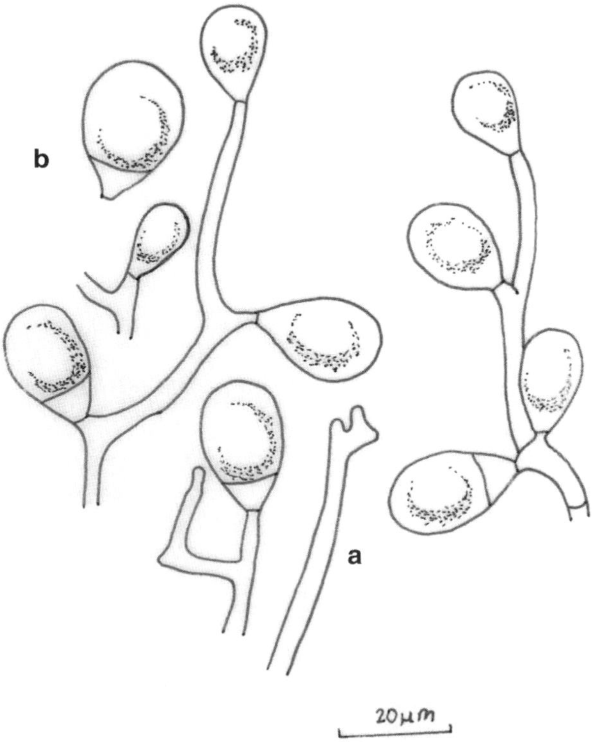

Fig. 3.36 *Arthrobotrys perpasta*. **a** conidiophore; **b** conidium. Bar = 20 μm

Distribution: China (Hubei, Xizang, Yunnan), Cuba, Germany (Berlin), USA (Maryland)

Material examined: W7230–3–06, isolated from field soil in Wuhan, Hubei in February 2001 by Xuefeng Liu *et al*; XZA–9, isolated from forest soil in Xizang, in August 2000 by Minghe Mo; YMF1.00035, YMF1.00600, isolated from forest soil in Jiuzhaigou, Xichuan in April 2004 by Xuefeng Liu. Permanent slide: W7230–3–06.

Notes: The conidia of *A. polycephala* slightly resemble those of *A. musiformis* in general shape, but differ in having larger and more abundant septate. The fungus was initially assigned to *Dactylaria* based on the plural septation of its conidia (Drechsler 1937). However, it was then transferred to *Arthrobotrys* based on trapping devices, nodose and repeatedly capitate forms of conidiophores (Rifai 1968). (Fig. 3.37)

Arthrobotrys pseudoclavata (Z.Q. Miao & Xing Z. Liu) J. Chen, L.L. Xu, B. Liu & Xing Z. Liu, Fungal Diversity 26 (1): 123 (2007)
≡ *Dactylella pseudoclavata* Z.Q. Miao & Xing Z. Liu, Can. J. Bot. 81: 453 (2003)

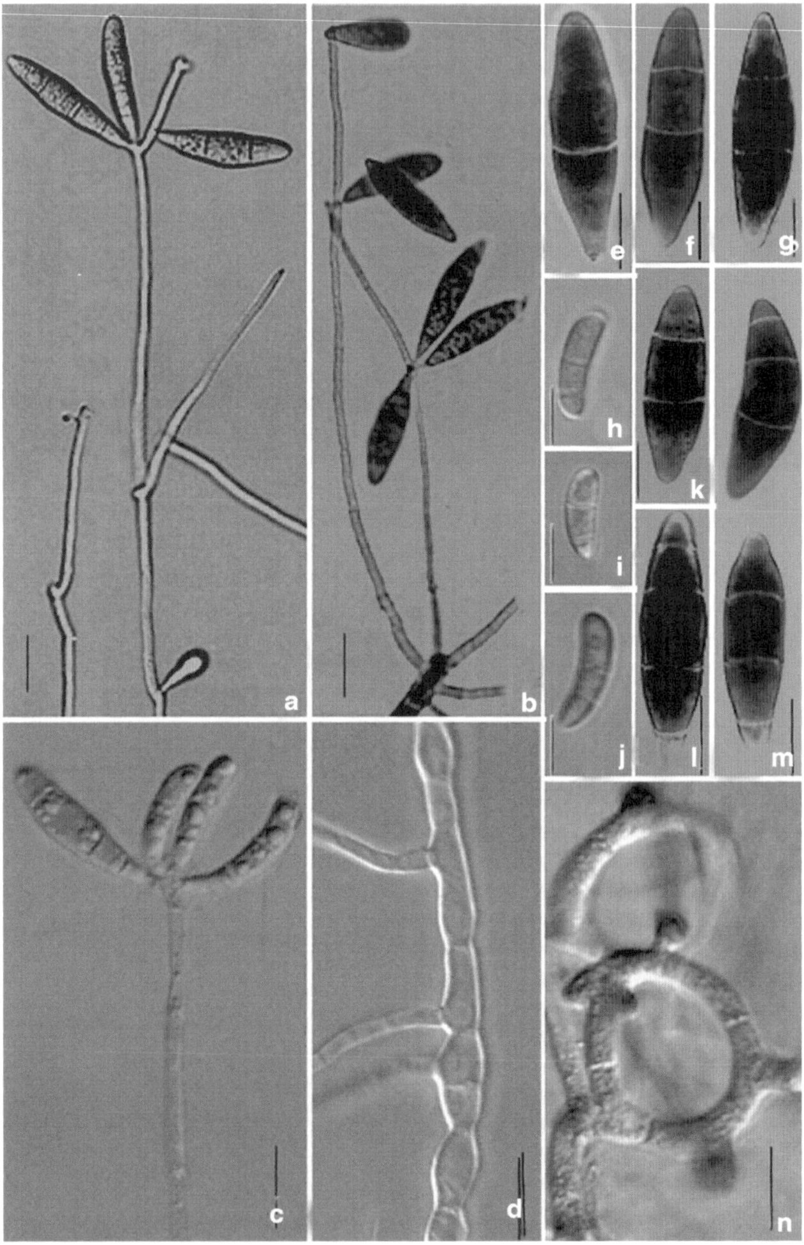

Fig. 3.37 *Arthrobotrys polycephala*. **a–c** conidiophore; **e–m** conidia; **d** chlamydospores; **n** adhesive network. Bars = 10 μm; Strain number: YMF1.00035

Characteristics: Colonies on CMA, slightly brown on top and light brown to yellowish below, effuse extending to a diameter of 7.5 cm at 25 °C within 7 days. Mycelium spreading, vegetative hyphae hyaline, septate, branched, prostrate, 2.5–7 μm wide. Conidiophores erect, simple or occasionally branched, hyaline, septate, 150–300 μm long, 6–7.5 μm wide at the base, gradually tapering upwards to a width of 4–5.5 μm at the apex. The conidiogenous cell produced 1–4 conidia holoblastically. Conidia 30–45 × 8–11 μm, obclavate, round to blunt at distal end, constricted into bottleneck shape at the base, with zero or one septum, which was generally at lower centre or at the septum. Some conidia broke into two parts at the septum. Chlamydospores were abundant in aged culture on PDA, intercalary or catenulate, yellowish to brown, globate or subglobate. Capturing nematodes by means of three dimensional adhesive networks.

Distribution China (Beijing)

Material examined: HMAS 84442, isolated from field soil in Haiding, Beijing in February 1998 by Zuoqing Miao.

Notes: *A. pseudoclavata* is characterized by obclavate, 0–1-septate conidia with round distal end and bottleneck-shaped base, simple conidiophores bearing 1–4 conidia at apex, and adhesive networks as the nematode-trapping device. These characters can distinguish it from other *Arthrobotrys* species. (Fig. 3.38)

Arthrobotrys psychrophila (Drechsler) M. Scholler, Hagedorn & A. Rubner, Sydowia 51 (1): 104 (1999)

≡ *Dactylaria psychrophila* Drechsler, Mycologia 36 (2): 161 (1944)
= *Genicularia psychrophila* (Drechsler) Rifai, Reinwardtia 7 (4): 367 (1968)
= *Geniculifera psychrophila* (Drechsler) Rifai, Mycotaxon 2: 216 (1975)
= *Golovinia psychrophila* (Drechsler) Mekht., Khishchnye Nematofagovye Griby—Gifomitsety (Baku): 151 (1979)
= *Monacrosporium psychrophilum* (Drechsler) R.C. Cooke & C.H. Dickinson, Trans. Br. mycol. Soc. 48: 622 (1965)

Characteristics: Mycelium spreading; vegetative hyphae hyaline, septate, mostly 2–6 μm wide. Conidiophores hyaline, erect, septate, mostly 150–500 μm high, 5–9 μm wide at the base, 2.5–4.5 μm wide at the apex, sometimes simple and sometimes bearing near the apex a branch (occasionally 2 branches) up to 35 μm long, often on giving rise terminally to 1–2 conidia elongating once or twice to produce 1 or 2 additional conidia, and thus frequently coming to bear 3 or 4 conidia in a loose head. Conidia hyaline, ellipsoidal or fusoid-ellipsoidal, rounded at the distal end, somewhat truncate at the proximal end, when developed under favorable conditions measuring mostly 46 to 70 μm (average 62.3 μm) in length and 21–29 μm (average 24.7 μm) in greatest width, containing from 1–5 cross-walls but mostly divided by 3 or cross-walls into 4–5 cells whereof one, as a rule– the penultimate cell usually in triseptate specimens and the median cell usually in quadriseptate specimens- greatly exceeds the others in size. Capturing nematodes by means of three dimensional adhesive networks.

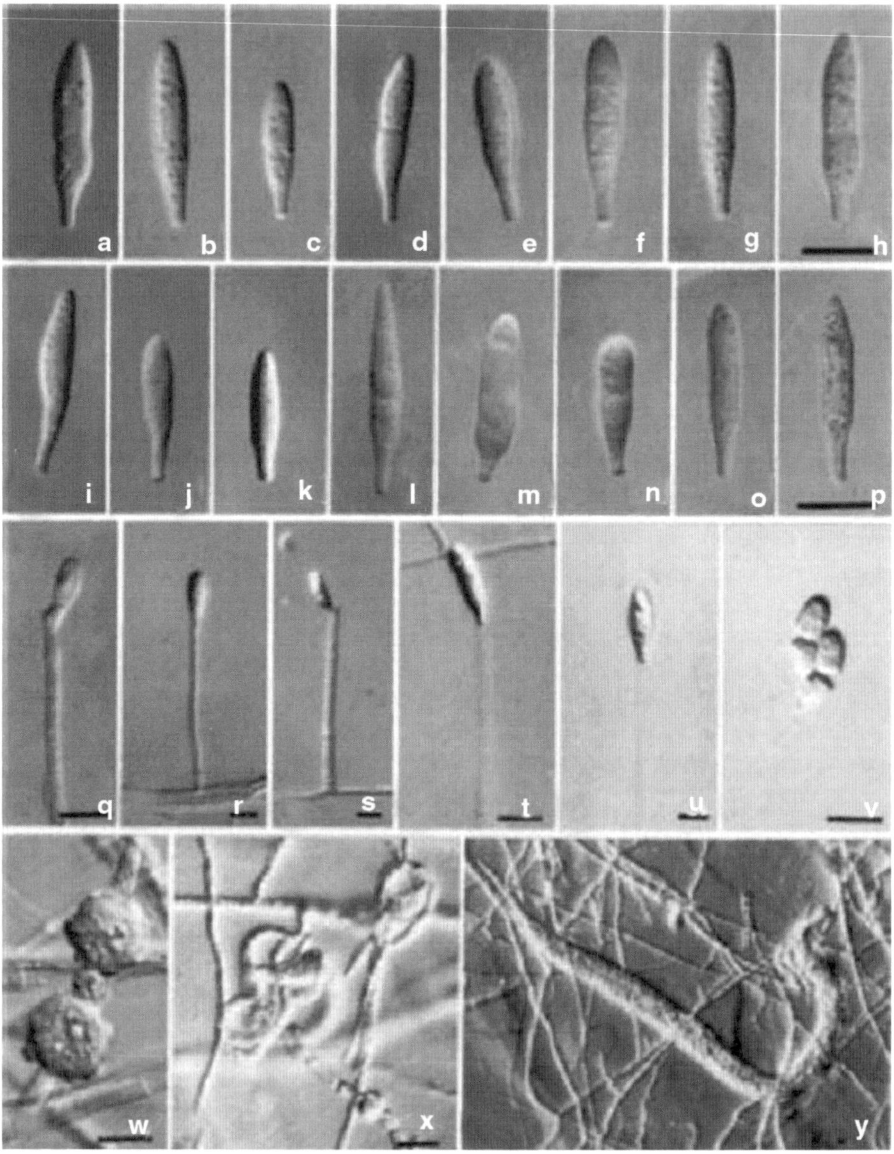

Fig. 3.38 *Arthrobotrys pseudoclavata.* **a–p** conidia; **q–s** conidiophore; **t–u** conidiophore without protoplasm; **v** conidia with breakpoint at septum; w chlamydospore; **x** adhesive network; **y** nematode trapped by adhesive network. Bars = 20 μm

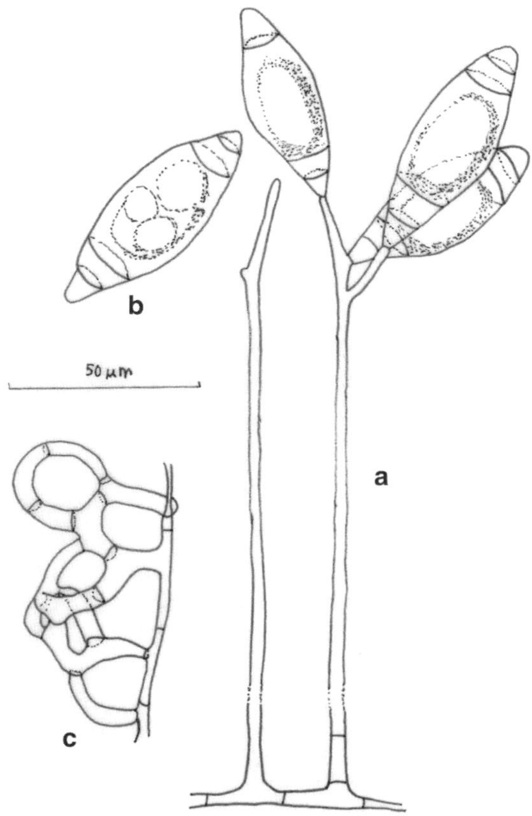

Fig. 3.39 *Arthrobotrys psychrophila.* **a** conidiophore; **b** conidium; **c** adhesive network. Bar = 50 µm

Sexual state: *Orbilia auricolor* (Bloxam ex Berk) Sacc., Syll. Fung., 8: 625,1889

Distribution: USA.

Notes: The description is based on the protologue. This species resembles *A. oudemansii*, but differs from the latter by mainly 3–4-septate conidia, while conidia of *A. oudemansii* are mainly 3-septate. (Fig. 3.39)

Arthrobotrys pyriformis (Juniper) Schenk, W.B. Kendr. & Pramer, Can. J. Bot. 55 (8): 984 (1977)
≡ *Dactylaria pyriformis* Juniper, Trans. Br. Mycol. Soc. 37: 437 (1954)
= *Dactylariopsis pyriformis* (Juniper) Mekht., Khishchnye Nematofagovye Griby—Gifomitsety (Baku): 119 (1979)

Characteristics: Colonies on CMA whitish. Mycelium spreading, vegetative hyphae hyaline, septate, branched, 2–5 µm wide. Conidiophores erect, hyaline, septate, 150–550 µm long, 4–8 µm wide at the base, tapering upwards to a width of 2.5–4 µm at the apex with denticle nodes; conidiophores usually continue development by repeated elongation and successive production of 2–3 additional denticle nodes, often containing 4–5 conidia. Conidia hyaline, elongate-pyriform, broadly rounded at the distal end, 17–38 × 6.5–11.5 μm, mostly 2–3-septate, occasionally 1-septate. Chlamydospores thick-walled, globose or ellipsoidal, 20–35 10–2 μm, yellowish with dense globuliferous content. Capturing nematodes by three dimensional adhesive networks.

Distribution: China (Yunnan), Canada (Quebec), Ecuador (Baños), Germany (Berlin), Netherlands (Noordoostpolder), UK (England)

Material examined: YMF1.00018, isolated from forest soil in Shennongngjia, Hubei in 2003 by Wei Zhou; Z125, isolated from forest soil in Xishuangbanna, Yunnan in 1994 by Ke-Qin Zhang. Permanent slide: LH31.

Notes: The specific epithet of *A. pyriformis* is based on the pyriform-shaped conidia. It resembles *Dr. brochopaga* in having 2–3-septate conidia, but differs in trapping devices and conidial shape. *A. pyriformis* produces elongate-oviform conidia and three dimensional adhesive networks while *Dr. brochopaga* produces cylindrical or elongate ellipsoidal conidia and constricting rings. Oorshot (1985) restudied the holotype (CBS204.83), and observed conidiophore branching, less expanded at the conidiogenous loci, and smaller conidia than in the original description (15–25 × 5–8 μm). The characters of our strain YMF1.00018 match well with the original description. (Fig. 3.40)

Arthrobotrys reticulata (Peach) M. Scholler, Hagedorn & A. Rubner, Sydowia 51 (1): 104 (1999)
≡ *Dactylella reticulata* Peach, Trans. Br. Mycol. Soc. 33 (1–2): 148 (1950)
= *Golovinia reticulata* (Peach) Mekht., Khishchnye Nematofagovye Griby-Gifomisety (Baku): 143 (1979)
= *Monacrosporium reticulatum* (Peach) R.C. Cooke & C.H. Dickinson, Trans. Br. Mycol. Soc. 48: 622 (1965)

Characteristics: Colonies on CMA whitish. Mycelium spreading, scanty, vegetative hyphae hyaline, septate, branched, 4–5 μm. Conidiophores hyaline, erect, septate, simple, only infrequently branched at the base, 250 μm long, bearing a single conidium on the apex. Conidia hyaline, ellipsoid, with one of the central cells being largest, 50–65 × 20–25 μm, mostly 4-septate, sometimes 5-septate. Chlamydospores present in aged cultures, in chains or clusters, globose to subglobose. Capturing nematodes by means of three dimensional adhesive networks.

Distribution: China (Anhui, Guizhou, Hebei, Shanxi), UK (near Harpenden Common)

Material isolated: FJ46, isolated from forest soil in Fanjing Mountain, Guizhou in 1994 by Ke-Qin Zhang;

Notes: Our isolate matches the original description (Peach, 1950). This species can be distinguished from similar species by its fusiform and 5-septate conidia. (Fig. 3.41)

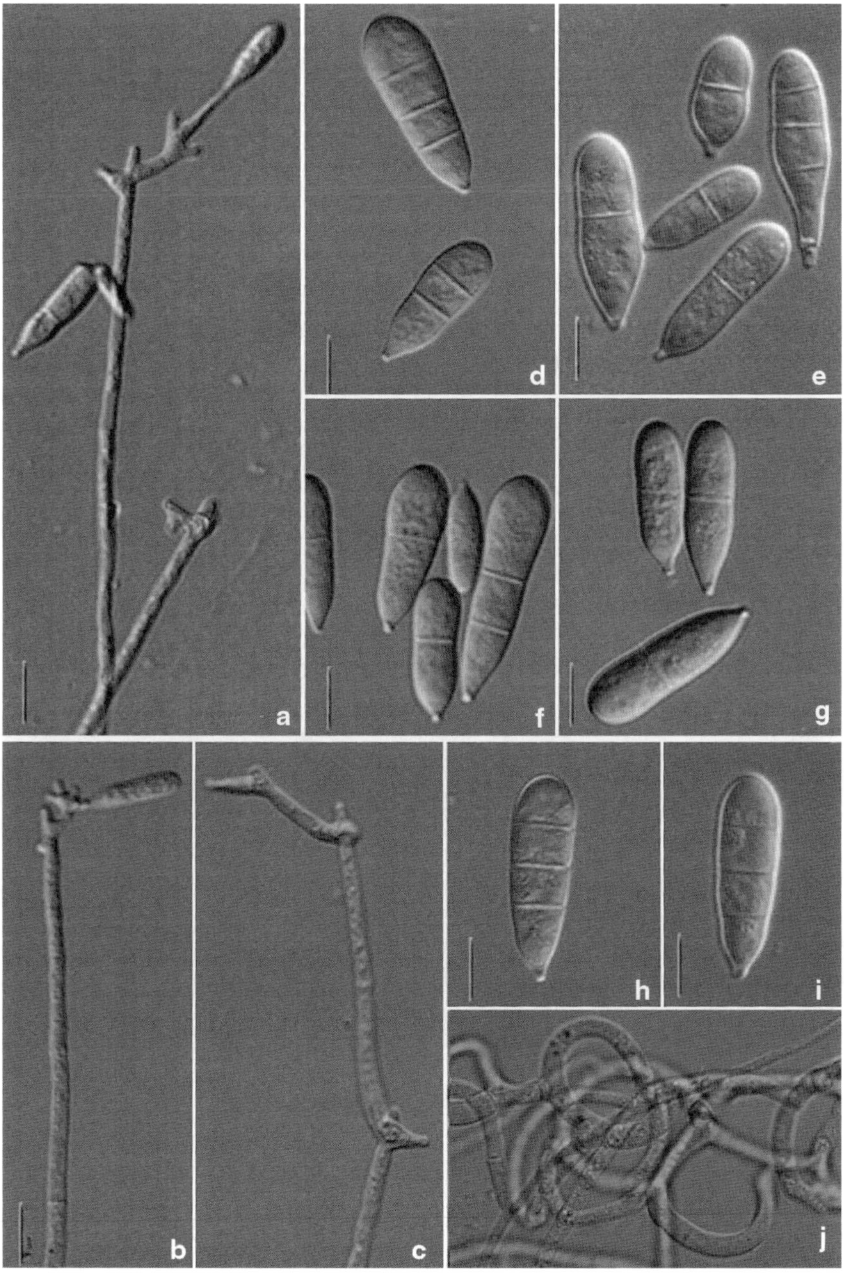

Fig. 3.40 *Arthrobotrys pyriformis.* **a–c** conidiophore; **d–i** conidia; **j** adhesive network. Bars = 10 μm; Strain number: YMF1.00018

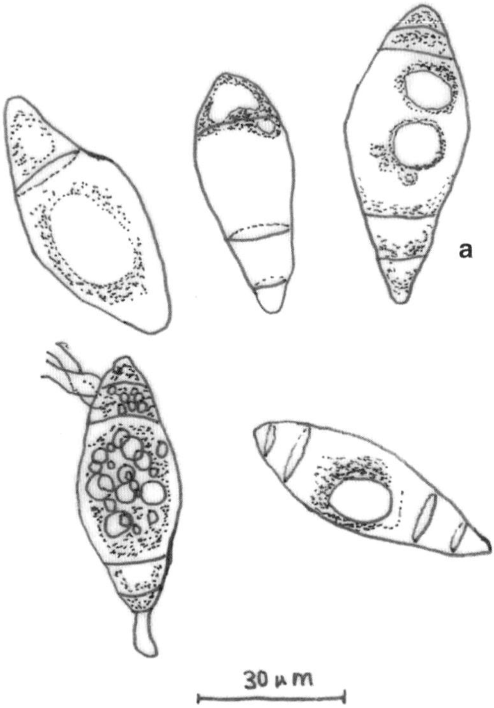

Fig. 3.41 *Arthrobotrys reticulata*. **a** conidia. Bar = 30 µm

Arthrobotrys robusta Dudd., Trans. Br. mycol. Soc. 34 (4): 598 (1952) [1951]

Characteristics: Colonies on CMA whitish, rapidly growing and extending to a diameter of 7.5 cm at 25 °C within one week. Mycelium spreading, vegetative hyphae hyaline, septate, branched, mostly 2–7 µm wide. Conidiophores erect, septate, branched, 200–300 µm long, 2–2.5 µm wide at the base, tapering upwards to a width of 1–2 µm at the slightly expanded apex, where on forming a compact of conidia (up to 6). Conidia oblong-pyriform, hyaline, 20–27.5 (24.4) × 7.5–12.5 (10.8) µm, 1-septate at the centre of spore. Chlamydospores not observed. Capturing nematodes by means of three dimensional adhesive networks.

Distribution: China (Anhui, Beijing, Guizhou, Hebei, Heilongjiang, Jilin, Liaoning, Neimeng, Shanxi, Yunnan), Canada (Quebec), Ecuador (Baños), Germany (Berlin), Netherlands (Noordoostpolder), UK (England).

Material examined: Z241, ZX41, isolated from forest soil in Guizhou and Shanxi in 1991 by Ke-Qin Zhang; ZS–3–2, isolated from forest soil in Xishan, Yunnan in June 1999 by Lu Cao; DQ4–8–2, isolated from soil in Xundian, Yunnan in September 2002 by Jing Zhang. Permanent slide: XS–3–2.

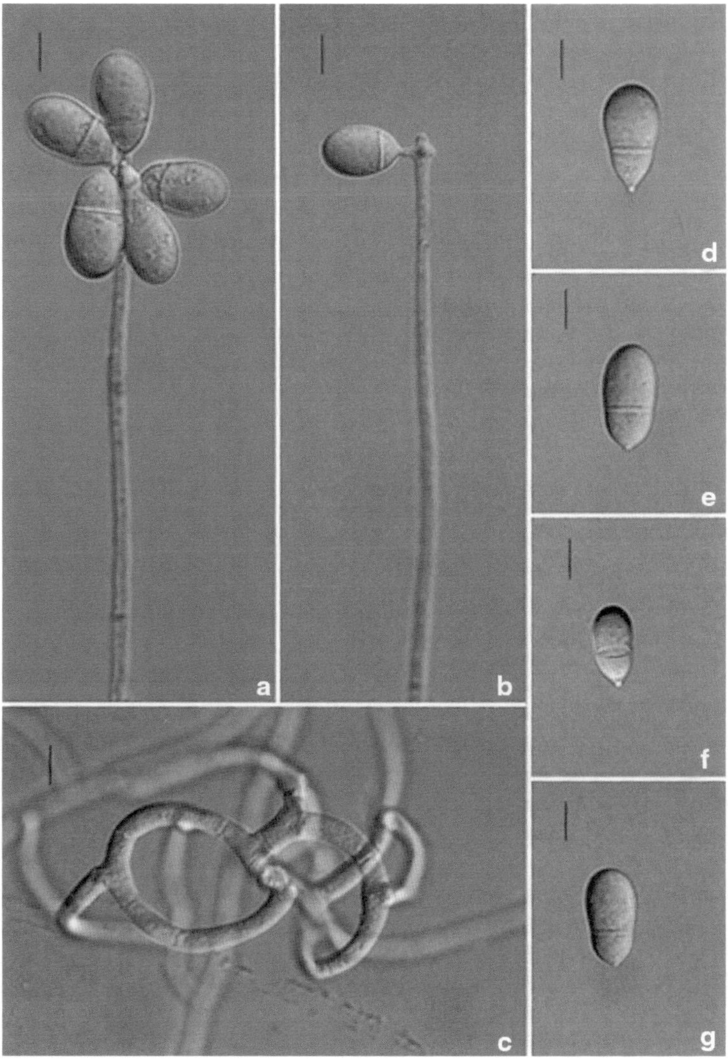

Fig. 3.42 *Arthrobotrys robusta*. **a–b** conidiophore; **c** adhesive network; **d–g** conidia. Bars = 10 μm; Strain number: DQ4-8-2

Notes: *A. robusta* resembles *A. superba* and *A. oligospora* in conidia, but differs in the freely branched conidiophores and in the absence of nodal development of conidia in whorls down the conidiophores. *A. robusta* also resembles *A. cladodes* in its general habit, differing from the former in both size of conidia and the absence of enlarged storage hyphae. (Fig. 3.42)

Arthrobotrys rutgeriense (R.C. Cooke & Pramer) Z.F. Yu, **comb.nov**
≡ *Monacrosporium rutgeriense* R.C. Cooke & Pramer, [as '*rutgeriensis*'] Phytopathology 58: 544, 1968
MB 804793

Characteristics: Colonies on CMA white. Vegetative hyphae hyaline, septate, branched. Conidiophores hyaline, erect, simple, septate, 67.5–350 μm long, 3–8 μm wide at the base, gradually tapering upwards to a width of 2–3 μm at the apex, bearing a single conidium. Sometimes conidiophores branched at the apex and two conidia were each borne singly and terminally on these branches, which up to 15 μm long. Conidia hyaline, globose to broadly turbinate, 27–47.5 (32.2) × 17.5–27.5 (22) μm, rounded at the distal end and tapered proximally to a narrow, truncate at the base, 2–3-septate. Capturing nematodes by means of three dimensional adhesive networks.

Distribution: China (Guizhou, Xizang, Yunnan), USA

Material examined: ⑨–13, isolated from field soil in Huaxi, Guizhou in 1996 by Ke-Qin Zhang; XZM–9, isolated from soil in Xizang in August 2000 by Minghe Mo; YMF1.00040, isolated from soil in Baoshan, Yunnan in October 2002 by Jing Zhang. Permanent slide: ⑨–13

Notes: *A. rutgeriense* resembles *A. eudermata* in conidial shape and septation, but differs markedly from it in conidial size, although there is a slight overlap between the two species. *A. rutgeriense* can be distinguished from *A. cookedickinson* in having predominantly 2–3-septate conidia in contrast to mainly 2-septate conidia in the latter species. (Fig. 3.43)

Arthrobotrys salina (R.C. Cooke & C.H. Dickinson) M. Scholler, Hagedorn & A. Rubner, Sydowia 51 (1): 104 (1999)
≡ *Monacrosporium salinum* R.C. Cooke & C.H. Dickinson, Trans. Br. Mycol. Soc. 48: 626 (1965)
= *Golovinia salinum* (R.C. Cooke & C.H. Dickinson)Mekht., Khishchnye Nematofagovye Griby—Gifomitsety (Baku): 145 (1979)

Characteristics: Single conidium produce from apex of conidiophores; conidia fusiform to ellipsoid, 32.5–52.5 × 12.5–17.5 μm, 0–4-septate, mainly 3-septate; chlamydospores not observed; capturing nematodes by means of three dimensional adhesive networks.

Distribution: UK.

Notes: The description is based on the protologue. This species varies in shape, septa and size of conidia, which result in difficulty of species identification. (Fig. 3.44)

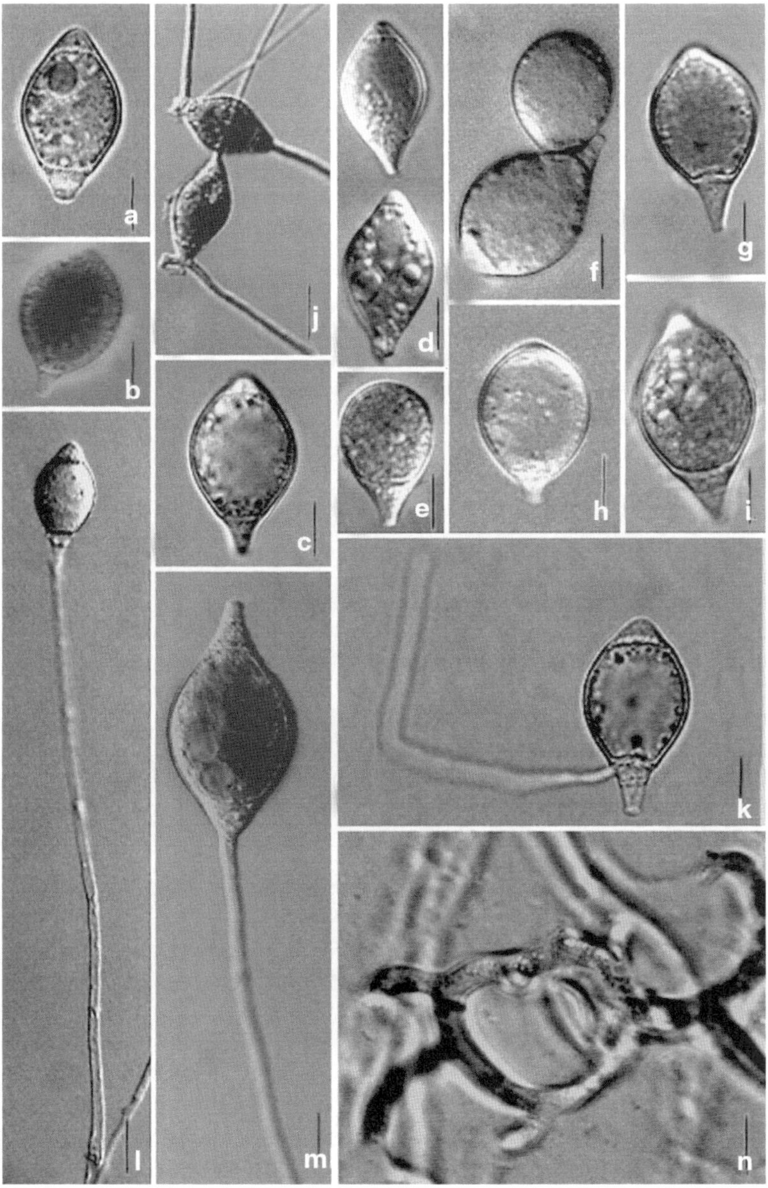

Fig. 3.43 *Arthrobotrys rutgeriense.* **a–j** conidia; **k** germinating conidia; **l–m** conidiophore; **n** adhesive network. Bars = 10 μm; Strain number: YMF1.00040

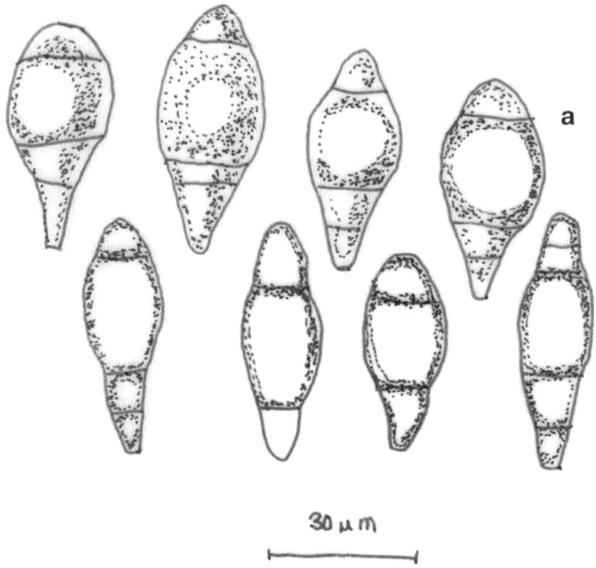

Fig. 3.44 *Arthrobotrys salina*. **a** conidia. Bar = 30 µm

Arthrobotrys scaphoides (Peach) S. Schenck, W.B. Kendr. & Pramer, Can. J. Bot. 55 (8): 984 (1977)
≡ *Dactylaria scaphoides* Peach, Trans. Br. Mycol. Soc.35 (1): 19 (1952)
= *Monacrosporium scaphoides* (Peach) Xing Z. Liu & K.Q. Zhang, Mycol. Res. 98 (8): 865 (1994)
= *Woroninula scaphoides* (Peach) Mekht, Khishchnye Nematofagovye Griby-Gifomitsety (Baku): 113 (1979)

Characteristics: Mycelium spreading, growing slowly, reaching 30 mm in diameter at 28 °C after 12 days. Vegetative hyphae hyaline, septate, 4–6 µm wide. Conidiophores erect, hyaline and septate, unbranched, 4.5–5.6 µm wide at base and tapering to a width of 4–4.5 µm, at a distance of 80–200 µm from the base producing 1–6 conidia, occasionally up to 10 conidia, in a loose capitate arrangement, then following repeated elongation often giving successively rise to up to 3 additional conidial clusters by branching at or just below slightly swollen warted nodules, producing more or less geniculate conidiophores up to 365–430 µm long. Conidia hyaline, 36.6–79.3 (57) × 11–17.5 (14) µm (Chinese strain), (50–)60–80 (–86) × (13–)14–15 (–16) µm (Dutch strain, both vital state in water), fusiform, not or slightly curved, 1–6-septate, mainly 2–3-septate, the proportion of conidia with 1, 2, 3, 4, 5 and 6 septa is 1.3, 48.8, 37.5, 10, 1.3 and 1.3 % respectively (Chinese strain), 2–3 (–4)-septate (Dutch strain), central cell mostly distinctly longer and wider than other cells. Three dimensional adhesive networks observed when nematodes added. Chlamydospores not observed in cultures.

Distribution: China, Netherlands

Material examined: YMF1.01895: Gansu Province, Jiuquan City, alt. 2450 m, from soil, under *Malus asiatica* Nakai (*Rosaceae*), from a private plantation. VIII.2006, YMF 1.01895, permanent slide culture (YMF 1.01895); H.B. 6972: The Netherlands: Zeeland, old harbor facing the Hertoginpolder, Verdronken Land van Saeftinghe, most eastern part, alt. 0 m, on previous year's leaves of *Scirpus maritimus* L. (*Cyperaceae*), 23.III.2001,G. Van Ryckegem (H.B. 6972b, dry specimen, associated with *A*. oudemansii M. Scholler et al. and its sexual state).

Notes: Within *A. scaphoides*, the Chinese strain shows a closer relationship to strain H.B. 6972: the ITS rDNA sequence analysis demonstrates that there is 99.6% similarity between them (2 nucleotide variance in ITS1-5.8s-ITS2 region), but 98.7% between the Chinese strain and CBS 226.52 (type strain, 7 nucleotide variances), and 99% between the strain from *Scirpus* and the type strain from *Typha* (5 nucleotide variance). Between the Chinese strain of *A. scaphoides* and other morphologically close species, the similarity is 84.2% to *A. microscaphoides* Xing Z. Liu & B.S. Lu, 84.9% to *A. thaumasia* (Drechsler) S. Schenck et al., (CBS 322.94), and 85.7% to *A. gampsospora*. (Fig. 3.45)

Arthrobotrys shahriari (Mekht.) M. Scholler, Hagedorn & A. Rubner, Sydowia 51 (1): 104 (1999)
≡ *Candelabrella shahriari* Mekht, Mycol. Res. 101: 334 (1997)

Characteristics: Colony growing rapidly, aerial mycelium initially white, becoming pale cream. Conidiophores erect, simple or branched and the apex candelabrum-like with 3–9 branches each bearing a conidium at the apex. The conidiophores rarely recommenced growth after the group of conidia had been produced and a second head was then formed above the first. Conidia elongate-obovoid or elongate-ellipsoidal, straight or sometimes curved, rounded distally and tapering proximally to a truncate base, 1-septate, 33.5–57 × 11–15.5 μm. Globose terminal chlamydospores also formed and adhesive traps formed rarely in pure culture. Capturing nematodes by means of three dimensional adhesive networks.

Distribution: Azerbaijan (Tabriz)

Notes: The description is based on the protologue. This species can be distinguished from other similar species by the candelabrum-like apex of the conidiophores and 1-septate, elongate-obovoid or elongate-ellipsoidal conidia. (Fig. 3.46)

Arthrobotrys shizishanna (X.F. Liu & K.Q. Zhang) J. Chen, L.L. Xu, B. Liu, Fungal Diversity 26 (1): 124 (2007)
≡ *Dactylella shizishanna* X.F. Liu & K.Q. Zhang, Fungal Diversity 14: 104 (2003)

Characteristics: Colonies on CMA whitish, slow growing and extending to a diameter of 3.5 cm within 15 days at 25°C. Aerial hyphae scant, hyaline, septate, branching, commonly 2.5–3.7 μm wide. Conidiophores growing from mycelium on the substratum, single, erect, rarely branched, 35–200 μm long, 1.5–2.5 μm wide at

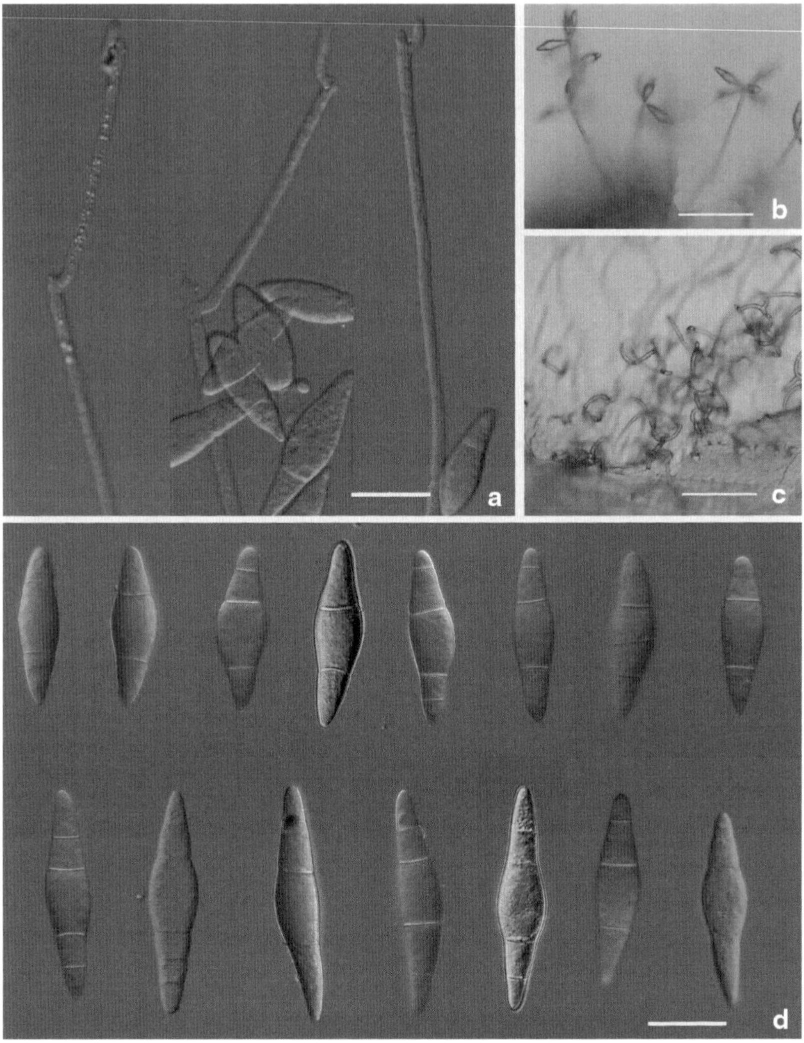

Fig. 3.45 *Arthrobotrys scaphoides*. **a** conidiophores; **b** conidiophores with conidia; **c** adhesive network; **d** conidia. Bars: **a, d** = 10 μm, **b, c** = 50 μm. Strain number:YMF1.01895

the base, tapering upwards gradually to a distal width of 0.5–1 μm, bearing a single conidium. Conidia hyaline, clavate, gradually narrowing at the basal end, obtuse at the distal end, straight or sometimes slightly curved, sometimes constricted at the septa, 2–9-septate, mostly 3–7-septate, 22.5–73.8 (50.6) × 5–10 (6.6) μm. The proportion of conidia with 3, 4, 5, 6 and 7 septa is 10, 30, 33, 18 and 6 %, respectively. Chlamydospores present in aged cultures. Capturing nematodes by means of three dimensional adhesive networks.

Distribution: China (Hubei)

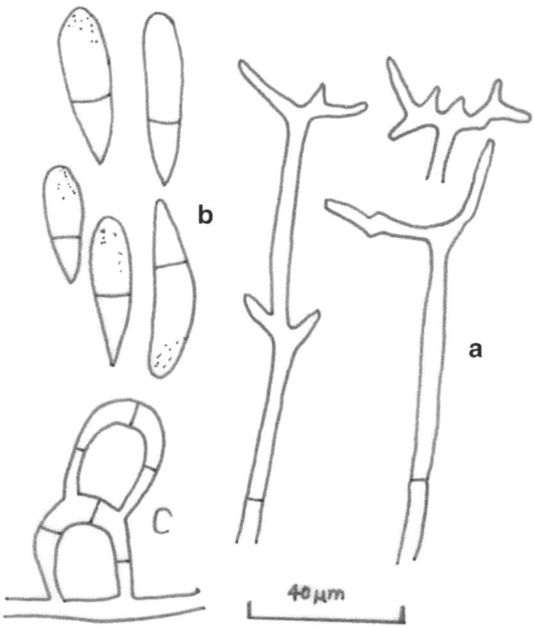

Fig. 3.46 *Arthrobotrys shahriari*. **a** conidiophore; **b** conidia; **c** adhesive network. Bar = 40 μm

Material examined: YMF1.00022, isolated from field soil in Shizishan, Hubei in February 2001 by Xuefeng Liu.

Notes: *A. shizishanna* is easily distinguished other *Arthrobotrys* species because having the most septate conidia. (Fig. 3.47)

Arthrobotrys sinensis (Xing Z. Liu & K.Q. Zhang) M. Scholler, Hagedorn & A. Rubner, Sydowia 51 (1): 104 (1999)
≡ *Monacrosporium sinense* Xing Z. Liu & K.Q. Zhang, Mycol. Res. 98: 863 (1994)

Characteristics: Colonies growing on CMA rapidly, attaining a diameter of 6 cm within 7 days at 25 °C, with sparse aerial hyphae. Vegetative hyphae hyaline, septate, branched, mostly 2.5–9.5 μm wide. Conidiophores hyaline, simple, erect, septate, occasionally branched, mostly 200–500 μm long, 4.5–5.5 μm wide at the base, tapering to 2.5–3 μm wide at the apex, bearing a single conidium. Conidia hyaline, subsphaerical to obovoid, 1–3-septate, 37 % of conidia 3-septate, 42 % 2-septate and 21 % 1-septate, 23.5–30 (27.6) × 17–25 (20) μm. Chlamydospores sphaerical to ellipsoidal, terminal to intercalary in chains, mostly 20–24 × 17–25 μm. Capturing nematodes by means of three dimensional adhesive networks.

Distribution: China (Anhui, Jiangxi, Xizang, Yunnan)

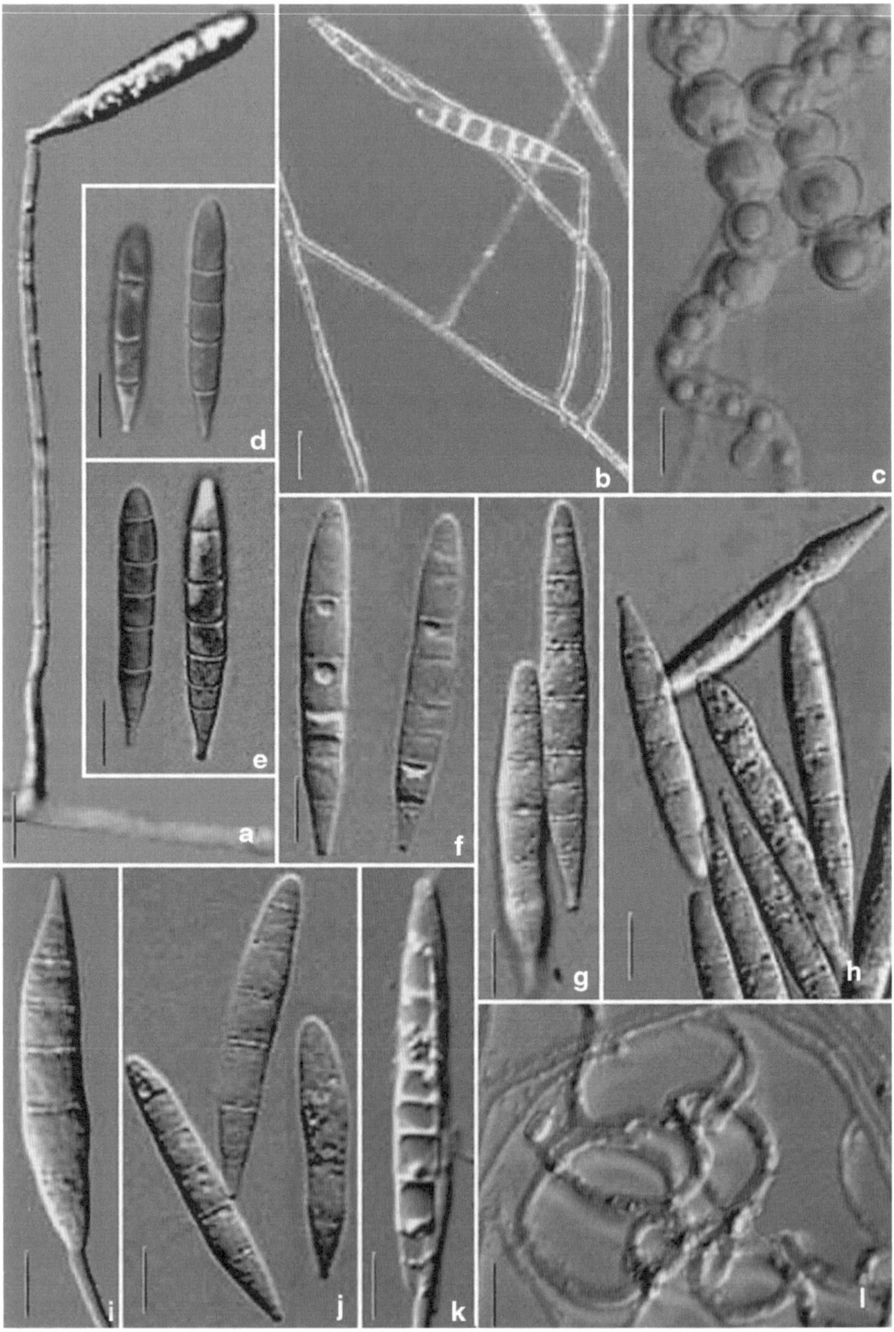

Fig. 3.47 *Arthrobotrys shizishana.* **a–b** conidiophore; **c** chlamydospore; **d–k** conidia; **l** adhesive network. Bars = 10 μm; Strain number: YMF1.00022

Material examined: YMF1.00025, isolated from field soil in Huaxi, Guizhou in 1996 by Ke-Qin Zhang; XZM–10, isolated from soil in Xizang in August 2000 by Minghe Mo. Permanent slide: CK9–1

Notes: This fungus closely resembles *A. cookedickinson* and *A. indica* in its relatively small conidia and conidial shape, but differs in size of conidia. In *A. cookedickinson,* conidia are 32.5–50 × 14–23 μm, in *A. indica* conidia are 22–30 × 14–20 μm, while in *A. sinense,* conidia are 25–30.5 × 15–18 μm. (Fig. 3.48)

Arthrobotrys sphaeroides (Castaner) Z.F. Yu, **comb. nov.**
≡ *Monacrosporium sphaeroides* Castaner, Am. Midl. Nat. 80: 282 (1968)
MB804795

Characteristics: Colonies on CMA white, growing rapidly. Mycelium spreading, scanty, vegetative hyphae hyaline, septate, branched, mostly 2–5 μm wide. Conidiophores hyaline, erect, simple, septate, 200–460 μm long, 5–8 μm wide at the base, gradually tapering upwards to a width of 2.5–3 μm at the apex, bearing a single conidium. Conidia hyaline, sphaeroid to obovoid, 20–44 (32) × 17–25 (20.4) μm, 0–3-septate. Chlamydospores in chains or clusters. Capturing nematodes by means of three dimensional adhesive networks.

Distribution: China (Guizhou, Xizang, Yunnan), USA

Material examined: YMF1.00030, isolated from field soil in Huaxi, Guizhou in 1996 by Ke-Qin Zhang; YMF1.00539, isolated from soil in Baoshan, Yunnan in September by Jing Zhang. Permanent slide: ④–5–1.

Notes: This fungus mostly resembles *A. cookedickinson.* In *A. cookedickinson,* conidiophores are simple or branched, and conidia are clavate to turbinate and 32.5–50 × 14–23 μm. In *A. sphaeroides* conidiophores are always simple, conidia are sphaeroid to obovoid and 28–42 × 19–29 μm. Comparisons between *A. cookedickinson* and *A. sphaeroides* indicate that the latter is sufficiently different to be considered a distinct species. (Fig. 3.49)

Arthrobotrys superba Corda, Pracht-Fl. Eur. Schimmelbild.: 43 (1839)
≡ *Didymozoophaga superba* (Corda) Soprunov & Galiulina, Mikobiology, Reading 20: 493 (1951)

Characteristics: Colonies on CMA whitish to reddish, rapidly growing. Mycelium dense, spreading, vegetative hyphae hyaline branched, septate. Conidiophores hyaline, smooth, 6–9-septate, rarely branched, erect, 180–600 μm long, 2.5–7.5 μm wide, apex swelling, bearing 20–39 conidia. Conidia elliptical, 1-septate at the centre of spore, slightly constricted at the septum, broadly rounded at the distal end, 7.5–27.5 (15.8) × 5–10.5 (6.6) μm. Capturing nematodes by means of three dimensional adhesive networks.

Distribution: China (Anhui, Guizhou, Hainan, Hubei, Jilin, Tianjing, Xizang, Yunnan), cosmopolitan.

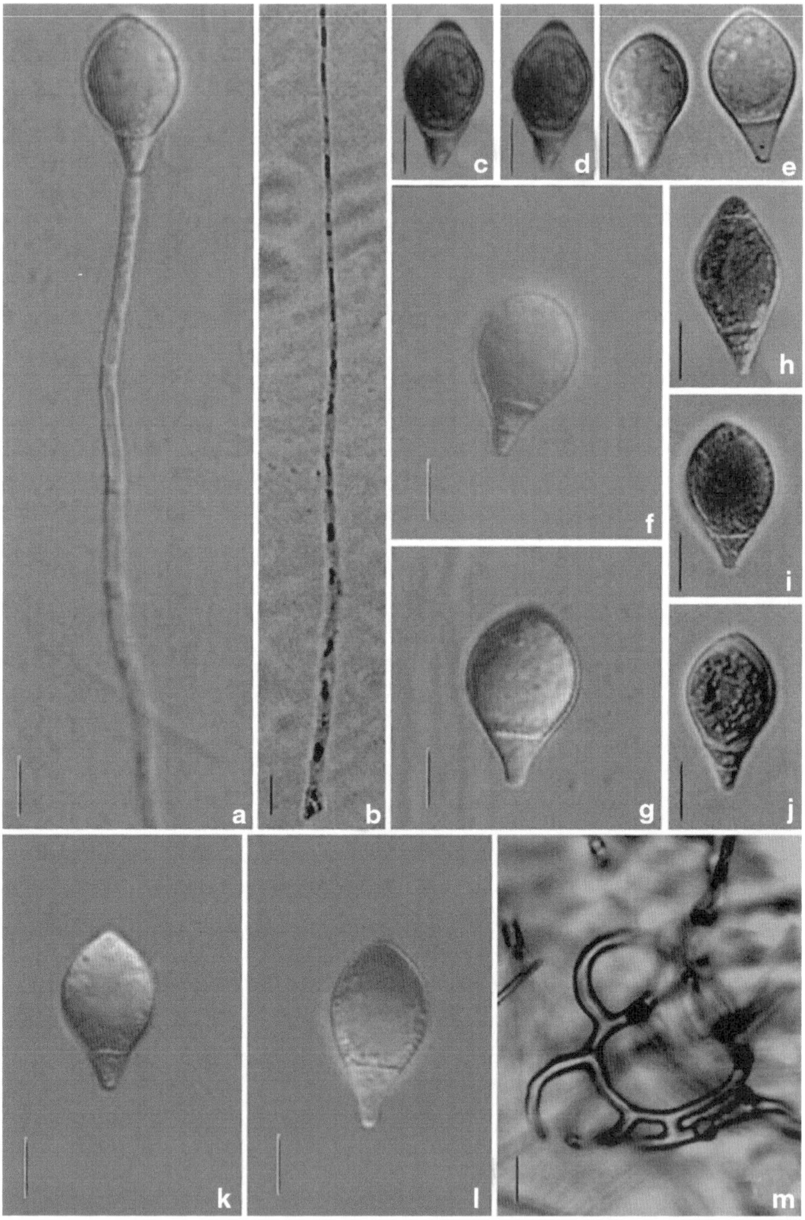

Fig. 3.48 *Arthrobotrys sinensis.* **a–b** conidiophore; **c–l** conidia; m adhesive network. Bars = 10 μm; Strain number: YMF1.00025

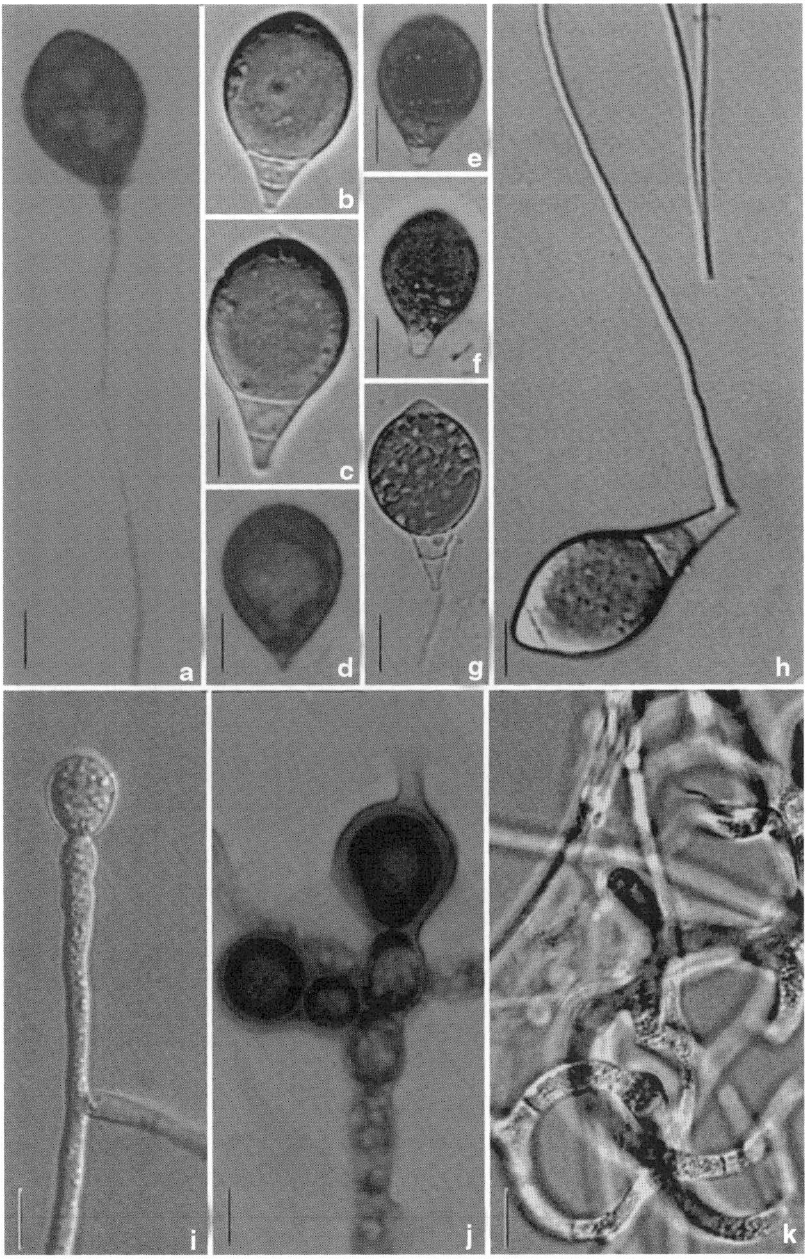

Fig. 3.49 *Arthrobotrys sphaeroides*. **a–g** conidia; **h** conidiophores with conidia; **i–j** chlamydospore; **k** adhesive network. Bars = 10 μm; Strain number: YMF1.00030

Material examined: Z19, isolated from field soil in Huaxi, Guizhou in 1996 by Ke-Qin Zhang; JLCB–1, isolated from forest soil in Changbai Mountain, Jilin in 1996 by Ke-Qin Zhang; LT–10, isolated from soil in Huaguoshan, Lincang in 1999 by Yanju Bi; J–1qz07, isolated from soil in Liqizhuang, Tianjing in 2000 by Wenpeng Li; XZA–10, isolated from pasture soil in Xizang in August 2000 by Minghe Mo; YMF1.00016, DQ4–2, isolated from forest soil in Xundian, Yunnan in September 2002 by Jing Zhang. Permanent slide: ⑨–19.

Notes: *Arthrobotrys superba* was established as the type species of *Arthrobotrys* (Corda 1839), which is the asexual state of *Orbilia fimicola* (Pfister 1994). *Arthrobotrys superba* resembles *A. conoides* and *A. oligospora* in most morphological characters except in conidia shape and the position of septum. In *A. superba* conidia are elliptical and 1-septate at the centre. In *A. conoides* conidia are elongate obconical, 1-septate about one third from the basal end, constricted at the septum. In *A. oligospora* conidia are obovoid, 1-septate near the base of the spore, and constricted at the septum. (Fig. 3.50)

Arthrobotrys tabrizica (Mekht.) M. Scholler, Hagedorn & A. Rubner, Sydowia 51 (1): 104 (1999)
≡ *Nematophagus tabrizicus* Mekht, Mycol.Res.102: 683 (1998)

Characteristics: Colony growing rapidly, velvety, initially white, changing through pale pink to intensive pink after 2 wk. Conidiophores simple, erect, branched, curved, with several swollen to distinctly denticulate conidiogenous regions forming close conidial heads; these proliferating to form additional conidiogenous regions. Conidia elongate-ovate to elongate-doliform or ellipsoidal, 1–2 (–3)-septate, 28.5–56 × 11.5–22.5 μm, with a pronounced protuberant base. Ellipsoidal to cylindrical intercalary chlamydospores present. Capturing nematodes by means of three dimensional adhesive networks.

Distribution: Azeraijan (Tabriz)

Notes: The description is based on the protologue. This species produces simple or branched conidiophores with several conidial clusters. Conidia are 1–3-septate; 2-septate conidia are more dominant in the original isolation plate than when the fungus was grown on malt agar. (Fig. 3.51)

Arthrobotrys thaumasia (Drechsler) S. Schenck, W.B. Kendr. & Pramer, Can. J. Bot. 55 (8): 984 (1977)
≡ *Dactylaria thaumasia* Drechsler, Mycologia 29 (4): 522 (1937)
= *Candelabrella thaumasia* (Drechsler)Rifai, Reinwardtia 7 (4): 369 (1968)
= *Golovinia thaumasia* (Drechsler) Mekht., Mikol. Fitopatol. 1: 276 (1967)
= *Monacrosporium thaumasium* (Drechsler) de Hoog & van Oorschot [as '*thaumasia*'], Stud. Mycol. 26: 120 (1985)

Characteristics: Colonies on CMA white to reddish. Mycelium spreading, vegetative hyphae hyaline, septate, branched, 2–8 μm wide. Conidiophores hyaline, erect,

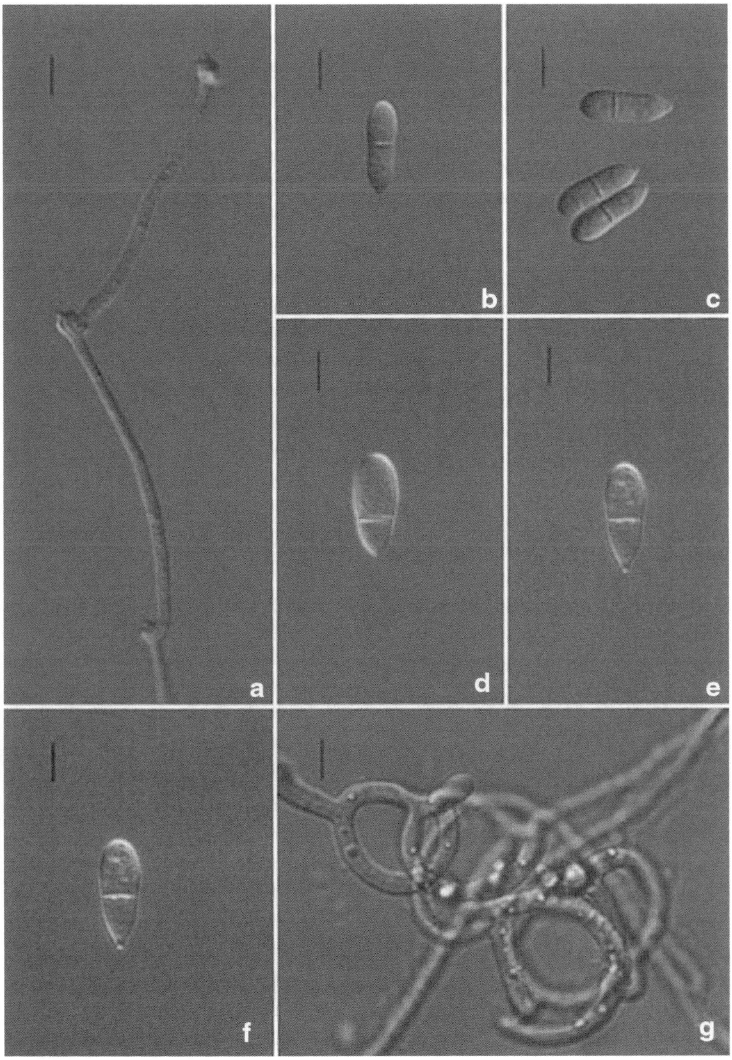

Fig. 3.50 *Arthrobotrys superba*. **a** conidiophore; **b–f** conidia; **g** adhesive network. Bars = 10 μm; Strain number:YMF1.00016

septate, 195–460 μm long, 4–8 μm wide at the base, gradually tapering upwards to a width of 2–4 μm at the apex, simple or often somewhat branched near the apex, and there borne on blunt sterigmata, mostly 2–10 μm long and 2–3 μm wide, usually 3–15, rarely up to 25 conidia in loose capitate arrangement. Conidia hyaline, top-shaped, rounded at the apex, tapering towards the frequently somewhat protruding truncate base, 30–60 (36.2) × 15–30 (20.2) μm, 1–4-septate, mostly 3-septate. Chlamydospores yellow, globose or ellipsoidal. Capturing nematodes by means of three dimensional adhesive networks.

Fig. 3.51 *Arthrobotrys tabrizica*. **a** conidiophore; **b** conidia; **c** adhesive network. Bar = 25 μm

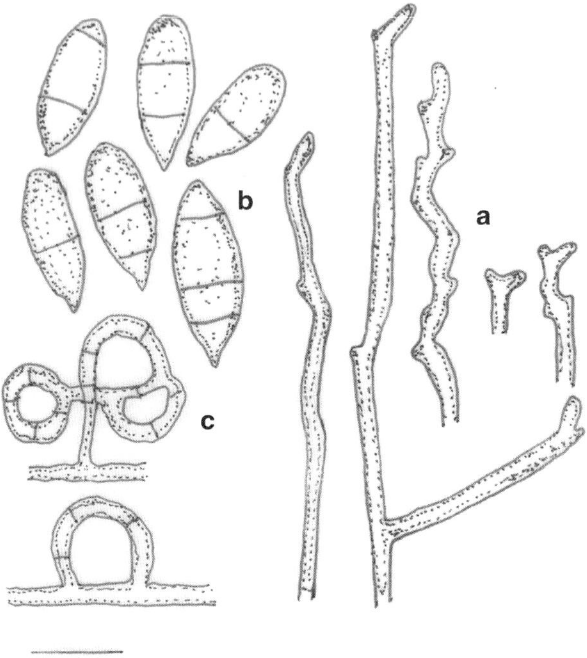

Distribution: China (Anhui, Beijing, Guangdong, Guizhou, Hebei, HuBei, Hunan, Jiangsu, Xizang, Yunnan, Zhejiang), Baukina Faso (Houet), Germany (Berlin–Dahlem), Netherlands (Schipborg), Thailand, UK (England), USA (Norfolk)

Material examined: L3–5, isolated from soil in Licang, Yunnan in 1999 by Yanju Bi; XZM–11, isolated from soil in Xizang in August 2000 by Minghe Mo; YMF1.00026, YMF1.00547, YMF1.00586, isolated from soil in Baoshan, Yunnan in October 2002 by Jing Zhang; Sjz3.11.20, isolated from soil in Huaxi, Guiyang in 1996 by Ke-Qin Zhang. Permanent slide: Sjz3.11.20

Notes: *Arthrobotrys thaumasia* resembles *A. elegans* in conidial shape and number of septa, but differs in conidiophores and conidial length. The conidia of *A. thaumasia* are longer than those of *A. elegans*. Moreover, in *A. thaumasia*, conidiophores branch near the apex, bearing blunt sterigmata with 3–15 conidia in a loose capitate arrangement, while in *A. elegans* conidiophores bear a single conidium at the apex, or occasionally form a branch near the apex with a conidium. (Fig. 3.52)

Arthrobotrys vermicola (R.C. Cooke & Satchuth.) Rifai, Reinwardtia 7 (4): 371 (1968)
≡ *Dactylaria vermicola* R.C. Cooke & Satchuth., Trans. Br. Mycol. Soc. 49: 27 (1966)
= *Nematophagus vermicola* (R.C. Cooke & Satchuth.) Mekht., Khishchnye Nematofagovye Griby—Gifomitsety (Baku): 105 (1979)

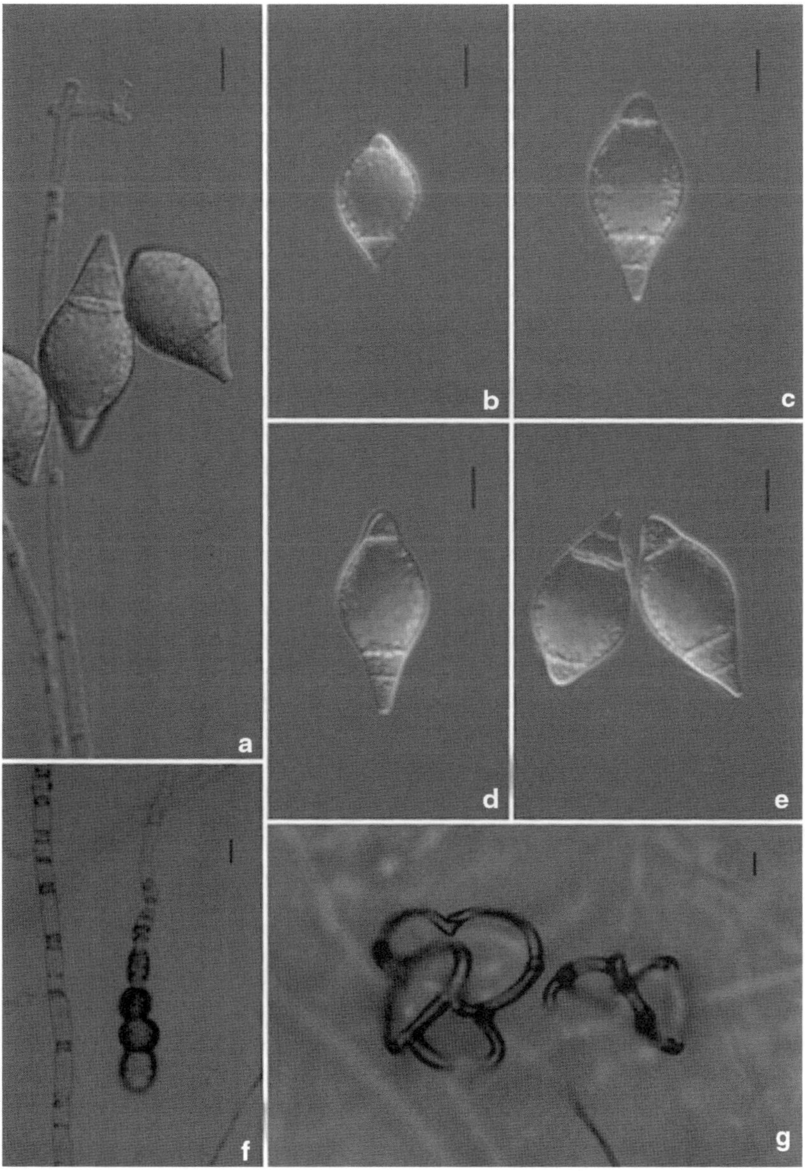

Fig. 3.52 *Arthrobotrys thaumasia*. **a** conidiophore; **b–e** conidia; **f** chlamydospore; **g** adhesive network. Bars = 10 μm; Strain number: YMF1.00547

Characteristics: Colonies on CMA whitish. Mycelium spreading, vegetative hyphae hyaline, septate, branched. Conidiophores erect, unbranched, septate, 140–470 μm long, forming 3–6 conidia in a loose capitate arrangement. Each conidium is formed on a small protuberance. The conidiophores often recommence growth after the first group of conidia had been produced, and a second head is then formed about 75 μm above the first. This process is repeated so that three to four groups of conidia are produced on each conidiophore, the latter then becoming somewhat geniculate. Conidia elongate-ellipsoidal to broadly fusiform, hyaline, 25–50 × 17.5–25 μm, bluntly rounded distally and tapering proximally to a narrow truncate base, 1–3-septate; proportion of conidia with 1, 2 and 3 septa is 4, 64 and 32 %, respectively. In 2-septate conidia the median cell, and in 3-septate conidia the penultimate cell, is usually larger than the others. Very rarely a 4-septate conidium was observed. Capturing nematodes by means of three dimensional adhesive networks.

Distribution: Cosmopolitan

Material examined: Z296, Z290, Z279–1, isolated from soil in Beijing, and Guiyang, Guizhou during 1991–1992 by Ke-Qin Zhang; YNBN–2, isolated from forest soil in Xishuangbanna, Yunnan in 1996 by Ke-Qin Zhang; J31–1, isolated from soil in Jianshui, Yunnan in 1999 by Yanju Bi; BS3–35, isolated from soil in Baoshan, Yunnan in October 2002 by Jing Zhang; YMF1.00533, YMF1.00534, YMF1.00554, from soil in Dehong, Yunnan in July 2003 by Hong Luo. Permanent slide: BS3–35.

Notes: *Arthrobotrys vermicola* resembles *A. guizhouensis* (Zhang 1994) in the conidial shape and trapping devices, but differs in conidia septa and conidiophore form. In *A. vermicola*, conidia are mostly 2–3-septate, produced along the conidiophores, with 3–6 conidia being located at each denticle node of the unbranched conidiophores in a loose capitate arrangement. In *A. guizhouensis*, conidia are mostly 1-septate, produced at the irregular expanded apex of the branched conidiophores. (Fig. 3.53)

Arthrobotrys yunnanensis M.H. Mo & K.Q. Zhang, Fungal Diversity 18: 109 (2005)

Characteristics: Colonies growing rapidly on CMA and attaining 6 cm diameter in 5 days at 28°C and mycelia spreading at the rate of 0.5 cm per 24 h, conidiophores and conidia are produced after 4 days. Mycelium scanty, spreading, vegetative hyphae hyaline, septate and branched, mostly 2–4 μm wide. Conidiophores hyaline, erect, septate, simple, frequently 60–200 μm long, 2–5 μm wide at the base and 1.5–2.4 μm at the apex, producing 1–5 conidia singly from conidiogenous loci on conspicuous denticles at and near the apex. Conidia hyaline, elongate ellipsoid-cylindrical or slightly clavate, broadly rounded at the apex, rounded to truncate at the narrowed base, sometimes constricted gradually at the distal part of conidia, usually aseptate, occasional 1-septate (<5 %) at the centre, 17.5–32.5 (22.6) × 2.75–7.5 (5.5) μm. Chlamydospores sphaerical to ellipsoidal, intercalary. Capturing nematodes by means of three dimensional adhesive networks.

Distribution: China (Yunnan)

Material examined: MF1.00593, isolated from soil in Yuliang, Yunnan in August 2003 by Minghe Mo.

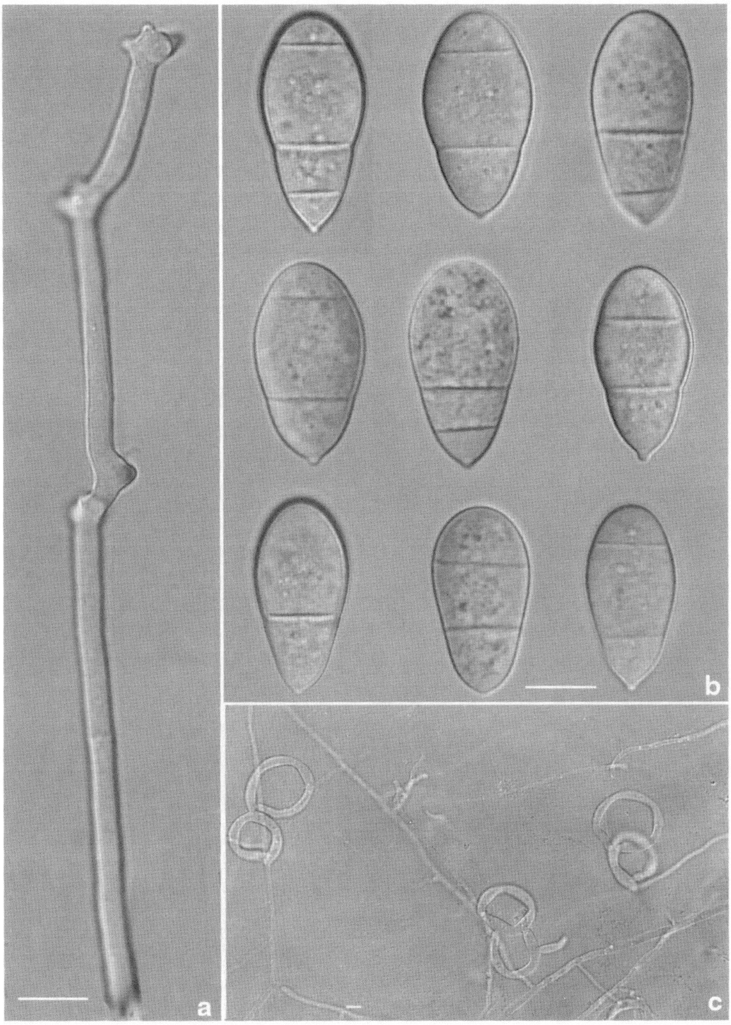

Fig. 3.53 *Arthrobotrys vermicola*. **a** conidiophore; **b** conidia; **c** adhesive network. Bars = 10 μm; Strain number: YMF1.00534

Notes: *Arthrobotrys yunnanensis* differs from the three species of *Arthrobotrys* which form aseptate, or occasionally 1-septate conidia, *A. anomala*, *A. amerospora* and *A. botryospora*. *Arthrobotrys yunnanensis* is characterized by predominantly elongate ellipsoid-cylindrical or slightly clavate, non-septate conidia which are borne on distinct and long denticles. *Arthrobotrys yunnanensis* and *A. anomala* have similar conidial shapes, but differ in the size of conidia. The conidia of *A. anomala* form one septum only when detached, while in the former species they are aseptate when detached. *Arthrobotrys yunnanensis* is easy to distinguish from *A. botryos-*

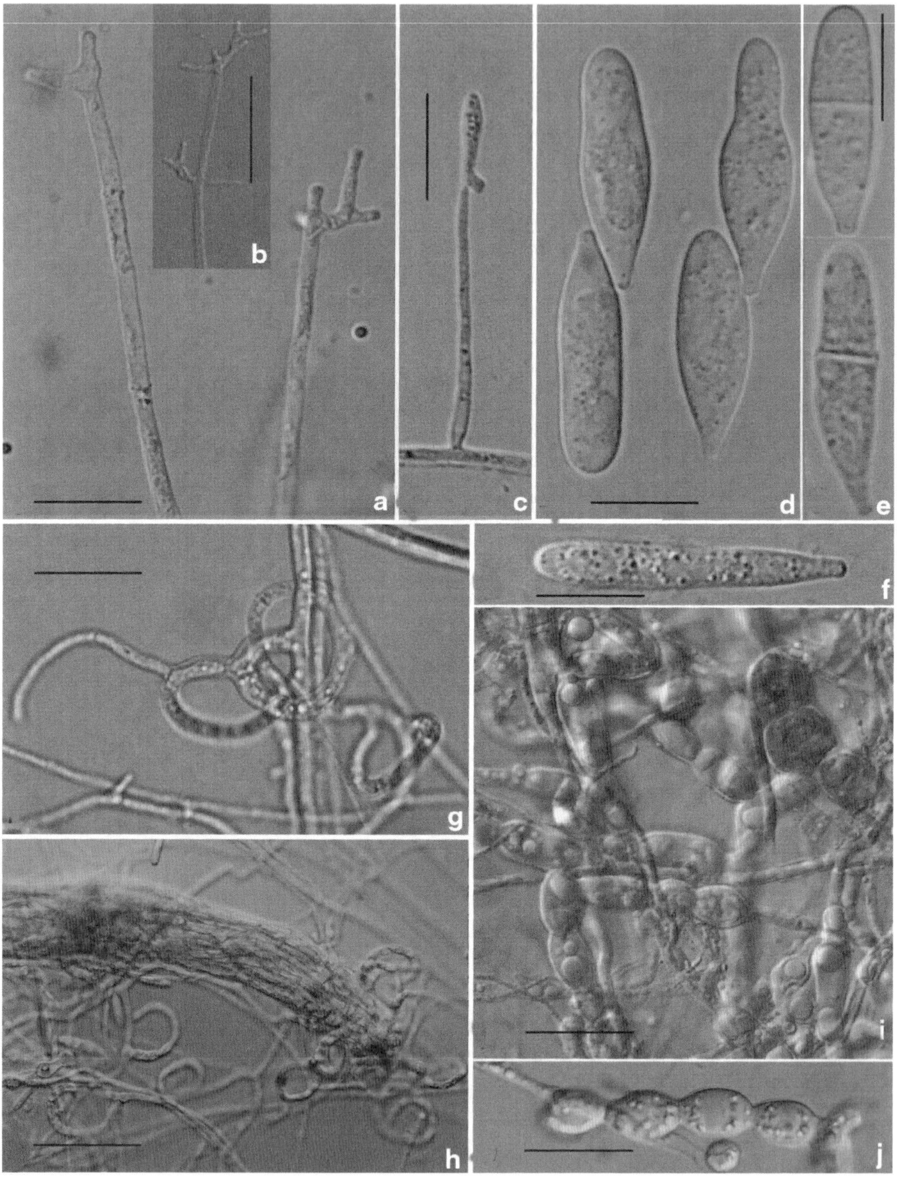

Fig. 3.54 *Arthrobotrys yunnanensis.* **a–c** conidiophore; **d–f** conidia; **g–h** adhesive network; **i–j** chlamydospore. Bars = 10 μm; Strain number: YMF1.00593

pora and *A. amerospora* by their different conidia and conidiophores (modification at apex). Furthermore, unlike *A. botryospora*, *A. yunnanensis* does not capture nematodes by adhesive hyphae, and unlike *A. amerospora*, which only produces aseptate conidia, *A. yunnanensis* occasionally produces 1-septate conidia. (Fig. 3.54)

Species of the genus *Dactylellina*

Dactylellina M. Morelet, emend. M. Scholler, Hagedorn & A. Rubner
Type species: *Dactylellina leptospora* (Drechsler) M. Morelet-Bull. Soc. Sci. Nat. Archéol. Toulon Var 178: 6 (1968)
≡ *Dactylella leptospora* Drechsler, Mycologia 29: 507 (1937)
= *Dactylosporium* Mekht. (1967) [non Harz 1871] (nom. illeg., Art. 53 ICBN)
Type species: *Dactylosporium leptosporum* (Drechsler) Mekht., Mikol. Fitopatol. 1: 278 (1967)
≡ *Dactylella leptospora* Drechsler, Mycologia 29: 507 (1937).
= *Kafiaddinia* Mekht. (1978)
Type species: *Kafiaddinia fusarispora* Mekht., Mikol. Fitopatol.12: 8 (1978).
= *Laridospora Nawawi* (1976)
Type species: *Laridospora appendiculata* (Anastasiou) Nawawi—Trans. Br. Mycol. Soc. 66: 344 (1976)
≡ *Dactylella appendiculata* Anastasiou—Pacif. Sci. 18: 202 (1964)

Characteristics: Mycelium slow-growing. Hyphae septate, branching, hyaline. Saprotrophic or predatory by trapping nematodes or other animals by means of stalked adhesive knobs, sometimes in combination with stalked three-celled non-constricting rings. Traps generally spontaneously formed. Conidiophores mostly simple; branches, when present, often near the apex, apex mostly simple without modifications or sometimes with short denticles. Conidia either formed singly or in clusters on the apex of the conidiophore, holoblastic, hyaline, 1–7 (15) septate, mostly cylindrical, ellipsoidal, fusiform or spindle-shaped, rarely clavate or obconical. Microconidia and microconidiophores very rarely formed. Chlamydospores absent. Sexual state belonging to the genus *Orbilia* Fr. (*Helotiales*).

Key to Species of *Dactylellina*

1. Trapping organs adhesive knobs or knobs and non-constricting rings 5
1. Trapping organs others 2
2. Trapping organs simple two-dimension networks .... ***D. gephyropaga***
2. Trapping organs adhesive branches, verruca, adhesive knobs and networks 3
3. Trapping organs adhesive branches .... ***D. cionopaga***
3. Trapping organs adhesive verruca, knobs and networks.................................... 4
4. Trapping organs adhesive verruca.. ***D. robustum***
4. Trapping organs knobs and networks... ***D. arcuata***
5. Trapping organs adhesive knobs and non-constricting rings 6
5. Trapping organs adhesive knobs 10
6. Near apex of conidiophores producing short branches, with conidia in strikingly loose capitate arrangement.. ***D. candidum***

6. Conidia singly grows at apices of conidiophores, or at apices of subapical lateral branches 7
7. Conidia elongate-fusiform to cylindrical, with 5-15 septa .... ***D. leptospora***
7. Conidia fusiform, mostly 4 or 5-septate 8
8. Conidia 3–6-septate, mostly 4 septate, largest cell not obvious
.... ***D. sichuanensis***
8. Conidia with largest cell 9
9. Conidia 2–4-septate, mostly 4-septate .... ***D. lysipaga***
9. Conidia 3–8-septate, mostly 5-septate .... ***D. illaqueata***
10.Adhesive unstalked, growing out to form branches, conidia with mostly 3-4 septa .... ***D. phymatopaga***
10. Adhesive stalked or with short stalk 11
11.Conidiophores with apical cluster of conidia or as well as intercalary cluster, conidia arranged into capitate heads 12
11. Conidiophores unbranched, or occasionally with very short subapical lateral branches, conidia singly arranged 18
12. Conidiophores with apical cluster of conidia as well as intercalary cluster
.... ***D. entomopaga***
12. Conidiophores with apical cluster of conidia.. 15
13. Conidiophores branched, which apex with denticles or sterigmata................ 14
13. Conidiophores unbranched 18
14. Conidia obovoid to elongate-cylindrical, 0–4-septate, mostly 3-septate
.... ***D. hertziana***
14. Conidia ovoid to clavate, 1-septate... ***D. ferox***
15.Conidiophore of the candelabrelloid type .... ***D. huisuniana***
15. Apex of conidiophores with sterigmata, without intercalary cluster of conidia
.... 16
16. Conidia cylindrical.. ***D. haptospora***
16. Conidia not cylindrical... 17
17. Conidia fusiform, mostly 3-septate, adhesive knobs joined together to form loops .... ***D. lobata***
17. Conidia elongate-fusiform or clavate, straight or curve, mostly 6–8-septa
.... ***D. daliensis***
18. Conidia typically fusiform 21
18. Conidia fusiform to other type 19
19. Distally end of conidia bluntly rounded... 20
19. Conidia fusiform to cylindrical, mostly 4–7-septate...................... ***D. formosana***
20. Conidia obconical to clavate, distally end bluntly rounded, 3-septate
.... ***D. asthenopaga***
20. Conidia ellipsoid to fusiform, 2-4-septate .... ***D. mammillata***
21. Conidia 1–3-septate, often with thorn-like appendage .... ***D. appendiculata***
21.Conidia with more than 2 septa 22
22. Shape of conidia varies on CMA, with multiform conidia............. ***D. mutabilis***

22. Conidia mostly fusiform 23
23. Conidia 4–9-septate, mainly 6–7-septate .... ***D. multiseptatum***
23. Conidia mostly 4–septate or 2–3-septate 24
24. Conidia mostly 2—septate.. 25
24. Conidia mostly 4–septate 26
25. Conidia 3–4-septate, mostly 2–3-septate, 29.5-48.7 × 10.7-15.4 μm
.... ***D. drechsleri***
25. Conidia 3–5-septa, mostly with 3 septa, with shorter stalk adhesive knobs
.... ***D. quercus***
26. Conidia relatively long, conidia 1–6-septate, mostly 4-septate, 56-97 × 8.5-16
....***D. copepodii***
26. Conidia about 50 μm long.. 27
27. Conidia 3–5-septate, mostly 4-septate, 33-50 (43.7) × 7.4-13.3 (10.7)
....***D. parvicolla***
27. Conidia 2–4-septate, mostly 4 septate, 37.5-62 × 8.7-19.3...........***D. ellipspspora***

Accepted Species of *Dactylellina*

Dactylellina appendiculata (Anastasiou) M. Scholler, Hagedorn & A. Rubner, Sydowia 51 (1): 110 (1999)
≡ *Dactylella appendiculata* Anastasiou, Pacif. Sci. 18: 202 (1964)
= *Monacrosporium tentaculatum* A. Rubner &W. Gams, Stud. Mycol.39: 97 (1996)
= *Laridospora appendiculata* (Anastasiou) Nawawi, Trans. Br. Mycol. Soc. 66 (2): 344 (1976)

Characteristics: Conidia fusiform, often with thorn on it, 20–59 μm long, 10–24 μm wide, 1–3-septate. Capturing nematodes by adhesive hyphae.

Distribution: Russia

Notes: This species has not been reported from other regions since it was first reported from Russia. The type strain cannot be located, but we still treat it as an accepted species. (Fig. 3.55)

Dactylellina arcuata (Scheuer & J. Webster) Ying Yang & Xing Z. Liu, Mycotaxon 97: 158 (2006)
≡ *Dactylella arcuata* Scheuer & J. Webster, Mycol. Res. 94: 720 (1990)
= *Arthrobotrys arcuata* (Scheuer & J. Webster) Yan Li, in Li, Hyde, Jeewon, Cai, Vijaykrishna & Zhang, Mycologia 97 (5): 1042 (2005)
= *Gamsylella arcuata* (Scheuer & J. Webster) M. Scholler, Hagedorn & A. Rubner, Sydowia 51 (1): 108 (1999)
= *Monacrosporium arcuata* (Scheuer & J. Webster) A. Rubner, Stud. Mycol.39: 49 (1996)

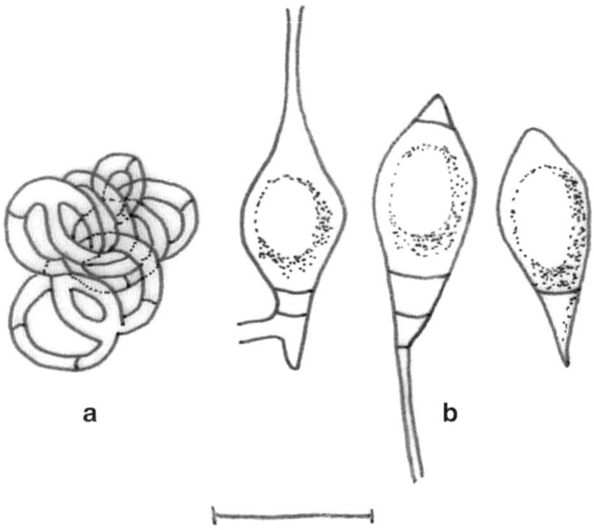

Fig. 3.55 *Dactylellina appendiculata.* **a** adhesive hyphae; **b** conidia. Bar =30 μm

Characteristics: Mycelium colourless, growing rapidly on 0.1% malt extract agar. Conidiophores colourless, single, septate, erect, bearing only one conidium at the apex or a bundle of 2–8 conidia on denticles or short branches. Occasionally an additional 'whorl' of conidia formed around the central of conidiophores. Conidia colourless, usually 3-septate, fusiform, widest below the centre (30–)35–54 × 4–6 μm; end cells longer than central cells. Occasionally 2 septa conidia with cells of about equal length were found. Capturing nematodes by adhesive anastomotic branches.

Distribution: UK.

Notes: The description is based on the protologue. This species differs from other species of *Dactylellina* by its special trapping device. (Fig. 3.56)

Dactylellina asthenopaga (Drechsler) M. Scholler, Hagedorn & A. Rubner, Sydowia 51 (1): 110 (1999)
≡ *Dactylella asthenopaga* Drechsler, Mycologia 29: 498 (1937)
= *Dactylariopsis asthenopaga* (Drechsler) Mekht., Mikol Fitopatol, 1: 279 (1967)
= *Monacrosporium asthenopaga* (Drechsler) A. Rubner, Stud. Mycol. 39: 50 (1996)

Characteristic: Conidiophores hyaline, septate, erect, 100–200 μm, mostly 125–175 μm high, 2.5–4 μm wide at the base, tapering gradually upwards to a width of approximately 1.5 μm, mostly unbranched and terminating in a single conidium, but occasionally giving off one or two branches some distance below the apex, and then bearing two or three conidia. Conidia hyaline, obconical or clavate, truncate at the narrow proximal end, broadly rounded at the distal end, 20–46 μm, mostly 26–36 μm (average 31.5 μm) long, 6.5–9.5 μm (average 8.2 μm)wide, containing

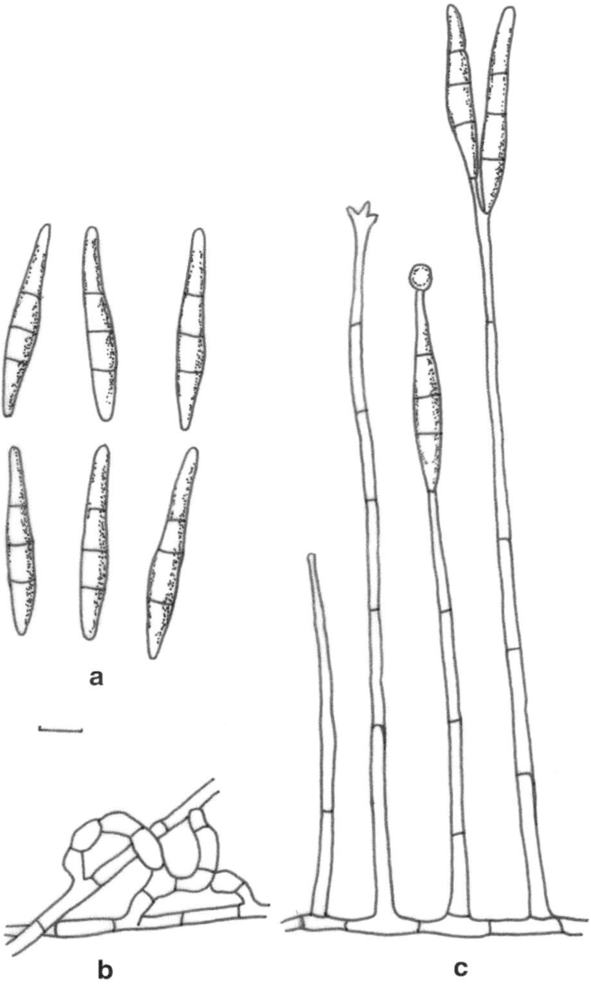

Fig. 3.56 *Dactylellina arcuata*. **a** conidia; **b** adhesive branches; **c** conidiophore. Bar = 20 μm

1–5-septa, but most often and most typically containing 3 septa, the basal of the 4 cells then delimited averaging about 8 μm in length, the antepenultimate cell about 7.5 μm, the penultimate cell about 9.8 μm, and the apical cell about 6.2 μm. Capturing nematodes by adhesive knobs.

Distribution: Australia, Netherlands, UK.

Notes: The description is based on the protologue. Within *Dactylellina* species which have adhesive knobs as trapping device, *Da. asthenopaga* is characterized by rarely unbranched conidiophores, and obconical or clavate conidia. (Fig. 3.57)

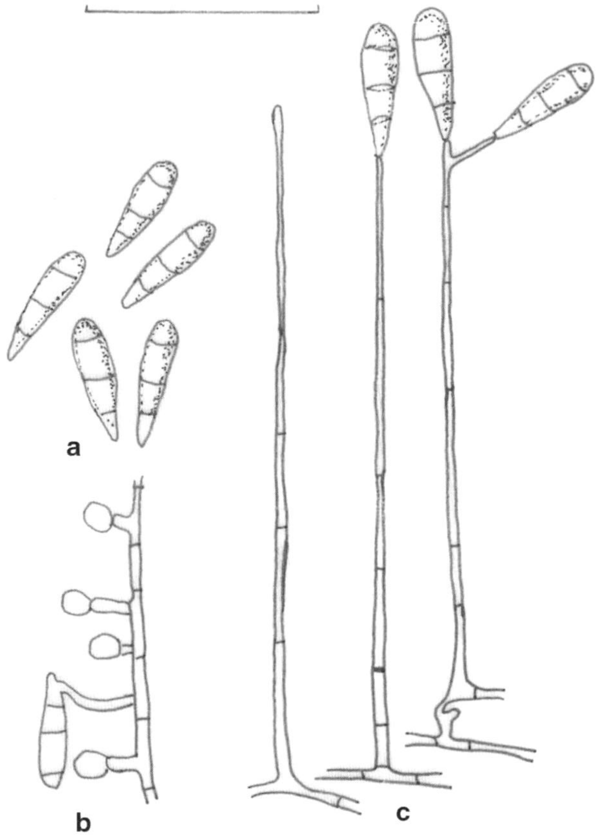

Fig. 3.57 *Dactylellina asthenopaga.* **a** conidia; **b** germinating conidia with adhesive knobs; **c** conidiophore. Bars = 50 μm

Dactylellina candidum (Nees) Yan Li, in Li, Hyde, Jeewon, Cai, Vijaykrishna & Zhang, Mycologia 97 (5): 1042 (2005)
≡ *Dactylium candidum* Nees, Syst Pilze Schw 58, 1816: Fries Syst. mycol. 3: 44, 829
= *Arthrobotrys candida* (Nees) S. Schenck, W.B. Kendr. & Pramer. Can. J. Bot. 55 (8): 982 (1977)
= *Candelabrella candida* (Nees) Rifai, Reinwardtia 7 (4): 369 (1968)
= *Dactylaria candida* (Nees) de Hoog, Stud. Mycol. 26: 102 (1886)
= *Dactylaria candida* Nees, Syst. Pilze (Würzburg): 58 (1816) [1816–17]
= *Monacrosporium candidum* (Nees) Xing Z. Liu & K.Q. Zhang, Mycol. Res. 98 (8): 864 (1996)
= *Dactylaria haptotyla* Drechsler, Mycologia 42: 48, 1950
= *Monacrosporium yunnanense* K.Q. Zhang, Xing Z. Liu & L. Cao, Mycol Res 100: 275 (1996)
≡ *Dactylellina yunnanensis* (K.Q. Zhang, Xing Z. Liu & L. Cao) M. Scholler, Sydowia 51: 108 (1999)
= *Monacrosporium chiuanum* Xing Z. Liu & K.Q. Zhang, Mycol. Res. 98: 863, 1994

Characteristics: Colonies hyaline initially and turned whitish or faintly pink after 15 days of incubation on CMA and extending a diameter of 4 cm at 25 °C within 10 days. Mycelium hyaline, scanty, vegetative hyphae septate and branched.

Conidiophores hyaline, erect, 5–7-septate, branched, 100–335 μm long, 2–3.7 μm wide at the base, gradually tapering upwards to a distal width of 1–2 μm, initially bearing a single terminal conidium, later often forming 2–12 (mostly 3–5) short branches near the apex with conidia in strikingly loose capitate arrangement. Conidia hyaline, narrow spindle-shaped or fusoid, gradually narrowing at the basal end, truncate at the base, 2–5-septate, mainly 4-septate, 27.5–57.5 (35) × 7.5–12.5 (9) μm. The proportion of conidia with 2, 3, 4 and 5 septa is 2, 10, 81 and 7%, respectively. Chlamydospores not observed in cultures. The fungus produces two types of trapping organ: non-constricting ring and adhesive knobs.

Distribution: worldwide.

Material examined: D9-1, D12-1, isolated from soil in Wuding, Yunnan in 1999 by Yanju Bi; YMF1.00543, isolated from forest foil in Lijiang, Yunnan in October 2002 by Jing Zhang; YMF1.00036, isolated from soil in Zhongdian, Yunnan in 1999 by Lu Cao; YMF1.000579, isolated from forest soil in E' mei Mountain, Sichuan in 2002 by Jinshi Deng. Permanent slide: DH5-5.

Notes: Rubner (1996) treated *Monacrosporium candidum*, *M. haptotylum* and *M. sclerohyphum* as conspecific under the name *M. haptotylum*. *M. yunnanense* K.Q. Zhang, X.Z. Liu & L. Cao is similar to *M. candidum* in having spindle-shaped, mostly 4-septate conidia, and branched conidiophores, bearing 3–10 conidia in a loose capitate arrangement. The only difference being that *M. candidum* and *M. yunnanense* form both stalked adhesive knobs and non-constricting rings, while *M. haptotylum* and *M. sclerohyphum* form only stalked adhesive knobs. These four taxa are phylogenetically closely related receiving high support, based on the rDNA, β-tubulin and combined datasets and this grouping agrees well with the morphological characters of the taxa (Li et al. 2005). These four taxa can therefore be considered as conspecific and should be named *Da. candidum* based on our results.

No material of *M. chiuanum* could be obtained to study the relationship with other related species based on molecular analyses. Comparing the morphological characters of *M. chiuanum* from original description with *M. candidum*, *M. yunnanense*, *M. haptotylum* and *M. sclerohyphum*, the difference among them is only in the 4–5 septa and occasionally branched conidiophores, which were not the crucial distinctive features to definite a new species. *M. chiuanum* was also regarded as a synonym of *M. haptotylum* by Rubner (1996). Based on the discussion above, we treat it as a synonym of *Da. candidum*.

(Fig. 3.58)

Dactylellina cionopaga (Drechsler) Ying Yang & Xing Z. Liu, Mycotaxon 97: 158 (2006)
≡ *Dactylella cionopaga* Drechsler, Mycologia 42: 30 (1950)

Characteristics: Mycelium sparse, hyphae, septate, 2–5 μm wide; Conidiophores hyaline, erect, 2–11 septate, 170–330 μm high, 5–7 μm wide at the base, gradually

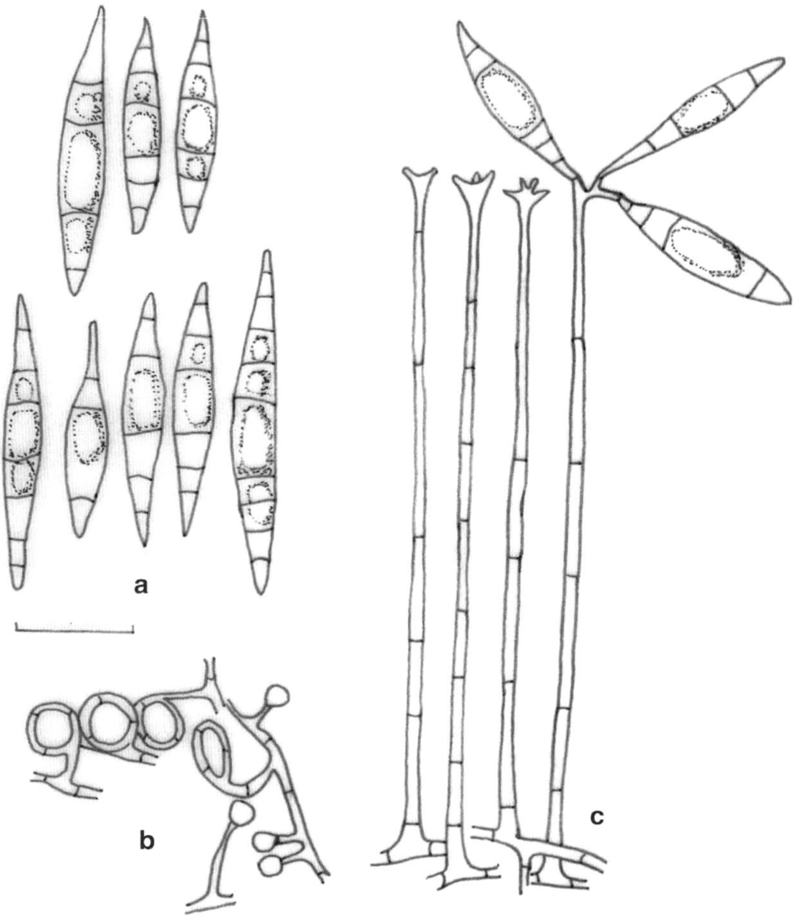

Fig. 3.58 *Dactylellina candidum.* **a** conidia; **b** adhesive knobs and non-constricting ring; **c** conidiophore. Bar = 20 μm

tapering upwards, 3 μm at the apex, sometimes branching and bearing a conidium at the apex of each branch. Conidia were broad fusoid, rounded at the apex, truncate at the base, 2–6-septate, mainly 3–4-septate, 35–60 × 13–21 μm. Chlamydospores not observed. Capturing nematode by adhesive branches.

Distribution: China (Yunnan, Sichuan), USA.

Material examined: TJ-1qz08, isolated from soil in Liqizhuang, Tianjing in 2000 by Wenpeng Li; YMF1.00569, isolated from soil in Deqin, Yunnan in September 2002 by Jing Zhang;YMF1.00580, isolated from forest soil in Jiuzhaigou, Sichuan in 2002 by Jingshi Deng;Hn2-1-3, isolated from field soil in Hainan in 2002 by Ke-Qin Zhang.

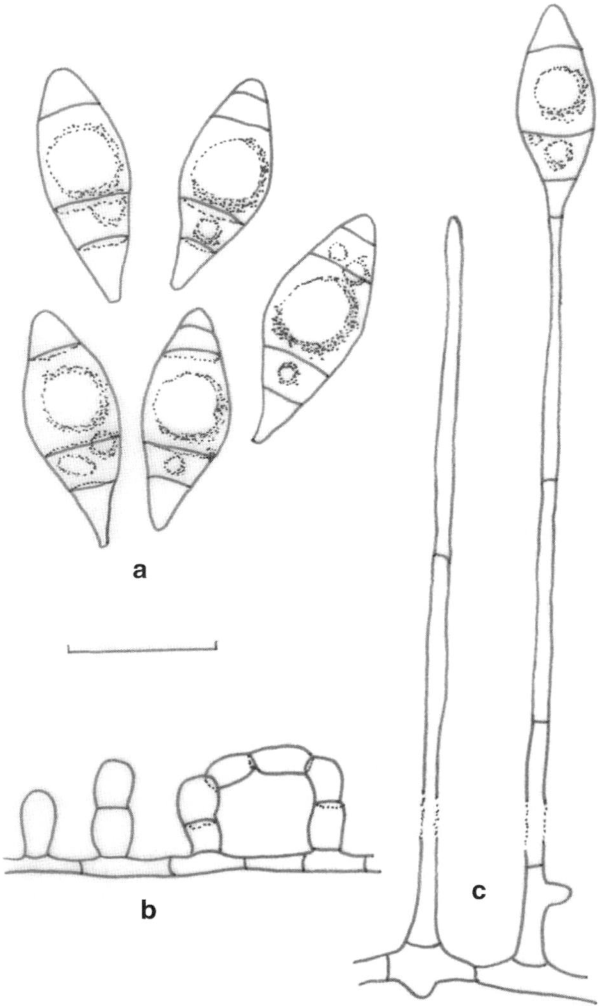

Fig. 3.59 *Dactylellina cionopaga.* **a** conidia; **b** adhesive branches; **c** conidiophore. Bar = 30 μm

Notes: Among the species of nematode-trapping fungi, only *Da. cionopaga* and *Da. gephyrophaga* form both unstalked adhesive knobs and simple two-dimension networks. Drechsler distinguished *Da. cionopaga* from *Da. gephyrophaga* by unstalked adhesive knobs, which seldom link to a scalariform network, while they commonly fuse in the latter. Conidiophores were found to be generally shorter, conidia noticeably longer in *Da. cionopaga* than in *Da. gephyrophaga*. Conidia of *Da. cionopaga* show a conspicuous vacuole and vary in their septation more than in *Da. gephyrophaga*. (Fig. 3.59)

Dactylellina copepodii (G.L. Barron) M. Scholler, Hagedorn & A. Rubner, Sydowia 51 (1): 110 (1999)
≡ *Dactylella copepodii* G. L. Barron, Can. J. Bot. 68: 692 (1990)
= *Monacrosporium copepodii* (G. L. Barron) A. Rubner, Stud. Mycol. 39: 54 (1996)

Characteristics: Conidiophores erect, slender, hyaline, septate, unbranched, and 160–360 μm tall, 2–2.5 μm wide for most of their length and somewhat wider 3.5 (4) μm at the base; Each conidiophore bears a solitary, terminal conidium at its apex. Conidia fusiform, broadest in the central region, and tapering to a narrow connective tissue at the base and an acutely rounded apex at the distal end, 1–6-septate, mainly 4 septate, 56–97 × 8.5–16 μm. Capturing nematode by adhesive terminal cells growing on branches.

Distribution: New Zealand

Notes: The description is based on the protologue. Among species of *Dactylellina*, only *D. copepodii* can trap copepods and is characterized by long branches with adhesive knobs at the apex. (Fig. 3.60)

Dactylellina daliensis H.Y. Su, Mycotaxon 105: 314 (2008)

Characteristics: Colonies white on CMA, growing rapidly, obtaining a diameter of 6 cm at 25 °C within 7 days. Mycelium sparse, hyphae hyaline, septate, branched, 2–4-septate, 21–101 μm high, 1.8–2.5 μm wide at the base, gradually tapering upwards to a width of 1.5–1.7 μm at the apex where forming 1–7 apical conidia. Conidia hyaline, elongate-fusoid or clavate, straight or somewhat curved, round at the distal end, truncate at the base, 1–13 (mainly 6–8) septate, 28.7–51.5 (40.8) × 3.2–5.2 (4.2) μm. Chlamydospores not observed in culture. Trapping nematodes by stalked adhesive knobs and non-constricting rings.

Distribution: China (Yunnan)

Notes: The description is based on the protologue. This species resembles only two other species in the *Dactylellina* genus: *Da. leptospora*, *Da. haptospora*. Both *Da. daliensis* and *Da. leptospora* produce elongate-fusoid conidia and capture nematodes by the same kinds of trapping devices. However, unlike *Da. daliensis*, which forms terminal cluster of conidia, *Da. leptospora* mainly bears a single terminal conidium, occasionally producing an additional conidium on a short branch attached to the main axis near the conidiophores apex. Furthermore, the conidial of *Da. daliensis* are shorter than those of *Da. leptospora*. Both *Da. daliensis* and *Da. haptospora* can bear terminal conidial clusters and form stalked adhesive knobs as trapping devices. However, *Da. daliensis* differs from *Da. haptospora* by producing an additional type of trapping device, in non-constricting rings. (Fig. 3.61)

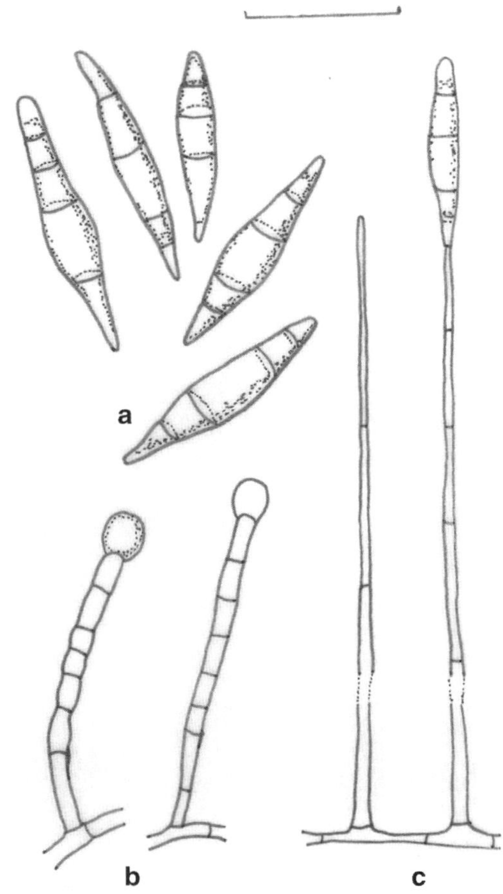

Fig. 3.60 *Dactylellina copepodii.* **a** conidia; **b** adhesive knobs; **c** conidiophore. Bar = 25 μm

Dactylellina drechsleri (Tarjan) M. Scholler, Hagedorn & A. Rubner, Sydowia 51 (1): 110 (1999)
≡ *Dactylella drechsleri* Tarjan, Mycopathologia 14: 143 (1961)
= *Monacrosporium drechsleri* (Tarjan) R.C. Cooke & C.H. Dickinson, Trans. Br. Mycol. Soc. 48 (4): 623 (1965)
= *Golovinia drechsleri* (Tarjan) Mekht., Mikol. Fitopatol. 1: 276 (1967)

Characteristics: Mycelium usually profuse, hyaline, slowly spreading; vegetative hyphae septate, 4 μm (3.2–5.3 μm) wide. Conidiophores hyaline, erect, sparingly septate, 145 μm (93–184 μm) long, each terminating in single conidium but often forming one or two branches, well below apex, also bearing solitary conidia. Conidia hyaline, somewhat fusiform, usually tapering conical with finely truncate base, but broadly obtuse to bluntly rounded at distal end, 29.5–48.7 (39.6) × 10.7–15.4 (13.1), 2–4-septate, mainly 3-septate. Capturing nematodes by stalked adhesive knobs, knobs 8–10 (9) × 7.2–9.6 (8.3) μm.

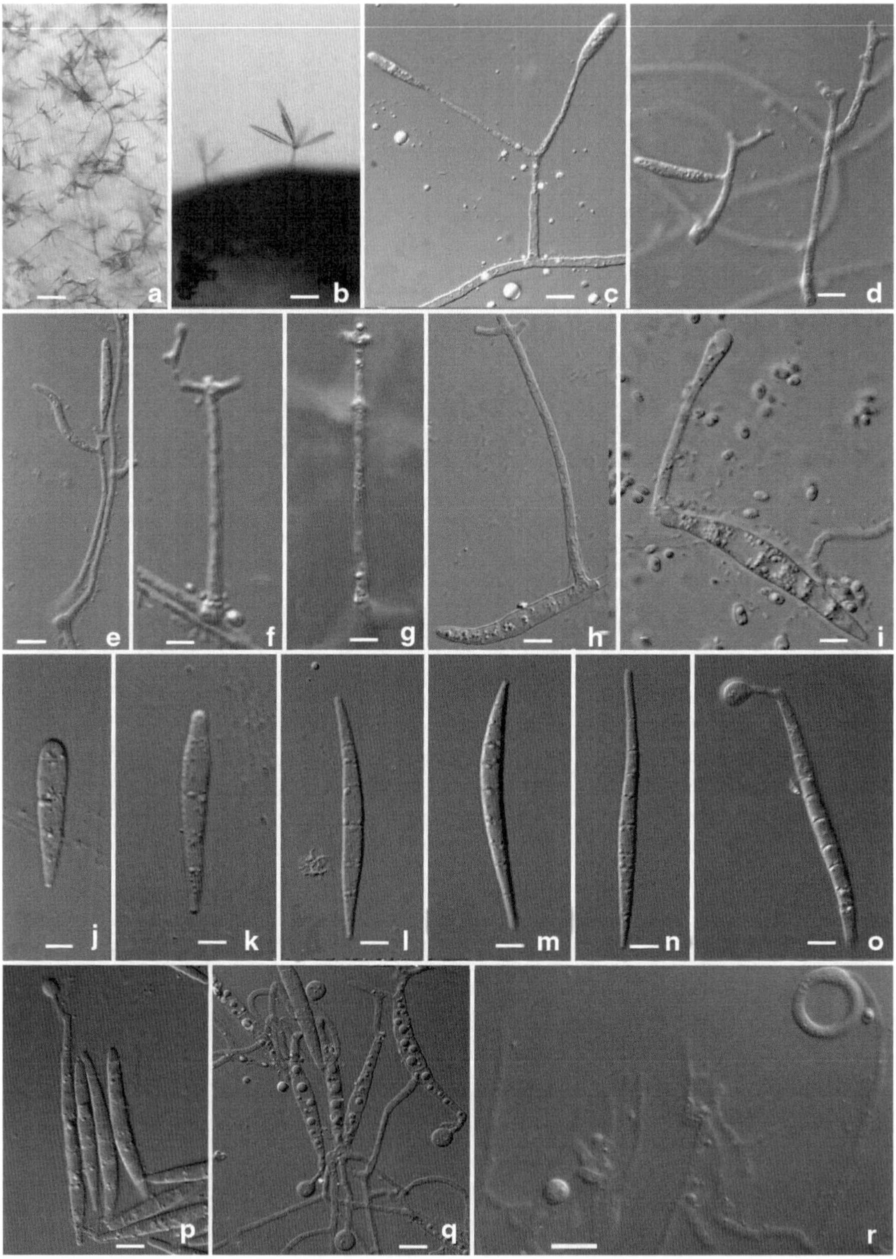

Fig. 3.61 *Dactylellina daliensis.* **a–i** conidiophores; **j–p** conidia; **q** adhesive knobs; **r** non-constricting rings. Bars: **a** = 50 μm; **b** = 25 μm; **c, e** = 12.5 μm; **d, f–r** = 5 μm

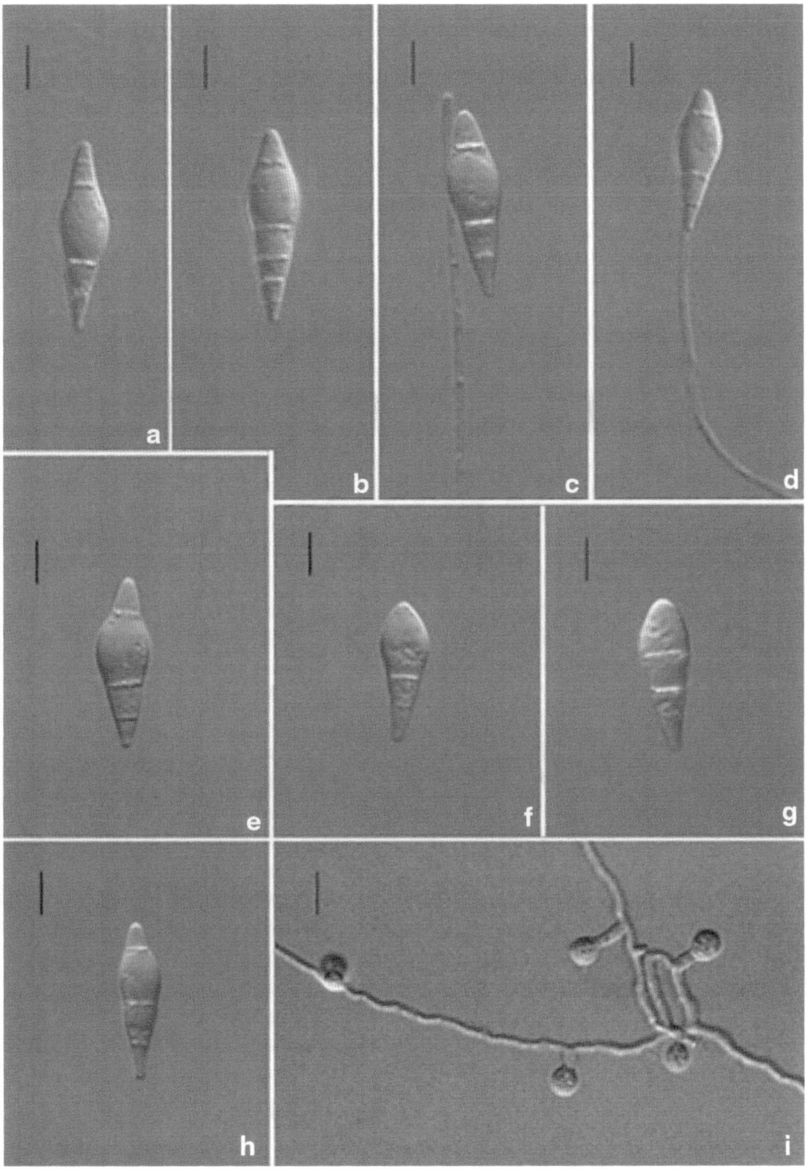

Fig. 3.62 *Dactylellina drechsleri.* **a–b, e–h** conidia; **c–d** conidiophore; **i** adhesive knobs. Bars = 10 μm; Strain number:YMF1.00573

Distribution: USA.

Notes: The description is based on the protologue. Among *Dactylellina* species with adhesive knobs as trapping devices, *Da. drechsleri* can be recognized by its 3 septate conidia, while others are usually 4 septate. (Fig. 3.62)

Dactylellina ellipsospora (Preuss) M. Scholler, Hagedorn & A. Rubner, Sydowia 51 (1): 110 (1999)
= *Dactylella ellipsospora* (Preuss) Grove, J. Bot. Lond. 24: 200 (1886)
≡ *Menispora ellipsospora* Preuss in Sturm, Deutschl. Fl. Abt. 3:H.8:T.47 (1848)
= *Monacrosporium ellipsosporum* (Preuss) R.C. Cooke & C.H. Dickinson, Trans. Brit. Mycol. Soc., 48: 623 (1965)

Characteristics: Colonies on CMA white to reddish, slow growing. Mycelium spreading, scanty, vegetative hyphae hyaline, septate, branched, mostly 2–3 μm wide. Conidiophores hyaline, erect, septate, unbranched, 100–330 μm long, 3–4.5 μm at the base, gradually tapering upwards to a width of 2–2.4 μm at the apex, bearing a single conidium. Conidia broadly spindle-shaped, hyaline, 40–57.5 (48.3) × 10–17.5 (13) μm, 2–4-septate, mostly 4-septate, the central cell being the largest, rounded on the apex. Chlamydospores not observed. Capturing nematodes by means of adhesive knobs.

Distribution: Cosmopolitan.

Material examined: Z29, isolated from forest soil in Tianzhu, Guizhou in 1988 by Zhang Ke-Qin ZJHZ-1, isolated from field soil in Hangzhou, Zhejiang in 1999 by Ke-Qin Zhang; XZM-3, isolated from forest soil in Xizang in August 2000 by Minghe Mo; YMF1.00032, isolated from soil in Ruili, Yunnan in October 2002 by Jing Zhang; 5102-1, isolated from field soil in Wuhan, Hubei in February 2001 by Xuefeng Liu.

Notes: *Da. ellipsospora* resembles *Da. lysipaga* in conidial characters, but differs from the latter in trapping devices. *Da. ellipsospora* produces adhesive knobs, while *Da. lysipaga* produces both adhesive knobs and non-constricting rings. The material examined almost matches the original description. (Fig. 3.63)

Dactylellina entomopaga (Drechsler) M. Scholler, Hagedorn & A. Rubner, Sydowia 51 (1): 110 (1999)
≡ *Arthrobotrys entomopaga* Drechsler, Mycologia 36: 392 (1944)
= *Arthrobotrys pauca* J. S. McCulloch, Trans. Br. Mycol. Soc. 68: 176 (1977)
= *Dactylariopsis entomopaga* (Drechsler) Mekht., Khishchnye Nematofagovye Griby—Gifomitsety (Baku): 122 (1979)

Characteristics: Mycelium spreading, colourless, septate, 2–3 μm wide. Conidiophores erect, colourless, meagerly septate, 75–175 μm tall, 3–4.5 μm wide at the base, about 2.5 μm wide farther upwards, often somewhat inflated at the top from which given off 3–8 simple or branched sterigmata of 2–7 μm long, whereon borne collectively 3–10 conidia in loose capitate arrangement; additional conidial clusters often being produced following renewed axial elongation. Conidia colourless, cylindrical or somewhat clavate, 15–28 × 4.5–5.5 μm, broadly rounded at the apex, often minutely pedicellate below, 1-septate, the 2 cells not pronouncedly unequal as a rule, even though the lower cell is often slightly longer than the upper one. Capturing nematodes by stalks ellipsoidal adhesive knobs measuring 8–13 × 4.5–8 μm.

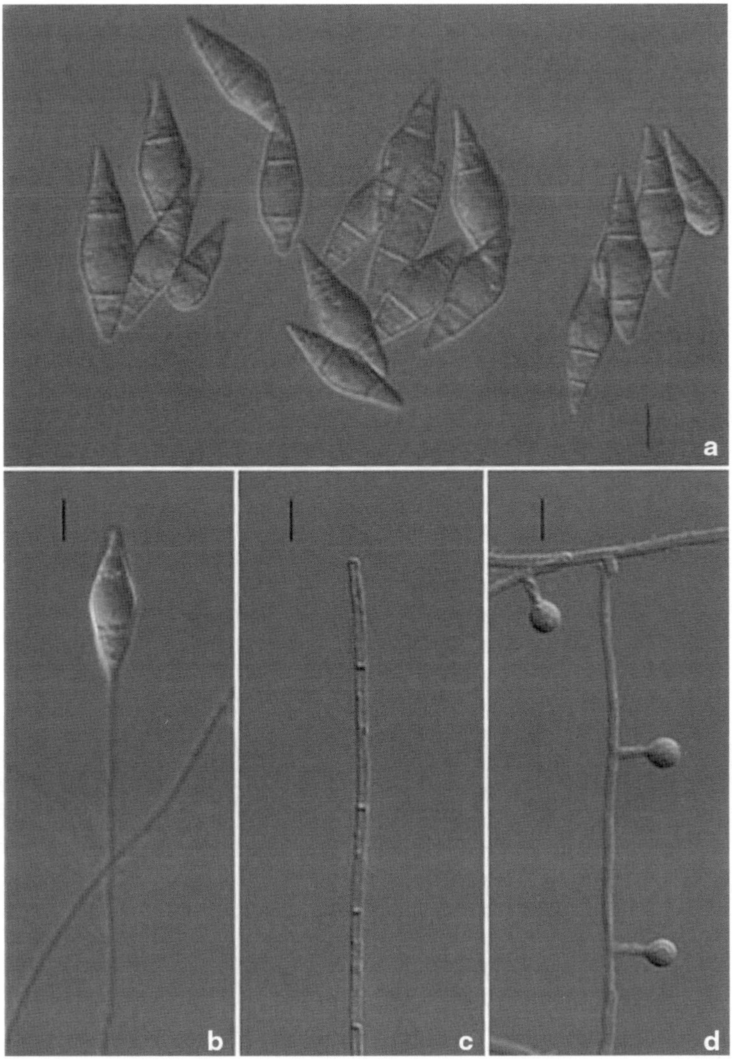

Fig. 3.63 *Dactylellina ellipsospora*. **a** conidia; **b–c** conidiophore; **d** adhesive knobs. Bars = 10 μm; Strain number: YMF1.00032

Distribution: Australia, USA.

Notes: The description is based on the protologue. Among *Dactylellina* species with adhesive knobs as trapping devices, this species is characterized by loose capitate arranged conidia and repeated extended conidiophores. (Fig. 3.64)

Dactylellina ferox (Onofri & S. Tosi) M. Scholler, Hagedorn & A. Rubner, Sydowia 51 (1): 110 (1999)

≡ *Arthrobotrys ferox* Onofri & S. Tosi, Mycotaxon 44: 446 (1992)

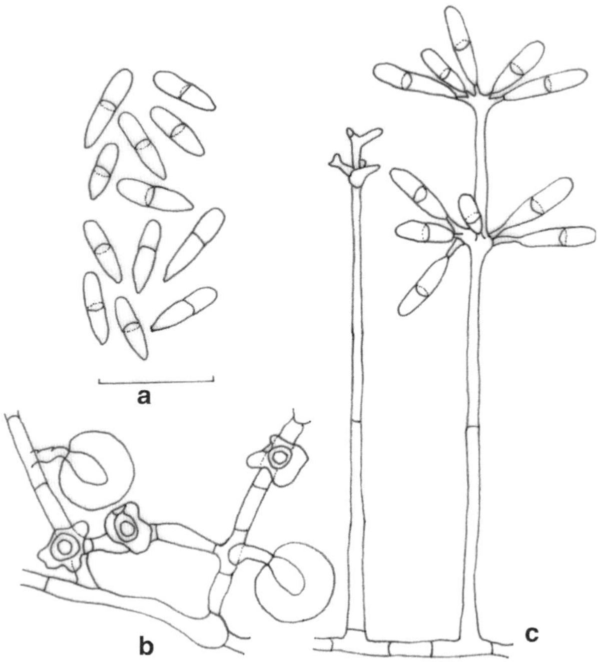

Fig. 3.64 *Dactylellina entomopaga*. **a**. conidia; **b**. adhesive knobs; **c**. conidiophores. Bar = 30 μm

Characteristics: Colonies on Czapek yeast white to pale pink-orange, mycelium hyaline; repent and aerial hyphae, 4.5–6.5 μm wide, septate at intervals of 22–35 (–45) μm. Conidiophores macronematous, mononematous, erect, septate, often branched, (33.5)44–144 (–466) μm long, 5–7 μm wide at the base and 3–4 μm farther upwards, producing (2–)4 (–10)conidia on 2–4.5 μm long denticles. Conidia hyaline, obovoidal to clavate, 1-septate, slightly constricted at the septum which is usually in the centre, (13–)15–18 (24.5) × (5–)6–8 (–9) μm. Predatory on springtails (*Gressittacantha terranova*) by stalks adhesive knobs measuring (18–)20–25 (–31) × (16.5–)17–21 (–21.5) μm.

Distribution: Italy

Notes: The description is based on the protologue. The trapping device of this species is hard to induce in pure culture. Its adhesive knobs are 3–4 times larger than other species. Because living cultures are not available, the species may be doubtful. (Fig. 3.65)

Dactylellina formosana (J.Y. Liou, G.Y. Liou & Tzean) M. Scholler, Hagedorn & A. Rubner, Sydowia 51 (1): 110 (1999)
≡ *Dactylella formosana* J. Y. Liou, G. Y. Liou & Tzean. Mycol. Res. 99: 751 (1995)

Characteristics: Colonies thin, white to orange white, 3.5–4.2 cm diameter in 1 wk. Mycelium hyaline, septate, branched, 2–4 μm wide. Conidiophores erect,

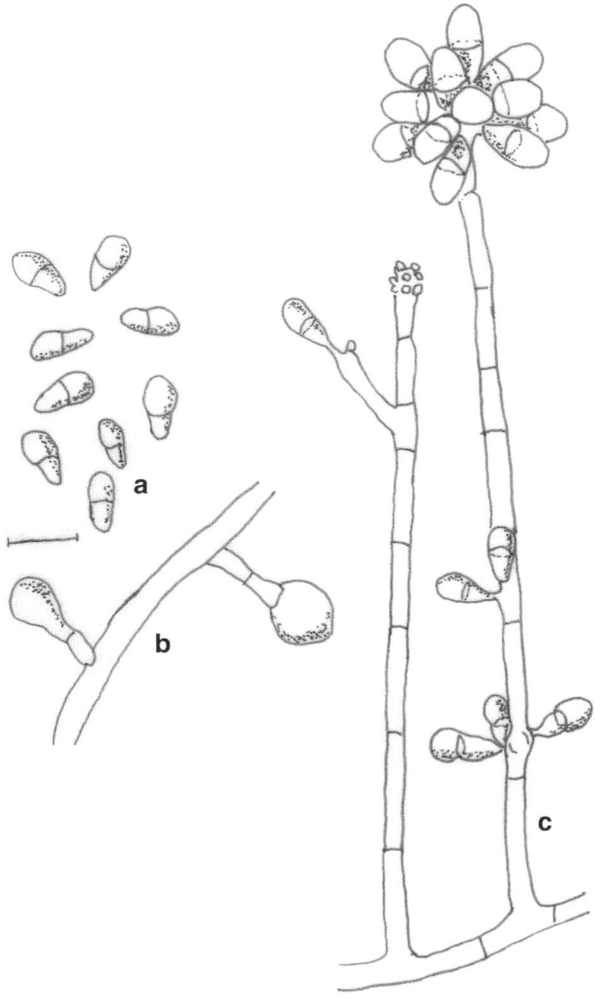

Fig. 3.65. *Dactylellina ferox*. **a**. conidia; **b**. adhesive knobs; **c**. conidiophores. Bars = 20 μm

unbranched, short, septate, hyaline, 17–90 μm long, 2.5–4.5 μm wide at the base, tapering to subterminal denticle. Conidia hyaline, fusoid, cylindrical, straight or slightly curved, (2)4–7 (8)-septate, 30–70 × 5–7.5 μm, tapering towards both ends. Chlamydospores absent. Capturing nematodes on adhesive knobs.

Distribution: China (Taiwan)

Notes: The description is based on the protologue. Morphologically *Da. formosana* resembles *Da. copepodii* and *Da. leptospora*. They share common features such as the ability to capture micro fauna, producing fusoid, cylindrical, multiseptate, solitary apical conidia on unbranched conidiophores. However, in contrast to *Da. formosana*, *Da. leptospora* has thinner, longer, more frequently septate

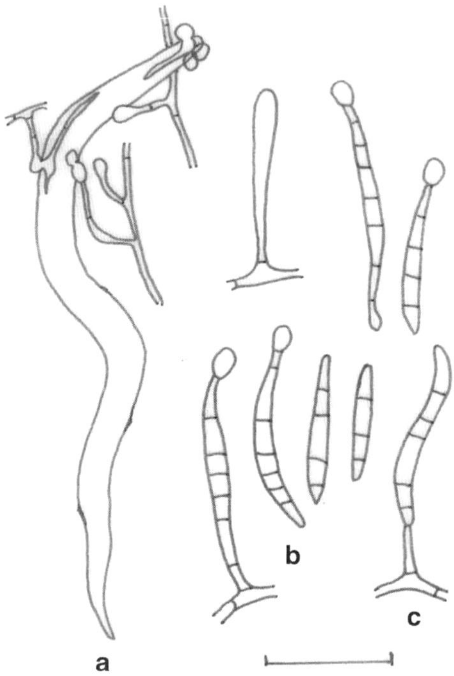

Fig. 3.66 *Dactylellina formosana*. **a**. adhesive knobs on hypha trapping nematodes; **b**. conidia; **c**. conidiophore. Bars = 30 μm

conidia, which produce secondary conidia when germinating, and trap nematodes by adhesive knobs and also by non-constricting rings (Drechsler 1937). Although both *Da. copepodii* and *Da. formosana* prey micro fauna by adhesive knobs, the former traps copepods as compared to nematodes. *Da. copepodii* also has conidia with larger intercalary cells and secondary adhesive knobs which arise from germinated conidia. These characteristics readily separate it from *Da. formosana* (Barron 1990). (Fig. 3.66)

Dactylellina gephyropaga (Drechsler) Ying Yang & Xing Z. Liu, Mycotaxon 97: 158 (2006)
≡ *Dactylella gephyropaga* Drechsler, Mycologia 29: 512 (1937)
= *Arthrobotrys gephyropaga* (Drechsler) Yan Li, in Li, Hyde, Jeewon, Cai, Vijaykrishna & Zhang, Mycologia 97 (5): 1042 (2005)
= *Gamsylella gephyropaga* (Drechsler) M. Scholler, Hagedorn & A. Rubner, Sydowia 51 (1): 110 (1999)
= *Golovinia gephyropaga* (Drechsler) Mekht., Mikol Fitopatol.1: 276 (1967)
= *Monacrosporium gephyropaga* (Drechsler) Subram., [as 'gephyropagum'] J. Indian bot. Soc.42: 293 (1964) [1963]

Characteristics: Colonies on CMA whitish, slow growing and extending to a diameter of 4 cm at 25 °C within ten days. Mycelium scanty, spreading, vegetative hyphae hyaline, septate, branched, mostly 2–5 μm wide. Conidiophores hyaline, erect, commonly unbranched, 2–11-septate, 75–360 μm long, 5–7 μm wide at the base, gradually tapering upwards to a width of 3 μm wide at the apex, bearing a single conidium, sometimes after repeated elongation forming one or two additional conidia. Conidia hyaline, mostly broadly spindle-shaped or somewhat clavate, broadly rounded at the distal end, but at the narrower proximal end tapering towards the bluntly truncate base, commonly 30–67.5 (45) × 10–23.5 (16) μm, 2–6-septate, mostly 3–4-septate. Chlamydospores not observed. Two types of trapping devices observed in culture: adhesive branches and simple two dimension networks.

Distribution: China (Anhui, Beijing, Guizhou, Hainan, Hebei, Taiwan, Tianjing, Yunnan), Portugal (Quinta de Sao Pedro), Netherlands (Valthernond), Spain (Tenerife)

Material examined: TJ-1qz08, isolated from soil in Liqizhuang, Tianjing in 2000 by Wenpeng Li; YMF1.00569, isolated from soil in Deqin, Yunnan in September 2002 by Jing Zhang; YMF1.00580, isolated from forest soil in Jiuzhaigou, Sichuan in 2002 by Jingshi Deng; Hn2-1-3, isolated from field soil in Hainan in 2002 by Ke-Qin Zhang. Permanent slide: Hn2-1-3.

Notes: Among the species of *Dactylellina*, only *Da. cionopaga* and *Da. gephyrophaga* form adhesive branches and simple two-dimension networks, but *Da. cionopaga* seldom form simple two dimension networks. For other differences between both species refer to *Da. cionopaga*. (Fig. 3.67)

Dactylellina haptospora (Drechsler) M. Scholler, Hagedorn & A. Rubner, Sydowia 51 (1): 110 (1999)
≡ *Dactylaria haptospora* Drechsler, Mycologia 32: 459 (1940)

Characteristics: Mycelium spreading; the vegetative hyphae hyaline, septate, 1.3–4.5 μm wide. Conidiophores hyaline, erect, septate, 50–150 μm high, 1.7–2.7 μm wide at the base, very gradually tapering upwards to a width of about 1.5 μm, then terminating often in a slightly branched more or less geniculate part formed by repeated elongation following successive apical production of conidia up to 15 in number in loosely capitate arrangement. Conidia hyaline, elongate-cylindrical, 35–60 × 2.2–3.7 μm, more rarely containing 5–8 cross-walls. Capturing nematodes by unicellular adhesive knobs and non-constricting rings. Adhesive subsphaerical or prolate ellipsoidal, 6–10 × 5–8.5 μm.

Distribution: China (Guizhou, Yunnan), USA.

Notes: The description is based on the protologue. This species differed from other *Dactylellina* species by elongate cylindrical and very narrow conidia. (Fig. 3.68)

Dactylellina hertziana (M. Scholler & A. Rubner) M. Scholler, Hagedorn & A. Rubner, Sydowia 51 (1): 111 (1999)
≡ *Arthrobotrys hertziana*, M. Scholler & A. Rubner, Mycol.Res.103: 764 (1999)

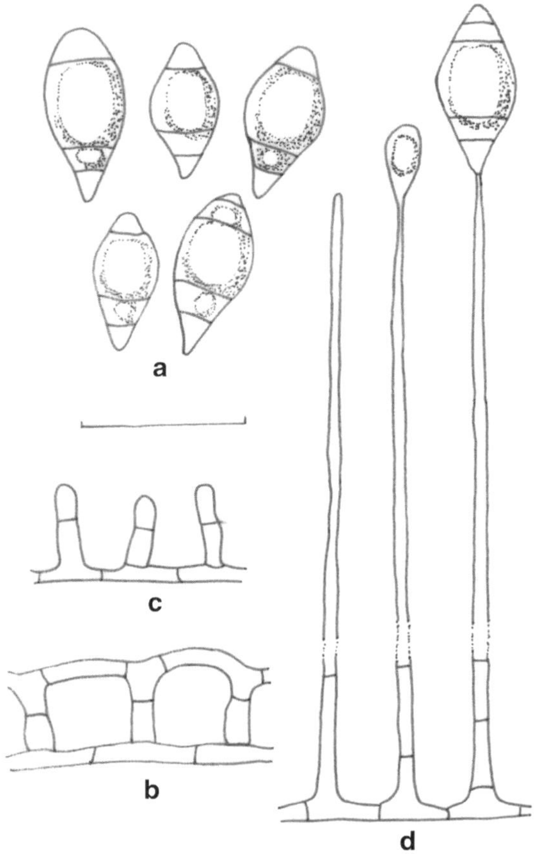

Fig. 3.67 *Dactylellina gephyropaga.* **a**. conidia; **b**. adhesive branches; **c**. adhesive network; **d**. conidiophore. Bars = 40 μm

Characteristics: Colonies and reverse whitish to pink, slow-growing, attaining less than 8 cm diameter in 17 days at 25 °C, producing aerial mycelium in concentric rings only on nutrient-rich media. Odor lacking. Morphological characteristics on CMA/2; Hyphae 3–5 μm wide, hyaline, septate, branched, often anastomosing. Conidiophores (150)–170–400 (–440) μm high, 3–5 μm wide in the lower part, erect, septate, mostly branched near the apex or at the bottom, branches slightly bent upwards near the branch base, then clearly erect. Fertile tips with weakly swollen nodes forming up to seven conidia on denticles of 1.5–2.5 (–4) μm long. Conidia hyaline, obovoid to ellipsoid-cylindrical, equi- or mostly inequidistantly (0–)3 (–4) septa, (16–)16.5–22–27.5 (–28.5) × (7.5–)7.5–9.5–11.5 (–12) μm, l/w ratio = 2.5. Nematodes are trapped by stalked or unstalked adhesive knobs. Chlamydospores not seen.

Distribution: Spain

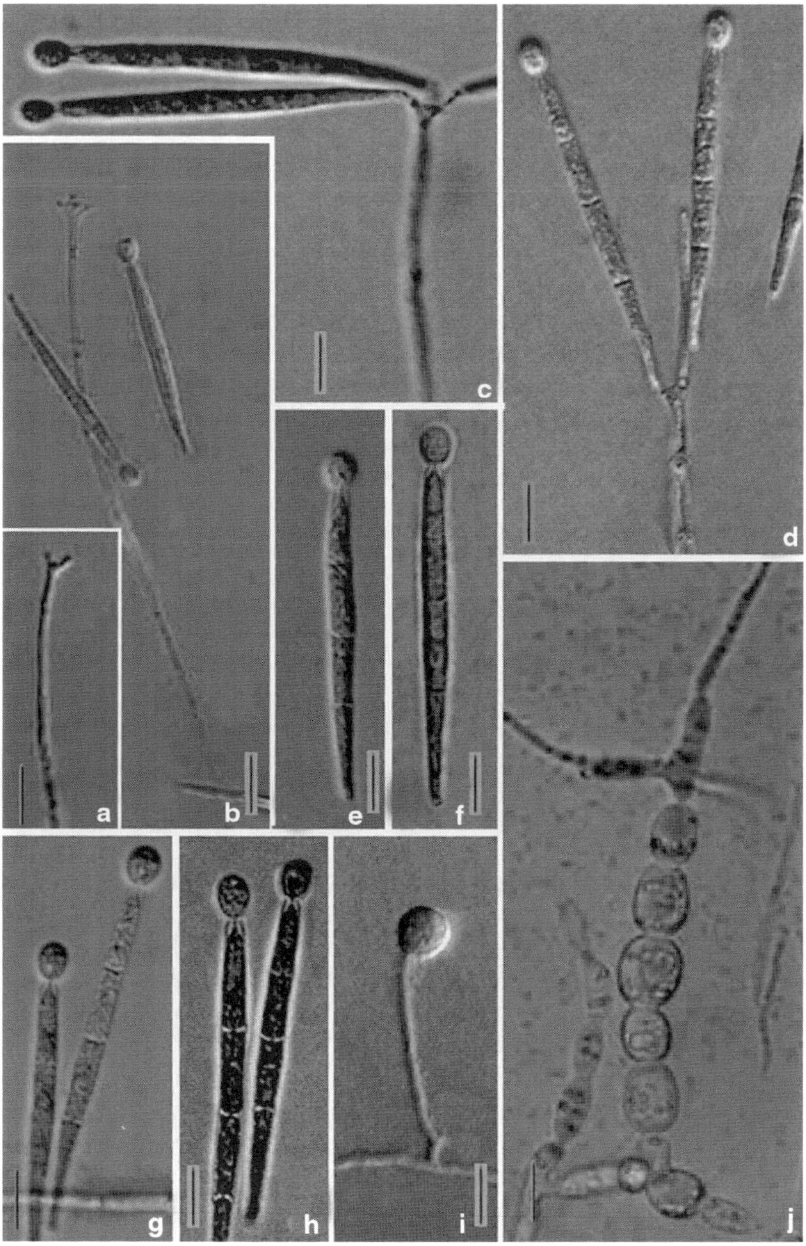

Fig. 3.68 *Dactylellina haptospora*. **a–d** conidiophore; **e–h** conidia; **i** adhesive knobs; **j** chlamydospores. Bars = 10 µm; Strain number: YMF1.00560

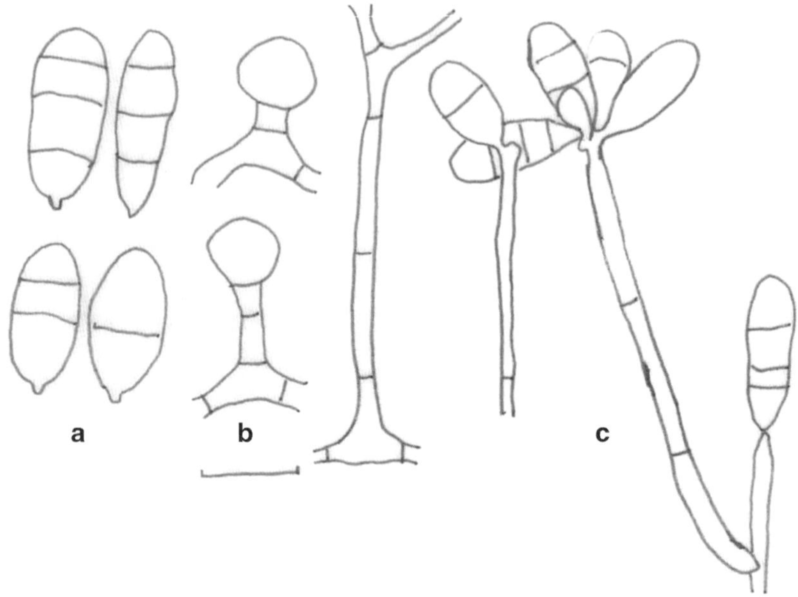

Fig. 3.69 *Dactylellina hertziana*. **a**. conidia; **b**. adhesive knobs; **c**. conidiophores. Bar = 15 μm

Notes: The description is based on the protologue. This species is characterized by branched conidiophores and obovoid to ellipsoid-cylindrical conidia of loose capitate arrangement. (Fig. 3.69)

Dactylellina huisuniana (J.L. Chen, T.L. Huang & Tzean) M. Scholler, Hagedorn & A. Rubner, Sydowia 51 (1): 111 (1999)
≡ *Dactylella huisuniana* J.L. Chen, T.L. Huang & Tzen, Mycol. Res. 102: 1269 (1998)

Characteristics: Colonies on oat meal agar (OMA) 35–40 mm in 17 d at 25 °C, effuse, velvety, very pale orange; reverse very pale orange white to pale orange. Mycelium mostly immersed, branched, septate, smooth, subhyaline to pale orange in mass, 1–4.5 μm wide;. Conidiogenous cells terminal or subterminal, lateral, with denticles up to 9 μm long, 1.5 μm wide, with sympodial proliferations, scattered and often branched, elongate, candelabrelloid at the apex in age, bearing 1–6 conidia. Conidia predominantly fusiform, 3-septate, occasionally clavate-fusoid and 2-septate or 4-septate, smooth, hyaline, 29.5–41 × 4.5–7.5 μm, the distal second cell the widest, base attenuated, truncated, 1–1.5 μm wide. Trapping nematodes by ovoid to subglobose adhesive knobs of 5–7 μm diameter.

Distribution: China (Taiwan)

Notes: The description is based on the protologue. This species is characterized by sympodial proliferations conidiophores with candelabrelloid apex. (Fig. 3.70)

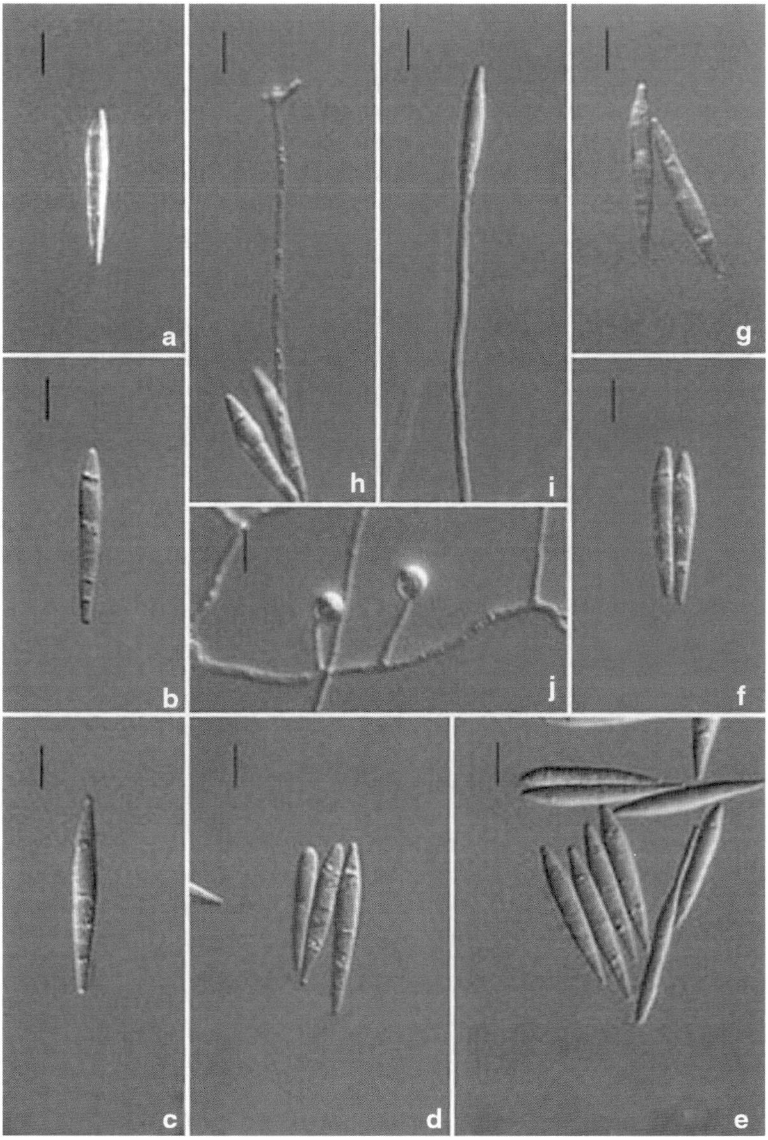

Fig. 3.70 *Dactylellina huisuniana*. **a–g** conidia; **h–i** conidiophore; **j** adhesive knobs. Bars = 10 μm; Strain number: YMF1.00574

Dactylellina illaqueata D.S. Yang & M.H. Mo, in Yang, Chen, Huang, Mo & Zhang, Mycotaxon 94: 215 (2005)

Characteristics: Mycelium scanty, spreading, vegetative mycelium colourless, septate, mostly 1.5–2.5 μm wide. Conidiophores colourless, erect, unbranched, often 95–250 μm high, 2.2–2.6 μm wide at base, and gradually tapering upwards to a width of 1.8–2.1 μm at the apex, bearing a single conidium on the apex, occasionally two conidia. Conidia colourless, elongate fusiform, narrowly obtuse at the distal end, truncate at the base, the central cell swelling obviously, 25.5–117.5 (66.5) × 5.5–15.2 (14.1) μm, 3–8-septate, mainly 5-septate. The proportion of conidia with 3,4,5,6,7, and 8 septa accounts for 9.1, 15.2, 63.6, 6.1, 3 and 3%, respectively. When induced with nematodes, the fungus produced non-constricting rings and stalked adhesive knobs. Chlamydospores sphaerical to ellipsoidal, intercalary.

Distribution: China (Yunnan)

Notes: *Da. illaqueata* is mainly characterized by its 5-septate conidia singly borne on the unbranched conidiophores and this species resembles *Da. yunnanensis* and *Da. lysipaga* in conidial shape. However, *Da. yunnanensis* usually forms short denticles at the apex of conidiophores and bears 2–5 conidia, and *Da. lysipaga* produces the conidia mainly with 2–4-septate. In comparison with *Da. leptospora*, the conidia of *Da. illaqueata* usually have a wider central cell (average 14.1 μm) than that of *Da. leptospora* (4–5.8 μm). In addition, conidia of *Da. leptospora* have more septa (5–15) than those of *Da. illaqueata* (3–8, mainly 5 septa). (Fig. 3.71)

Dactylellina leptospora (Drechsler) M. Morelet, Bull. Soc. Sci. nat. Arch. Toulon & du Var 178: 6 (1968)
≡ *Dactylella leptospora* Drechsler, Mycologia 29: 507 (1937)
= *Dactylosporium leptospora* (Drechsler) Mekht., Mikol. Fitopatol. 1: 277 (1967)
= *Monacrosporium leptosporum* (Drechsler) A. Rubner, Stud. Mycol. 39: 79 (1996)

Characteristics: Colonies on CMA whitish, rapidly growing. Mycelium spreading, scanty, vegetative hyphae hyaline, septate, mostly 1.2–3 μm, rarely up to 4 μm wide. Conidiophores hyaline, erect, septate, 75–250 μm long, 2.5–5 μm wide at the base, tapering upwards very slightly, 1–2 μm wide at the apex, mostly simple and bearing a single terminal conidium, but occasionally bearing an additional conidium on a short branch attached to the main axis near its apex. Conidia hyaline, mostly straight, elongate fusoid or cylindrical, 30–92 (64) × 4–6 (4.5) μm, 4–15-septa, after falling off, often giving rise individually to 1 or 2 globose knobs mostly on scarcely modified distal prolongations, more rarely on lateral stalks; often too, producing on slightly branched conidiophores, 50–125 μm long, 2–3 μm wide at the base and 1.3–1.8 μm wide at the apex, secondary conidia mostly 10–35 × 4–5.8 μm, 3–8-septate. Two types of trapping devices were observed in culture: stalked adhesive knobs and non-constricting rings.

Distribution: China (Beijing, Guizhou, Hubei, Xizang, Yunnan), India, Russian, USA

Material examined: SJt3.9.68, isolated from field soil in Huaxi, Guizhou in 1996 by Ke-Qin Zhang; XZD-2, XZD-3, isolated from soil in Xizang in August 2000 by Minghe Mo; DL1-1-1, isolated from forest soil in Dali, Yunnan in September 2002

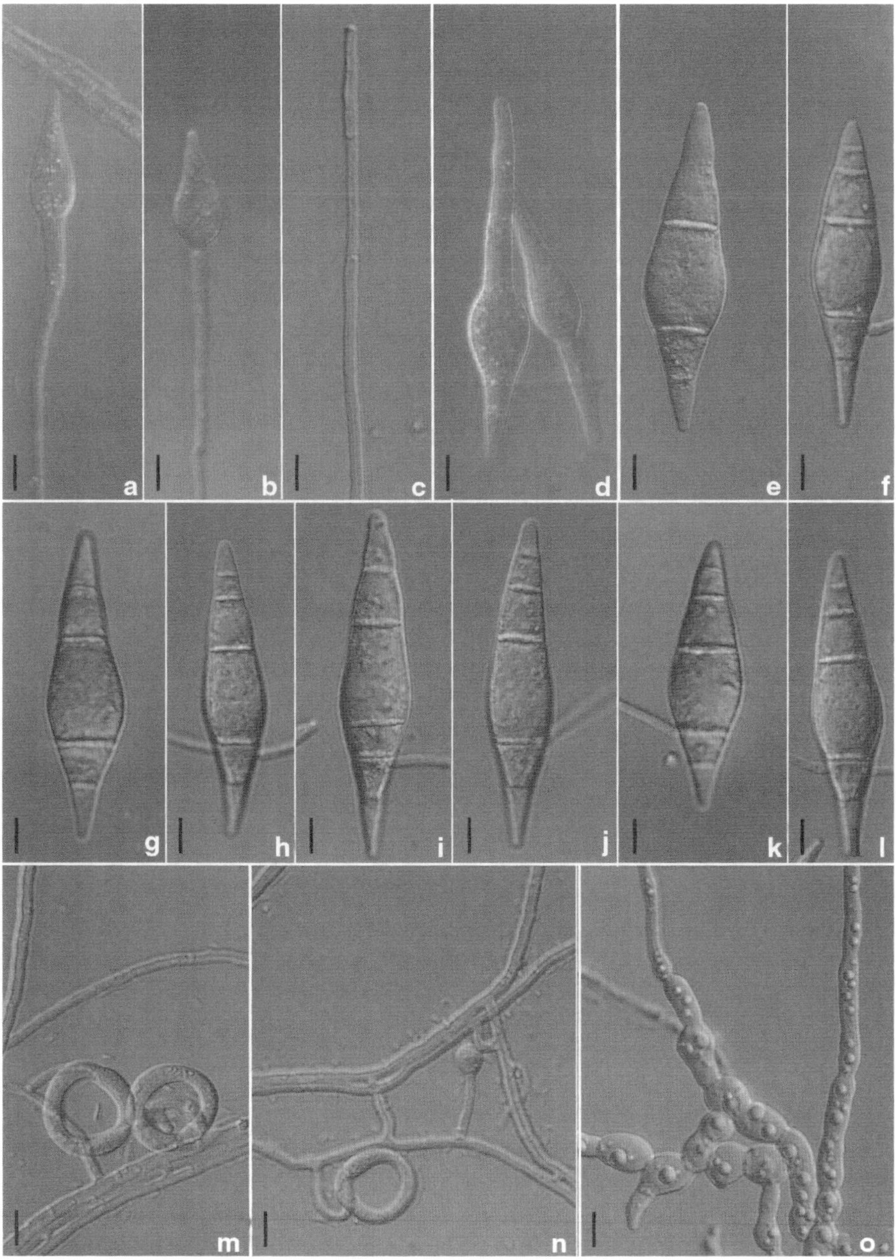

Fig. 3.71 *Dactylellina illaqueata*. **a–c** conidiophore; **d–l** conidia; **m–n** adhesive knobs and non-constricting rings; **o**. chlamydospore. Bars = 10 μm

by Jing Zhang; YMF1.00114, isolated from soil in Kunming, Yunnan in 2002 by Lu Cao; J34-1, J34-2, isolated from forest soil in Jianshui, Yunnan in 1999 by Yanju Bi; YMF1.00115, YMF1.00117, YMF1.00042, YMF1.00562, isolated form soil in Dali and Lijiang, Yunnan in September 2002 by Jing Zhang. Permanent slide: STt3.9.68.

Notes: The fungus is easy to distinguish from other species because of its elongate-fusiform to cylindrical and greater number of conidial septa. *Dactylaria dasguptaii* and *Kafiaddinia fusariispora* were treated as synonym of *Dactylaria leptospora* because of similar trapping devices and conidial characteristics (Rubner, 1996). All the characteristics of examined material matched well with original distribution of *Da. leptospora*. (Fig. 3.72)

Dactylellina lobata (Dudd.) Yan Li, in Li, Hyde, Jeewon, Cai, Vijaykrishna & Zhang, Mycologia 97 (5): 1042 (2005)
≡ *Dactylella lobata* Dudd., Trans. Br. mycol. Soc. 34 (4): 489 (1951)
= *Gamsylella lobata* (Dudd.), M. Scholler, Hagedorn & A. Rubner, Sydowia 51 (1): 108 (1999)
= *Monacrosporium lobata* (Dudd.) A. Rubner, Stud. Mycol.39: 80 (1996)

Characteristics: Mycelium septate, occasionally branched, 2–5 μm wide. Conidiophores erect, single, 250 μm high. Conidia fusiform, mainly with 3 septa, 32–54 × 8–12 μm, each at the apex of an erect conidiophore. Trapping nematodes by adhesive unstalked knobs, which sometimes joined together to form loops.

Distribution: UK.

Notes: The description is based on the protologue. In this species, knobs often form a string, which resemble adhesive branches of *Da. cionopaga*, but the adhesive cell of the latter is rectangle. In addition, conidia of the two species are different. (Fig. 3.73)

Dactylellina lysipaga (Drechsler) M. Scholler, Hagedorn & A. Rubner, Sydowia 51 (1): 111 (1999)
≡ *Dactylella lysipaga* Drechsler, Mycologia 29: 503 (1937)
= *Golovinia lysipaga* (Drechsler) Mekht., *Mikol. Fitopatol.*1: 277 (1967)
= *Monacrosporium lysipagum* (Drechsler) Subram., J. Indian Bot. Soc. 42: 293 (1963)

Characteristics: Colonies on CMA whitish. Mycelium spreading, scanty, vegetative hyphae hyaline, septate, branched, mostly 1.5–3.5 μm wide. Conidiophores hyaline, septate, erect, simple, 125–760 μm long, 3–5 μm wide at the base, gradually tapering upwards to a width of 1–2 μm, bearing a single conidium, but occasionally producing a second conidium on a short branch attached to the main axis some distance below the apex. Conidia hyaline, sometimes obovoid-fusoid, but much more frequently and more typically rather symmetrically fusoid, somewhat acutely rounded at the apex, truncate at the narrow base, 2–4-septate, mostly with 3–4 septa, 27.5–85 (40.7) × 9–17.5 (11) μm. Capturing nematodes by means of adhesive knobs and non-constricting rings.

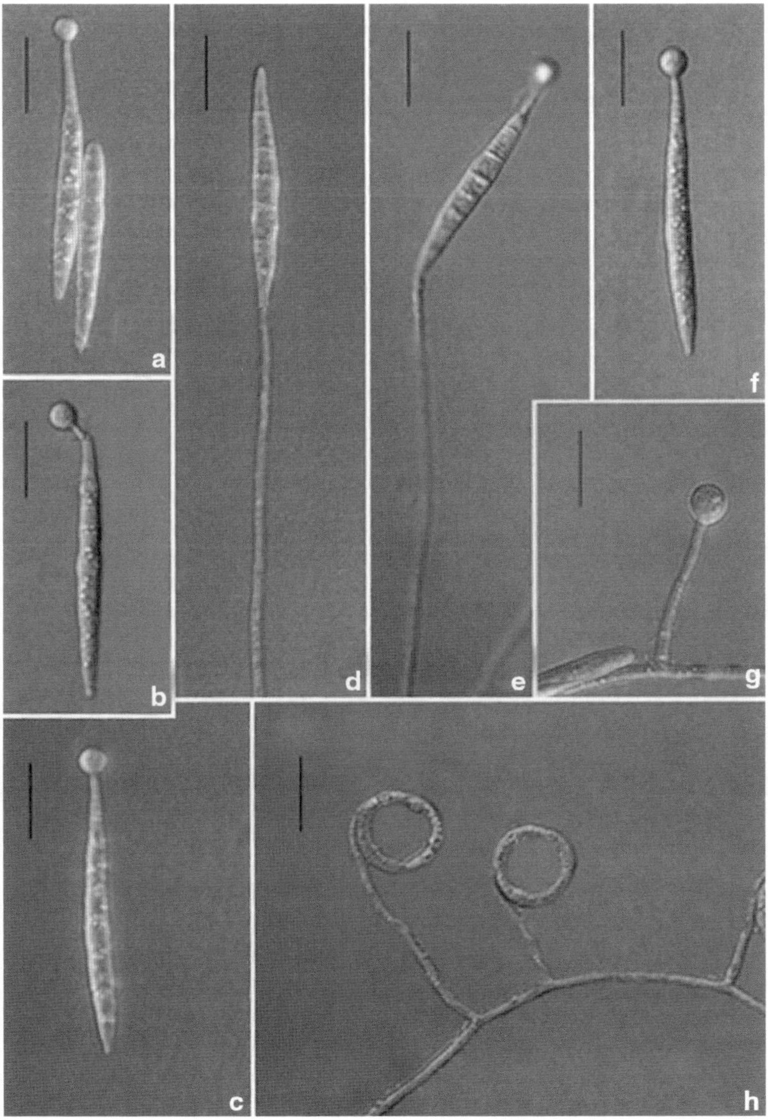

Fig. 3.72 *Dactylellina leptospora.* **a–c**, **f** conidia; **d–e** conidiophore; **g** adhesive knobs; **h** non-constricting rings. Bars = 10 μm; Strain number: YMF1.00042

Distribution: China (Beijing, Guizhou, Hebei, Yunnan), Portugal (Azores)

Material examined: Z23, isolated from field soil in Tianzhu, Guizhou in 1988 by Ke-Qin Zhang; YMF1.00535, isolated from soil in Baoshan, Yunnan in October 2002 by Jing Zhang; P7.1.1, isolated from forest soil in Wenshan, Yunnan in July 2002 by Zhiwei Zhao. Permanent slide: P7.1.1

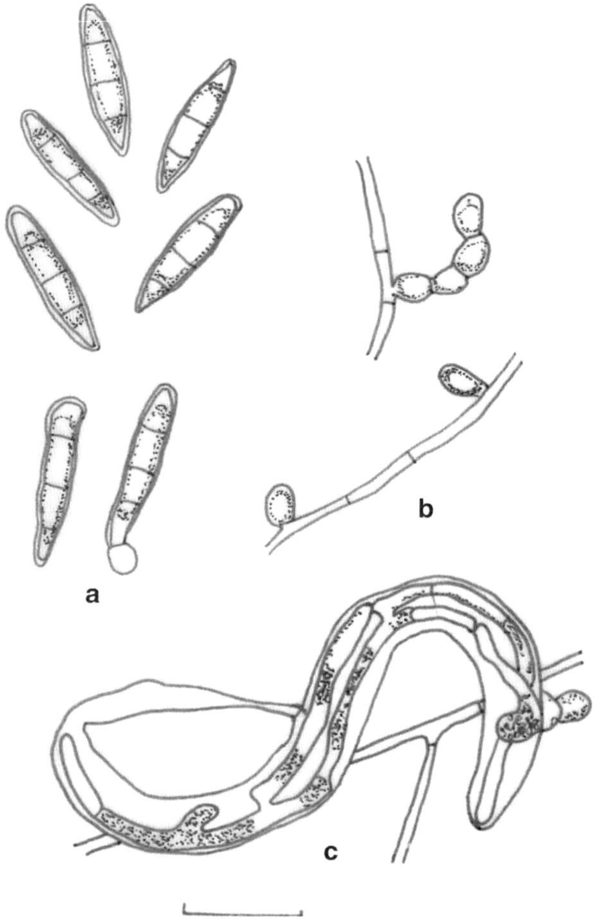

Fig. 3.73 *Dactylellina lobata.* **a**. conidia; **b**. adhesive knobs; **c**. hypha infect nematodes and produce adhesive knobs. Bar = 30 μm

Notes: *Da. lysipaga* resembles *Da. ellipsospora* in conidial shape, septate, but the conidia of the former are smaller than those of the latter. Moreover, *Da. lysipaga* forms both adhesive knobs and non-constricting rings, while *Da. ellipsospora* forms only adhesive knobs. *Da. lysipaga* also resembles *Da. candidum* in conidial shape and trapping devices, but differs in conidiophores. In *Da. lysipaga*, conidiophores simple, bear a single conidium at the apex, occasionally forming additional one next the apex. While in *Da. candidum*, conidiophores form short branches near the apex with conidia in strikingly loose capitate arrangement. The conidia of present isolate are longer than in the description of Subramanian [28–55 (40.7) × 9–14 (11.6) μm] (1963). (Fig. 3.74)

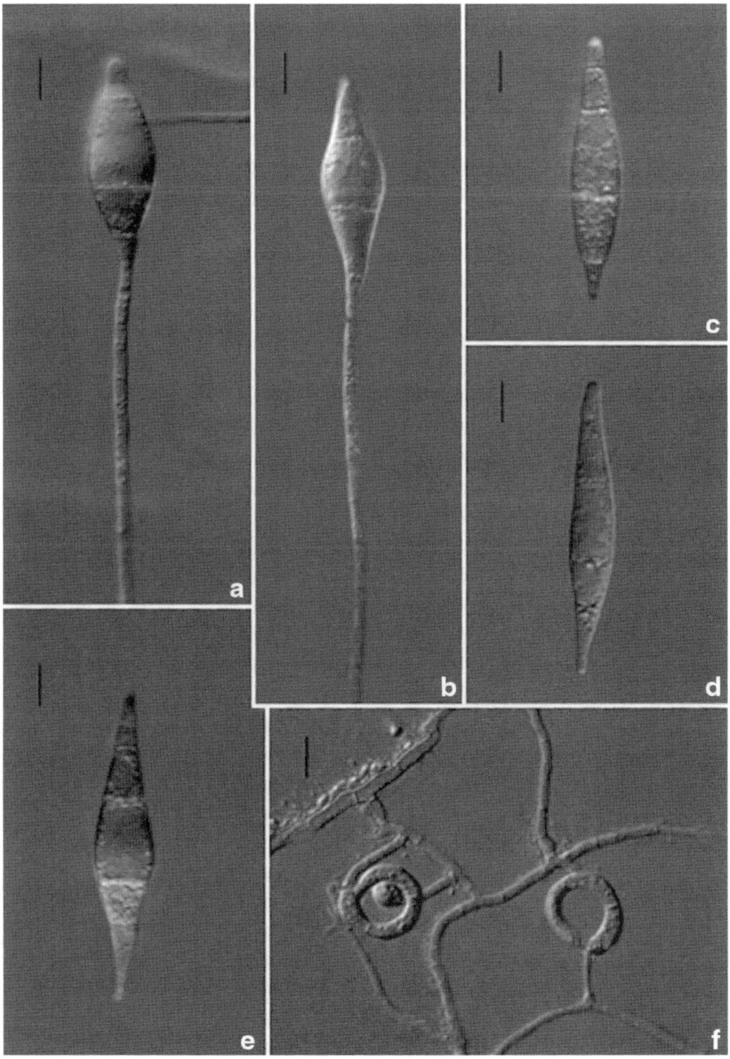

Fig. 3.74 *Dactylellina lysipaga.* **a–b** conidiophore; **c–e** conidia; **f** adhesive knobs and non-constricting rings. Bars = 10 μm; Strain number: YMF1.00535

Dactylellina mammillata (S.M. Dixon) M. Scholler, Hagedorn & A. Rubner, Sydowia 51 (1): 111 (1999)

≡ *Dactylella mammillata* S.X. Dixon, Trans. Br. Mycol. Soc. 35: 144 (1952)

= *Golovinia mammillata* (S.X. Dixon) Mekht., Khishchnye Nematofagovye Griby—Gifomitsety (Baku): 159 (1979)

= *Monacrosporium mammillata* (S.X. Dixon) R.C. Cooke & C.H. Dickinson[as 'mammillata'] Trans. Br. Mycol. Soc. 48 (4): 622 (1965)

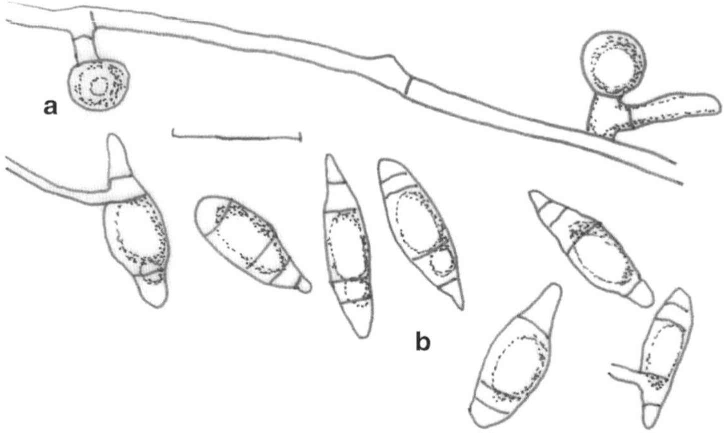

Fig. 3.75 *Dactylellina mammillata*. **a**. adhesive knobs; **b**. conidia. Bar = 30 µm

Characteristics: Mycelium scanty; hyphae hyaline, septate. Conidiophores 160–280 µm high, conidia borne singly on the apex of conidiophores. Conidia ellipsoidal to fusiform, obtuse rounded at the apex, 3–4-septate, 30–60 × 9–17 µm. Capturing nematodes by stalks adhesive knobs 6.5–9 × 6.5–7 µm.

Distribution: Canada, Germany, UK.

Notes: This species is most similar to *Da. ellipsospora*, but differs in having a rounded conidia apex and 3–4-septate conidia, while *Da. ellipsospora* most often has 4-septate, occasionally 3 or 2 septate conidia. (Fig. 3.75)

Dactylellina mutabilis (R.C. Cooke) M. Scholler, Hagedorn & A. Rubner, Sydowia 51 (1): 111 (1999)
≡ *Monacrosporium mutabile* R. C. Cooke [as '*mutabilis*'], Trans. Br. Mycol. Soc. 53: 318 (1969)
= *Golovinia mutabilis* (Cooke) Mekht., Khishnchnye Nematofagovye Griby–Gifomitesty (Baku): 161 (1979)

Characteristics: Conidiophores simple, septate, erect, bearing a single conidium at the apex. Conidia ellipsoidal to fusiform, 50–60 × 16–17.5 µm, with 4 septa, with the median cell markedly larger than the rest. Conidia taper to a narrowly rounded cell distally and proximally to a narrowly truncate base. In pure culture on cornmeal agar conidia are markedly different in size and shape from those on the nematode-infested substrate, being 37.5–50 × 12.5–15 µm. In addition, about 14% of conidia were 3 septate. Curved, malformed spores were frequent. Capturing nematodes by stalks adhesive knob about 5 µm in diameter.

Distribution: Samoan Islands

Notes: The description is based on the protologue. Among species of *Dactylellina* with 4 septate conidia, *Da. mutabilis* can be distinguished from other species by densely crowded, stalked, adhesive knobs. (Fig. 3.76)

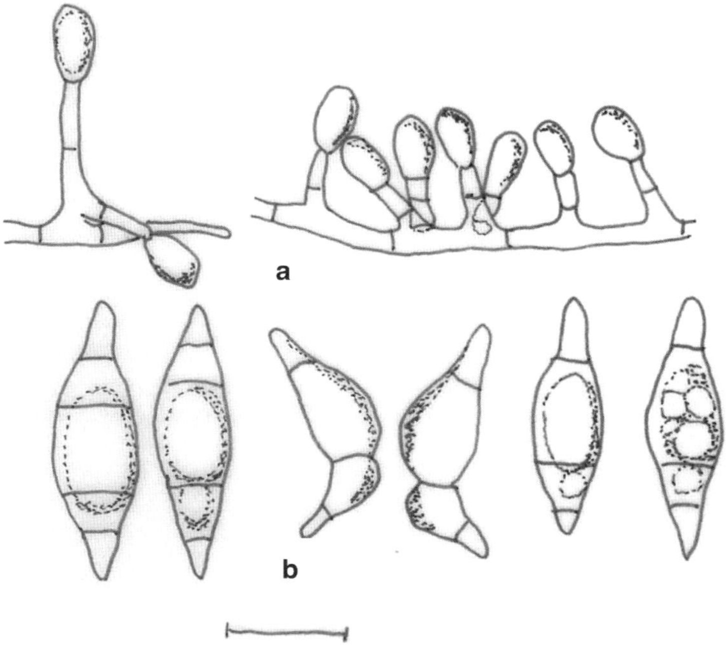

Fig. 3.76 *Dactylellina mutabilis*. **a**. adhesive knobs; **b**. conidia. Bar = 20 μm

Dactylellina multiseptata (H.Y Su & K.Q. Zhang) Z.F. Yu, **comb. nov.**
≡ *Monacrosporium multiseptatum* H.Y. Su & K.Q. Zhang. Mycotaxon 92: 193 (2005)
MB 804794

Characteristics: Colonies on CMA plates growing quickly, attaining 6 cm diameter within 7 days at 25 °C. Mycelium spreading, vegetative hyphae hyaline, septate and branched, mostly 1.8–2.5 μm wide. Conidiophores erect, simple, septate, 170–260 μm long, 2.5–3 μm wide at the base, gradually tapering upwards to a width of 1.7–2 μm at the apex, bearing a single conidium, occasionally two conidia. Conidia hyaline, elongate fusiform to straight clavate, sometimes apparently curved, narrowly round at the distal end, tapering towards the narrow truncate at the base, somewhat constricted at septa, 67.5–132.5 (91.6) × 3.8-17.5 (15.5) μm, 4–9-septate, mainly 6–7- septate. The proportion of conidia with 4, 5, 6, 7, 8 and 9 septa is 1.8, 3.6, 23, 45, 25.1 and 2 % respectively. Chlamydospores sphaerical to ellipsoidal, intercalary, about 5–8 × 4–6 μm. In the presence of nematodes the fungus forms stalked adhesive knobs which are unicellular, subsphaerical. The knobs usually 11 μm long, 10 μm wide, with a stalk of 20–67.5 μm.

Distribution: China (Yunnan)

Material examined: YMF1.00127, isolated from soil in Yunnan in 2003 by HongYan Su.

Notes: *Da. multiseptata* resembles but distinctly differs from two nematode-trapping fungi, *Da. candidum* and *Da. ellipsospora* in conidial shape and type of predacious device. *Da. multiseptata* produces larger conidia [67.5–132.5 (91.6) × 3.8–17.5 (15.5) μm] and more septa (usually 6–8) than *Da. candidum* [33–50 (43.7) × 7.4–13.3 (10.7) μm, mainly 4-sepate] and *Da. ellipsospora* [37.5–62 (48.3) × 8.7–19.3 (13) μm, often with 4 septa]. Additionally, *Da. multiseptata* does not form denticles on the apex of conidiophores, while *Da. candidum* and *Da. ellipsospora* often produce 2–5 short branches near the apex. (Fig. 3.77)

Dactylellina parvicolla (Drechsler) Yan Li, in Li, Hyde, Jeewon, Cai, Vijaykrishna & Zhang, Mycologia 97 (5): 1042 (2005)
≡ *Dactylella parvicollis* Drechsle, Sydowia 15: 13 (1962)
= *Gamsylella parvicollis* (Drechsler) M. Scholler, Sydowia 51: 109 (1999)
= *Golovinia parvicollis* (Drechsler) Mekht., Khishchnye nematofagovye Griby-Gifomitsety: (Baku)162 (1979)
= *Monacrosporium parvicolla* (Drechsler) R.C. Cooke & C.H. Dickinson [as '*parvicollis*'], Trans Br Mycol Soc 48 (4): 622 (1965)

Characteristics: Mycelium colourless, sterile hyphae branched, septate, 1.5–3.8 μm wide. Conidiophores colourless, erect, mostly 130–290 μm long, 3–4.5 μm wide at the base, tapering gradually upwards, 1.8–2.2 μm wide at the apex bearing a single conidium. Conidia colourless, commonly spindle-shaped, rounded at the distal end, narrowly truncate at the base, 35–45 × 8–14 μm, 2–4- septate, mainly 4-septate. Capturing nematodes by globose or obovoid or prolate ellipsoid stalks adhesive knobs measuring 6.5–11 × 5.5–9 μm.

Distribution: China (Beijing, Fujian, Hubei, Shandong, Xizang, Yunnan), Ecuador (National Parc Cotopaxi), Germany (Berlin-Dahlem, Berlin), USA (California, Colorado).

Material examined: YMF1.00028, YMF1.00029, YMF1.00546, isolated from field soil in Wuhan, Hubei in 2001 by Xuefeng Liu; XZM-8, isolated from forest soil in Xizhang in August 2000 by Minghe Mo.

Discussion: *Da. parvicolla* can be distinguished from related species of *Dactylellina* by its 3–5 septate (mostly 4-septate), spindle-shaped conidia. The conidial dimensions of the present isolate is wider than those of the type strain (29–47 × 6–10.5 μm). (Fig. 3.78)

Dactylellina phymatopaga (Drechsler) Yan Li, in Li, Hyde, Jeewon, Cai, Vijaykrishna & Zhang, Mycologia 97 (5): 1042 (2005)
≡ *Dactylella phymatopaga* Drechsler, Mycologia 46: 775 (1954)
= *Gamsylella phymatopaga* (Drechsler) M. Scholler, Sydowia 51: 109 (1999)
= *Golovinia phymatopaga* (Drechsler) Mekht., Khishchnye nematofagovye Griby-Gifomitsety (Baku): 165 (1979)
= *Monacrosporium phymatopagum* (Drechsler) Subram.[as '*phymatophagum*'], J. Indian bot. Soc 42: 293 (1963)

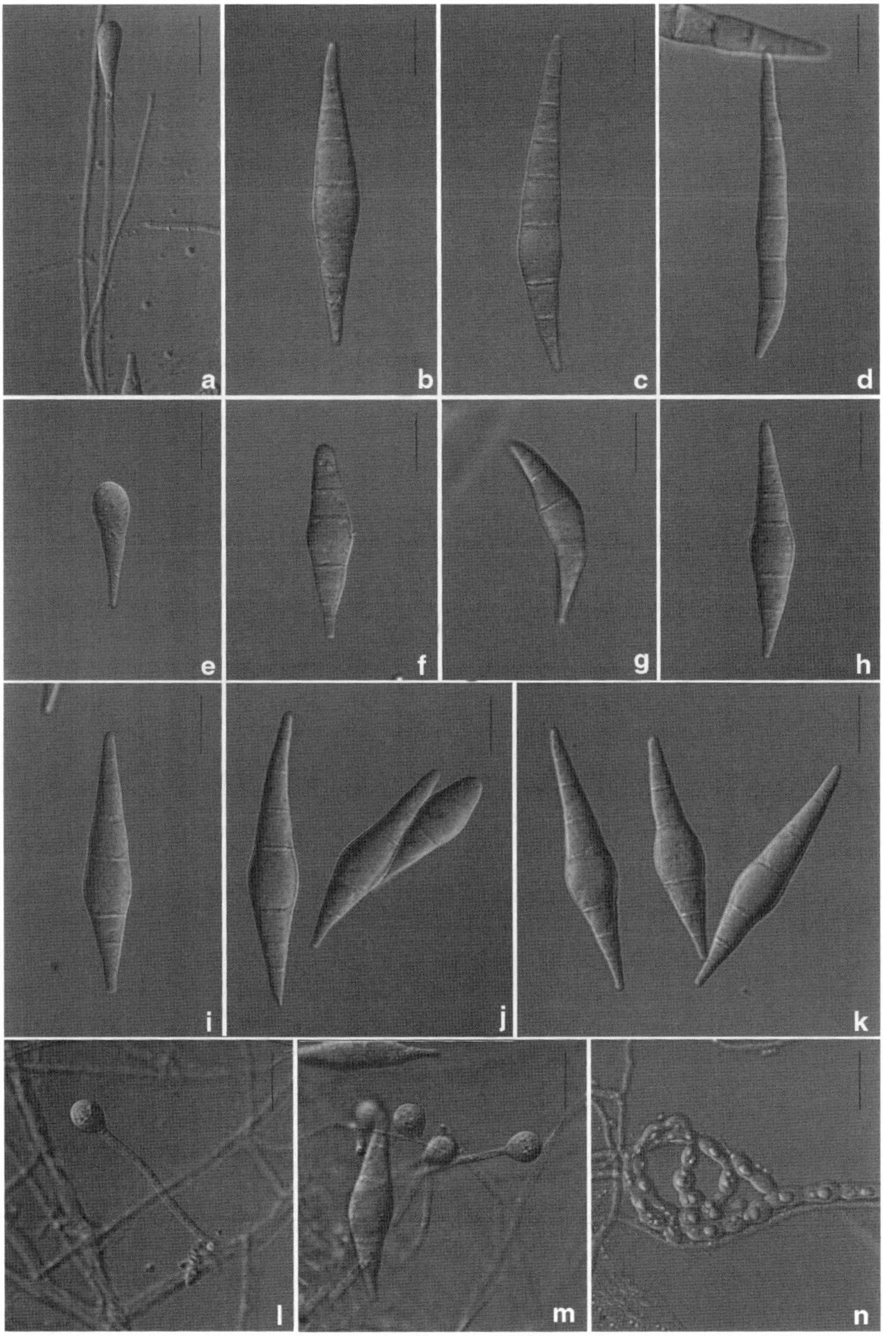

Fig. 3.77 *Dactylellina multiseptata.* **a** conidiophores; **b–k** conidia; **l–m** adhesive knobs; **n** chlamydospores. Bars = 10 μm

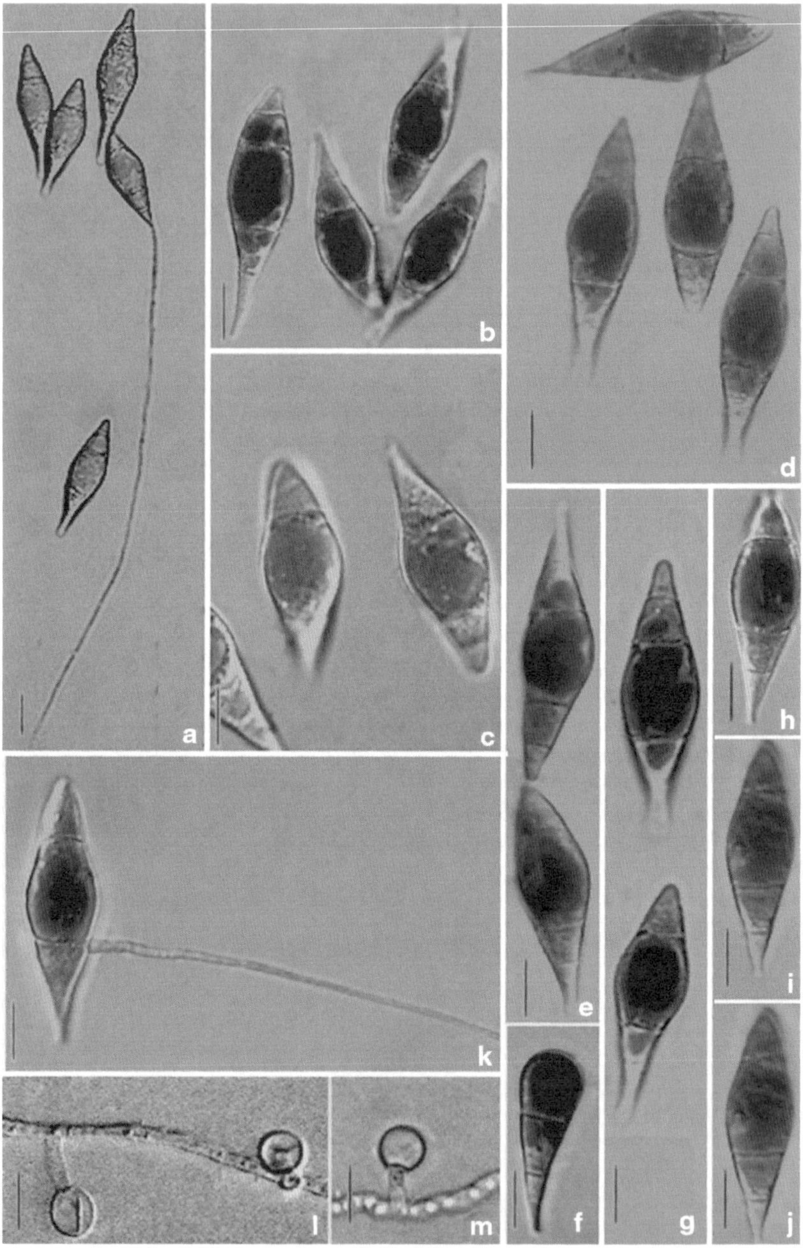

Fig. 3.78 *Dactylellina parvicolla.* **a** conidiophore; **b–k** conidia; **l–m** adhesive knobs. Bars = 10 μm; Strain number: YMF1.00029

Characteristics: Mycelium scanty; vegetative hyphae hyaline, branched, septate at moderate intervals, 1.8–3.5 μm wide. Conidiophores hyaline, erect, septate, 150–500 μm long, 5–9 μm wide at the base, gradually tapering upwards to a width of 2.5–4.5 μm at the apex, bearing a single conidium, occasionally forming 1–2 short branches near the apex with conidia. Conidia hyaline, spindle-shaped, truncate and rather narrow at the base, rounded at the distal end, 46–71 (62.3) × 21–29 (24.7) μm, 1–5-septate, mostly 3–4-septate. Capturing nematodes by means of unstalked adhesive knobs, which can grow out to form branches.

Distribution: China (Beijing, Hebei, Jilin, Sichuan, Yunnan), Germany (Berlin Dahlem, Berlin), India, Netherlands (Naaldwijk), Sweden.

Material examined: N4-1, isolated from forest soil in Jiuzhaigou, Sichuan in April 2003 by Xuefeng Liu; YMF1.1413, isolated from forest soil in Dehong, Yunnan in 2003 by Hong Luo. Permanent slide: N4-1

Notes: This species closely resembles *Da. ellipsospora* with its solitary conidia, which are spindle-shaped and 4 septa; however, conidia are much longer in *Da. ellipsospora*. The trapping devices of *Da. phymatopaga* are not the typical adhesive knobs like those of *Da. ellipsospora*, which are obovoid rather than globose and can grow out to form adhesive branches. (Fig. 3.79)

Dactylellina quercus Bin Liu, Xing Z. Liu & W.Y. Zhuang [as '*querci*'], FEMS Microbiol. Lett. 245 (1): [99] (2005)

Characteristics: Colonies colourless on PDA or CMA, reached to 60–65 mm in diameter on PDA and 45 mm CMA after 20 days culture at 23–25 °C. Aerial mycelium sparse, hyphae hyaline, septate, branched, 2.5–3.5 μm wide. Conidiophores mostly 130–180 μm high, 5–6.5 μm wide at the base, 1.5–2.5 μm at the apex, sometime branched near the apex, bearing a single conidium. Conidia were commonly spindle-shaped, slightly rounded at the distal end, narrowly truncate at the base, 25–32.5–40 (–50) × 8–9.5–12 μm, with 3–5 and mainly 3-septate, but the central cell is not larger than the others. Chlamydospores not observed. Trapping nematodes by means of stalked knobs (0–5.5 μm), knobs produced frequently on the nutritional agar plates even without challenging with nematodes.

Sexual state: Orbilia querci

Distribution: China (Beijing, Huairou County)

Notes: This species resembles *Da. drechsleri*, *Da. ellipsospora* and *Da. haptotyla* in the shape of conidia and adhesive knobs, but differs in that it lacks the largest cell in conidia and in having shorter stalks of adhesive knobs. The sequence divergence of the ITS1 region between *Da. quercus* and the other knobs forming species tested is 23.8–33.4, which separates *Da. quercus* from its related species. (Fig. 3.80)

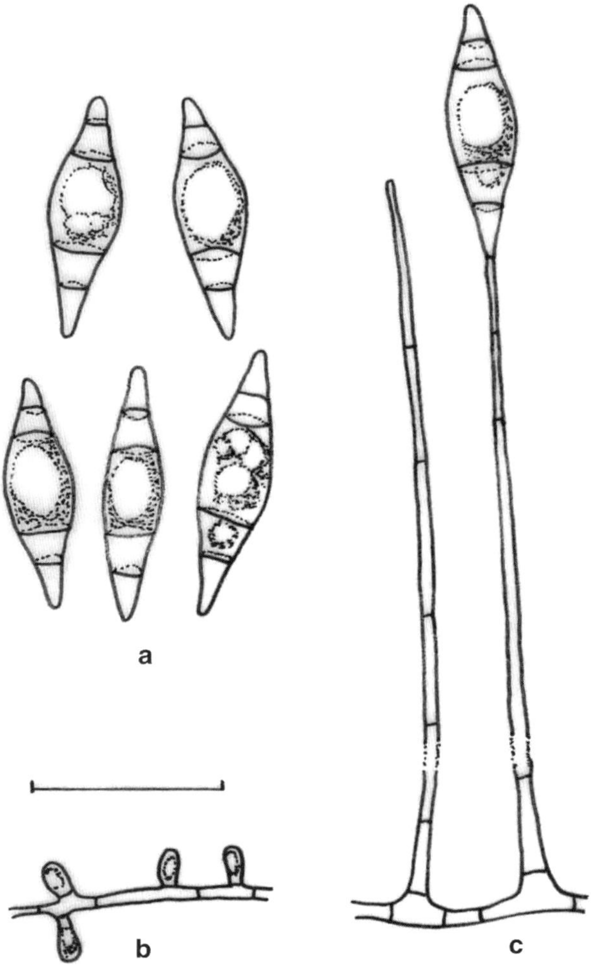

Fig. 3.79 *Dactylellina phymatopaga*. **a**. conidia; **b**. adhesive knobs; **c**. conidiophores. Bar = 40 μm

Dactylellina robusta (J.S. McCulloch) Yan Li, in Li, Hyde, Jeewon, Cai, Vijaykrishna & Zhang, Mycologia 97 (5): 1043 (2005)
≡ *Monacrosporium robustum* J. S. McCulloch, Trans. Br. Mycol. Soc. 68: 177 (1977)
= *Gamsylella robusta* (J.S. McCulloch) M. Scholler, Hagedorn & A. Rubner, Sydowia 51: 109 (1999)

Characteristics: Colonies on CMA white, slow growing. Mycelium spreading, scanty, vegetative hyphae hyaline, septate, branched. Conidiophores hyaline, erect, septate, unbranched, 150–450 μm long, 5–8 μm wide at the base and tapering to 3–4 μm wide at the apex, bearing a single conidium. Conidia hyaline, spindle-shaped, 68–85 × 20–30 μm, 3–5-septate. Chlamydospores not observed. Capturing nematodes by means of stout unicellular unstalked adhesive knobs, 16–24 × 9–10 μm.

Distribution: China (Yunnan), West Australia.

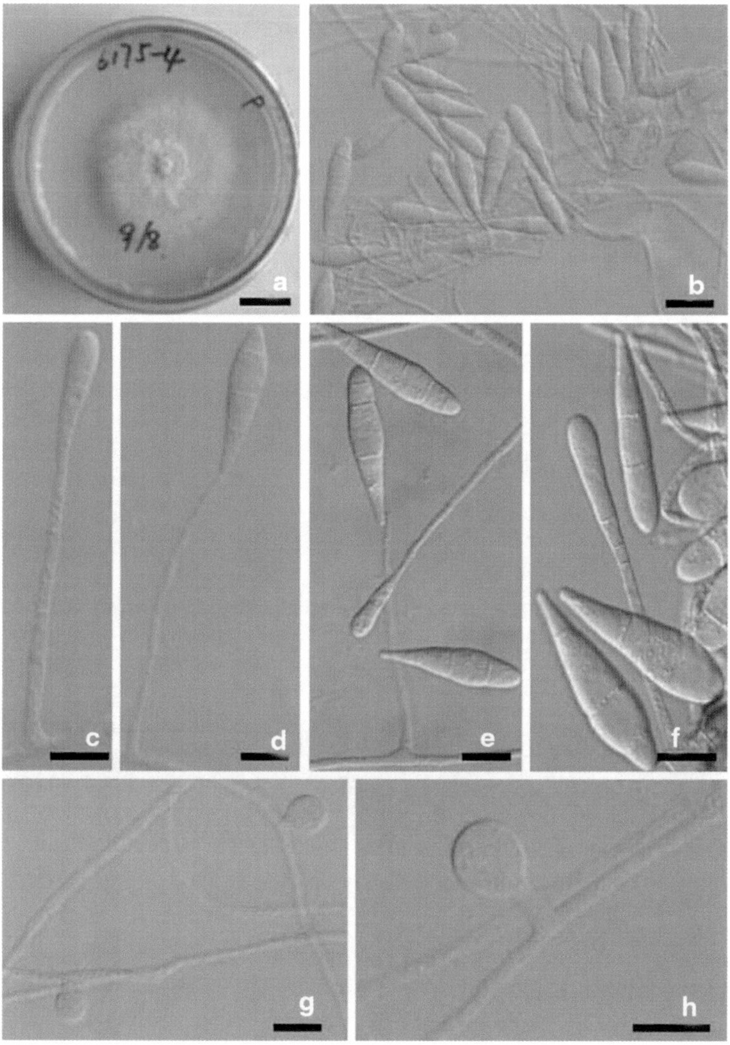

Fig. 3.80 *Dactylellina quercus.* **a** colony on PDA; **b–f** conidiophores and conidia, note the single conidium produced on each apex of the conidiophores; **g–h** stalked knobs. Bars: **a**=2 cm; **b**=20 μm; **c–h**=10 μm

Material examined: YMF1.1413, isolated from forest soil in Ruili, Yunnan in September 2003 by Hong Luo.

Notes: *Da. robusta* resembles *Da. phymatopaga* (Drechsler) Y. Li (2005), in trapping devices and spore shape, but differs in the larger size of the hyphae, traps and conidia. *Da. robusta* is easily recognized by its large conidia and its characteristic trapping devices, which just are unicellular adhesive knobs. In *Da. phymatopaga*, trapping devices can grow out to form adhesive branches. (Fig. 3.81)

Dactylellina sichuanensis Yan Li, K.D. Hyde & K.Q. Zhang, in Li, Jeewon, Hyde, Mo & Zhang, Mycol. Res. 110 (7): 792 (2006)

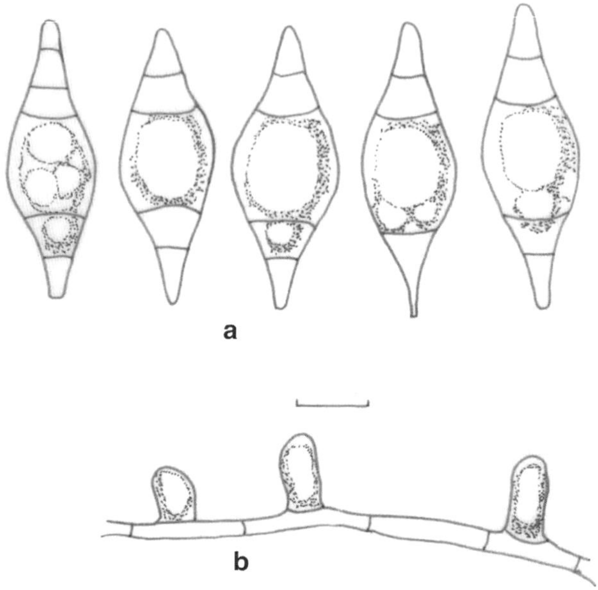

Fig. 3.81 *Dactylellina robusta*. **a**. conidia; **b**. adhesive branches; Bar = 20 µm

Characteristics: Colonies on PDA villiform, whitish, extending to 5 cm diameter at 25 °C, 7 cm diameter at 28 °C and not growing at −4 °C or 35 °C within 7 d. The reverse side of the media is of the same colour. Colonies on CMA whitish, slow-growing, and extending to a diameter of 4 cm at 25 °C, 6 cm 28 °C and not growing at −4 °C or 35 °C within 10 d. Mycelium spreading, scanty, vegetative hyphae hyaline, septate, branched, mostly 2.5–5 µm wide. Conidiophores hyaline, erect, simple, septate, 139–198.5 µm (168.5 µm) high, 2–2.5 µm wide at the base, gradually tapering upwards to a width of 0.5–1 µm at the apex, bearing one single conidium. Conidia 35–82.5 µm (53.5 µm) × 7.5–17.5 µm (13 µm), hyaline, spindle-shaped, straight or sometimes slightly curved, acutely narrowing towards and truncate at the base, the central cell the largest, (3–)4 (–6) septa; proportion of conidia with 3,4,5 and 6 septa are 14, 76, 9 and 1 %, respectively. Two types of trapping devices were observed in culture: stalked adhesive knobs, 20–50 µm long, unicellular, subsphaerical or prolate ellipsoidal; non-constricting rings approximately circular, composed of three subequal arcuate cells. Chlamydospores were not observed in culture.

Distribution: China (Sichuan)

Notes: This species is distinct in having two types of trapping devices and larger conidia from similar species within the same genus. *Da. sichuanensis* is morphologically similar to *Da. lysipaga*, which also possesses two types of trapping devices, simple conidiophores and spindle-shaped conidia. However, the main differences between these two are the size of the conidia and the number of conidial septa. *Da. sichuanensis* produces larger conidia than *Da. lysipaga*. Only 2–4 septa are present in *Da. lysipaga* whereas *Da. sichuanensis* is usually characterized by more than four septa. (Fig. 3.82)

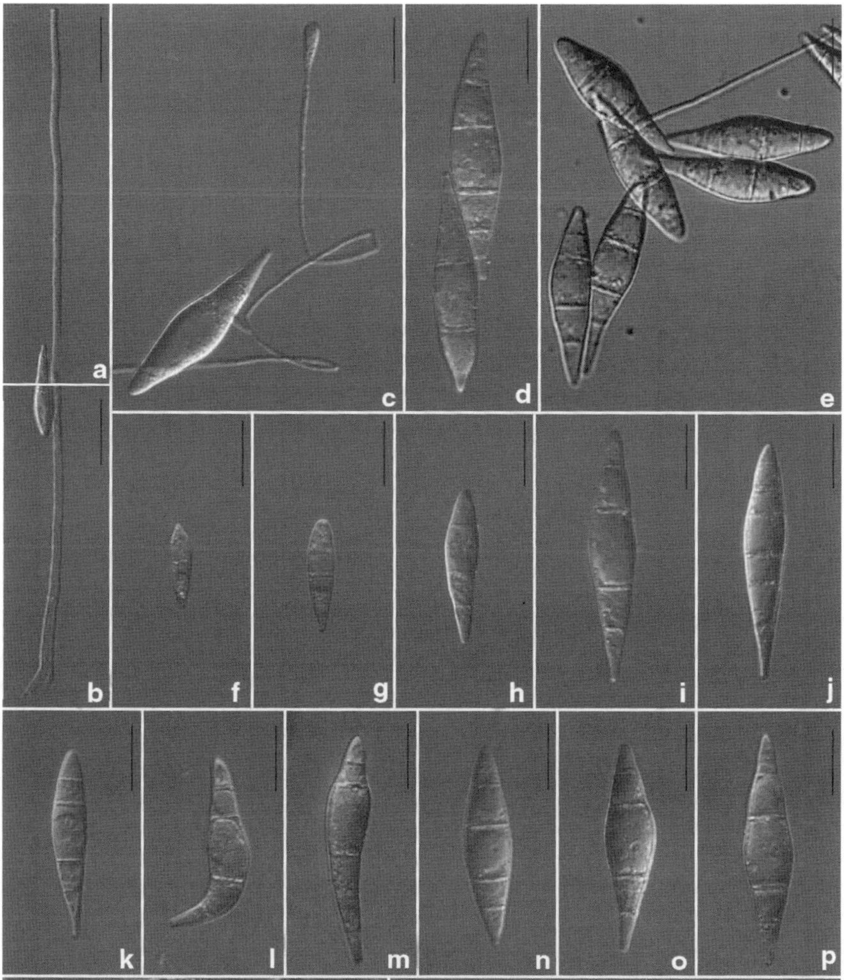

Fig. 3.82 *Dactylellina sichuanensis.* **a–b** conidiophore; **c–p**, **r** conidia; **q** adhesive knobs and non-constricting rings. Bars = 20 μm; Strain number: YMF1.00234

Species of the Genus *Drechslerella*

Drechslerella Subram. emend. M. Scholler, Hagedorn & A. Rubner
Type species: *Drechslerella acrochaeta* (Drechsler) Subram.–J. Ind. Bot. Soc. 42: 299 (1963)
≡ *Dactylella acrochaeta* Drechsler—Mycologia 44: 541 (1952)
= *Dactylariopsis* Mekht. 1967
Type species: *D. brochopaga* (Drechsler) Mekht.—Mikol. Fitopatol. 1: 278 (1967)
≡ *Dactylaria brochopaga* Drechsler—Mycologia 29: 517, 1937
= *Golovinia* Mekht. 1967
Type species: *Golovinia bembicodes* (Drechsler) Mekht.—Mikol. Fitopatol.1: 257 (1967)
≡ *Dactylella bembicodes* Drechsler—Mycologia 29: 491 (1937)

Characteristics: Mycelium slow-growing. Hyphae septate, branching, hyaline. Conidiophores simple or rarely branched, apex mainly simple without modifications, rarely with short denticles or with geniculate proliferations. Conidia mainly formed singly at the apex of the conidiophores, rarely in clusters. Conidia holoblastic, hyaline, one to four septate, mainly ellipsoidal, obovoidal, spindle-shaped, or top-shaped, rarely clavate, or cylindrical. Microconidia and microconidiophores rarely observed. Chlamydospores, when present, intercalary and in chains, thick-walled, sphaerical to ovoid, yellow pigmented. Sexual state, when known, belonging to the genus *Orbilia* Fr. (*Helotiales*). Trapping nematodes by means of spontaneously formed three-celled, stalked, constricting rings; saprotrophic capacities weak.

Key to Species of *Drechslerella*

1. Conidiophores usually with swollen tips and denticles, bearing apical clusters of conidia 2
1. Tips of conidiophores bearing a single conidium, occasionally with very short subapical lateral branches 5
2. Conidia elongate-obovoid, mostly 1-septate, occasionally 2-septate .... ***D. anchonia***
2. Conidia other types 3
3. Conidia digitiform 4
3. Conidia elongated-ellipsoid, 0–1-septate, mostly 1-septate .... ***D. yunnanensis***
4. Conidia cylindrical to clavate, mostly 3-septate .... ***D. brochopaga***
4. With two types of conidia, macroconidia clavate, 1-septate .... ***D. dactyloides***
5 Conidiophores of the genicularioid type, conidia ellipsoid .... ***D. effusa***
5. Conidiophores unbranched or occasionally with subapical lateral branches 6
6. Conidia ellipsoid or broadly obovoid 7

6. Conidia spindle-shaped 9
7. Conidia broadly obovoid, with 1-septate at the base .... ***D. polybrochum***
7. Conidia ellipsoid or wide digitiform 8
8. Having two types of conidia, macroconidia ellipsoid, 2-septate ...***D. heterospora***
8. Conidia ellipsoid or wide digitiform, straight or curved, 1–3-septate .... ***D. stenobrocha***
9. Conidia 2-septate 10
9. Septa of conidia more than 2 septa 11
10. Conidia 26.5–45 × 14–18.5 μm, bearing an appendage at conidia distal apex .... ***D. acrochaetum***
10. Conidia 25–52.5 (33.2) × 12.5–29 (17.3) μm, conidiophores with globose apex .... ***D. doedycoides***
11. Conidia mostly 3-septate 12
11. Conidia mostly 4-septate 13
12. Conidia narrower, 42.5–62.5 (47) × 15–22.5 (16.9) .... ***D. inquisitor***
12. Conidia 2–4-septate, mostly 3-septate, 40–57.5 (51) × 15.5–35 (24.6) .... ***D. aphrobrocha***
13. Conidia 36–43.2 (40) × 16.8–21.6 (20.5) .... ***D. bembicodes***
13. Conidia 2–5-septate, 45.6–55.2 (49.5) × 16.8–21.6 (19.8).......***D. coelobrochum***

Accepted Species of *Drechslerella*

Drechslerella acrochaeta (Drechsler) Subram., J. Indian Bot. Soc. 42: 299 (1964) [1963]
≡ *Dactylella acrochaeta* Drechsler, Mycologia 44 (4): 541 (1952)
= *Golovinia acrochaeta* (Drechsler) Mekht., Khishchnye Nematofagovye Griby—Gifomitsety (Baku): 138 (1979)
= *Monacrosporium acrochaetum* (Drechsler) R.C. Cooke, Trans. Br. Mycol. Soc. 50 (2): 317 (1967)

Characteristics: Colonies on CMA rather slow growing and attaining a diameter of 1 cm within one week at 25 °C. Hyphae hyaline, septate, branched. Conidiophores hyaline, erect, septate, simple, 110–320 μm long, 2–4 μm wide, bearing a single conidium at the apex. Conidia hyaline, broadly spindle-shaped, rounded at the distal end and truncate at the base, 1–2-septate, 26.5–45 × 14–18.5 μm, bearing distally an appendage approximate alignment with the spore axis. Trapping nematodes by means of constricting rings.

Distribution: China (Guizhou, Hebei, Jilin, Yunnan), USA

Material examined: 95X9–1, isolated from forest soil in Xishuangbanna Yunnan in June 1994 by Xingzhong Liu.

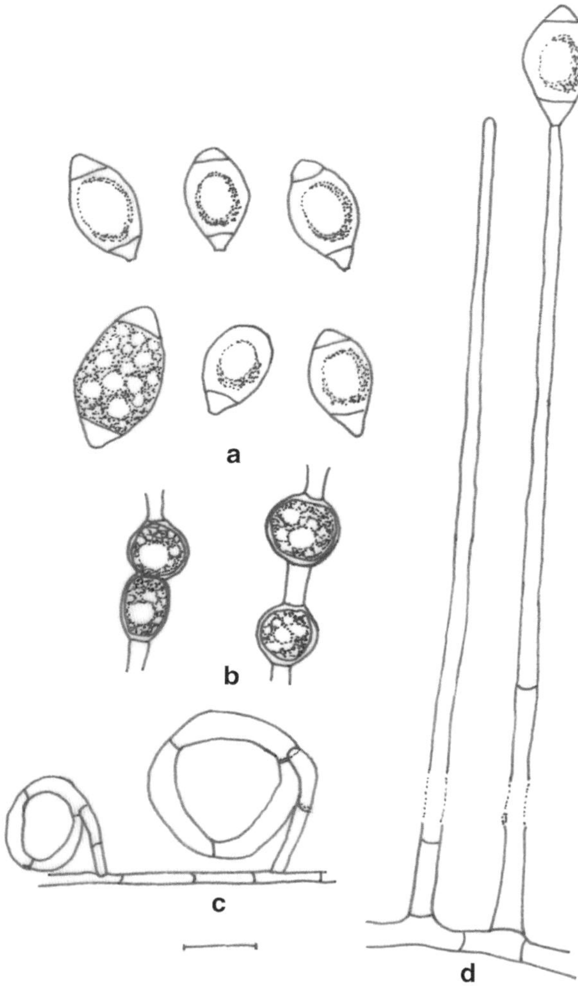

Fig. 3.83 *Drechslerella acrochaeta*. **a**. conidia; **b**. chlamydospore; **c**. constricting rings; **d**. conidiophore. Bar = 30 μm

Notes: The conidial appendage was observed in culture in our isolate. Conidia shape, septa and size of examined material matched well with original description (Drechsler 1952). (Fig. 3.83)

Drechslerella anchonia (Drechsler) M. Scholler, Hagedorn & A. Rubner, Sydowia 51 (1): 99 (1999)
= *Arthrobotrys dactyloides* Drechsler, Mycologia 29 (4): 486 (1937)
= *Arthrobotrys anchonia* Drechsler, Mycologia 46: 762 (1954)
≡ *Dactylariopsis dactyloides* (Drechsler) Soprunov [as '*dactyloidea*'], Predacious fungi-Hyphomycetes and their use in the control of pathogenic nematodes: 142 (1958)
= *Nematophagus anchonius* (Drechsler) Mekht., Khishchnye Nematofagovye Griby-Gifomitsety, (Baku): 108 (1979)

Characteristics: Mycelium spreading, scanty, vegetative hyphae hyaline, septate, branched, 2–5 μm. Conidiophores hyaline, erect, 3–8-septate, simple, 350–500 μm long, 4–6 μm wide at the base, tapering gradually upwards to a width of 2.5–3.5 μm at the apex, there bearing 3–8 conidia in radiating capitate arrangement; conidia hyaline, elongate obovoid, rarely bi-septate, usually 1-septate with the proximal cell commonly 8 to 18 μm (average 11.5) long and the distal cell commonly 17–27 (average 23.5) μm long, 29–43 (35) × 15–19 (16.8) μm, capturing nematodes by means of constricting rings.

Distribution: USA

Material examined: No live culture was obtained, so the description is based on the protologue.

Notes: Among species of *Drechslerella*, there are four species with conidia in a radiating capitate arrangement at the apex of conidiophores, including *Dr. anchonia*, *Dr. brochopaga*, *Dr. dactyloides*, *Dr. yunnanensis*. Within these species, *Dr. anchonia* can be distinguished from the other three species by elongate obovoid and 1-septate conidia. (Fig. 3.84)

Drechslerella aphrobrocha (Drechsler) M. Scholler, Hagedorn & A. Rubner, Sydowia 51 (1): 99 (1999)
≡ *Dactylella aphrobrocha* Drechsler, Mycologia 42: 20 (1950)

Characteristics: Colonies on CMA slow growing and extending a diameter of 2.6 cm within 7 days at 25 °C. Mycelium spreading, scanty, vegetative hyphae hyaline, septate, branched, mostly 1.7–3.7 μm. Conidiophores hyaline, erect, 2–5-septate, simple, 162–350 μm long, 3–7.5 μm wide at the base, tapering gradually upwards to a width of 2.5–5 μm at the apex, bearing a single conidium and sometimes a second one on a short perpendicular branch below the apex. Conidia hyaline, broadly spindle-shaped, the central cell the largest, rounded at the distal end and somewhat truncate at the base, 2–4-septate, mostly 3-septate, 40–57.5 (51) × 15.5–35 (24.6) μm, bearing appendages occasionally. Chlamydospores not observed. Capturing nematodes by means of constricting rings, which consist of three cells, outer diameter 20–27.5 μm, inner diameter15–17.5 μm.

Distribution: Carolina, China (Beijing, Guizhou, Jiling, Yunnan), India, North USA

Material examined: YMF1.00119, isolated from forest soil in Lijiang, Yunnan in September 2002 by Jing Zhang. Permanent slide: LJ2–11–②

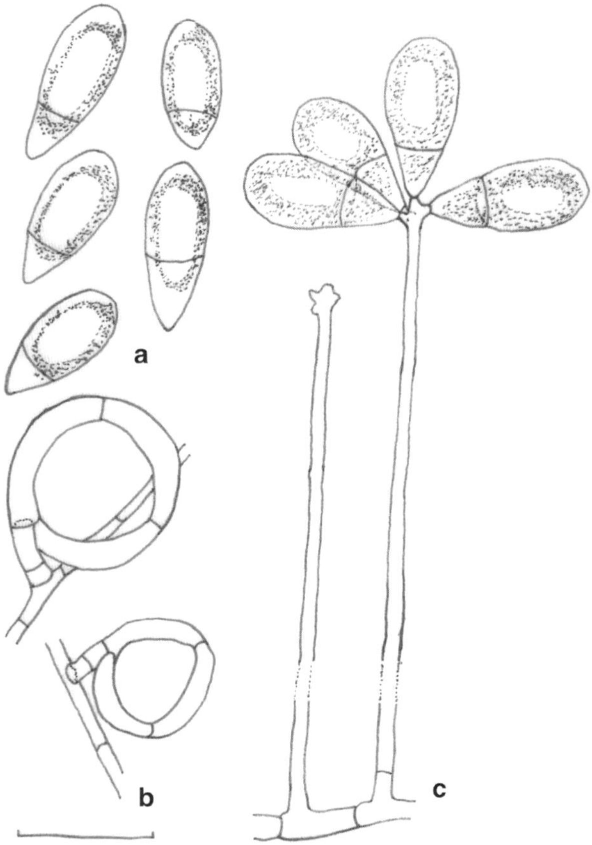

Fig. 3.84 *Drechslerella anchonia*. **a**. conidia; **b**. constricting rings; **c**. conidiophore. Bar = 30 μm

Notes: The conidia of our isolate matched the original description (41–55 × 17–26 μm) (Drechsler 1950) and is slightly larger than Subramanian (1963) description (46.7 × 21.9 μm). Conidial length and width of *Dr. aphrobrocha* is very similar to *Dr. coelobrocha*, but Drechsler (1950) emphasized the greater conidial length and the protracted apical and basal cells of the latter (Rubner 1996). (Fig. 3.85)

Drechslerella bembicodes (Drechsler) M. Scholler, Hagedorn & A. Rubner, Sydowia 51 (1): 99 (1999)
≡ *Dactylella bembicodes* Drechsler, Mycologia 29 (4): 491 (1937)
= *Monacrosporium bembicodes* (Drechsler) Subram. J. Indian Bot. Soc.42: 293 (1964) [1963]
= *Golovinia bembicodes* (Drechsler) Mekht., Mikol. Fitopatol. 1: 275 (1967)

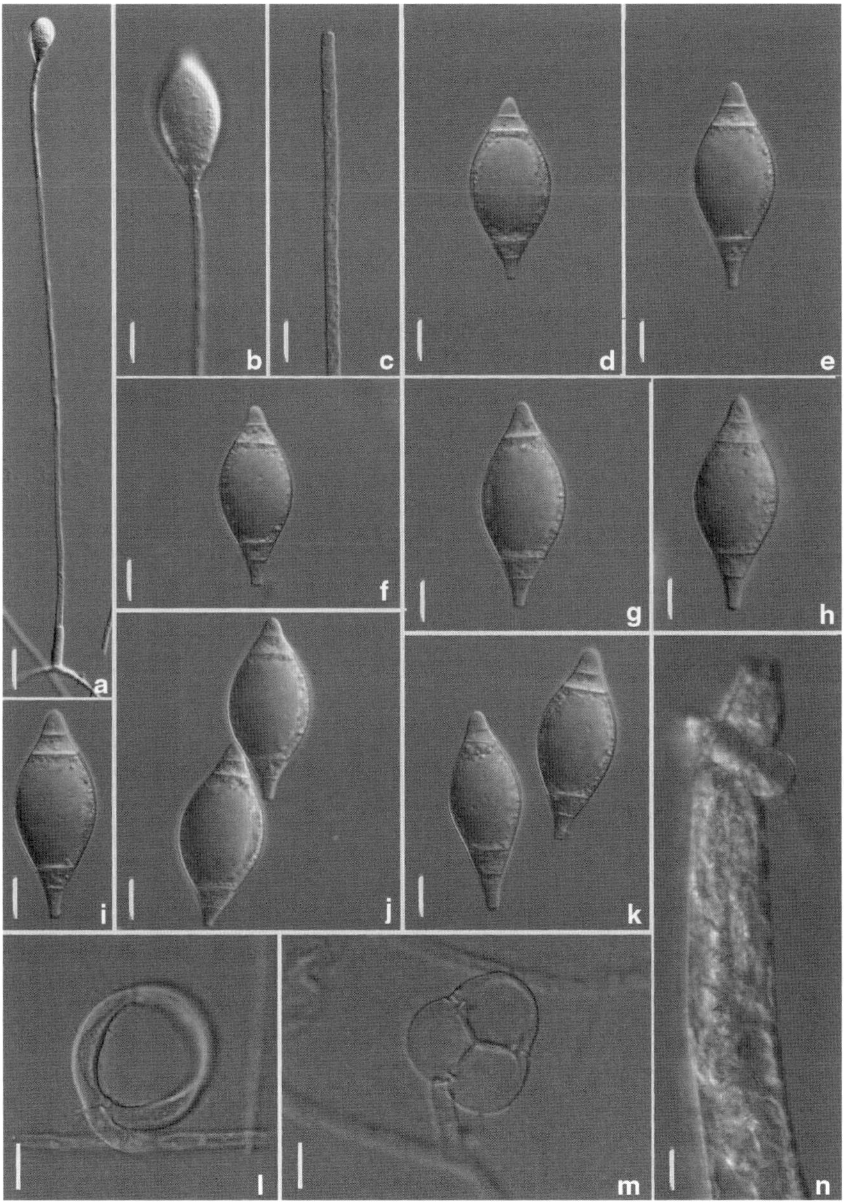

Fig. 3.85 *Drechslerella aphrobrocha.* **a–c** conidiophore; **d–k** conidia; **l–m** constricting rings; **n** nematode trapped by constricting. Bars = 10 μm; Strain number: YMF1.00119

Characteristics: Mycelium spreading, vegetative hyphae hyaline, septate, mostly 2–5 μm wide. Conidiophores hyaline, erect, septate, 250–450 μm long, 5–7.5 μm wide at the base, tapering gradually upwards to a width of 2–3 μm at the apex, where bear a single conidium. Conidia hyaline, broadly spindle-shaped, broadly rounded at the apex, tapering towards the slightly protruding truncate base, 36–43.2 (40) × 16.8–21.6 (20.5) μm, 3–4-septate, mostly 4-septate. Microconidia developing on conidiophores with several (3–7) subterminal denticles, obovoid, 10–19 × 4–5 μm, usually 1-septate. Chlamydospores not observed. Trapping nematodes by means of constricting rings.

Distribution: China (Guizhou), India, Netherlands (Schovenhorst near Putten), New Zealand, USA

Material examined: YMF1.01429, isolated from soil in Kuankuoshui, Guizhou in October 2003 by Yu'e Hao. Permanent slide: YMF1.1429D.

Notes: The conidia size of our isolate is slightly smaller than in the original description [34–48 (42) × 16–23 (20) μm] (Drechsler 1937). (Fig. 3.86)

Drechslerella brochopaga (Drechsler) M. Scholler, Hagedorn & A. Rubner, Sydowia 51 (1): 99 (1999)
= *Arthrobotrys gracilis* (Dudd.) S. Schenck, W.B. Kendr & Pramer, Can. J. Bot. 55 (8): 983 (1977)
= *Candelabrella brochopaga* (Drechsler) Subram., Kavaka 5: 95 (1977) [1978]
≡ *Dactylaria gracilis* Dudd., Trans. Br. Mycol. Soc.34 (2): 194 (1951)
= *Dactylariopsis brochopaga* (Drechsler) Mekht, Mikol. Fitopatol. 1: 278 (1967)
= *Dactylariopsis gracilis* (Dudd.) Khishchnye Nematofagovye Griby—Gifomitsety (Baku): 117 (1979)
≡ *Dactylella brochopaga* Drechsler, Mycologia 29: 517 (1937)

Characteristics: Colonies on CMA initially whitish then turning reddish. Mycelium spreading; vegetative hyphae hyaline, septate, branched. Conidiophores erect, single, septate, 77.5–582.5 μm high, mostly 2.5–5 μm wide at the base, gradually tapering upwards to a width of 2–3 μm at the apex, there bearing on short blunt sterigmata 2–12 μm, mostly 3–8 conidia in radiating capitate arrangement or less often and less typically producing up to 15 conidia in more scattered, irregularly racemose arrangement. Conidia 20–45 (30) × 5–12.5 (6) μm, hyaline, straight of slightly curved, cylindrical or elongate ellipsoidal, broadly rounded at the distal end, usually tapering noticeably towards the somewhat truncate base, 1–3-septate, mainly 3-septate. Capturing nematodes by means of constricting rings.

Distribution: China (Beijing, Guizhou, Hainan, Hebei, Jiangsu, Jilin, Yunnan), USA (Florida)

Material examined: Hn38–2–4, isolated from field soil in Haikou, Hainan, China in August 2002 by Ke-Qin Zhang; MW13-5, isolated from forest soil in Mengla, Yunnan, China in 1999 by Yanju Bi; YMF1.00008, isolated from field soil in Simao, Yunnan in September 2000 by Yanju Bi; BS3-3, YMF1.00557, isolated from soil in Baoshan, Yunnan in October 2002 by Jing Zhang. Permanent slide: Hn38-3-4

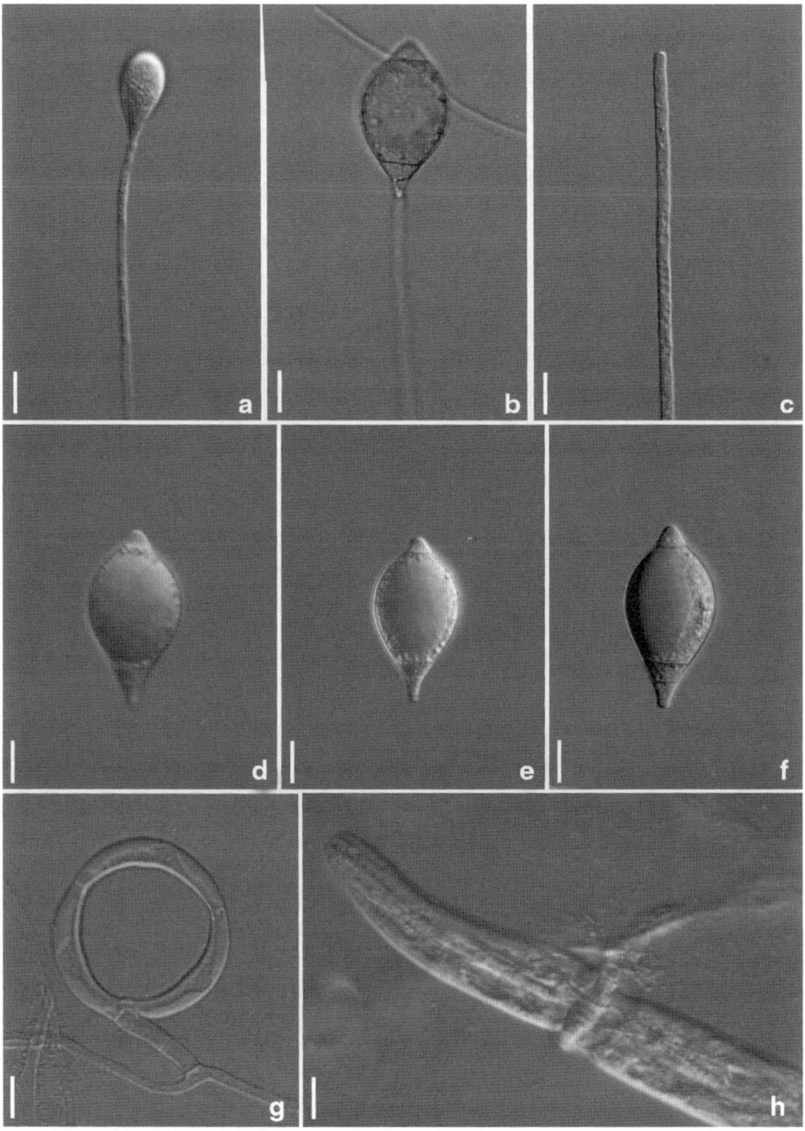

Fig. 3.86 *Drechslerella bembicodes*. **a–c** conidia and conidiophore; **d–f** conidia; **g** constricting rings; **h** nematode trapped by constricting. Bars = 10 μm; Strain number: YMF1.01429

Notes: This species resembles *Dr. dactyloides* in conidia shape, but it mainly has 3 septa while the later often has 1 septum. It occasionally forms conidial traps to trap nematodes including *Acrobeles*, *Acrobeloides*, *Cephalobus*, *Diplogaster*, *Diploscapter*, *Picctus*, *Rhabditis* and *Mononchus*. The conidial trap is smaller in diameter than the constricting ring. Dackman & Nordbring (1992) indicated that conidial traps were formed as an adaption to a poor environment. (Fig. 3.87)

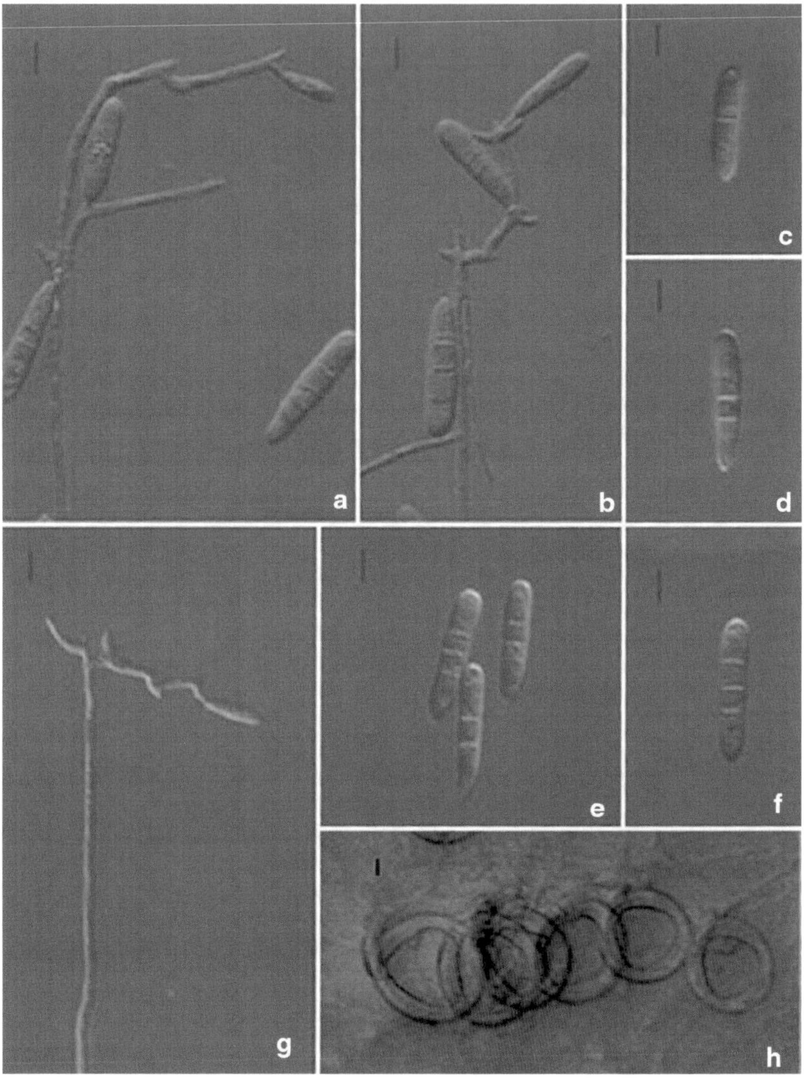

Fig. 3.87 *Drechslerella brochopaga*. **a–b**, **g** conidiophore; **c–f** conidia; **h** constricting rings. Bars = 10 µm; Strain number: YMF1.00008

Drechslerella coelobrocha (Drechsler) M. Scholler, Hagedorn & A. Rubner, Sydowia 51 (1): 99 (1999)

≡ *Dactylella coelobrocha* Drechsler, Mycologia 39: 17 (1947)

= *Monacrosporium coelobrochum* (Drechsler) Subram., J. Indian Bot. Soc. 42: 293 (1963)

= *Golovinia coelobrocha* (Drechsler) Mekht., Khishchnye Nematofagovye Griby—Gifomitsety (Baku): 141 (1979)

Characteristics: Mycelium spreading, vegetative hyphae hyaline, septa, mostly 2–3 μm. Conidiophores hyaline, erect, containing 1–3 septa, mostly in the basal portion, usually unbranched, commonly 180–470 μm long, 3.5–6 μm wide at the base, tapering gradually upwards to a width of 1.5–3 μm at the apex, bearing a single conidium, and sometimes on renewed elongation forming a second conidium. Conidia hyaline, mostly broadly fusiform and strongly ventricose in the centre, 2–5-septate, mostly 4-septate, 45.6–55.2 (49.5) × 16.8–21.6 (19.8) μm, appendages sometimes wholly absent, but at other times present in numbers from 1 to 5, usually arising from the short parabasal cell at right angles to the spore axis, hyaline, filiform, straight, rather rigid. Capturing nematodes by means of constricting rings.

Distribution: China (Guizhou), USA

Material examined: YMF1.01480, isolated from soil in Kuankuoshui, Guizhou in October 2003 by Yu'e Hao.

Notes: The conidial size of *Dr. coelobrocha* is very similar to *Dr. aphrobrocha*, but Drechsler (1950) emphasized the greater conidial length, the protracted apical and basal cells in *Dr. coelobrocha*. This species resembles *Da. ellipsospora* in conidial shape and septa, but the latter traps nematodes by means of adhesive knobs. Compared with original description, the present isolate has slightly smaller conidia. (Fig. 3.88)

Drechslerella dactyloides (Drechsler) M. Scholler, Hagedorn & A. Rubner, Sydowia 51 (1): 99 (1999)
≡ *Arthrobotrys dactyloides* Drechsler, Mycologia 29 (4): 486 (1937)
= *Arthrobotrys anchonia* (Drechsler), Mycologia 46 (6): 762 (1954)
= *Dactylaria dactyloides* (Drechsler) Soprunov [as 'dactyloidea'], Hyphomycetes and their use in the control of pathogenic nematodes: 142 (1958)
= *Dactylariopsis dactyloides* (Drechsler) Mekht., Khishchnye Nematofagovye Griby—Gifomitsety (Baku): 120 (1979)
= *Nematophagus anchonius* (Drechsler) Mekht., Khishchnye Nematofagovye Griby—Gifomitsety (Baku): 108 (1979)

Characteristics: Colonies on CMA whitish. Mycelium spreading, vegetative hyphae, branched, septa, mostly 2–5 μm wide. Conidiophores hyaline, septate, erect, mostly 230–650 μm long, 5–6 μm wide at the base, tapering gradually upwards to a width of 1.5–2 μm at the apex, bearing on sterigmata 1–5 μm long and 2–3 μm wide, forming 4–13 conidia usually in a single loose head, more rarely in two somewhat distinct clusters. Conidia 35–51.5 (42.1) × 6.5–8 (7.5) μm, hyaline, usually elongate ellipsoidal or somewhat digitiform, straight or slightly curved, tapering noticeably from the broadly rounded wider distal end, towards the narrower truncate basal end, single septum near the centre of the spore, but occasionally becoming older and shorter, measuring as much as 25–29 × 10–13 μm, then often 1–2-septate, spindly with an inflated central cell greatly exceeding the end cells in size. Capturing nematodes by means of constricting rings.

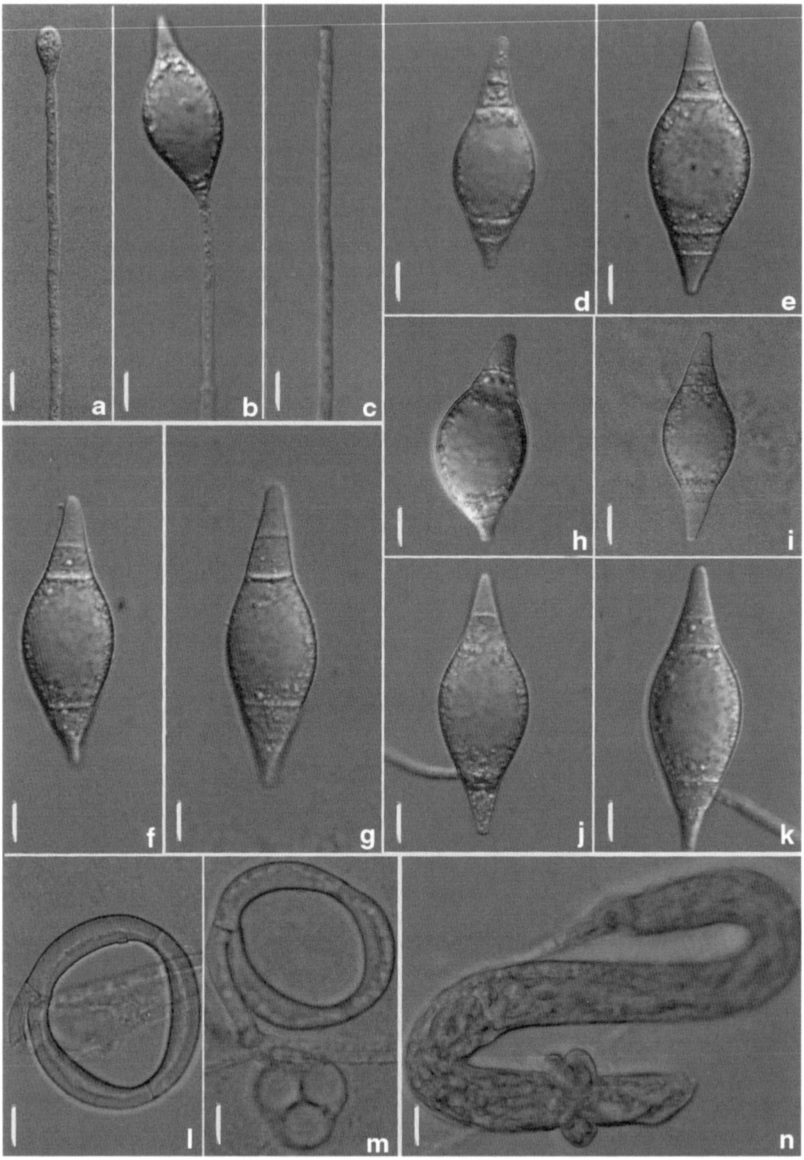

Fig. 3.88 *Drechslerella coelobrocha*. **a–c** conidia and conidiophore; **d–k** conidia; **l–m** constricting rings; **n** nematode trapped by constricting. Bars = 10 μm; Strain number: YMF1.01480

Distribution: Canada, China (Beijing, Guizhou, Hebei, Taiwan, Tianjing, Yunnan, Xizhang), Germany (Berlin, Bonn, Jena), South Africa (North West Province), USA (Maryland).

Material examined: Z258, Z285, Z279, H055, H056, H057, isolated from forest soil in Guizhou and Beijing, in 1992 by Ke-Qin Zhang; YMF1.00131, isolated from field soil in Xizang in 2000 by Minghe Mo; YMF1.00031, YMF1.00555, isolated from field soil in Xundian, Yunnan in March 1999 by Lu Cao; TJ–gzz05, isolated from soil in Gaozhuangzi, Tianjin in 2000 by Wenpeng Li; LJ2–9, isolated from soil in Lijiang, Yunnan in September by Jing Zhang. Permanent slide: D104.

Notes: *Drechslerella dactyloides* resembles *Arthrobotrys musiformis* and *A. javanica* in conidia shape, but differs in modification of conidiophores, conidia size and trapping–devices. *Drechslerella dactyloides* forms constricting-rings, while *Arthrobotrys musiformis* and *A. javanica* form adhesive three-dimensional networks. In *Drechslerella dactyloides*, the apex of conidiophores bears a simple and short sterigmata. In *Arthrobotrys musiformis* and *A. javanica*, the apex of conidiophores often bears branched and long sterigmata. *Drechslerella dactyloides* can form conidia traps, which are smaller than constricting rings formed from hyphae, especially in the presence of nematodes. Higgins and Pramer (1967) described the process on conidia trap in detail. In addition, this species is a good material to control *Meloidogyne javanica* (Stirling et al. 1998). (Fig. 3.89)

Drechslerella doedycoides (Drechsler) M. Scholler, Hagedorn & A. Rubner, Sydowia 51 (1): 99 (1999)
≡ *Dactylella doedycoides* Drechsler, Mycologia 32 (4): 454 (1940)
= *Golovinia doedycoides* (Drechsler) Mekht., Khishchnye Nematofagovye Griby—Gifomisety (Baku): 136 (1979)
= *Monacrosporium doedycoides* (Drechsler) R.C. Cooke & C.H. Dickinson, Trans. Br. Mycol. Soc. 48: 622 (1965)

Characteristics: Colonies on CMA whitish. Mycelium spreading, vegetative hyphae hyaline, septa, branched, 2–4 μm wide. Conidiophores hyaline, erect, 2–6-septate, 200–550 μm long, 5–8 μm wide at the base, tapering gradually upwards to a width of 2–4 μm wide, then expanding abruptly into a knob-like apex 3–5 μm wide, whereon single conidium are borne. Conidia hyaline, somewhat top-shaped, tapering noticeably towards the abruptly or somewhat concavely truncate base, 1–3-septate, mostly 2-septate, 25–52.5 (33.2) × 12.5–29 (17.3) μm. Capturing nematodes by means of constricting rings composed of three cells, out diameter 22–40 μm, inner diameter 10–20 μm.

Distribution: China (Beijing, Guizhou, Yunnan), Portugal (Quinta de Sao Pedro), UK

Material examined: YMF1.00553, isolated from soil in Lijiang, Yunnan in September 2002 by Jing Zhang. Permanent slide: LJ4-1

Notes: The knob-like apex of fertile branches is the crucial feature to distinguish it from other species. Our isolate has the knob-like tips to the conidiophores as in the original description (Cooke and Dickinson, 1965). (Fig. 3.90)

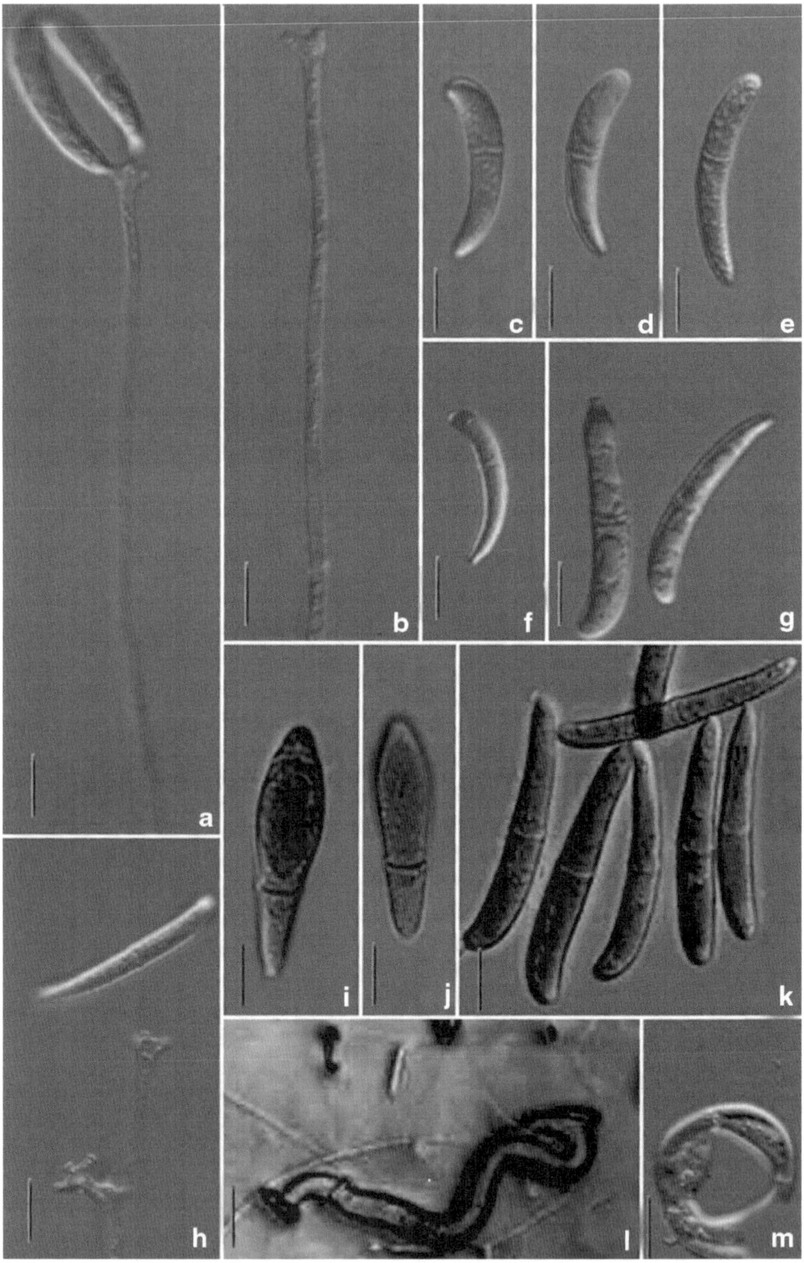

Fig. 3.89 *Drechslerella dactyloides.* **a–b**, **h** conidiophore; **c–g**, **i–k** conidia; **l** nematode trapped by constricting; **m** constricting rings. Bars = 10 μm; Strain number: YMF1.00555

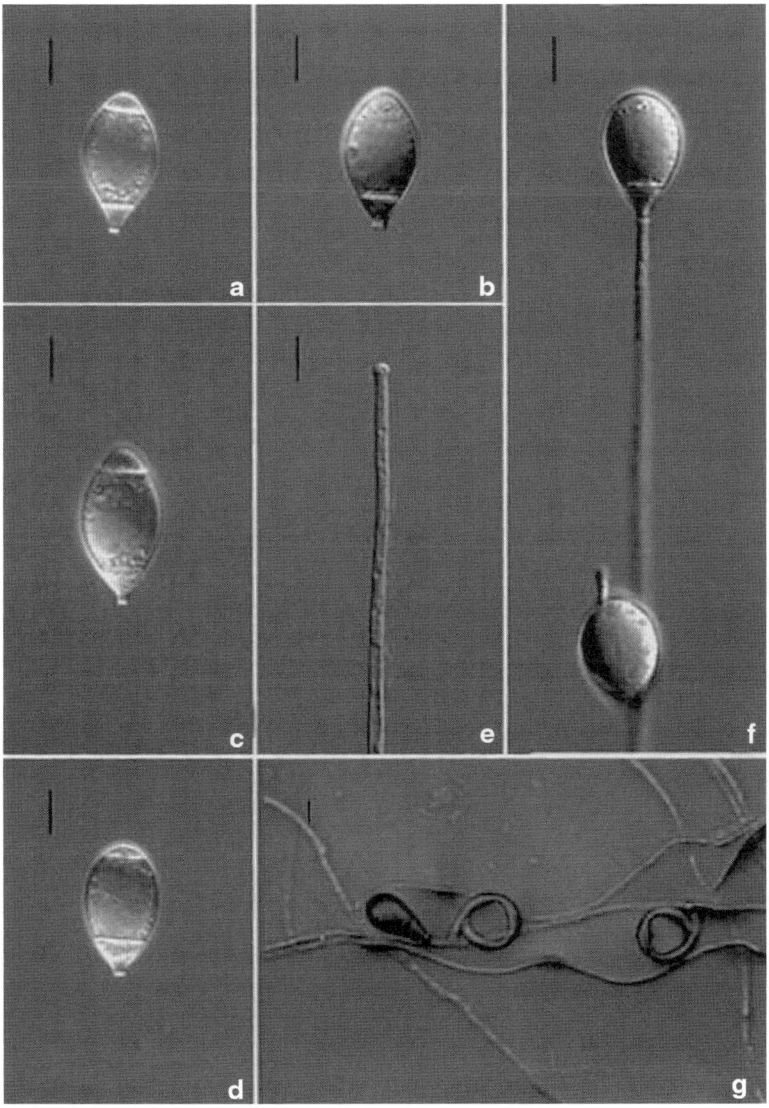

Fig. 3.90 *Drechslerella doedycoides.* **a–d** conidia; **e–f** conidiophore; **g** constricting rings. Bars = 10 μm; Strain number: YMF1.00553

Drechslerella effusa (Jarow.) M. Scholler, Hagedorn & A. Rubner, Sydowia 51 (1): 99 (1999)
= *Arthrobotrys effusa* (Jarow.) S. Schenck, W.B. Kendr. & Pramer, Can. J. Bot. 55: 982, 1977
≡ *Dactylaria effusa* Jarow., Bull. Acad. Polon. Sci., Cl. II. Sér. sci. biol. 16: 773 (1968)
= Geniculifera effusa (Jarow.) Oorschot, Stud. Mycol. 26: 93 (1985)
= *Monacrosporium effusum* (Jarow.) Xing Z. Liu & K.Q. Zhang, Mycol. Res. 98: 864 (1994)

Characteristics: Colonies on CMA forming a cobweb-like, gray-like coating. Mycelium spreading, scanty, delicate, vegetative hyphae hyaline, branched, thread-like, septate, 2–4 µm wide. Conidiophores straight, erect, simple, septate, 220–400 µm long, 5–7.5 µm wide at the base, gradually tapering upwards to a width of 2.5 µm at the apex, initially with a single apical spore, later bearing a loose head consisting of 2–12 (or more) conidia, supported by short wide, set apart sterigmata. Conidia ellipsoid, widely rounded at the distal end, and widely obtuse, somewhat truncate at the base, 32.5–45 (38.9) × 17.5–25 (21.4) µm, 1–2-septa. The proportion of conidia with 1 and 2 is 25 and 75 % respectively. Microconidia formed from aging hyphae in old culture, clavate, 18–26 × 9–10.5 µm, 0–1-septum.Capturing nematodes by means of constricting rings consisted of three cells, 25–30 µm in outer diameter, 15–20 µm in inner diameter, stalks 7.5–10 µm long, and 5 µm wide.

Distribution: China (Sichuan), Canada (Manitoba)

Material examined: YMF1.00583, isolated from forest soil in Jiuzhaigou, Sichuan in April 2003 by Xuefeng Liu. Permanent slide: N58–10.

Notes: *Drechslerella effusa* resembles *Dr. heterospora* and *Dr. doedycoides* in conidial shape, but differs in conidial size (13–20 × 35–47 µm, and 19.7 × 30–41.4 µm, respectively). Moreover, microconidia present in *Dr. effusa*, but absent in *Dr. doedycoides*, while microconidia show somewhat cylindrical shape in *Dr. heterospora*. (Fig. 3.91)

Drechslerella heterospora (Drechsler) M. Scholler, Hagedorn & A. Rubner, Sydowia 51 (1): 100 (1999)
≡ *Dactylella heterospora* Drechsler, Mycologia 35: 347 (1943)

Characteristics: Colonies on CMA whitish, slow growing and extending to a diameter of 2.9 cm within 7 days at 25 °C. Mycelium spreading, scanty, vegetative hyphae hyaline, septate, branched, mostly 2–4 µm. Conidiophores hyaline, erect, septate, simple, occasionally branched, 150–450 µm long, 5–6 µm wide at the base, gradually tapering upwards to a width of 2.5–5 µm at the apex, bearing a single conidium. Conidia hyaline, prolate ellipsoidal, slightly protuberant at the abruptly truncate base, rounded at the distal end, 1–2-septate, 17.5–45 (34) × 17.5–25 (20.4) µm. Microconidiophores often 15–25 µm long, with a cylindrical, curved and somewhat allantoid conidium, 23–40 (31.3) × 5.3–8 (6.8) µm, one septate. Chlamydospores regularly formed, yellowish, in chains. Capturing nematodes by means of constricting rings composed of three cells, 25–35 µm in outer diameter and 17.5–22.5 µm in inner diameter.

Distribution: Chian (Jilin, Yunnan), USA

Material examined: YMF1.00550, isolated from forest soil in Deqin, Yunnan in September 2002 by Jing Zhang. Permanent slide: DQ6–1

Notes: This fungus can be distinguished from other *Drechslerella* species by its prolate ellipsoidal conidia and slightly protuberant at the abruptly truncate base. (Fig. 3.92)

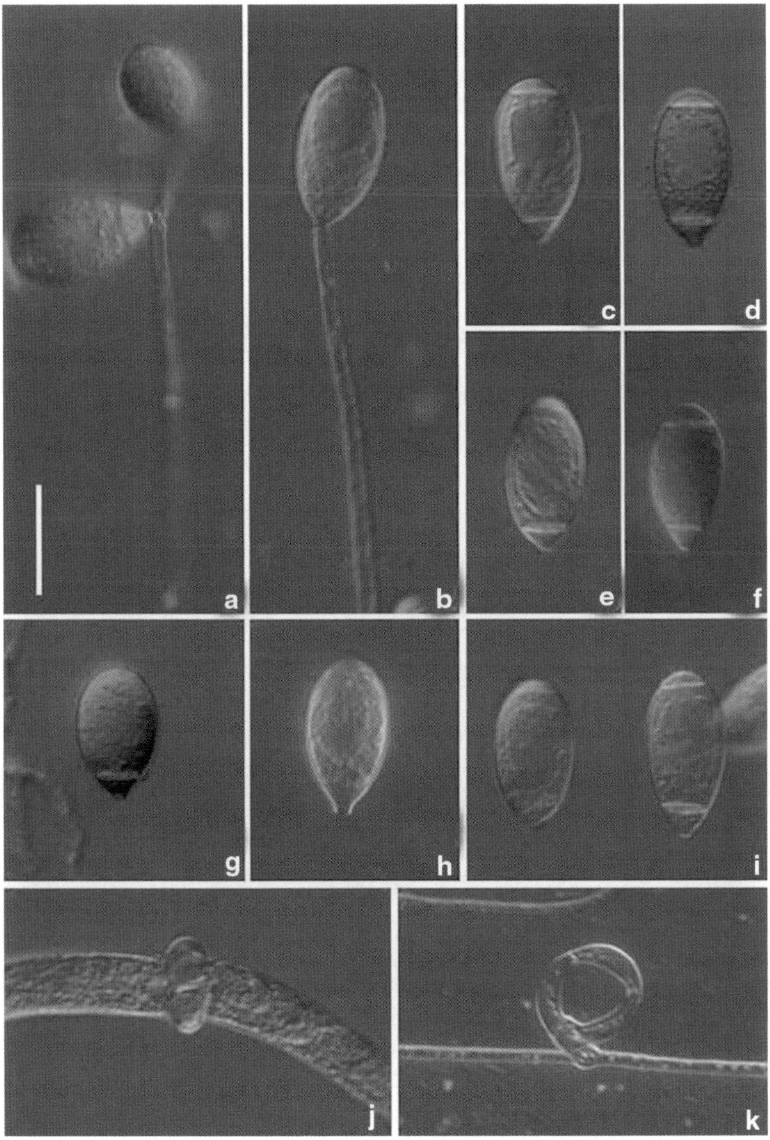

Fig. 3.91 *Drechslerella effusa*. **a–b** conidiophore; **c–i** conidia; **j** nematode trapped by constricting; **k** constricting ring. Bar = 20 μm; Strain number: YMF1.00583

Drechslerella inquisitor (Jarow.) M. Scholler, Hagedorn & A. Rubner, Sydowia 51 (1): 100 (1999)

≡ *Dactylella inquisitor* Jarow. [as '*inquistor*'], Acta. Mycologica, Warszawa 7: 4 (1971)

Characteristics: Mycelium spreading, vegetative hyphae hyaline, septate, branched, 1.8–2.7 μm wide. Conidiophores hyaline, erect, septate, 193–390 μm

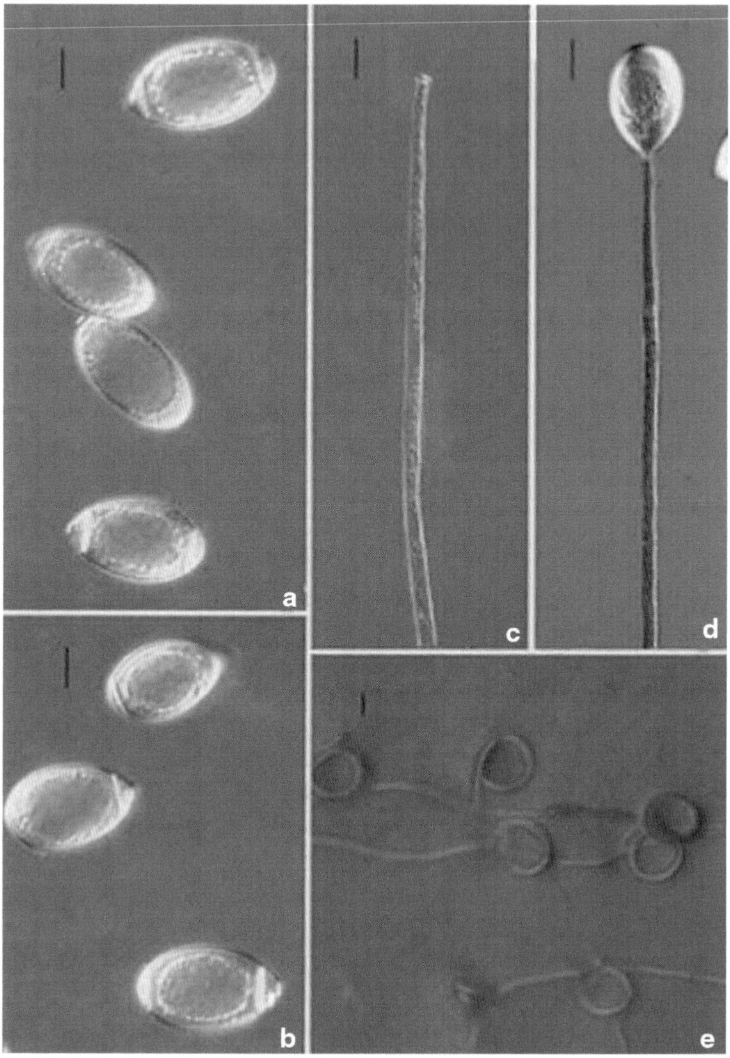

Fig. 3.92 *Drechslerella heterospora*. **a–b** conidia; **c–d** conidiophore; **e** constricting rings. Bars = 10 μm; Strain number: YMF1.00550

long, 5–7.5 μm wide at the base, gradually tapering upwards to a width of 2–4.5 μm at the apex, bearing one single conidium. Conidia hyaline, spindle-shaped, 42.5–62.5 (47) × 15–22.5 (16.9) μm, 1–4-septate, mostly 3-septate. Chlamydospores not observed. Capturing nematodes by means of constricting rings.

Distribution: China (Yunnan), Poland

Material examined: YMF1.00585, isolated from soil in Lushui, Yunnan in September 2002 by Jing Zhang. Permanent slide: NJ5–43–3

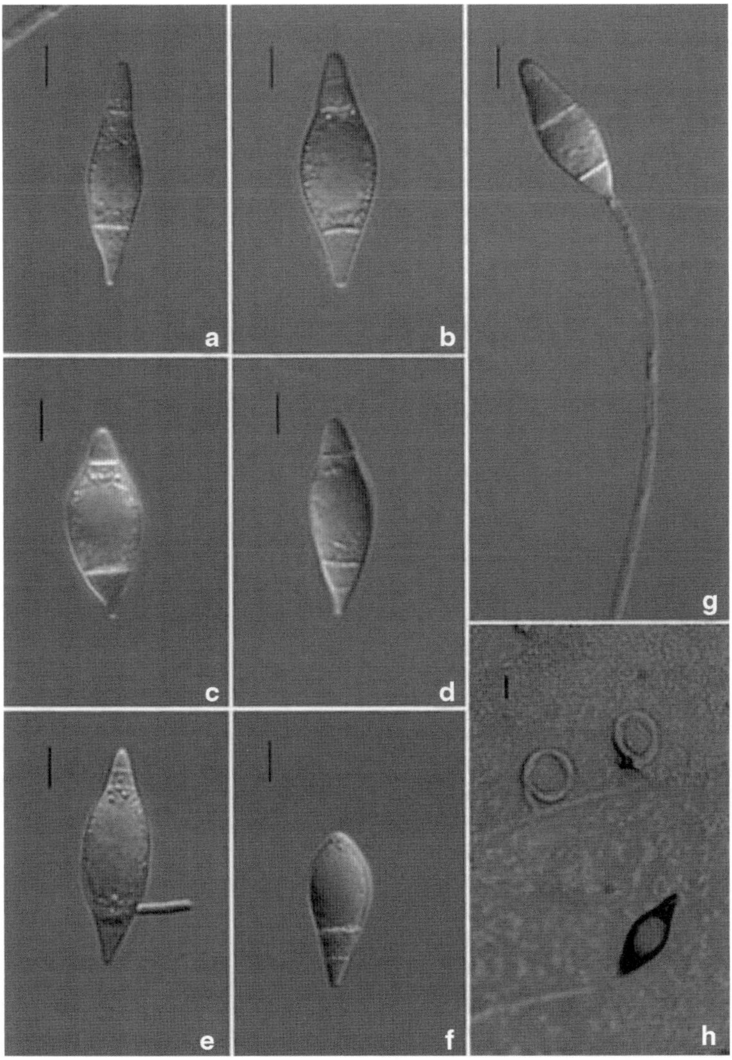

Fig. 3.93 *Drechslerella inquisitor.* **a–f** conidia; **g** conidiophore; **h** constricting rings. Bars = 10 μm; Strain number: YMF1.00585

Notes: *Drechslerella inquisitor* resembles *Dr. bembicodes* in having spindle-shaped and mostly 3 septate conidia. However, the conidia of *Dr. inquisitor* are much longer and narrower than those of *Dr. bembicodes*. Our isolate is smaller than the original description (Jarowaja 1971) [50–65 (55) × 18–24 (19) μm], and occasionally formed 1–2 short branches near the apex of the conidiophores. (Fig. 3.93)

Drechslerella polybrocha (Drechsler) M. Scholler, Hagedorn & A. Rubner, Sydowia 51 (1): 100 (1999)
= *Dactylella polybrocha* (Drechsler) Mekht., [as'*polybrochum*']Dokl. Akad. Nauk Aerb. SSR, 20: 70 (1964)

Fig. 3.94 *Drechslerella polybrocha*. **a**. conidia; **b**. constricting ring; **c**. conidiophore. Bars = 25 μm

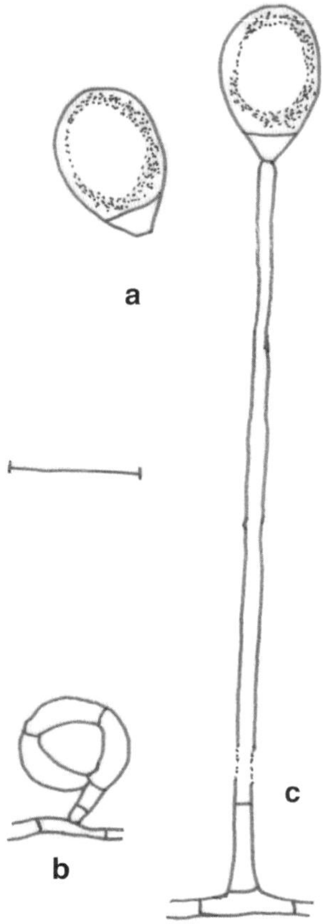

= *Golovinia polybrocha* (Drechsler) Mekht., Khishchnye, Nematofagovye Griby Gifomitsety, 137 (1979)
≡ *Monacrosporium polybrochum* (Drechsler) Subram., Kavaka 5: 96 (1977)[1978]
= *Orbilia tenebricosa* (Svrček) Baral, in Yu, Zhang, Qiao, Baral, Weber & Zhang, Mycotaxon 96: 167 (2006)
= *Patinella tenebricosa* Svrček, česká Mykol. 31 (3): 135 (1977)
= *Trichothecium polybrochum* Drechsler, Mycologia 29 (4): 536 (1937)

Characteristics: Mycelium spreading, vegetative hyphae hyaline, septate, 2 to 4.5 μm wide; conidiophores hyaline, septate, erect, 275–400 μm long, 7 μm wide at the base, gradually tapering upwards to a width of 3 μm at the apex, bearing one single conidium. Conidia hyaline, broadly obovoid, 35 × 24 μm, divided by a single septum into a small obconical basal cell about 6.5 μm long, and a much larger distal cell, about 28.5 μm long, often enveloped in a hyaline layer of mucus 2 to 3 μm thick. Capturing nematodes by means of constricting rings.

Distribution: USA

Notes: The description is based on the protologue. This species differs from other species in having 1 septate, broadly obovoid conidia, which is often enveloped in a hyaline layer of mucus when attached on conidiophores. (Fig. 3.94)

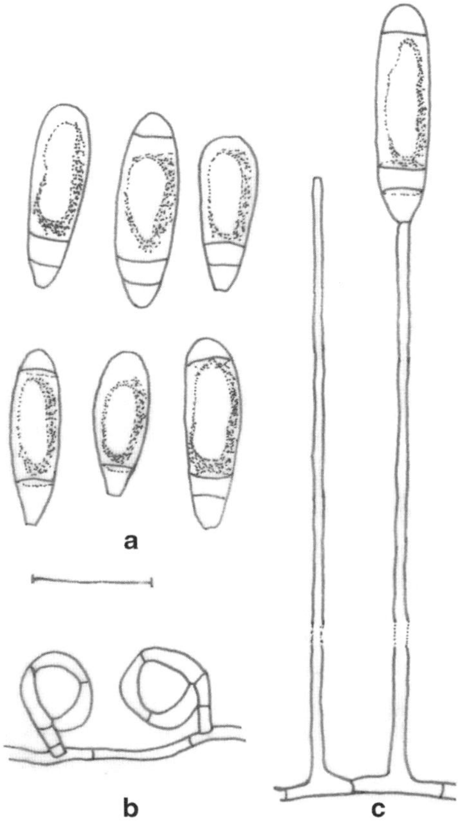

Fig. 3.95 *Drechslerella stenobrocha*. **a**. conidia; **b**. constricting rings; **c**. conidiophore. Bars = 30 μm

Drechslerella stenobrocha (Drechsler) M. Scholler, Hagedorn & A. Rubner, Sydowia 51 (1): 100 (1999)
≡ *Dactylella stenobrocha* Drechsler, Mycologia 42: 10 (1950)

Characteristics: Mycelium scanty; vegetative hyphae hyaline, septate, branched 1.7–3.7 μm wide. Conidiophores hyaline, erect, 425–550 μm long, 4.5–6.5 μm wide at the base, gradually tapering upwards to a width of 2.5–3 μm at the apex, bearing a single conidium. Conidia hyaline, elongate ellipsoidal or broadly finger-shaped or sometimes slightly clavate, straight or slightly curved, somewhat rounded to truncate at the narrowed base, broadly rounded at the apex, 34–56.5 × 12.5–16.5 μm, 1–3-septate. Chlamydospores not present. Capturing nematodes by means of constricting rings.

Distribution: China (Hebei), USA (Wyoming)

Notes: The description is based on the protologue. This species can be distinguished from other species of *Drechslerella* by its ellipsoid or wide digitiform conidia. (Fig. 3.95)

Drechslerella yunnanensis Z.F. Yu & K.Q. Zhang, Mycotaxon 110: 254 (2009)

Characteristics: Colonies white, growing slowly on CMA medium, attaining less than 25 mm diam. in 10 days at 25 °C. Vegetative hyphae hyaline, septate, 3.5–4 µm wide, aerial mycelium sparse, hyaline, septate, branched, 2.5–4 µm wide. Conidiophores hyaline, septate, erect, unbranched or occasionally branched below, 60–100 (–220) µm, 3.8–4.2 µm wide at the base, tapering gradually upwards to 1.5–2 µm near the apex, with 2–7 (–10) denticles of 2.3–4.2 µm long, each bearing one conidium in a capitate or racemose arrangement. Conidia hyaline, straight, elongate ellipsoidal, rounded at the apex, with a small truncate protuberance at the base, 7.8–12.9 (–17.8) × 3.3–4.2 µm, (0–)1-septate, proportion of aseptate conidia 17%. Nematodes are captured by three-celled constricting rings. In the non-constricted state, the outer diameter is 18–21.9 µm, the inner diameter is 12–13.6 µm.

Sexual state: *Orbilia* cf. *orientalis.*

Distribution: China (Yunnan)

Material examined: YMF1.01863, isolated from *Orbilia* cf. *orientalis*, collected in Yimen, Yunnan in August 2006 by Z. F. Yu.

Notes: Within species of *Drechslerella* where conidia are arranged in a cluster at the apex of conidiophores, *Dr. yunnanensis* can be distinguished from the other three species by the size and shape of conidia. Details are as follows: *Dr. brochopaga* has much larger conidia (26–46 × 5–9 µm) with more (mainly 3) septa (Drechsler 1937), *Dr. anchonia* has obovoid, much larger conidia 29–43 (–35) × 15–19 (–16.8) µm with a single septum in the lower third (Drechsler 1954). Conidia of *Dr. dactyloides* are narrowly ellipsoidal or somewhat digitiform, often very slightly curved, and there are more conidia in the apex of conidiophores and conidia are much larger 32–48 × 7–9.5 µm (Drechsler 1937). (Fig. 3.96)

Asexual State-Sexual State Connections of Nematode-Trapping Fungi

Fungi are classified primarily based on the structures associated with sexual reproduction, which tend to be evolutionarily conserved. However, many taxa reproduce only asexually, which cannot easily be placed in a classification based on sexual characters, while some produce both asexual and sexual states. The International Code of Botanical Nomenclature (Vienna Code) (McNeill et al., 2006) previously permitted mycologists to give asexually reproducing fungi (asexual states) separate names from their sexual states (sexual states). When names are available for both the asexual and sexual states of the same taxon, the holomorph usually took the sexual state name. However, where several asexual states were associated with the same sexual morph, this nomenclature could not be followed. Although dual

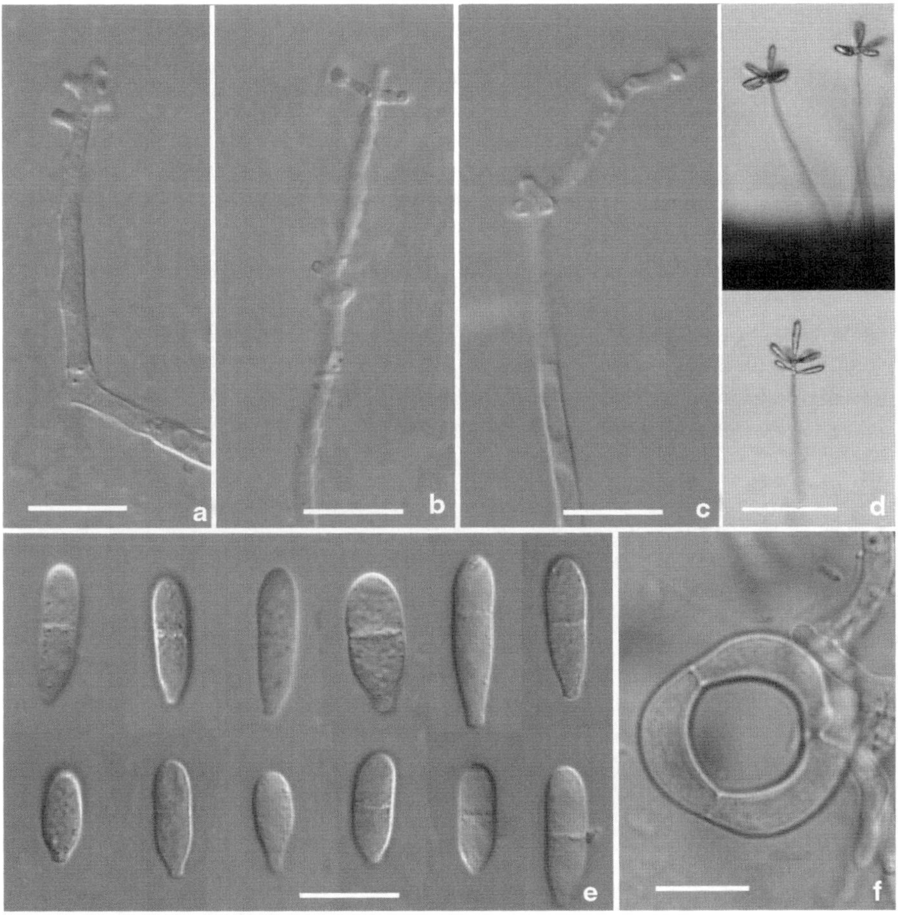

Fig. 3.96 *Drechslerella yunnanensis*. **a–d** conidiophore; **e** conidia; **f** constricting rings. Bars: **a–c**, **e–f** = 10 μm, **d** = 50 μm; Strain number: YMF 1.01863

nomenclature has persisted for decades, the concept of permitting separate names for asexual states of fungi with a pleomorphic life-cycle has been an issue of debate since the phenomenon was first recognized in the mid-19th century. More recently, it was deemed that one fungus can only have one name and this is now being put into practice (Taylor, 2011). After 1 January 2013, one taxon can only have one name, the system of permitting separate names to be used for sexual and asexual states is no longer permissible (Hawksworth, 2011). The topic presented here focuses only on the asexual—sexual state connections, but not with name validity, as the latter is still being debated.

Drechsler (1937) was possibly the first to report an asexual state connected to a sexual state in nematode-trapping fungi. He obtained apothecia of an *Orbilia*-

like species in a contaminated culture of *Arthrobotrys superba* when challenged with nematodes. The apothecia were flesh-coloured, 0.5-0.8 μm diameter, somewhat stalked and with a prominent marginal border and the cylindrical asci were 29-32 × 3 μm. Ascospores were 5 × 1.3 μm, hyaline and tear-shaped. This sexual state was similar to *Orbilia fimicola* Jeng & Krug in morphology. Zachariah (1983) obtained *Orbilia*-like apothecia when crossing an auxotrophic strain (D30A) of *Drechslerella dactyloides* with a prototrophic strain (D31P). However, the apothecia were sterile; no asci were formed. Since that time, at least 14 nematode-trapping asexual species have had credible connections established with *Orbilia* species and one with *Patinella tenebricosa* (Table 3.1). In the following, detailed descriptions of 12 asexual nematode-trapping fungi states connected to a sexual state are presented and most illustrated. Two of the connections identified by Pfister (1997), i.e. *Drechslerella doedycoides*—*Orbilaria* sp., and *Drechslerella polybrochum*—*Patinella tenebricosa*, lack taxonomic descriptions.

Arthrobotrys cladodes Drechsler var. macroides Drechsler
Sexual morph: *Orbilia auricolor* (A. Bloxam ex Berk.) Sacc.
Reference: Pfister and Liftik (1995)

Asexual morph (Culture No. 90): Cultures of germinated ascospores on MEYE (Lilly and Barnett, 1951), white, becoming pale tan, forming hyphal strands on the surface and cottony patches, remaining white. On CMA growth sparse, remaining white. Conidiophores erect, branching, not proliferating; each branch terminating in a loose whorl of conidia. Conidia blastic, hologenous, 18–20 × 6–8 μm, elongate-ellipsoid, with a single septum, the cells nearly equal, without constrictions at the septa. Conidia basipetally produced at several closely spaced loci on a sometimes inflated, often distorted, terminal area of the conidiogenous cell; secession schizolytic, leaving a slight protuberant, unthickened, open denticle with a slight frill. Conidiogenous cell determinant, persistent. Cultures induced by nematodes (*Cephalobus* sp.) normally producing adhesive networks within 12 h.

Sexual morph (Specimen No. 90): Ascomatal specimens were collected by Pfister and Liftik in July, 1994, on decorticated wood from a swampy area, Purgatory, Tract, Westwood, Massachusetts. Ascomata up to 1 mm diameter, yellow to orange, pulvinate to turbinate; asci 8-spored, 30–40 × 3–4 μm, cylindric, J-, tapered towards the base and often forked, at the apex truncate without an obvious pore; ascospores curved, narrowly clavate, broad and rounded at one end, narrowing to an acute point at the other end, non-septate containing a single inclusion which stains intensely in cresyl blue, 8–12.5 × 0.9–1.5 μm; paraphyses branching below, abruptly swelling to become capitate above reaching a diameter of 3-3.5 μm; excipulum of angular cells.

Arthrobotrys cystosporia
Sexual morph: *Orbilia auricolor* (A. Bloxam ex Berk.) Sacc.

Asexual morph (Culture No. SBP88, Fig. 3.97): Colonies colourless, spreading on CMA, reaching 5 cm at 25 °C after 10 days. Vegetative hyphae hyaline, septate,

Table 3.1 Asexual nematode-trapping fungi in *Orbiliaceae* connected to a sexual state

Asexual states	Sexual state	Trapping device	References
Arthrobotrys cladodes Drechsler var. *macroides*	*Orbilia auricolor*	Adhesive networks	Pfister and Liftik, Mycologia,1995, 87 (5): 684–688
A. cystosporia	*O. auricolor*	Adhesive networks	Su et al., Mycotaxon, 2013
A. oligospora Fresen. var. oligospora	*O. auricolor*	Adhesive networks	Pfister and Liftik, Mycologia,1995, 87 (5): 684–688
A. psychrophilum	*O. auricolor*	Adhesive networks	Rubner 1996, Studies in Mycology, 39: 1–134
A. yunnanensis	*O. auricolor*	Adhesive networks	Mo et al., Fungal Diversity, 18: 107–115
A. superba	*O. fimicola*	Adhesive networks	Pfister, Mycologia, 1994, 86 (3): 451–453
A. nonseptata	*Orbilia* sp.	Adhesive networks	Li et al., Mycotaxon, 2009, 109: 247–254
A. vermicola	*O. blumenaviensis*	Adhesive networks	Qiao et al., Mycological Progress 2011, DOI 10.1007/s11557-011-0744-3
Dactylellina parvicolla	*O. cunninghamii*	Adhesive knobs	Liu et al., 2002, The 3rd Asia-Pacific Mycological Congress on Biodiversity and Biotechnology, pp. 75 (abstract).
Da. quercus	*O. quercus*	Adhesive knobs	Liu et al., FEMS Microbiology Letters, 2005a, 245: 99–105
Drechslerella brochopaga	*O. orientalis*	Constricting rings	Yu et al., Mycotaxon, 2006, 96: 163–168
Dr. doedycoides	*Orbilia* sp.	Constricting rings	Pfister, Mycologia, 1997, 89 (1): 1–23
Dr. polybrochum	*Patinella tenebricosa*	Constricting rings	Pfister, Mycologia, 1997, 89 (1): 1–23
Dr. yunnanensis	*O.* cf. *orientalis*	constricting rings	Yu et al., Mycotaxon, 2009a, 110: 253–259

3.2–5.4 μm wide. Aerial mycelium sparse. Conidiophores hyaline, erect, simple, septate, 60–200 μm high, 2.5–5.5 μm wide at the base, gradually tapering upwards to a width of 1.8–2.5 μm at the apex. Conidia produced at the apex, sometimes in a cluster of 2–5 conidia. Primary conidia hyaline, broadly clavate or broadly turbinate to obovoid, 26–38.5 (31.4) × 13–22 (20) μm, with 2–3 septa. The second conidiophores from the primary conidia colourless, erect, simple, septate, 35–140 μm high, 2–4.5 μm wide at the base and 1.8–2.5 μm at the apex, producing 3–7 secondary conidia on conspicuous denticles loci at near the apex of conidiophores. The secondary conidia hyaline, elongate ellipsoid-cylindrical or slightly clavate, broadly rounded at the apex, rounded truncate at the narrowed base, 26–38.5 (31.4) × 13–22 (20) μm, 0–1-septate. Chlamydospores ellipsoidal. Trapping nematodes by three-dimensional adhesive networks. (Fig. 3.97)

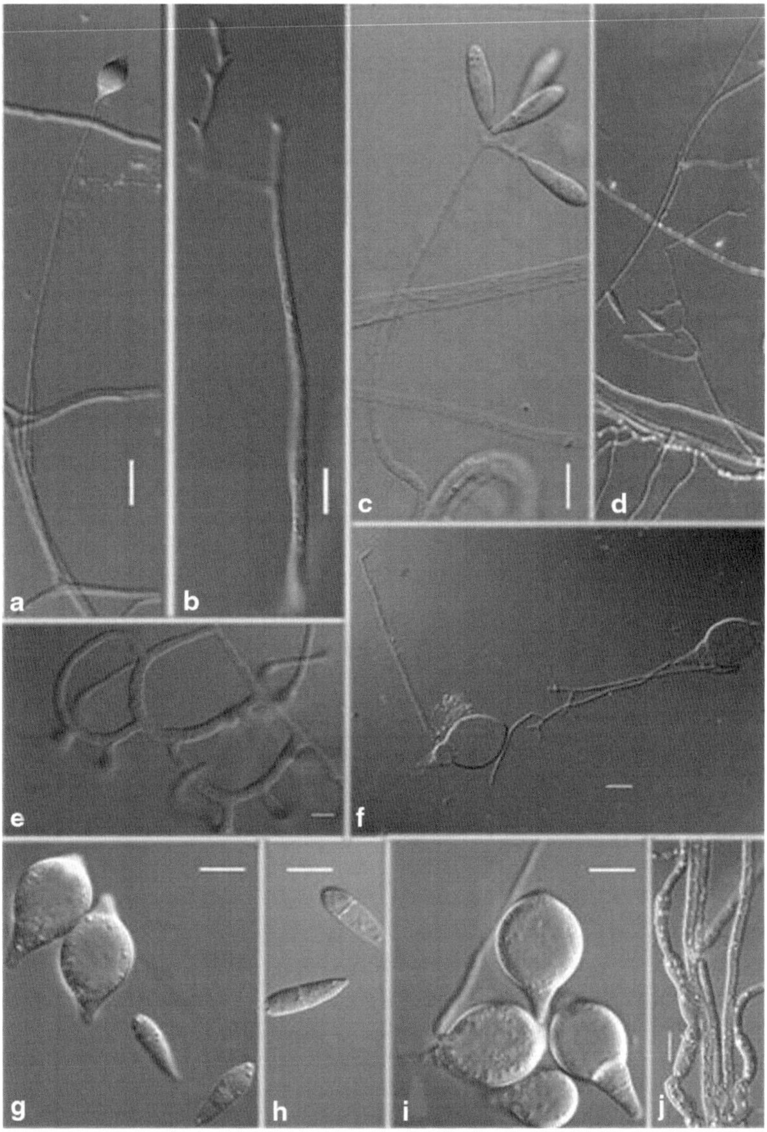

Fig. 3.97 *Arthrobotrys cystosporia* **a–b**, **d** conidiophores; **c**, **h** secondary conidia; **e** adhesive networks; **g–i** conidia; **j** chlamydospores. Bars: **a**, **d**, **f** = 20 μm; **b**, **c**, **e**, **g–i** = 10 μm. Strain number: SBP88

Sexual morph (Specimen No. SBP88, Fig. 3.98): Soils samples were collected from Binchuan County, Yunnan, China on March 21, 2008 by HongYan Su. Subsamples of 2–5 g were spread on CMA plates and stored at 28°C. After incubation for 10 days, apothecia of an *Orbilia* sp. were observed on the soil granules, and later also on other areas of the plates. Apothecia superficial, sessile, pale cream. Disc 0.4–0.9 mm diameter, smooth, plane, margin even. Ectal excipulum composed from base to margin of globose or subglobose cells, 5–12 μm diameter, with thin or slightly thickened walls. Asci cylindric-clavate, tapered and often forked at the base, apex rounded or truncate-rounded, 8-spored, 35.5–39 (37) × 3.5–4.7 (4.1) μm in living state. Ascospores banana-shaped or narrowly clavate, medium curved, one end narrower, broadest above and slightly trapped to the rounded at proximal end, non-septate, containing an elongate tear-shaped spore body at the broader end, 7.5–11 (8.6) × 0.9–1.2 μm. Paraphyses hyaline, cylindric-clavate, often branched below, septate in the lower part, 1.4–1.8 (1.6) μm diameter, slightly expanded to 2.3–3 μm at the apex, which is not encrusted. (Fig. 3.98)

Arthrobotrys oligospora Fresen. var. *oligospora*
Sexual morph: *Orbilia auricolor* (A. Bloxam ex Berk.) Sacc.
Reference: Pfister and Liftik (1995)

Asexual morph (Culture Nos. 45 and 55): Cultures of germinated ascospores on MEYE at first white, becoming rose within 1 wk, growth partially cottony with hyphal strands forming across the surface of the medium. On CMA growth is uniform but sparse, remaining white. Conidiophores erect, with whorls of conidia. Conidia blastic, hologenous, 14–28 × 10–14 μm, obovoid, 1-septate, often constricted at the septum, the proximal cell smaller than the distal cell. Conidia basipetally produced, at several closely spaced loci, on somewhat inflated areas of the conidiogenous cell; secession is schizolytic, leaving protuberant, open, unthickened denticles with frills. Conidiogenous cell are indeterminate, persistent, with hologenous sympodial, presumptive proliferation. Cultures with nematodes (*Cephalobus* sp.) produced the trap device of adhesive networks within 12 h.

Sexual morph (Specimen Nos. 45 and 55): Ascomatal specimens were obtained from a crustose lichen on a rock which had been incubated in a moist chamber. Taxonomic characteristics of specimen Nos. 45 and 55 were similar as that of specimen No. 90 according to description of Pfister and Liftik (1995).

Arthrobotrys psychrophilum
Sexual morph: *Orbilia auricolor* (A. Bloxam ex Berk.) Sacc.
Reference: Rubner (1996)

Asexual morph (Culture Nos. CBS 319.94): The strain was isolated form soil collected from Ecuador (Province Cotopaxi, north of Latacunga, hacienda "La Cienega"). Cultures grew well on CMA medium. Macroconidia were formed after one week on mostly geniculate conidiophores, microconidia appeared later, especially from germinating macroconidia (and also on the fruit bodies). Conidiophores of macroconidia simple, 250–310 μm high, bearing up to 6 conidia. Macroconidia spindle-shaped and mostly 3-septate, with a ratio of length to width of 2.48, 47–52–59.5 × 17.5–21–22.5 μm. Microconidia ellipsoidal to obovoidal and mostly 1-septate, 15-18-22 × 5-6.5-7.5 μm. Chlamydospores in chains, thick-walled, smooth and

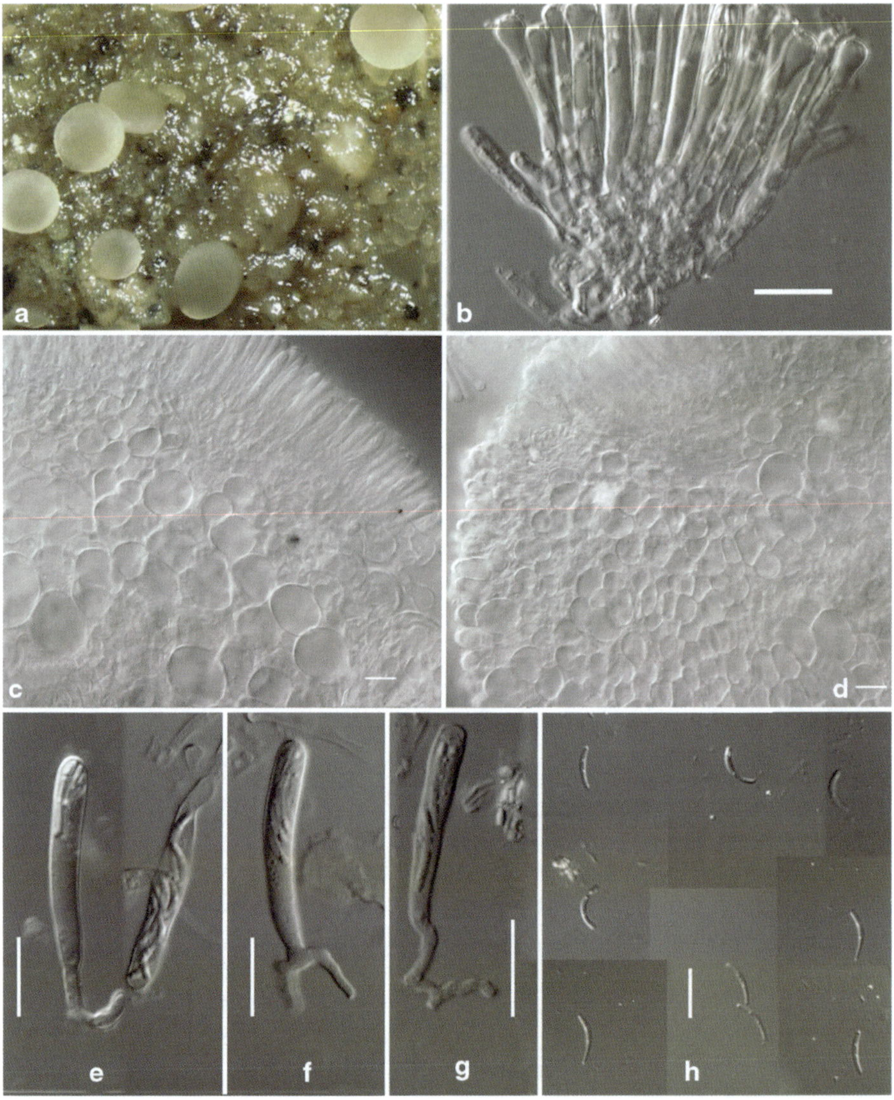

Fig. 3.98 *Orbilia auricolor* **a** fresh apothecia; **b** Paraphyses; **c–d** cell morphology of ectal excipulum; **e–g** asci; **h** alive ascospores. Bars = 10 μm. Specimen number: SBP88

yellow. When adding the nematode species *Turbatrix aceti*, the fungus produced adhesive networks.

Sexual morph (Specimen No. A.R.9312): Culture of CBS 319.94 on CMA medium amended with oat husks produced apothecia after two months. Mature apothecia deviating from normal *Orbilia* apothecia by strongly convex warted hymenia. Asci 45–58 × 3.5–4.2 μm, ascospores 7.5–13 × 1–1.3 μm. An ellipsoid refractive spore body is attached to the spore apex by a thin thread and 3–5 of the lower spores are inversely oriented within asci. Paraphyses slightly inflated at apex. Cortical cells of ectal excipulum terminated by 2–6 μm high glassy caps.

Arthrobotrys superba Corda
Sexual morph: *Orbilia fimicola* Jeng & Krug.
Reference: Pfister (1994)

Deer dung was collected from Berkshire County in western Massachusetts and was placed in a moist chamber. After $2^1/_2$ mouths at room temperature and in natural light, the ascomata discovered on dung was used to project ascospores and obtain the asexual state culture. Pfister (1994) confirmed that taxonomic morphology of the sexual state agreed with the original description of *Orbilia fimicola* (Jeng and Krug, 1977) and he identified the asexual state as *Arthrobotrys superba* (Drechsler, 1937), but morphological details on both sexual state and asexual state had not been described.

Arthrobotrys yunnanensis M.H. Mo & K.Q. Zhang
Sexual morph: *Orbilia auricolor* (A. Bloxam ex Berk.) Sacc.
Reference: Mo et al. (2005)

Asexual morph (Culture No. YMF1.00593, HT1.00593, Fig. 3.99): Colony derived from single ascospore, whitish and slow growing on CMA. Mycelium spreading, scanty; vegetative hyphae hyaline, septate, branched, except for occasional storage filaments that densely filled with protoplasm and up to 11 μm wide measuring mostly 2–7 μm in diameter. Conidiophores hyaline, erect, septate, more or less branched, 68–236 μm long, 2.5–5 μm wide at the base, tapering gradually upwards to a width of 2–3 μm below the irregularly expanded, globose or somewhat coralloid apex whereon are borne 10–20 conidia in usually dense capitate arrangement. Conidia hyaline, ellipsoid or elongate obovoid, mostly 10–20 (17.5) × 5–8 (6.2) μm, 1–septate approximately near in the centre of the spore. Chlamydospores not present. Cultures on CMA produce adhesive networks when nematodes were added (*Panagrellus redivivus*). (Fig. 3.99)

Sexual morph (Specimen No. Mo MH O 002, Fig. 3.100): Wet soil samples were from Mt Xiaobailong, Yiliang, Yunnan, China collected by MingHe Mo on 15 August 2003. Subsamples of 2–5 g were spread on CMA plates and stored at room temperature (about 20–28°C). After incubation for 20 days, apothecia were observed on the soil granules, and later also on other areas of the plates. Apothecia superficial, sessile, pale cream. Disc 0.3–0.8 mm diameter, smooth, plane, margin even. Ectal excipulum composed from base to margin of globose or subglobose cells, 6–12 μm diameter, with thin or slightly thickened walls. Asci cylindric-clavate, tapered and often forked at the base, apex rounded or truncate-rounded, 8-spored, 30–45 × 3.5–5 μm in living state. Ascospores banana-shaped or narrowly clavate, medium curved, one end narrower, broadest above and slightly trapped to the rounded proximal end, non-septate, containing an elongate tear-shaped spore body

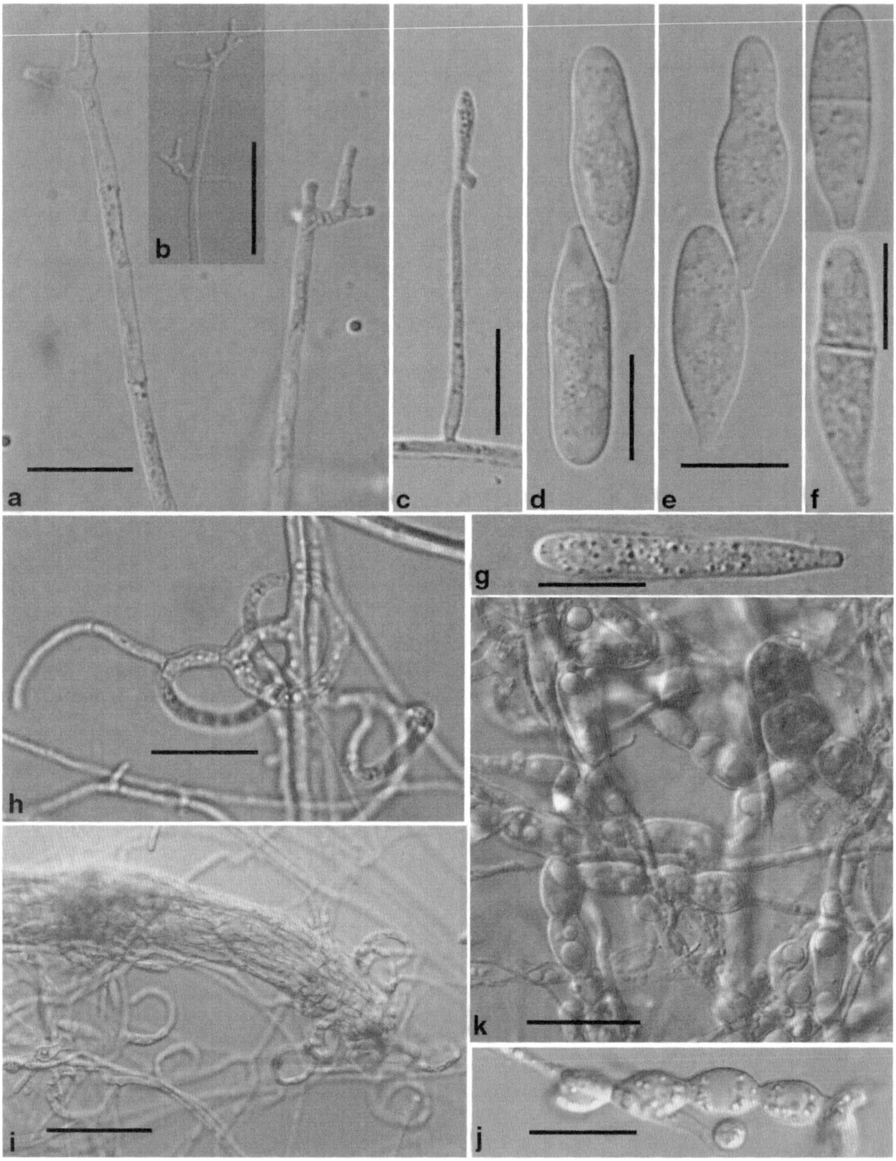

Fig. 3.99 *Arthrobotrys yunnanensis* **a**–**c** conidiophores with short denticles; **c** an immature conidium attached to a conidiophore; **d**–**f** elongate-ellipsoid-cylindrical or slightly clavate conidia; **g** adhesive networks; **h** trapped nematode in adhesive networks; **i**–**j** sphaerical to ellipsoidal chlamydospores. Bars: **a**, **d**–**g**, **i**–**j** = 10 μm; **b**, **c**, **h** = 20 μm. Strain number: YMF1.00593

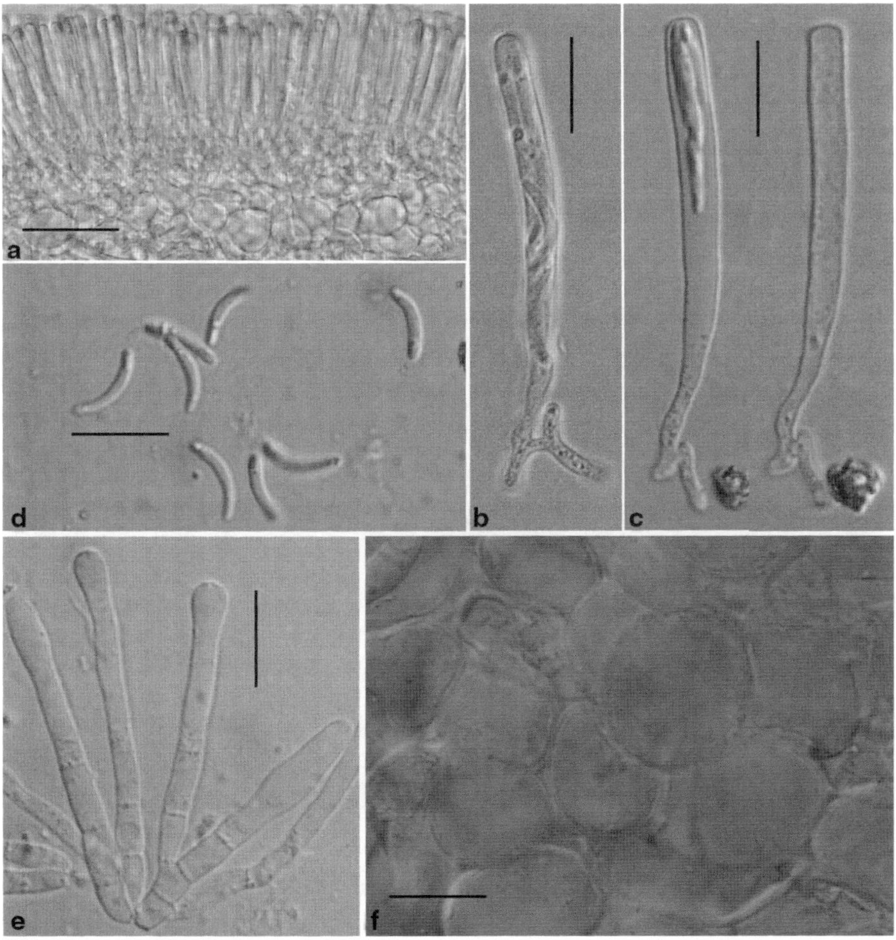

Fig. 3.100 *Orbilia auricolor* **a** cluster of asci and paraphyses; **b** eight living ascospores projected from an ascus; **c** ascospores within the dead ascus; **d** a turgescent living ascus (left) and an empty ascus (right) after projection of spores; **e** paraphyses; **f** globose to subglobose ectal excipulum cells. Bars: **a** = 40 μm; **b–d** = 10 μm; **e–f** = 7 μm. Specimen number: Mo MH O 002

at the broader end, 9 × 1.4 μm when shooting from asci and still in living state. Paraphyses hyaline, cylindric-clavate, often branched below, septate in the lower part, 1.5–2 μm diameter, slightly expanded to 2.5–3.5 μm at the apex, not encrusted at apex. (Fig. 3.100)

Arthrobotrys nonseptata Z.F. Yu, S.F. Li & K.Q. Zhang
Sexual morph: *Orbilia* sp.
Reference: Li et al. (2009)

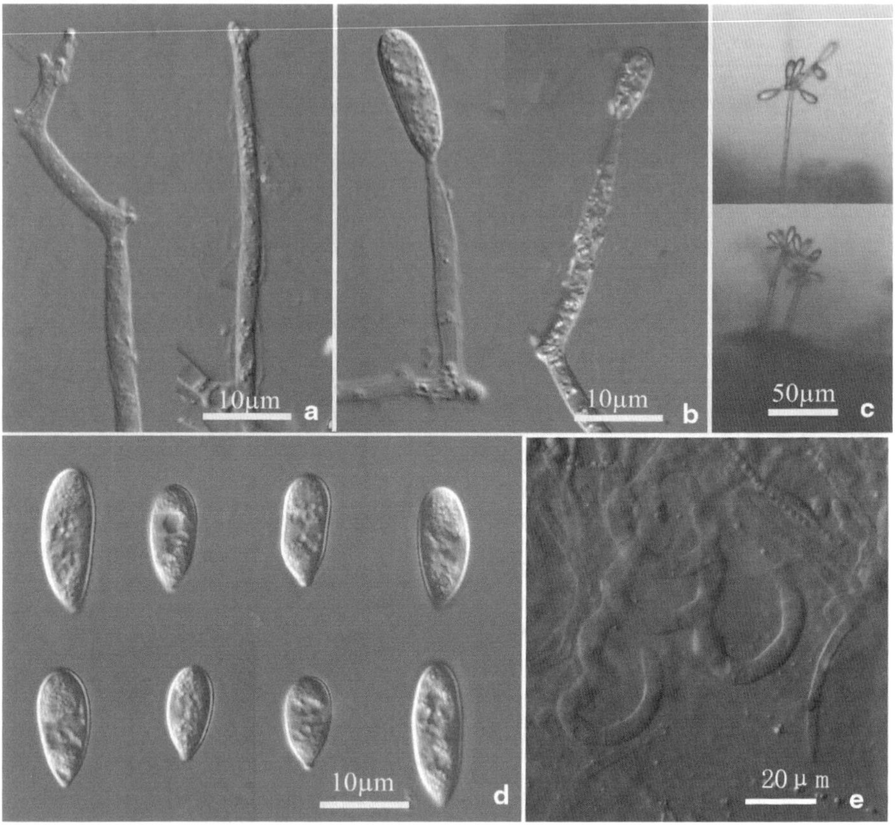

Fig. 3.101 *Arthrobotrys nonseptata* **a** conidiophores with short denticles; **b** conidium attached to a conidiophore; **c** conidiophores bearing conidia in clusters; **d** conidia; **e** adhesive networks. Strain number: YMF1.01852

Asexual morph (Culture No. YMFT 1.01852, Fig. 3.101): Cultures were derived ascospores. Colonies slow-growing on CMA medium, attaining less than 35 mm diameter in 10 days at 25 °C. Vegetative hyphae hyaline, septate, 3.5–4 μm wide, aerial mycelium sparse, hyaline, septate, branched, 2.5–4 μm wide. Conidiophores erect, septate, unbranched, 40–120 μm high, 2–4 μm wide in the lower part, 1.5–2 μm wide at the apex, producing 3–10 conidia from retrogressive conidiogenous loci on conspicuous denticles at and near the apex. Conidia hyaline, non-septate, 11–16.8 × 5–6.6 μm, elongate ellipsoid, constricted at the base by forming a small truncate protuberance. Cultures on CMA produce adhesive networks when nematodes added (*Panagrellus redivivus*). (Fig. 3.101)

Sexual morph (Specimen No. YMFT 1.01852, Fig. 3.102): Fresh fruit bodies were collected by Y. Zhang on 18 August 2006 from decaying bark of a broad-leaved tree in forest dominated by *Cyclobalanopsis glaucoides* Schottky and *Pinus armandii* Franch., in Dalongkou Park of Yimen County, Yunnan Province, China.

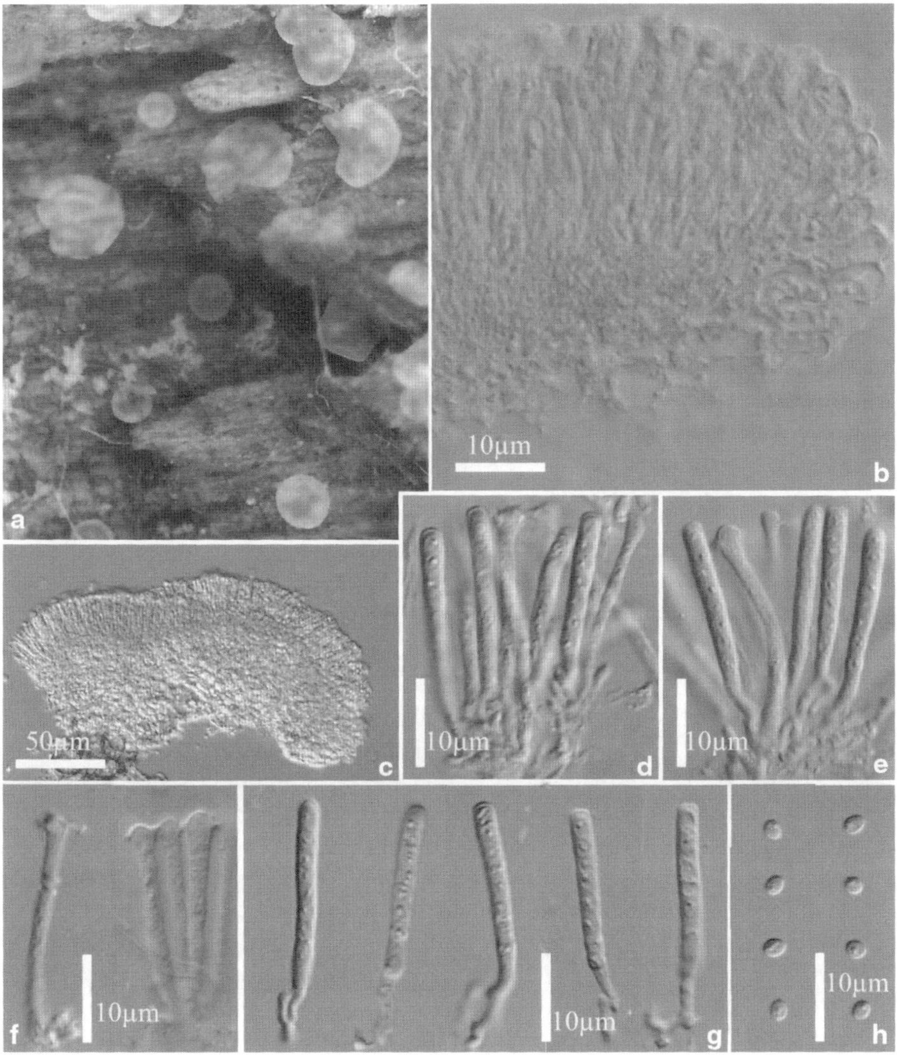

Fig. 3.102 *Orbilia* sp **a** apothecia; **b** cells of ectal excipulum; **c** vertical section of part apothecia; **d–e** cluster of dead asci and paraphyses with living spores; **f** paraphyses; **g** asci; **h** ascospores. Specimen number: YMFT 1.01852

The authors confirmed that their specimen belongs to an unidentified species of the genus *Orbilia* and presented a morphological plate, but did not provide a taxonomic description. (Fig. 3.102)

Arthrobotrys vermicola (R.C. Cooke & Satchuth.) Rifai
Sexual morph: *Orbilia blumenaviensis* (Henn.) Baral & E. Weber
Reference: Qiao et al. (2012)

Asexual morph (Culture No. YMF 1.03002, Fig. 3.103): Colonies growing rapidly on CMA medium, attaining 40 mm diameter in 6 days at 28 °C. Mycelium spreading, vegetative hyphae hyaline, septate and branched, mostly 3–6 μm wide. Conidiophores colourless, produced on the mycelium growing at the fringe of the plate, appressed or erect, branched, septate; 150–300 μm high when erect, often 700–1,500 μm when appressed, 4–8 μm wide at the base and 3.5–5 μm at the apex, often recommencing growth after the first group of conidia had been produced and a second head is then formed about 100 μm above the first. This process was repeated until 3–4 (–7) whorls of conidia were produced on a single conidiophore, each group with 2–8 conidia. Two types of colourless conidia were simultaneously formed which mainly differ in their length-width ratio but also in the number of septa: Type a with 1-septate, pear-shaped to obovoid, sometimes elongate ellipsoidal conidia being broadly rounded at the apex, rounded-truncate at the narrowed base, sometimes slightly gradually attenuated at the proximal end, (16.4–)20.4–28.2 × (9.8–)13.5–15.3 (–19) μm; Type b with 1–2 septa, elongate cylindric-clavate, obconical conidia being broadly rounded at the apex, (21.7–)30–34.5 (–39.2) × 12.7–14 μm. The proportion of conidial types a and b was 88 and 12 % respectively. Nematodes are trapped by three-dimensional adhesive networks. (Fig. 3.103)

Sexual morph (Specimen No. YMFT 1.03002, Fig. 3.104): Specimen was collected on decaying angiosperm wood fallen on the ground of a broad-leaved subtropical evergreen forest located in Xiushan Forest Park of Tonghai County, Yunnan Province, China, in July 2007, by S.F. Li and J.W. Guo. Apothecia 0.7–1.5 mm in diameter, superficial, with a distinct stalk up to 0.2–0.3 mm high, scattered to gregarious on decayed wood, smooth, margin even, yellow and translucent throughout when moist, pale brown when dry. Ectal excipulum composed of globose or subglobose cells, near base 20–43 × 15–38 μm, on flanks 15–30 × 12–25 μm. Asci cylindric-clavate, often forked at the base, apex medium truncate (rounded in side view), 8-spored, 28–39 (–45) × 2.6–3.7 (–4.4) μm in dead state. Ascospores falcate, medium to strongly curved, lower end distinctly narrowed, upper end only slightly so, non-septate, containing a spore body at the broader end, with 1–2 minute lipid bodies in the centre, (6.2–)8–11 (–12) × 1–1.2 μm when shot from asci and still in living state. Spore body rounded or elongate ellipsoid, 0.9–1.6 × 0.7–0.8 μm (1.5–2.5 μm including invisible connecting part). Paraphyses hyaline, cylindrical, with distinctly widened, lanceolate apex, septate below, 1.7–2.8 μm wide at widest point, not encrusted. (Fig. 3.104)

Dactylellina parvicolla
Sexual morph: *Orbilia cunninghamii*
Reference: Liu et al. (2002)

Asexual morph: Colonies reached to 25 mm in diameter on PDA after 10 days. Conidia sparse on the nutritionally rich media, spindle-shaped at maturity, 27.5–37.5 × 8–10 μm, mainly with 3–4 septa. Trapping nematodes by means of sessile knobs, occasionally the knob with a short stalk.

Sexual morph: Specimen was collected from Huai-rou County, Beijing, China on rotten wood of *Quercus* sp.. Apothecia 0.2–0.5 in diameter, asci 20–25 × 3–4 μm, sub-

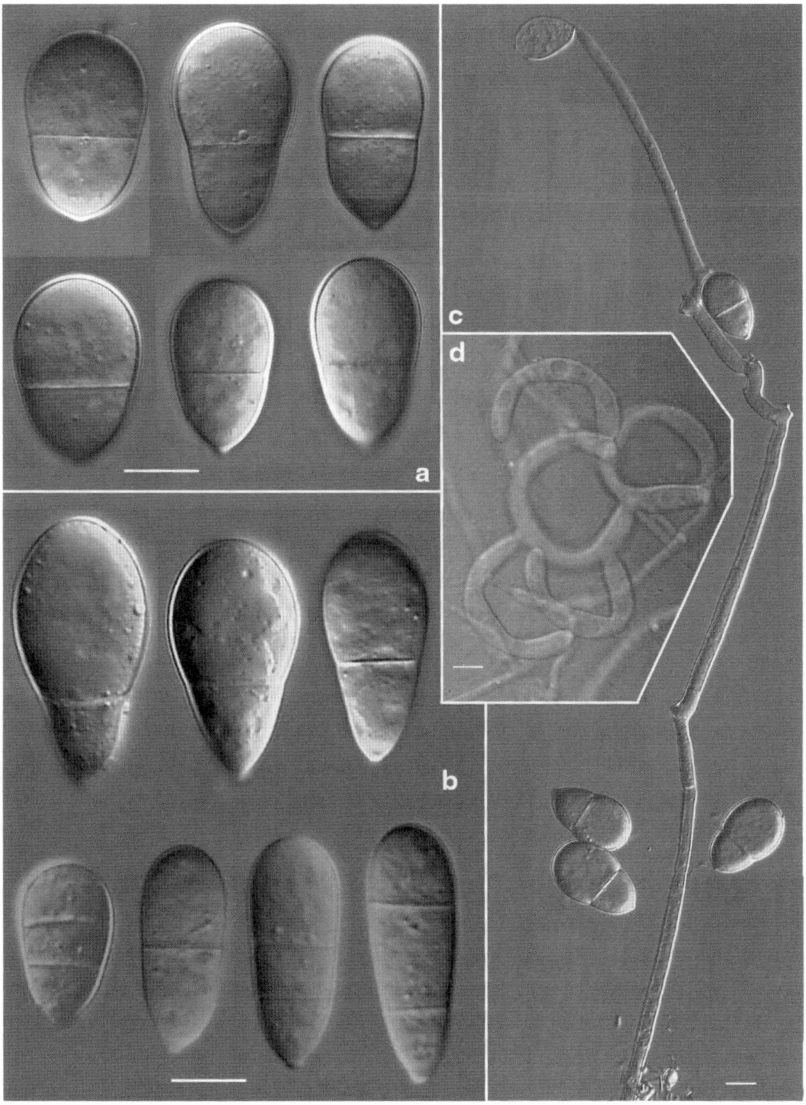

Fig. 3.103 *Arthrobotrys vermicola* **a–b** conidia; **c** conidiophore and conidia; **d** adhesive network. Bars = 10 μm. Strain number: YMF 1.03002

cylindrical to clavate, tapered and bifurcate at the base, ascospores 5–6 × 1–1.5 μm, rod-shaped to clavate.

Dactylellina quercus Bin Liu, Xing Z. Liu & W.Y. Zhuang
Sexual morph: *Orbilia quercus* Bin Liu, Xing Z. Liu & W.Y. Zhuang
Reference: Liu et al. (2005)

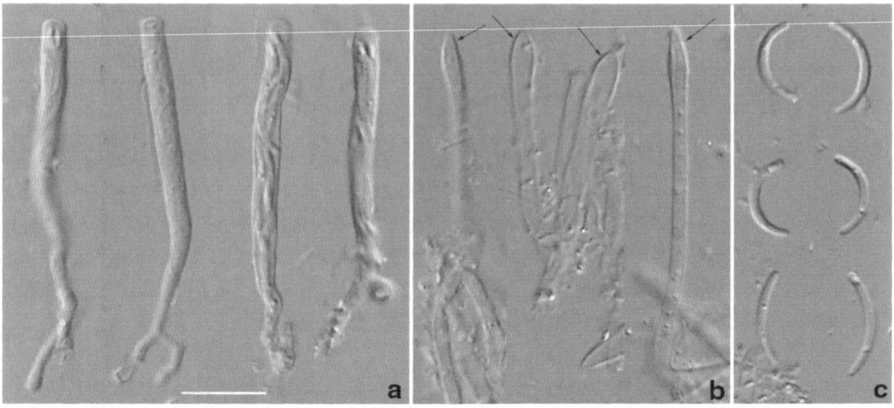

Fig. 3.104 *Orbilia blumenaviensis* **a** asci; **b** paraphyses; **c** ascospores. Bar = 10 μm. Specimen number: YMFT 1.03002

Asexual morph (Culture No. AS 3.6762, Fig. 3.105): Colonies colourless on PDA or CMA, reached to 60–65 mm in diameter on PDA and 45 mm on CMA after 20 days culture at 23–25°C. Aerial mycelium sparse, hyphae hyaline, septate, branched, 2.5–3.5 μm wide. Conidiophores mostly 130–180 μm high, 5–6.5 μm wide at the base, 1.5–2.5 μm at the apex, sometime branched near the apex, bearing a single conidium. Conidia were commonly spindle-shaped, slightly rounded at the distal end narrowly truncate at the base, 25–50 × 8–12 μm, with 3–5 and mainly 3-septa, but the central cell is not much larger than the others. Chlamydospores not observed. Trapping nematodes by means of stalked knobs (0–5.5 μm), knobs sphaerical to subsphaerical, 8–12 × 7.5–10 μm. Knobs produced frequently on the nutritional agar plates even without challenging with nematodes. (Fig. 3.105)

Sexual morph (Specimen No. HMAS 88781, Fig. 3.106): Specimen was collected from Labagoumen Forest Park of Huai-rou County, Beijing, China in July, 2002. Apothecia superficial on rotten wood, gregarious, sessile, 0.2–0.5 mm in diameter. Disc concave, smooth, translucent, whitish to pale yellow when fresh, brownish-yellow when dried. Ectal excipulum of *textura angularis*, 78–96 μm thick, cells isodiametric, 6–10 (–14) μm in diameter. Medullary excipulum of *textura intricata*, 27–35 μm thick. Subhymenium poorly developed. Asci cylindrical-clavate, narrower and tapered towards the base, sometime forked at the base, apex truncate to rounded, 18–30 × 2.5–3 μm. Ascospores hyaline, subcylindrical to cylindrical-clavate, straight or sometimes slightly curved, non-septate, usually overlapping tightly and biseriate within ascus, 5–6 × 0.8–1.2 μm in water, spore body (SB) tear-shaped to short rod-shaped, 1.1–1.8 × 0.7–1 μm. Paraphyses filiform with a clavate to capitate apex, hyaline, 18–22 μm long, 2–2.5 μm wide at apex and 1.5–2 μm wide below. (Fig. 3.106)

Drechslerella brochopaga (Drechsler) M. Scholler, Hagedorn & A. Rubner
Sexual morph: *Orbilia orientalis* (Raitv.) Baral
Reference: Yu et al. (2006)

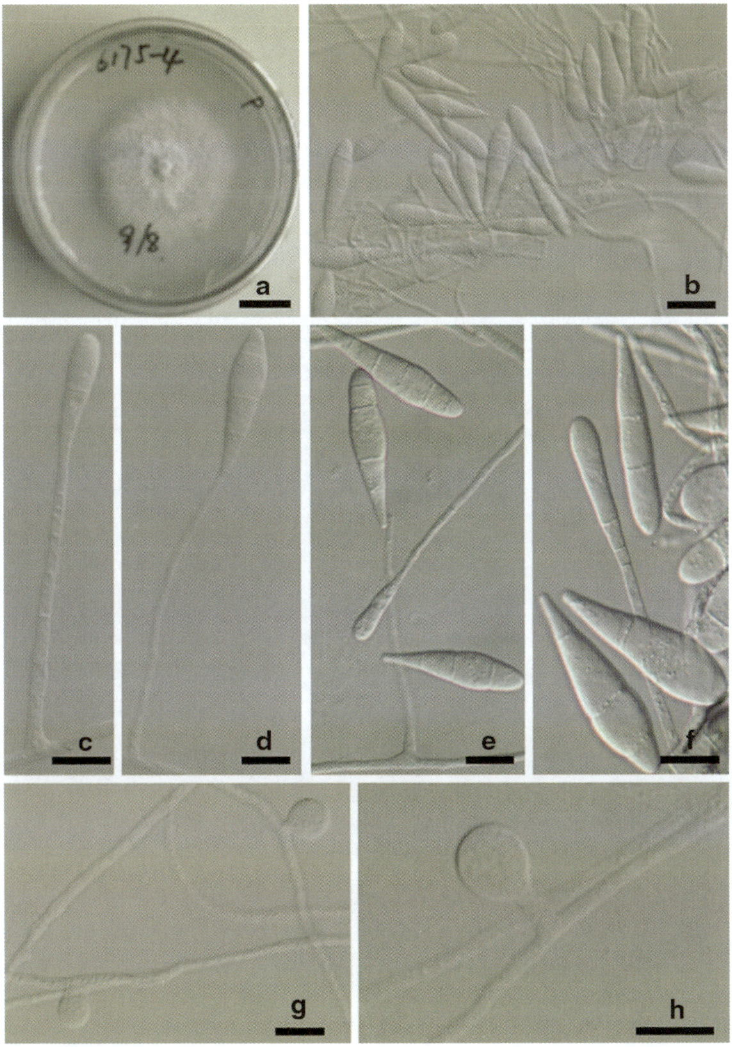

Fig. 3.105 *Dactylellina quercus* **a** colony on PDA; **b–f** conidiophores and conidia, note the single conidium produced on each apex of the conidiophores; **g–h** stalked knobs. Bars: **a** = 2 cm; **b** = 20 μm; **c–h** = 10 μm. Strain number: AS 3.6762

Asexual morph (Culture No. YMF1.01829, Fig. 3.107): Cultures were derived from ascospores. Colonies colourless, spreading on CMA, reaching 48 mm after 14 days at 25 °C. Vegetative hyphae hyaline, septate, 3.5–4 μm wide. Aerial mycelium sparse, hyaline, septate, branched, 2.5–4 μm wide. Conidiophores hyaline, erect, septate, mostly 380–430 μm high, 3.5–4 μm at the base, tapering upwards to a width of 2–2.5 μm at the apex, there bearing short blunt denticles 2–10 μm in

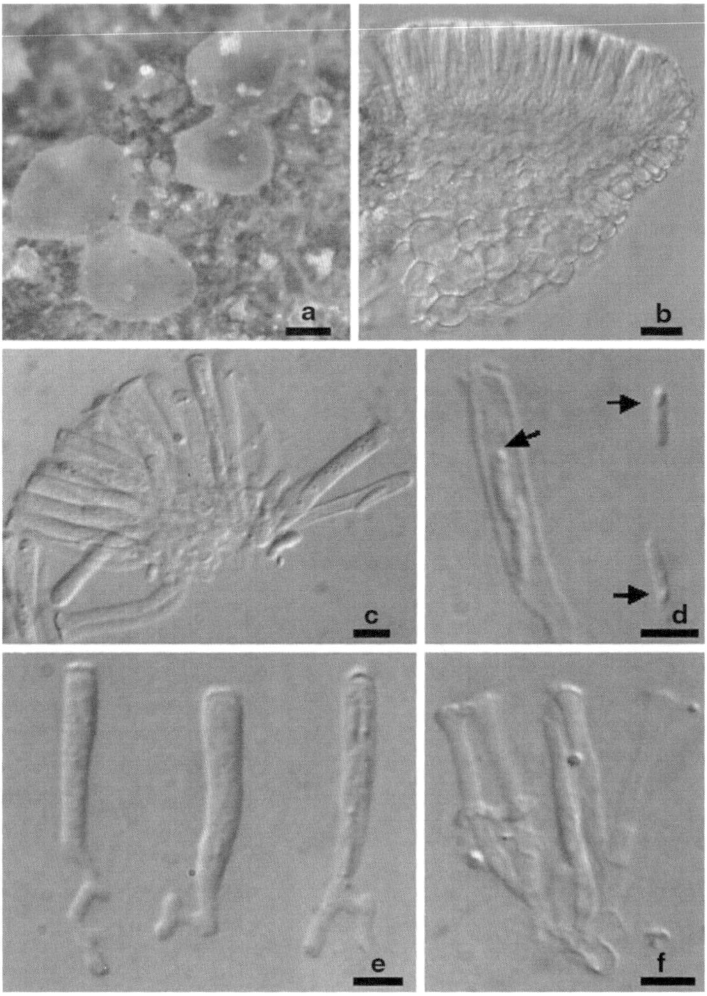

Fig. 3.106 *Orbilia quercus* **a** dried apothecia; **b** vertical section of an apothecium; **c–d** Asci and ascospores, *arrow* indicates the SB in living ascospore; **e** asci; **f** paraphyses. Bars: **a** = 100 μm; **b** = 10 μm; **c–f** = 5 μm. Specimen number: HMAS 88781

length, whereon bearing 3–10 conidia in radiating capitate arrangement, occasionally producing up to 13 conidia in more scattered, irregularly racemose arrangement. Conidia commonly cylindric-oblong, rounded at the distal end, shortly tapered at the somewhat truncate base, 20–36 × 5–7.5 μm, slightly to medium curved especially near distal end, with 1–3 septa, predominantly 3-septate, then terminal cells often exceeding the length of the central cells by a third or even a half. Cultures on CMA produce constricting rings adhesive networks when challenged with nematodes (*Panagrellus redivivus*). (Fig. 3.107)

Sexual morph (Specimen No. YMFT1. 01829, Fig. 3.108): Specimens were collected on decaying bark of broadleaved tree in a forest at 848 m altitude, located

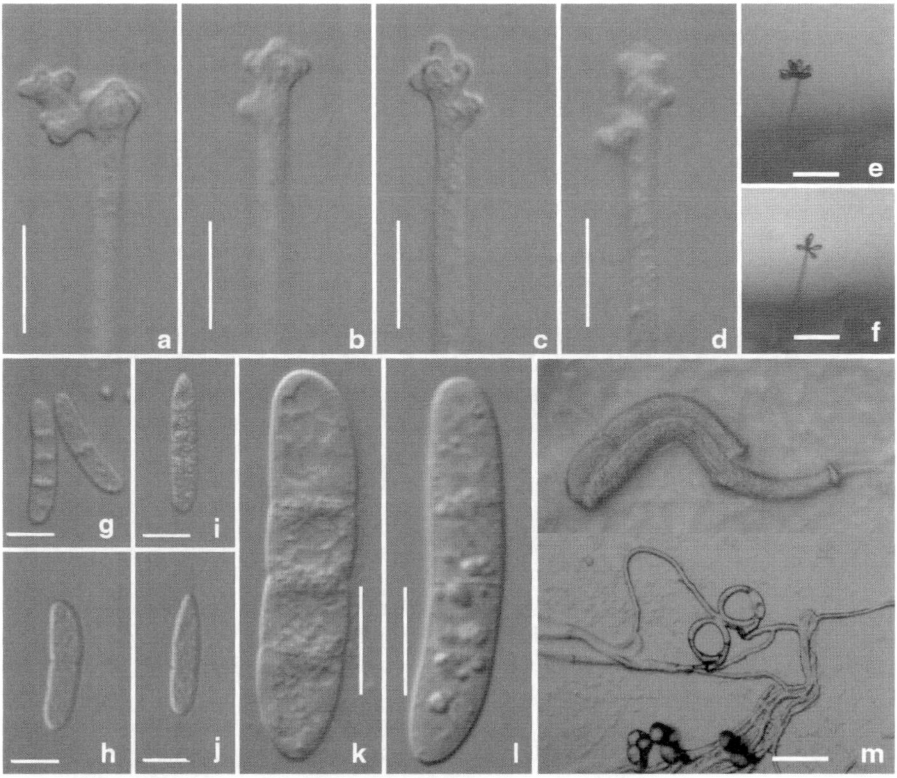

Fig. 3.107 *Drechslerella brochopaga* **a–f** conidiophores; **g–l** conidia; **m** constricting rings. Bars: **a–d**; **g–l** = 10 μm; **e–f**; **m** = 50 μm. Strain number: YMF 1. 01829

in Jinggangshan Park of Jian City, Jiangxi Province, China, in October 2005 by Y. Zhang. Apothecia 0.8–1.2 mm in diameter, superficial, with a distinct stalk up to 0.2–0.3 mm high, scattered to gregarious on decayed bark, white and translucent throughout when moist, pale brown when dry. Disc slightly concave to flat, margin even and smooth. Ectal excipulum of angular or globose cells, 8–13.8 μm in diameter, with thin or only slightly thickened walls, towards margin of *textura prismatica-angularis* oriented at a high angle, marginal cells tipped by glassy processes up to 30–40 × 3.5–4.5 μm, curved outwards, agglutinated to form rounded teeth. Medullary excipulum rather thick, of *texture globulosa* (*-prismatica*) with intermingled hyphae, subhymenium poorly developed. Asci 29.5–38 × 3.2–4.5 μm, 8-spored, lower (2)3–5 (6) spores inversely oriented (with spore body towards ascus base), cylindric, rounded or truncate at the apex (depending on side of view), tapered and often forked at the base (L, hor H-shaped). Ascospores hyaline, nonseptate, cylindric- ellipsoid, sometimes slightly tapered at lower end, 3–4 × 1.6–1.8 μm, with a refractive rod-shaped spore body (SB) at upper end in living mature ascospores, 1.2–1.4 × 0.3–0.5 μm in diameter. Paraphyses 1.3–2 μm wide, enlarged to 2.8–3.7 μm in diameter at the medium to rather strongly clavate or capitate apex,

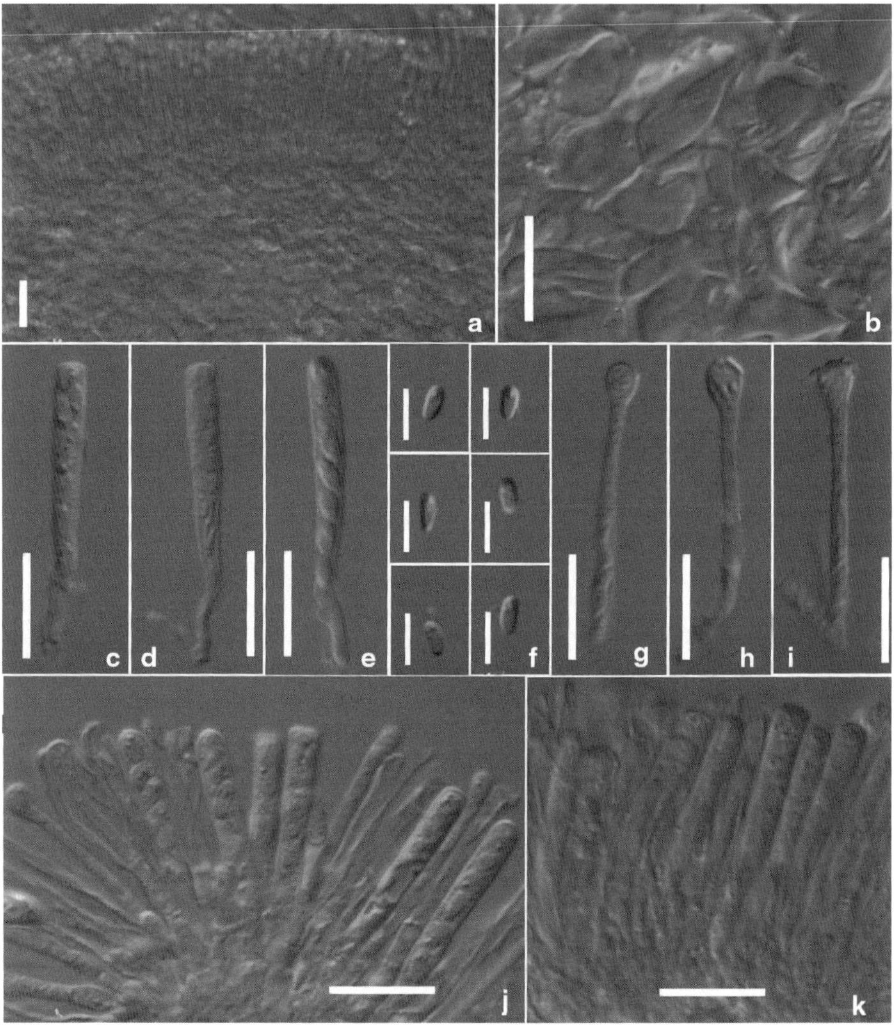

Fig. 3.108 *Orbilia orientalis* **a** hymenium; **b** cells of ectal excipulum; **c–e** dead asci; **f** living ascospores with globose spore bodies; **g–i** Dead paraphyses; **j–k** cluster of dead asci and paraphyses with living spores. Bar: **a–e**, **g–k** = 10 μm; **f** = 5 μm. Specimen number: YMFT 1. 01829

terminal cell much longer than lower cells, apex covered by a rough thin layer of exudates. (Fig. 3.108)

Drechslerella yunnanensis Z.F. Yu & K.Q. Zhang
Orbilia cf. orientalis (Raitv.) Baral.
Reference: Yu et al. (2009)

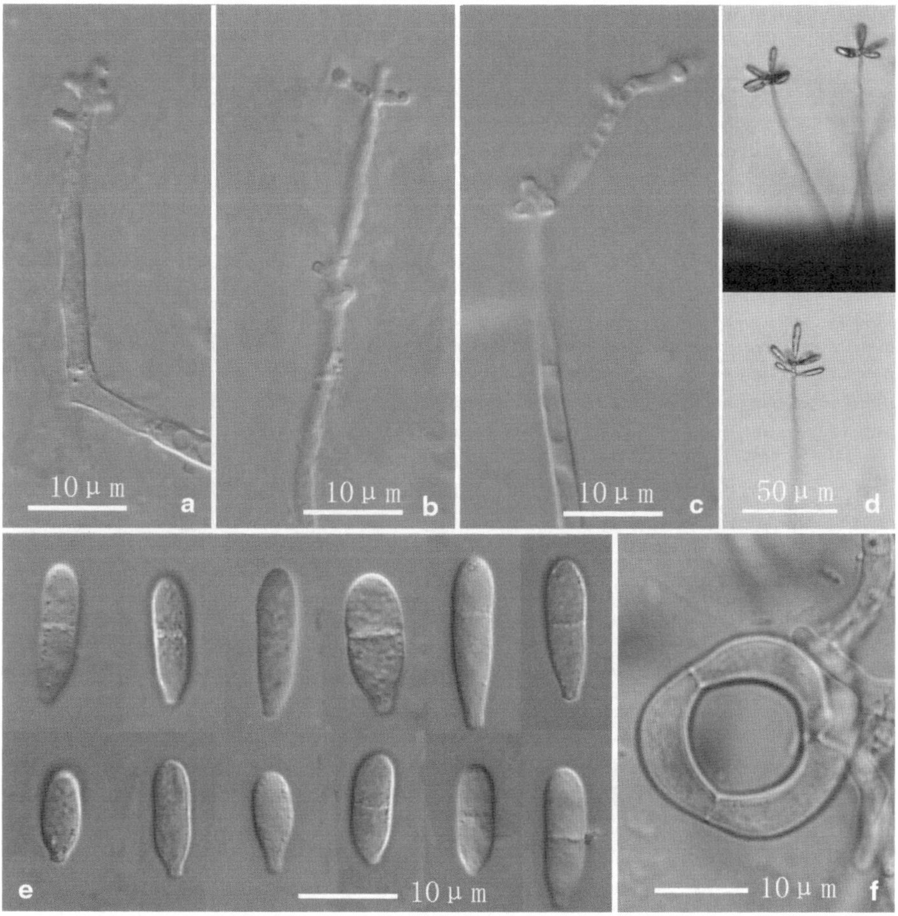

Fig. 3.109 *Drechslerella yunnanensis* **a–c** conidiophores with short denticles; **d** conidiophores bearing conidia in clusters; **e** conidia; **f** constricting rings. Bar: **a–c**, **e–f**=10 μm; **d**=50 μm. Strain number: YMF 1.01863

Asexual morph (Culture No. YMF1.01863, Fig. 3.109): Cultures were derived from ascospores. Colonies white, slow-growing on CMA medium, attaining less than 25 mm diameter in 10 days at 25 °C. Vegetative hyphae hyaline, septate, 3.5–4 μm wide, aerial mycelium sparse, hyaline, septate, branched, 2.5–4 μm wide. Conidiophores hyaline, septate, erect, unbranched or occasionally branched below, 60–100 (–220) μm, 3.8–4.2 μm wide at the base, tapering gradually upwards to 1.5–2 μm near the apex, with 2–7 (–10) denticles 2.3–4.2 μm long, each bearing one conidium in a capitate or racemose arrangement. Conidia hyaline, straight, elongate ellipsoidal, rounded at the apex, with a small truncate protuberance at the base, 7.8–12.9 (–17.8) × 3.3–4.2 μm, (0-)1 septate, proportion of aseptate conidia 17 %. Cultures on CMA produce the trap device of constricting rings when induced by nematodes (*Panagrellus redivivus*). (Fig. 3.109)

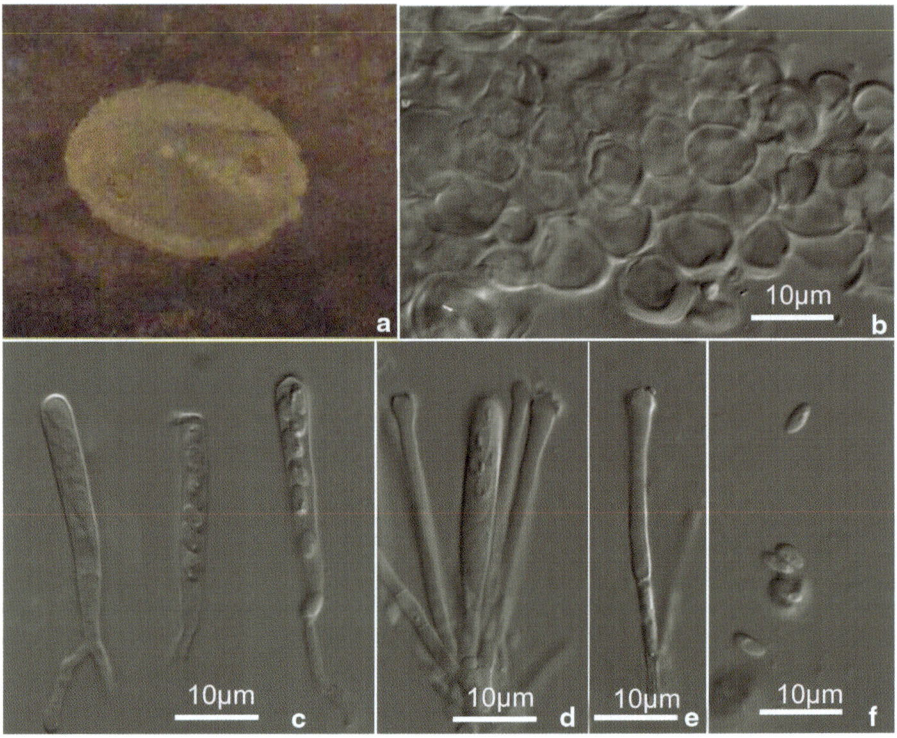

Fig. 3.110 *Orbilia cf. orientalis* **a** apothecium; **b** cells of excipulum; **c** asci; **d** cluster of asci and paraphyses; **e** paraphyses; **f** ascospores. Bar: **b**–**f** = 10 μm. Specimen number: YMFT 1.01863

The sexual morph (specimen No. YMFT 1.01863, Fig. 3.110) was collected as fresh fruit bodies by ZeFen Yu in August 2006 from decaying bark of a broad-leaved tree, in Dalongkou Park of Yimen County, Yunnan Province, China (N24°34, E101°00′, alt. 1580 m, coniferous-broadleaf forest dominated by *Cyclobalanopsis glaucoides* Schottky. and *Pinus armandii* Franch.). The sexual state is very similar to *Orbilia orientalis* in having cylindric-ellipsoid ascospores with a refractive, rod-shaped, laterally arranged spore body (SB) at the upper end. Also spore size is very close (protologue of *O. orientalis*: 3–4 × 1.5–2 μm, the specimen YMFT1.01863: 4.0–4.2 × 1.6–1.9 μm). The apothecial margin is only minutely crenulate and hardly bears any glassy processes, while in the *O. orientalis* type, the glassy processes are long and agglutinated, forming distinct teeth at the margin. (Fig. 3.110)

Conclusion

In the classification of nematode-trapping fungi, trapping devices are generally more informative than the morphologies of the conidia (shape, number and size of cells) and conidiophores (branching, modification of the apex). The latter characters were previously used in delimiting genera. We conclude that all orbiliaceous nematode-trapping fungi should be placed in the genera of *Arthrobotrys*, *Dactylellina* and *Drechslerella* based on their trap types. *Arthrobotrys* includes species producing adhesive networks, *Dactylellina* contain species forming stalked adhesive knobs, unstalked adhesive knobs, or non-constricting rings, while *Drechslerella* includes species producing constricting rings. Species lacking trap formation, especially members of the genus *Dactylella*, are excluded from the nematode-trapping fungi, although some share similar morphologies to species of *Dactylellina*.

A total of 96 species of orbiliaceous nematode-trapping fungi are described and illustrated in detailed, with 54 species from *Arthrobotrys*, 28 from *Dactylellina* and 14 from *Drechslerella*. Nematode-trapping fungi have been found in all parts of the world and in all climatic zones. Most of the nematode-trapping fungi annotated here were isolated from agricultural soils and forest soils, while some of them also inhabit freshwater. The majority of nematode-trapping fungi are fairly ubiquitous, with few species restricted geographically. Species forming adhesive networks are the dominant group.

Fourteen connections have been established between asexual nematode-trapping fungi and their *orbiliaceous* sexual state. Five asexual species are associated with *Orbilia auricolor* suggesting that *O. auricolor* is a species complex. It is difficult to distinguish between members of this complex based only morphological characters of the sexual state. Beside the morphological features, more molecular information, such as gene sequences of ITS, 28S rDNA, 5.8S rDNA and ß-tubulin, should be taken into account in the taxonomy of *Orbilia*.

References

Ahrèn, D., Ursing, B. M., & Tunlid, A. (1998). Phylogeny of nematode-trapping fungi based on 18S rDNA sequences. *FEMS Microbiology Letters, 158,* 179–184.

Anastasiou, C. J. (1964). Some aquatic fungi imperfecti from Hawaii. *Pacific Science, 18,* 202–206.

Barron, G. L. (Ed.). (1977). *The Nematode-Destroying Fungi. Topics in Mycology No. 1*. Canadian Biological Publications, Guelph.

Barron, G. L. (1979). Nematophagous fungi: A new *Arthrobotrys* with aseptate conidia. *Canadian Journal of Botany, 55,* 1371–1373.

Barron, G. L. (1990). A new predatory hyphomycete capturing copepods. *Canadian Journal of Botany, 68,* 691–696.

Barron, G. L., & Davidson, J. G. (1972). Nematophagous hyphomycetes: *Arthrobotrys anomala* sp. nov. *Canadian Journal of Botany, 50,* 1773–1774.

Castaner, D. (1968). *Monacrosporiumsphaeroides*, a new nematode-destroying hyphomycete from Iowa. The *American Midland. Naturalist, 80,* 280–283.

Chen, J. L., Huang, T. L., & Tzean, S. S. (1998). *Dactylella huisuniana*, a new nematode-trapping fungus from Taiwan. *Mycological Research, 102*(10), 1269–1273.

Chen, J., Xu, L. L., Liu, B., & Liu, X. Z. (2007a). Taxonomy of *Dactylella* complex and *Vermispora*. I. Generic concepts based on morphology and ITS sequences data. *Fungal Diversity, 26,* 73–83.

Chen, J., Xu, L. L., Liu, B., & Liu, X. Z. (2007b). Taxonomy of *Dactylella* complex and *Vermispora*. II. Generic concepts based on morphology and ITS sequences data. *Fungal Diversity, 26,* 85–126.

Chen, J., Xu, L. L., Liu, B., & Liu, X. Z. (2007c). Taxonomy of *Dactylella* complex and *Vermispora*. III. Generic concepts based on morphology and ITS sequences data. *Fungal Diversity, 26,* 127–142.

Chowdhry, P. N., & Bahl, N. (1982). Notes on Indian hyphomycetes—VII *Monacrosporiealla*. *Current Science, 20,* 894–896.

Cooke, R. C. (1969). Two nematode-trapping hyphomycetes, *Duddingtonia flagrans* gen. et. comb. nov. and *Monacrosporium mutabilis* sp. nov. *Transactions of the British Mycological Society, 53,* 315–318.

Cooke, R. C., & Dickison, C. H. (1965). Nematode-trapping species of *Dactylella* and *Monacrosporium*. *Transactions of the British Mycological Society, 48,* 621–629.

Cooke, R. C., & Pramer, D. (1968). *Monacrosporium rutgeriensis* sp. nov., a nematode-trapping hyphomycetes. *Phytopathology, 58,* 544–545.

Cooke, R. C., & Satchuthananthavale, V. (1965). Some nematode-trapping species of *Dactylaria*. *Transactions of the British Mycological Society, 49,* 27–32.

Corda, A. G. J. (1839). Pracht-Flora europaischer Schimmelbildungen, vol 43. Leipzig.

Dackman, C., & Norbring, B. (1992). Nematophagous fungi and their activities in soil. Soil Biochemistry 7: 95–130

Dixon, S. M. (1952). Predacious fungi from rotton-wood. *Transactions of the British Mycological Society, 35,* 144–148.

Dowsett, J. A., Reid, J., & Kalkat, R. S. (1984). A new species of *Dactylella*. *Mycologia, 76,* 563–566.

Drechsler, C. (1937). Some hyphomycetes that prey on free-living terricolous nematodes. *Mycologia, 29,* 447–552.

Drechsler, C. (1940). Three new hyphomycetes preying on free-living terricolous nematodes. *Mycologia, 32,* 448–470.

Drechsler, C. (1941). Four hyphomycetes destructive to nematodes and rhizopods. *Mycologia, 33,* 248–269.

Drechsler, C. (1943). A new nematode-capturing *Dactylella* and several related hyphomycetes. *Mycologia, 35,* 339–362.

Drechsler, C. (1944a). Three hyphomycetes that capture nematodes in adhesive networks. *Mycologia, 36,* 138–171.

Drechsler, C. (1944b). A species of *Arthrobotrys* that captures springtails. *Mycologia, 36,* 383–397.

Drechsler, C. (1947). A nematode-strangling *Dactylella* with broad quadriseptate conidia. *Mycologia, 39,* 5–18.

Drechsler, C. (1950). Several species of *Dactylella* and *Dactylaria* that capture free-living nematodes. *Mycologia, 42,* 1–79.

Drechsler, C. (1952). Another nematode-strangulating *Dactylella* and some related hyphomycetes. *Mycologia, 44,* 533–556.

Drechsler, C. (1954). Some hyphomycetes that capture eelworms in southern states. *Mycologia, 46,* 762–782.

Drechsler, C. (1962). Some clampless hyphomycetes predaceous on nematodes and rhizopods. *Sydowia, 15,* 9–25.

Duddington, C. L. (1951a). Two predacious hyphomycetes. Transactions of the British Mycological Society*, 34,* 598–603.

Duddington, C. L. (1951b). *Dactylella lobata*, predacious on nematodes. Transactions of the British Mycological Society*, 34,* 489–491.

Eriksson, O., Baral, H., Currah, R., Hansen, K., Kurtzman, C., Lassoe, T., & Rambold, G. (2003). Notes on ascomycete systematics Nos 3580-3623. *Myconet, 9,* 91–103.

Glockling, S., & Dick, M. (1994). *Dactylella megalobrocha*, a new species of nematophagous fungus with constricting ring. *Mycological Research, 98,* 845–853.

Grove, W. B. (1884). New or noteworthy fungi. *American Journal of Botany, 22,* 195–201.

Hagedorn, G., & Scholler, M. (1999). A re-evaluation of predatory orbiliaceous fungi. I. Phylogenetic analysis using rDNA sequence data. *Sydowia, 51,* 27–48.

Hao, Y. E., Luo, J., & Zhang, K. Q. (2005). A new aquatic nematode-trapping hyphomycete. *Mycotaxon, 89,* 235–239.

Hawksworth, D. L. (2011). A new dawn for the naming of fungi: Impacts of decisions made in Melbourne in July 2011 on the future publication and regulation of fungal names. *International Mycological Association Fungus, 2,* 155–162.

Higgins, M. L., & Pramer, D. (1967). Fungal morphogenesis: ring formation and closure by Arthrobotrys dactyloides. *Science, 155,* 345–346.

Jarowaja, N. (1968). *Dactylaria effusa* sp. nov. -na new nematode-trapping hyphomycetes. Bulletin de l'Academie Polonaise des Sciences. *Serie des Sciences Biologiques, 16,* 773–778.

Jarowaja, N. (1970). The genus *Arthrobotrys* Corda. *Acta Mycology, 6,* 337–406.

Jarowaja, N. (1971). *Dactylella inquisitor* sp. nov. a new nematode-strangulating fungus. *Acta Mycology, 7,* 3–6.

Jeng, R. S., & Krug, J. C. (1977). A new species of Orbilia from Venezuela. *Mycologia, 69,* 832–835

Juniper, A. J. (1954). *Dactylaria pyriformis* sp. nov. *Transactions of the British Mycological Society, 37,* 437–440.

Kuthubutheen, A. J., & Webster, J. (1985). *Arthrobotrys dendroides*, a new nematode-trapping synnematous *Arthrobotrys* from Malaysia. *Transactions of the British Mycological Society, 84,* 563.

Li, S. D., Miao, Z. Q., Zhang, Y. H., & Liu, X. Z. (2003). *Monacrosporium janus* sp. nov., a new nematode-trapping hyphomycetes parasitizing sclerotia and hyphae of *Sclerotinia sclerotiorum*. *Mycological Research, 107,* 888–894.

Li, T. F., Zhang, K. Q., & Liu, X. Z. (2000). *Taxonomy of nematophagous fungi (Chinese).* Beijing: Science Press.

Li, Y., Hyde, K. D., Jeewon, R., Cai, L., Vijaykrishna, D., & Zhang, K. Q. (2005). Phylogenetics and evolution of nematode-trapping fungi estimated from nuclear and protein coding genes. *Mycologia, 97,* 1034–1046.

Li, Y., Jeewon, R., Hyde, K. D., Mo, M. H., & Zhang, K. Q. (2006). Two new species of nematode-trapping fungi:relationships inferred from morphology, rDNA and protein gene sequence analyses. *Mycological Research, 110,* 790–800.

Li, S. F., Yu, Z. F., Zhang, Y., Guo, J. W., & Zhang, K. Q. (2009). *Arthrobotrys nonseptata*, a new asexual state from *Orbilia* sp. *Mycotaxon, 109,* 247–254.

Lilly, V. G., & Barnett, H. L. (1951). *Physiology of the Fungi* (pp. 464). New York: McGraw-Hill.

Liou, J. Y., Liou, G. Y., & Tzean, S. S. (1995). *Dactylella formosana*, a new nematode-trapping fungus from Taiwan. *Mycological Research, 99,* 751–755.

Liu, X. Z., & Lu, B. S. (1993). Two new species of *Monacrosporium* from China. *Mycosystema, 6,* 65–69.

Liu, X. Z., & Zhang, K. Q. (1994). Nematode-trapping species of *Monacrosporium* with special reference to two new species. *Mycological Research, 98,* 862–868.

Liu, X. Z., & Zhang, K. Q. (2003). *Dactylella shizishanna* sp. nov., from Shizi Mountain, China. *Fungal Diversity, 14,* 103–107.

Liu, B., Yang, Y., Liu, X. Z., & Zhuang, W. Y. (2002). *Monacrosporium parvicolla*, a knob-forming nematophagous hyphomycetes from *Orbilia cunninghamii*. In: The 3rd Asia-Pacific Mycological Congress on Biodiversity and Biotechnology, P. 75 (abstract).

Liu, B., Liu, X., & Zhuang, W. Y. (2005a). *Orbilia querci* sp. nov. and its knob-forming nematophagous asexual state. *FEMS Microbiology Letters, 245,* 99–105.

Liu, B., Liu, X., & Zhuang, W. Y. (2005b). A new species of *Hyalorbilia* and its asexual state from China. *Novartis Foundation Symposium, 81,* 145–155.

Luo, H., Mo, M. H., Huang, X. W., Li, X., & Zhang, K. Q. (2004). *Coprinus comatus*: A basidiomycete fungus forms novel spiny structures and infects nematodes. *Mycologia, 96,* 1218–1225.

Luo, H., Li, X., Pan, Y. B., Li, G. H., & Zhang, K. Q. (2006). Acanthocytes of *Stropharia rugosoannulata* function as a nematode-attacking device. *Applied and Environmental Microbiology, 72,* 2982–2987.

Luo, H., Liu, Y. J., Fang, L., Li, X., Tang, N. H., & Zhang, K. Q. (2007). *Coprinus comatus*: Damages nematode cuticles mechanically with spiny balls and produces potent toxins to immobilize nematodes. *Applied and Environmental Microbiology, 72,* 3916–3923.
McCulloch, J. S. (1977). New species of nematophagous fungi from Queensland. *Transactions of the British Mycological Society, 68,* 173–179.
McNeill, J., Barrie, F. R., Burdet, H. M., Demoulin, V., Hawksworth, D. L., Marhold, K., Nicolson, D. H., Prado, J., Silva, P. C., Skog, J. E., & Wiersema, J. H. (Eds.). (2006). International Code of Botanical Nomenclature (Vienna Code) adopted by the Seventeenth Interational Botanical Congress, Vienna, Austria, July 2005. (Regnum Vegetabile No. 146.) Ruggell: A.R.G. GantnerVerlag.
Mekhtieva, N. A. (1973). Species novae fungorum carnivororum generis *Arthrobotrys* Corda. *Nov Sist Niz Rast, 10,* 174–176.
Mekhtieva, N. A. (1979). Khishchnye Nematofagovye Griby—Gifomitsety (Predacious Nematophagous Hyphomycetes). A Kademiya Nauk Azerbaidzhanskoi SSR. Baku.
Mekhtieva, N. A. (1997). A new nematophagous fungus, *Candelabrella shahriari* sp. nov., from south Azerbaijan. *Mycological Research, 101,* 334.
Mekhtieva, N. A. (1998). New nematophagous fungi from Azerbaijan. *Mycological Research, 102,* 683–684.
Miao, Z. Q., Liu, X. Z., Li, S. D., & He, M. X. (2003). *Dactylella pseudoclavata* sp. nov., a new nematode-trapping fungus. *Canadian Journal of Botany, 81,* 1–5.
Mo, M. H., Huang, X. W., Zhou, W., Huang, Y., Hao, Y. E., & Zhang, K. Q. (2005a). *Arthrobotrys yunnanensis* sp. nov., the fourth asexual state of *Orbilia auricolor*. *Fungal Diversity, 18,* 107–115.
Mo, M. H., Zhou, W., Huang, Y., Yu, Z. F., & Zhang, K. Q. (2005b). A new species of *Dactylella* and its sexual state. *Mycotaxon, 91,* 185–192.
Morelet, M. (1968). De aliquibus in mycologia novitatibus. *Bulletin de la Société des Sciences Naturelles and d'Archéologie de Toulon and du Var, 178,* 6.
Nees, V., & Esenbeck, C. G. D. (1817). Das system der Pilze und Schwämme. Würzburg.
Onofri, S., & Tosi, S. (1992). *Arthrobotrys ferox* sp. nov., a springtail-capturing hyphomycete from continental antarctica. *Mycotaxon,* 445–451.
Oorshot, C. A. N. (1985). Taxonomy of the *Dactylaria* complex V. A review of *Arthrobotrys* and allied genera. In Hoog, G.S. DE (Ed.).*Taxonomy in the Dactylaria Complex, IV–VI, Studies in mycology* 26: 61–96.
Oudemans, C. A. (1885). Aanwinstenvoor de flora mycological van Nederland. *Ned Kruidk Archf Ser, 24,* 236–257.
Peach, M. (1950). Aquatic predacious fungi. *Transactions of the British Mycological Society, 33,* 148–153.
Peach, M. (1952). Aquatic predacious fungi. *Transactions of the British Mycological Society, 35,* 19–23.
Pfister, D. H. (1994). *Orbilia fimicola,* a nematophagous discomycete and its *Arthrobotrys* anamorph. *Mycologia, 86,* 451–453
Pfister, D. H., & Liftik, M. E. (1995). Two *Arthrobotrys* anamorphs from *Orbilia auricolor*. *Mycologia, 87,* 684–688.
Pfister, D. H. (1997). Castor, Pollux and life histories of fungi. *Mycologia* 89: 1–23.
Qiao, M., Zhang, Y., Li, S. F., Baral, H. O., Su, H. Y., Xu, J. P., Zhang, K. Q., & Yu, Z. F. (2012). *Orbilia blumenaviensis* and its *Arthrobotrys* asexual state. *Mycological Progress, 11,* 255–262.
Rifai, M. A. (1968). The hyphomycetes genus *Dactylaria* Sacc. *Reninwardtia, 7,* 357–374.
Rifai, M. A., & Cooke, R. C. (1966). Studies on some didymosporous genera of nematode-trapping hyphomycetes. *Transactions of the British Mycological Society, 49,* 147–168.
Rubner, A. (1996). Revision of predacious hyphomycetes in the *Dactylella-Monacrosporium* complex. *Studies in Mycology, 39,* 1–134.
Scheuer, C. H., & Webster, J. (1990). *Dactylella arcuata* sp. nov., a nematode-trapping hyphomycete. *Mycological Research, 94,* 718.

Schenck, S., Kendrick, W. B., & Pramer, D. (1977). A new nematode-trapping hyphomycetes and a revaluation of *Dactylaria* and *Arthrobotrys*. *Canadian Journal of Botany, 55,* 977–985.

Scholler, M., & Rubner, A. (1999). *Arthrobotrys hertziana* sp. nov. from the Canary Islands. *Mycological Research, 103,* 764–768.

Scholler, M., Hagedorn, G., & Rubner, A. (1999). A reevaluation of predatory orbiliaceous fungi. II. A new generic concept. *Sydowia, 51,* 89–113.

Scholler, M., Hagedorn, G., & Rubner, A. (2000). *Arthrobotrys oudemansii* comb. nov., a new name for a nematode-trapping fungus to avoid homonymy. *Sydowia, 52,* 59–60.

Siddiqui, Z. A., & Mahmood, I. (1996). Biological control of plant parasitic nematodes by fungi: A review. *Bioresource Technology, 58,* 229–239.

Soprunov, F. F. (1958). Kishchnyegriby-gifomitsetiiikhprimenenie v bor'be s patogennyminematodami (Predcious hyphomycetes and their application in the control of pathogenic nematodes). Akademiya Nauk Turkmenskoi SSR, Ashkhabad. (Translation published by S. Monson, Jerusalem).

Stirling G. R., Nicol J. M., & Reay, F. (1998). Advisory services for nematodes pests - operational guide. Rural Industries Research and Development Corporation Publication No. 99/41. Canberra. pp 120.

Su, H. Y., Li, Y., Zhang, K. Q., & Mo, M. H. (2005). *Monacrosporium multiseptatum*, a new species from China. *Mycotaxon, 92,* 193–196.

Su, H. Y., Hao, Y. E., Yang, X. Y., Yu, Z. F., Deng, J. S., & Mo, M. H. (2008). A new species of *Dactylellina* producing adhesive knobs and non-constricting rings to capture nematodes. *Mycotaxon, 105,* 313–316.

Su, H. Y., Liu, S. R., Li, Y. X., Cao, Y. H., Chen, M. H., & Yang, X. Y. (2011). *Arthrobotrys latispora*, a new nematode-trapping fungus from southwest China. *Mycotaxon, 117,* 29–36.

Subramanian, C. V. (1963). *Dactylella, Monacrosporium* and *Dactylina*. *Journal of Indian Botanical Society, 42,* 291–300.

Swe, A., Jeewon, R., Pointing, S. B., & Hyde, K. D. (2008). Taxonomy and molecular phylogeny of *Arthrobotrys mangrovispora*, a new marine nematode-trapping fungal species. *Botanica Marina, 51,* 331–338.

Tarjan, A. C. (1961). Growth characteristics of *Dactylella drechsleri* nov. sp., an adhesive-knobbed nematode-trapping hyphomycete from Florida. *Mycopath. Mycopathology Mycology Applied, 14,* 136–144.

Taylor, J. W. (2011). One Fungus = One Name: DNA and fungal nomenclature twenty years after PCR. *International Mycological Association Fungus, 2*(2), 113–120.

Tzean, S. S., & Liou, J. Y. (1993). Nematophagous resupinate basidiomycetous fungi. *Phytopathology, 83,* 1015–1020.

Yang, Y., & Liu, X. Z. (2005). *Dactylella coccinella* sp. nov., an asexual stateic species. *Mycotaxon, 91,* 127–132.

Yang, Y., & Liu, X. Z. (2006). A new generic approach to the taxonomy of predatory asexual stateic *Orbiliaceae* (Ascomycotina). *Mycotaxon, 97,* 153–161.

Yang, D. S., Chen, W. M., Huang, Y., Mo, M. H., & Zhang, K. Q. (2005). A new predatory fungus from China. *Mycotaxon, 94,* 215–217.

Yang, Y., Yang, E. C., An, Z. Q., & Liu, X. Z. (2007). Evolution of nematode-trapping cells of predatory fungi of the *Orbiliaceae* based on evidence from rRNA-encoding DNA and multiprotein sequences. *Proceedings of National Academy of Science of USA of America, 104,* 8379–8384.

Yu, Z. F., Zhang, Y., Qiao, M., Baral, H. O., Weber, E., & Zhang, K. Q. (2006). *Drechslerella brochopaga*, the asexual state of *Orbilia* (*Hyalinia*) *orientalis*. *Mycotaxon, 96,* 163–168.

Yu, Z. F., Qiao, M., Zhang, Y., Baral, H. O., & Zhang, K. Q. (2007a). *Orbilia vermiformis* sp. nov. and its asexual state. *Mycotaxon, 99,* 271–278.

Yu, Z. F., Zhang, Y., Qiao, M., & Zhang, K. Q. (2007b). *Orbilia dorsalia* sp. nov., the sexual state of *Dactylella dorsalia* sp. nov. *Cryptogamie Mycologie, 28*(1), 55–63.

Yu, Z. F., Qin, L., Zhang, Y., Qiao, M., Kong, Y. J., & Zhang, K. Q. (2009a). A new *Drechslerella* species isolated from *Orbilia* cf. *orientalis*. *Mycotaxon, 110,* 253–259.

Yu, Z. F., Kong, Y. J., Zhang, Y., Qiao, M., Guo, J. W., & Zhang, K. Q. (2009b). A new *Dactylella* species from *Orbilia alba*. *Journal of Microbiology, 47*(3), 265–269.

Yu, Z. F., Qiao, M., Zhang, Y., Qin, L., & Zhang, K. Q. (2011). *Pseudotripoconidium*, a new asexual state genus connected to *Orbilia*. *Mycologia, 103,* 164–173.

Zachariah, K. (1983). Ascocarp induction in a natural autotroph of a predatory fungus. *Canadian Journal of Botany, 62,* 3262–3266

Zhang, K. Q., Liu, X. Z., & Cao, L. (1994). A review of *Dactylella* and a new species. *Mycosystema, 7,* 111–118.

Zhang, K. Q., Liu, X. Z., & Cao, L. (1996a). Nematophagous species of *Monacrosporium* from China. *Mycological Research, 100,* 274–276.

Zhang, K. Q., Liu, X. Z., Cao, L., & Gao, R. H. (1996b). A new species *Arthrobotrys* from China. *Mycological Research, 100,* 527–530.

Zhang KQ, Liu XZ, Li TF. (Ed.). (2001). *Biology of nematophagous fungi.* Chinese Scientific & Technological Publication, Beijing, China.

Chapter 4
The Ecology of Nematophagous Fungi in Natural Environments

Ying Zhang, Ke-Qin Zhang and KD Hyde

Contents

Abstract Nematode-trapping fungi have extensively been studied both because of their unique predatory life history and because they are potential biocontrol agents of economically important plant- and animal- parasitic nematodes. Fundamental knowledge of the ecology of these fungi is therefore essential before the value of such biocontrol methods can be assessed. The ecology of nematode-trapping fungi is reviewed in this chapter. Topics dealt with include occurrence and habitats,

K.-Q. Zhang (✉) · Y. Zhang
Laboratory for Conservation and Utilization of Bio-Resources, and Key Laboratory of Microbial Diversity in Southwest China, Ministry of Education, Yunnan University, 650091 Kunming, Yunnan, China
e-mail: kqzhang1@ynu.edu.cn

KD. Hyde
Institute of Excellence in Fungal Research, Mae Fah Luang University, Chiang Rai, Thailand

School of Science, Mae Fah Luang University, Chiang Rai, Thailand

K.-Q. Zhang, K. D. Hyde (eds.), *Nematode-Trapping Fungi,* Fungal Diversity Research Series 23, DOI 10.1007/978-94-017-8730-7_4, © Mushroom Research Foundation 2014

geographical and seasonal distribution, quantification of abundance, and effects of soil conditions and nematode density on their distribution. Competition between nematodes and nematophagous fungi, the effect of fungistasis and extreme ecological factors such as heavy metals and salinity on these fungi, and their genetic diversity and speciation are included in this chapter.

Keywords *Arthrobotrys* • Competition • Genetic diversity • Habitats • Distribution • Nematodes • Fungistasis

Introduction

Fungi possess broad strategies to obtain nutrients; they may be saprobes (living on dead material), pathogens (causing diseases of plant and animal hosts), symbionts (having associations with other microbes, plants or animals) or predacious (trapping and utilizing living animals for their food). The diverse strategies enable fungi to colonize a broad range of habitats. The predacious and as such carnivorous habit is a fascinating life form in nutrient acquisition (Yang et al. 2012; Barron 1977; Pathak et al. 2012).

Nematode-trapping fungi, a monophyletic lineage within the *Orbiliales* (Ascomycota), have distinctive trapping devices, such as adhesive hyphae, adhesive knobs, adhesive networks, constricting rings, and non-constricting rings, for trapping nematodes (Li et al. 2005). These devices develop from mycelial extensions and enable the fungi to capture and consume nematodes in soil, leaf litter, and other substrates (Gray 1987). Nematode trapping fungi have been extensively studied both because of their unique predatory life history and because they are potential biocontrol agents of economically important plant- and animal- parasitic nematodes (Linford et al. 1938; Wolstrup et al. 1996; Jaffee et al. 1993; Jaffee et al. 1998; Morton et al. 2003; Vilela et al. 2012; Kumar et al. 2011; El-Borai et al. 2011; Liu et al. 2009; Shams Ghahfarokhi et al. 2004; Palma-Guerrero et al. 2008; Yeates et al. 2007; Singh et al. 2012; Paraud et al. 2012; Braga et al. 2011; Carneiro et al. 2011; Braga et al. 2013). A fundamental knowledge of the ecology of these fungi is, however, essential before the value of such biological control methods can be assessed. The ecology of these fungi has been reviewed by various authors (e.g., Gray 1987; Dackman et al. 1992; Barron 1977; Duddington 1954; Duddington 1951; Jaffee 2003). This chapter updates these reviews by considering various aspects of ecology, such as occurrence and habitats, geographical and seasonal distribution, quantification of abundance, and effects of soil conditions and nematode density on their distribution. Competition between nematode and nematophagous fungi, the effects of fungistasis and extreme ecological factors such as heavy metals and salinity on these fungi, and their genetic diversity and speciation are also addressed.

Habitats and Distribution

Habitats and Species Diversity

Nematode trapping fungi have a worldwide distribution (Gray 1987), and species diversity have been investigated in many regions including Canada (Mahoney and Strongman 1994), China (Liu and Zhang 1994; Liu and Zhang 2003; Zhang et al. 2011; Su et al. 2011), Cuba (Hidalgo-Diaz et al. 2000), Brazil (Vilela et al. 2012; de Almeida et al. 2012; Falbo et al. 2013; Vilela et al. 2013), Europe (Boag and Lopez-Llorca 1989; Kelly et al. 2009; Saxena 2008), India (Saxena and Mukerji 1991), Latin America (Rubner 1994), and South Africa (Durand et al. 2005). Nematophagous fungi can also be found in a wide range of habitats. They have been isolated most frequently from agricultural pasture, coastal vegetation and coniferous leaf litter (Gray 1983b; Kelly et al. 2009; Kumar et al. 2011; Liu et al. 2009; Saxena 2008; Shams Ghahfarokhi et al. 2004). Species diversity and population densities, however, varies in different substrates. Gray and Bailey (1985) found nematode trapping fungi at depths of a maximum of 35 cm in a deciduous woodland. A much greater species diversity of nematophagous fungi was present in the upper organic zones. Predators forming constricting rings, adhesive branches and adhesive knobs were restricted to the upper litter and humus layers. The net forming predators were isolated at all depths, although they were significantly more abundant in the lower mineral rich soils. Jaffee et al. (1998) found that the abundance and variety of nematode-trapping fungi were usually higher in organically managed plots than in conventional plots, while two species, *Arthrobotrys haptotyla* and *Arthrobotrys thaumasia* tended to be more numerous in conventional than in organic plots. Persmark and Jansson (1997) investigated the densities and species composition of nematode-trapping fungi in agricultural soils and in the rhizosphere of crop plants in Sweden; the densities were 5–20 times higher in pea rhizosphere than in root-free soil. *Arthrobotrys oligospora* was the most common species, in both root-free soil and the rhizosphere, but there were some differences in the species composition. Asexual nematode-destroying hyphomycetous fungi were isolated from three woodland types (poor sandy pine forest, mesotrophic beech forest and a sporadically inundated river bank) in Berlin, with the highest number of isolates in the beech forest, while the lowland forest yielded the highest number of species, but the lowest amount of isolates (Sünder and Lysek 1988). Kumar et al. (2011) recorded 17 species of nematophagous fungi from compost, cow dung manure, leaf litter and agricultural soils, they found substrate colonization by nematophagous fungi was higher in leaf litter, compost and cow dung manure, and *Arthrobotrys eudermatum*, *Arthrobotrys cladodes*, *Drechslerella brochopaga*, *Arthrobotrys oligospora* and *Drechslerella dactyloides* (synonym: *Arthrobotrys dactyloides*) had higher percentage of soil colonization. Nematode-trapping fungi were also isolated from seriously polluted environments such as metalliferous mines. The distribution was not restricted by high concentration of lead as greater species diversity was slightly positively correlated with the lead pollution levels ($r=0.29$) (Mo et al. 2006, 2008).

Besides terrestrial habitats, nematode-trapping fungi have been investigated in aquatic and some other unique environments. Hao et al. (2005) examined 1000 waterlogged soil samples from China and nematode-trapping fungi were found in shallow freshwater (at a water depth of 20 cm), but not in samples collected at the bottom of a 4 m deep lake. Hao et al. (2005) isolated 35 species of predacious fungi, and *Arthrobotrys oligospora, A. musiformis, A. thaumasia* and *A. longiphorum* were the most common species. Species with adhesive networks were the most frequently isolated taxa in shallow freshwater. Swe et al. (2008) found a new species of *Arthrobotrys* (*A. mangrovispora*) from decaying submerged wood in mangroves of Hong Kong. They also identified 16 species in mangrove habitats worldwide. The commonly encountered taxa were *Arthrobotrys oligospora* and *Arthrobotrys thaumasia*. Diversity and abundance of nematode-trapping fungi from decaying litter in terrestrial, freshwater and mangrove habitats were also compared, the species richness and diversity was higher in terrestrial than in freshwater and mangrove habitats. Taxa characterized by adhesive nets were more frequent in all habitats (Swe et al. 2009). To sum up, nematode-trapping fungi have wide distribution range in global terrestrial and aquatic environments.

Temporal Distribution

The distribution of asexual nematode-trapping hyphomycetous fungi also shows temporal (mainly seasonal) differences. The seasonal distribution of nematode-trapping fungi in the vicinity of apple and grape orchards in China was investigated by Miao and Liu (2003). Nematode-trapping fungi were abundant in late spring and early summer and less so in the autumn. *Drechslerella dactyloides* was the predominant species in all orchards, and *Arthrobotrys oligospora, Arthrobotrys conoides* and *Stylopage* sp. were frequently encountered.

The influence of season and altitude on the occurrence of nematode-trapping fungi in cattle faeces has also been investigated (Su et al. 2007). The predominant species from all three plateau pastures were *Arthrobotrys oligospora, A. musiformis, Dactylellina ellipsosporum* and *Arthrobotrys thaumasia*. Species with adhesive networks were the most frequently isolated. Overall, species diversity was negatively correlated with altitude and was different among seasons within the same site. Levels of diversity were highest in the summer, followed by autumn, spring, and winter. The influence of climate and time of deposition on the colonization of sheep faeces by nematode-trapping fungi was examined in Brazil by Saumell and Padilha (2000). Cow and heifer faecal samples were examined monthly for the presence of nematophagous fungi. Nematode-trapping fungi were present in the cow faeces during the dry months of August and September. They were also recovered at the beginning and middle of the rainy season: one isolate of *Arthrobotrys musiformis* in October, and one isolate of *Arthrobotrys gampsosporum* (synonym: *Monacrosporium gampsosporum*) and one unidentified fungus which produced adhesive buds in December. In the individual samples collected from heifers, fungi were present only in the months of September (end of dry season) and March (end of rainy season). A strong seasonal variation in species densities in the rhizosphere of barley, pea and white mustard was revealed in a study in Sweden, with the highest numbers occurring during the late summer and autumn months. In all, the species diversity and density of nematode-trapping fungi may correlate with seasonal variation.

Differences in Growth Rate and Competitive Saprobic Ability

Aside from their differences in distribution and seasonal variation of occurrence, different species of nematode-trapping fungi demonstrate wide differences in their growth rate, competitive saprobic ability, effect on free-living nematode populations and ability to produce traps spontaneously in the same habitat. Several studies have indicated that species producing adhesive reticulate traps tended to have the most rapid growth rates and the highest saprobic ability. Cooke (1963) indicated that the development of predaceous efficiency has been accompanied by a tendency to lose those characters associated with an efficient saprobic existence in the soil; which is rapid growth rate and good competitive saprobic ability. Observations by fluorescence microscopy and isolation have verified this hypothesis (Saxena and Lysek 1993). Several studies have indicated that *Arthrobotrys oligospora* has a strong ability to trap nematodes in a variety of conditions such as the presence of saprobic competitors (Quinn 1987), high carbon to nitrogen ratios in soils (Jaffee 2004b), or in badly polluted environments such as metalliferous mines (Mo et al. 2006). *Arthrobotrys oligospora* was among the most abundant of taxa in many of the so called stressful environments (Mo et al. 2006, 2008). In comparison to the same three-dimensional network producing species, *Arthrobotrys musiformis*, the isolates of *Arthrobotrys oligospora* has less demanding grow requirements and sporulates in different culture media and in two environments (B.O.D at 25±1 °C and the environment of the Laboratory) (Soares et al. 2009). The reason for its competitive advantage and the mechanism governing population dynamics is not understood (Persmark and Jansson 1997), but taxa with three-dimensional networks may have a better competitive saprobic ability. Such knowledge could help in designing strategies to ensure long-term survival and adaptation of nematode-trapping fungi in biological control methodologies.

Abiotic Factors Affecting Distribution and Diversity

In soil environments, the growth rate and competitive saprobic ability are strongly affected by several abiotic factors, such as organic matters, vitamins, biotin, abscisic acid, sodium nitroprusside, metals, temperature and pH.

Nematode-trapping fungi are frequently found in animal faeces or soils with abundant organic matter. The presence of farmyard manure results in increased amounts of organic matter, numbers of propagules of predatory and endoparasitic fungi, and numbers of bacteria and nematodes (Dackman et al. 1987). Carbohydrate nutrition of nematode-trapping fungi was assessed by Satchuthananthavale and Cooke (1967a). The affect of organic amendments on *Dactylellina haptotyla* and *Arthrobotrys oligospora* populations was assessed by soil cages (Jaffee 2004a). The population density and trap formation ability of *Dactylellina haptotyla* were most enhanced by the addition of a small quantity of alfalfa. The population density of *Arthrobotrys oligospora* however was most enhanced with addition of larger quantities of alfalfa. *Arthrobotrys oligospora* however, trapped fewer or no nematodes, regardless of the alfalfa amendment.

Trapping fungi can also respond positively to other organic substrates, such as lupine leaves, grass leaves, dead isopods, dead moth larvae, isopod faeces, deer faeces, shrimp shells, and powdered chitin (Nguyen et al. 2007). Positive responses of *A. oligospora* were directly related to the quantity of nitrogen added with each substrate, and those substrates that caused large increases in resident nematodes usually caused large increases in *Arthrobotrys oligospora*. Interestingly, low C:N ratios (dead isopods, lupine leaves, dead moth larvae) usually caused large increases in *Arthrobotrys oligospora* populations, whereas those with higher C:N ratios (isopod faeces, deer faeces, grass leaves) did not. An exception was chitin powder, which had a low C:N ratio, but which did not cause *Arthrobotrys oligospora* to proliferate.

Other factors such as vitamins, biotin, abscisic acid (ABA), sodium nitroprusside and metals amounts in soils also affects the growth of nematode-trapping fungi. *Arthrobotrys oligospora* and *Arthrobotrys thaumasia* which form adhesive networks, and *Drechslerella anchonia, D. dactyloides, D. bembicodes* and *D. doedycoides* which form constricting rings, require thiamine for growth (Satchuthananthavale and Cooke 1967b). With the exception of *Arthrobotrys thaumasia*, all species are deficient in biotin. The average number of traps (constricting rings) per colony and the percentage of nematodes (*Caenorhabditis elegans*) trapped were greatly increased by the addition of ABA, but greatly suppressed when sodium nitroprusside was added (Xu et al. 2011).

Rosenzweig and Pramer (1980) measured the impact of various concentrations of divalent cadmium, zinc, or lead on the growth and morphogenesis of seven species of nematode-trapping fungi. Cadmium was found to exhibit the greatest toxicity followed by zinc and lead. In most cases, inhibition of growth was directly correlated with a decreased capacity to form traps. The activity of the collagenase was less sensitive than was growth or trap formation to heavy-metal inhibition. Swe (2009) found that a strain of *Arthrobotrys musiformis* isolated from mangroves had a higher growth rate than terrestrial or freshwater strains at certain salinity levels. The mangrove strain grown on CMA with increasing levels of NaCl had longer, frequently non-septate mature conidia, but the numbers of trapping devices in all strains decreased when salinity levels increased.

Temperature and pH can also influence the radial growth and predatory activity of different isolates of nematode-trapping fungi. They could grow in medium with pH of 5.0 to 8.0 at 10–30 °C. The optimal conditions for fungal growth were pH 6.0–6.5 and 20–25 °C. Fernandez et al. (1999) found all nematode-trapping fungal isolates tested showed a higher growth rate at 20 °C. At 10 and 15 °C, all *Arthrobotrys flagrans* (synonym: *Duddingtonia flagrans*) isolates showed very similar patterns of radial growth at both constant and fluctuating temperatures. At 20 °C, they grew significantly faster at constant, than at fluctuating temperatures.

Competition Between Nematode-Trapping Fungi and Nematodes

In soil environments, nematode-trapping fungi and their target nematodes compete with each other to colonize a broad range of habitats. Fundamental studies on these tritrophic (plant, nematode and nematode-trapping fungi) or multitrophic (plant, soil microorganisms, nematode and nematode-trapping fungi) interactions under natural conditions are required from a fungal ecology perspective.

Suppression of Nematodes by Nematode-Trapping Fungi

The density of nematodes and nematode-trapping fungi in soil can interact with each other. Cooke (1962) reported that increasing the amount of sucrose added to soil increased the nematode population, but resulted in a decrease in predacious activity of the fungi. Experiments using nematode-free soil suggested that the presence of nematodes is necessary to initiate the formation of trapping organs, while fungi were incapable of maintaining an active predacious state in the absence of an organic energy source other than the nematodes. Koppenhofer et al. (1996) indicated that suppression of the entomopathogenic nematode, *Heterorhabditis hepialus*, was usually greater in the presence of background fungi than in pasteurized fungus free soil. The interaction between larvae of bovine gastrointestinal nematode parasites and nematode-trapping fungi has been studied in Denmark by Gronvold et al. (1993). High populations densities of *Arthrobotrys oligospora* gave rise to a significant reduction in the number of infective parasite larvae in the dung and surrounding herbage. Two nematophagous fungi, *Arthrobotrys oligospora* and *Dactylaria* sp., reduced the increase in nematode numbers due to organic enrichment in the untreated soil (Bouwman et al. 1996). Therefore, there is a trade-off balance of nematodes and nematode-trapping fungi in soil.

Persmark and Jansson (1997) found that a host threshold density exists for the nematophagous fungi in soils containing nematodes; the density of nematode-trapping fungi and the number of nematodes showed similar seasonal dynamics. The correlation between population density and trapping was further detected using soil cages by most probable number (MPN) procedures (Jaffee 2003), regardless of the fungus population density. *Arthrobotrys oligospora* and *A. eudermata* trapped few if any nematodes in soil, and consequently, trapping and fungus population density were not correlated. The correlation between population density and trapping was weak for *Drechslerella dactyloides,* but relatively strong for *Dactylellina ellipsospora* and *D. haptotyla*. Therefore the trade-off balance between nematodes and different nematode-trapping fungi varies in different fungal species.

Competition Between Different Fungal Species and Nematodes

Competition between nematode-trapping fungi and nematodes differs among species of fungi (Singh et al. 2013). Infection of individual nematodes by more than one fungus has been reported by Barron (1977). However, in the presence of hyphae of different fungal species, especially *Harposporium* sp., zoosporangia of the endoparasitic fungus *Catenaria* have less pathogenic ability. Gray (1983a) found both endoparasites and predators display a small degree of selectivity, and that any selection is most likely due to the anatomy of the host or prey and the mode of infection or capture of the fungus. In nematodes infected with the endoparasitic nematophagous fungus *Drechmeria coniospora* and the predatory fungus *Arthrobotrys oligospora*, the hyphae of the latter were usually dead or degenerated, indicating that the former is antagonist (Dijksterhuis et al. 1994). Jaffee (2000) found that the nematophagous fungus *Hirsutella rhossiliensis* was quite sensitive to biotic inhibition when formulated as pelletized hyphae, but insensitive as parasitized nematodes. When the same was done with the nematode-trapping fungus *Arthrobotrys haptotyla*, it however, exhibited the opposite trend. *Arthrobotrys haptotyla* was more sensitive to biotic inhibition when added to soil as fungus-parasitized nematodes than as pelletized hyphae. With the exception of *Hirsutella indica* (neutral to Myzocytium) and *Steinernema* sp. (neutral to Catenaria), El-Borai et al. (2011) indicated that the tested nematodes were repelled by activated *Arthrobotrys* species, but attracted to activated *Myzocytium* and *Catenaria* species. Competition between nematode-trapping fungi and nematodes can be detected in the soil environments and it has been carefully studied as a density-dependent parasitism.

Density-Dependent Parasitism

Competition between nematode-trapping fungi and nematodes can be qualified as density-dependent parasitism (Jaffee et al. 1989). Density-dependent suppression of hosts by parasites is defined as a regulation. The probability of a host being attacked by a parasite is often thought to be dependent on host density, which is a common phenomenon among biological organisms (Alexander 1981). Regulation of soil-borne nematodes by fungal and bacterial parasites had attracted little attention until the end of the 1980s.

Temporal density-dependent parasitism of the plant-parasitic nematode *Criconemella xenoplax* by *Hirsutella rhossiliensis* was investigated using spatial sampling (Jaffee et al. 1989). Patches of soil in which the nematode and fungus interacted were assumed to possess similar density-dependent dynamics. The relative shallowness of the increase or decrease and the large variation in percentage parasitism was not explained by nematode density, suggesting that *H. rhossiliensis* is a weak regulator of *Criconemella xenoplax* population density. This appeared however, not to be the case in consequent experiments. In a laboratory soil experiment, the proportion of parasitism increased to nearly 100% when a high nematode density was maintained, but declined to nearly 0% at low densities (Jaffee et al. 1992). In a field survey where many samples were collected at one time in a mature peach orchard, the percentage of nematodes parasitized correlated well with the nematode density (Jaffee et al. 1991). Substrates other than nematodes were therefore thought to be important to these parasitic fungi. This may be the case with other nematode trapping fungi since nematode-trapping fungi showed different increases in percentage parasitism as compared to host density (Jaffee et al. 1993). Recent greenhouse trials showed that parasitism of *Hirsutella rhossiliensis* OWVT-1 was strongly correlated to the density of the soybean cyst nematode (Liu et al. 2009). Parasitism of the soybean cyst nematode by *H. rhossiliensis* was dependent on fungal density, but independent of nematode density. These findings lead to the development of a new mathematical model which explains the complex competition between nematodes and nematode-trapping fungi (Yang et al., unpublished data).

Conidial-Trap Formation

The formation of trapping devices is a strong indicator that fungi and nematodes interact and compete with each other, suggesting that the nutrients needed by nematode-trapping fungi switch from saprotrophism to parasitism. Ananko and Teplyakova (2011) investigated the optimal concentrations of carbon and nitrogen sources such as sucrose, ammonium ions and tryptone to promote trap formation of the nematophagous fungus *Arthrobotrys flagrans* with the nematode *Panagrellus redivivus*. Persmark and Jansson (1997) found that the degree of competition required to induce conidial-trap formation was lower for more parasitic nematode-trapping fungi than for saprobic ones, indicating that conidial traps are partly a response to competition with other microorganisms for nutrients. Whether a third party, a chemical compound or another organism might be involved in the process when nematode-trapping fungi and nematodes interact and compete with each other, is not well understood.

Conidial traps were formed when conidia were allowed to germinate in cow dung (Dackman et al. 1992), fungistatic soil (Mankau 1980), rhizosphere soil or soil extracts (Persmark and Nordbring-Hertz 1997), and the formation of conidial traps was believed to be a response to nutrient deprivation due to strong nutrient competition between soil microorganisms. Recently, a bacterial species has been identified as important inducer of conidial traps, mycelial traps and conidia in the nematode-trapping fungus *Arthrobotrys oligospora* (Li et al. 2011a). Trap formation in *Arthrobotrys oligospora* may be stimulated by bacterial metabolites that are released into the environment, or by the direct contact between the fungus and the bacterium. In the natural soils, some native bacteria could also be nematophagous (Niu et al. 2010), employing fungi to trap and kill the nematodes which they also then utilize. The population dynamics of these three organisms needs further research.

Sensitivity to Fungistasis of Nematode-Trapping Hyphomycetes

Attempts to establish nematode-trapping fungi in agricultural soils for the purposes of biological control of pathogenetic nematodes are unlikely to meet with success all the time. There is a large possibility that the nematode-trapping fungi are sensitive to the fungistasis of the soil. Soil fungistasis was first described by Dobbs and Hinson (1953), and results in the suppression of either germination or growth of fungi and is widespread in most natural soils. The intensity of fungistasis varies depending on the physical and chemical properties of soils, as well as on soil microbial activity (Dobbs and Gash 1965; Lockwood 1977; Mondal and Hyakumachi 1998; Alabouvette 1999). Unfortunately, most nematode-trapping hyphomycetes could not germinate and grow normally in soils due to the soil fungistasis (Zhou and Mo 2002).

Two popular hypotheses have been proposed to account for the mechanism of soil fungistasis – nutrient-deprivation (Lockwood 1964, 1977) and antibiosis (Romine and Baker 1973). Nutrient-deprivation means that energy sources are decreased by strong competition by other soil microorganisms, resulting in lack of nutrients for the germinating fungal spores. Antibiosis emphasizes that soil fungistasis can be caused by antifungal compounds produced by other soil microorganisms. Recent developments suggest that specific groups of microorganisms might play an important role in the process of soil fungistasis (Alabouvette 1999; Ellis et al. 2000; De Boer et al. 2003).

Microbial community composition is also an important factor in determining soil fungistasis (De Boer et al. 2003) and the presence and antifungal activity of pseudomonads may be essential in this respect. Xu et al. (2004) assessed the effects of soils (soil direct fungistasis) and volatile compounds (VOCs) produced by natural soils (soil volatile fungistasis) on spore germination of several fungi. Some VOCs, trimethylamine, 3-methyl-2-pentanone, dimethyl disulfide, methyl pyrazine, 2,5-dimethyl-pyrazine, benzaldehyde, *N, N*-dimethyloctylamine and nonadecane, were produced by various fungistatic soils. They concluded that soil fungistasis and VOC fungistasis were significantly correlated ($P<0.001$). The microbial origins of volatiles were further elaborated by Zou et al. (2007); among the 1018 bacterial isolates tested, 328 were found to produce antifungal volatiles that could inhibit spore germination and mycelial growth of two nematicidal fungi, *Paecilomyces lilacinus* and *Pochonia chlamydosporia*. Li et al. (2008) characterized the soil bacteria producing non-volatile fungistatic substances and 24.67 % of total strains showed fungistatic activity towards the nematophagous fungi *Paecilomyces lilacinus* and *Trichoderma viride* by producing non-volatile substances. Three genera, *Arthrobacter, Bacillus*, and *Pseudomonas*, were the most frequently encountered groups. Li et al. (2011b) tested the ability to suppress the conidial germination of nine biocontrol fungal agents in 287 agricultural soil samples collected from 26 provinces or autonomous regions of China. Ten soil samples that have stronger fungistatic intensity (germination inhibition rate 68.3%) to the target fungi and the soil actinobacteria involved in fungistasis in soils were also evaluated. Of the 1,000 isolates from those soil samples, 345 actinobacteria exhibited fungistatic activity to conidial germination of *Trichoderma viride* and *Paecilomyces lilacinus* with germination inhibition rates higher than 10%. The actinobacteria involved in the soil fungistasis had a close phylogenetic relationship with the members of *Streptomycetaceae*, *Microbacteriaceae*, *Micrococcaceae*, and *Nocardiaceae*. The negative effects of fungistasis on fungal agents certainly requires attention.

The antibiosis hypothesis for soil fungistasis is increasingly being shown to be correct (Xu et al. 2004; Zou et al. 2007). Except in the case of soil microorganisms such as bacteria and actinobacteria, fungal species may play a key role in soil fungistasis, and other compounds such as ammonia, benzaldehyde, ethylene, benzylamine, and aluminium ion have been identified as inhibiting factors (Zou et al. 2007). How these compounds act and transmit inhibit signals to bio-control candidates are still unknown and further studies are needed.

Genetic Diversity and Speciation

Molecular techniques have increasingly been used to examine phylogenetic relationships among nematode-trapping *Orbiliales* (Rubner 1996; Ahren et al. 1998; Scholler et al. 1999; Li et al. 2005; Yang et al. 2007; Yang et al. 2012; Cho et al. 2008), but they have seldom been used to study the ecology of nematode-trapping fungi. Restriction fragment length polymorphism (RFLP) and DNA sequencing of the ITS region have revealed cryptic species among morphologically similar isolates of *Arthrobotrys* and *Monacrosporium* (Persson et al. 1996; Ahrén et al. 2004; Meyer et al. 2005; Zhang et al. 2011). Studies on nematode-trapping fungi species diversity have mostly use the culture-based, "sprinkle plate" method (Drechsler 1937; Duddington 1951; Gray 1983a; Jaffee et al. 1996; Li et al. 2000). Briefly, for each sample, 1–2 g soil/water is sprinkled onto sterile corn meal agar (CMA, 20 g cornmeal, 18 g agar, and water to the final volume of 1,000 ml), inoculated with the free-living nematodes and incubated at room temperature. After 1 month, nematode-trapping fungi are examined using a dissecting microscope. The estimation of species diversity and frequency could be conducted directly or indirectly by subcultures. However, sprinkle plates may be biased toward detecting species of nematophagous fungi better able to grow in culture (Bailey and Gray 1989). A major drawback is also reproducibility as many biological and ecological interactions in the natural environment may vary, making it difficult to compare between sprinkle plate data studies (Gray 1983b; Gray and Bailey 1985).

Smith and Jaffee (2009) suggested that the combined use of *Orbiliales*-specific primers and culture-based techniques may benefit future studies of nematode-trapping fungi. They developed *Orbiliales*-specific ITS and 28s rDNA PCR primers to directly detect nematode-trapping fungi without culturing and compared the results of molecular detection with those obtained using a culture-based method. Of the eight species of nematode-trapping *Orbiliales* detected with the culture-based assay, only three were detected with PCR, but the molecular assay detected 18 species of uncultured *Orbiliales*, many of which are closely related to nematode-trapping fungi and fungal parasites of nematode eggs. PCR assays using species-specific primers located in the ITS regions have been developed for the rapid and accurate identification of *Duddingtonia flagrans* (Kelly et al. 2008).

Arthrobotrys oligospora is one of the most widely distributed and studied members of the predacious fungi. Its broad ecological distribution and efficient trapping ability have made *A. oligospora* a strong candidate for the control of plant and animal parasitic nematodes. The asexual state *A. oligospora* has a corresponding sexual morph in *Orbilia auricolor* (Pfister 1995) and in the future one of these names will be selcted for use and the other will become a synonym. It should be noted here that three other morphologically different nematode-trapping hyphomycetes: *Arthrobotrys cladodes* var. *macroides*, *A. yunnanensis*, and *A. psychrophilum* have also been identified as asexual morphs of the *Orbilia auricolor* species-complex (Mo et al. 2005). There are up to 29 and 31 different kinds of possible pleomorphisms and holomorphisms, respectively. There are 15 different possible ways of reproductive duality i.e. existence of a sexual and an asexual phase (Hennebert 1987; Shenoy et al. 2007). The genetic structure of the asexual morph and its culturally connected asexual morphs are not well understood. Possible phenotypic plasticity in *Arthrobotrys cladodes* var. *macroides*, *A. oligospora*, *A. yunnanensis*, and *A. psychrophilum* should be considered. There is an indication of a significant phenotypic variation influenced by salinity-gradient in some of the members of *Arthrobotrys* (Swe 2009).

At present, the genetic divergence and relationships among these closely related species, or within any of the individual species are not well understood. In an earlier report, the genetic variation of *Arthrobotrys flagrans*, which has become a promising biocontrol agent of animal parasitic nematodes, was investigated within a worldwide collection of 22 isolates (Ahrén et al. 2004). They also analyzed 12 isolates of *A. oligospora* and found a higher level of genetic variation in *A. oligospora* than in *A. flagrans*. A neighbour-joining tree based on the AFLP data showed no clear association between genotype and geographical origin within *A. flagrans* strains. Furthermore, the AFLP data suggest that *A. flagrans* is mainly clonal and no recombination could be detected, not even within the same country. Genetic studies of nematode-trapping fungi are however, rather limited.

Ahrén et al. (2004) also suggested that the genetic variation in *Arthrobotrys oligospora* might be geographically structured. However, limited sampling has prevented a comprehensive analysis of the potential effects of various ecological factors on genetic variation in these or other nematode-trapping fungi. Zhang et al. (2011) collected and analyzed 97 *A. oligospora* strains from diverse geographic locations and ecological niches in China in order to examine the patterns of genetic divergence, dispersal, and reproductive behavior. The ecological niches included common agricultural areas and forests, but also other specialized niches such as pristine and polluted aquatic environments, as well as salty and slag-containing soils.Different amounts of polymorphisms among the three tested genes were identified, and all three genes include both broadly distributed alleles and geographically restricted ones, especially to those in western China such as Yunnan, Guizhou, Sichuan, Tibet and Qinghai provinces. The large number of unique alleles and genotypes as well as their limited geographic distributions suggested that *A. oligospora* may have a large effective population size and that they might have undergone genetic differentiation during its adaptation to different ecological environments in different geographic regions.

Many species of nematode trapping fungi are likely to be species complexes (Persson et al. 1996; Ahrén et al. 2004; Meyer et al. 2005). Phylogeographic analyses revealed that 97 strains of *Arthrobotrys oligospora* isolated from China are from a species complex with at least three divergent lineages (cryptic species). Furthermore, the nested clade phylogeographic analyses (NCPA) revealed a significant geographical association in the distribution of the genetic variation of *A. oligospora* populations, at least one of the cryptic species experienced allopatric fragmentation in their history. In addition, there was significant geographic structuring with unambiguous evidence for localized recombination within two of the three lineages in nature.

Up to now, molecular ecology and intra-specific population studies on nematode-trapping fungi have been very limited. More extensive sampling of this kind of fungi using traditional sprinkle methodology and molecular detection from more broad areas is needed. Issues include modes of reproduction in nature, patterns of gene flow, and the relationships between genetic variation and ecological parameters which can reveal species boundaries and sometimes lead to revelation of new species. For example, an understanding of spatial and temporal patterns of genetic variation of nematode-trapping fungi could help design a sustainable biocontrol application strategy (Zhang et al. 2013).

References

Ahrén, D., Ursing, B. M., & Tunlid, A. (1998). Phylogeny of nematode-trapping fungi based on 18S rDNA sequences. *FEMS Microbiology Letters, 158,* 179–184.

Ahrén, D., Faedo, M., Rajastiekar, B., & Tunlid, A. (2004). Low genetic diversity among isolates of the nematode-trapping fungus *Duddingtonia flagrans*: Evidence for recent worldwide dispersion from a single common ancestor. *Mycological Research, 108,* 1205–1214.

Alabouvette, C. (1999). *Fusarium* wilt suppressive soils: An example of disease suppressive soil. *Australasia Plant Pathology, 28,* 57–64.

Alexander, M. (1981). Why microbial predators and parasites do not eliminate their prey and hosts. *Annual Review Microbiology, 35,* 113–133.

Ananko, G., & Teplyakova, T. (2011). Factors responsible for transition of the *Duddingtonia flagrans* carnivorous fungus from the saprotrophic to the zootrophic nutrition type. *Microbiology (Reading, England), 80,* 188–193.

Bailey, F., & Gray, N. F. (1989). The comparison of isolation techniques for nematophagous fungi from soil. *Annals of Applied Biology, 114,* 125–132.

Barron, G. L. (Ed). (1977). *The nematode-destroying fungi.* [Topics in Mycology No. 1.]. Guelph: Canadian Biological Publications.

Boag, B., & Lopez-Llorca, L. V. (1989). Nematodes and nematophagous fungi associated with cereal fields and permanent pasture in Eastern Scotland. *Crop Research, 29,* 1–10.

Bouwman, L. A., Hoenderboom, G. H., van der Maas K. J., & de Ruiter P. C. (1996). Effects of nematophagous fungi on numbers and death rates of bacterivorous nematodes in arable soil. *Journal of Nematology, 28,* 26–35.

Braga, F. R., Araújo, J. V., Araujo, J. M., Tavela, A. O., Ferreira, S. R., Freitas Soares, F. E., Benjamin, L. D. A., & Frassy, L. N. (2011). Influence of the preservation period in silica-gel on the predatory activity of the isolates of *Duddingtonia flagrans* on infective larvae of cyathostomins (Nematoda: Cyathostominae). *Experimental Parasitology, 128,* 460–463.

Braga, F. R., Araújo, J. V., Soares, F. E. F., Araujo, J. M., Tavela, A. O., de Carvalho L. M., de Mello I. N. K., de Paula A. T., Lelis, R., & Queiroz, J. H. (2013). Interaction of the nematophagous fungus *Duddingtonia flagrans* on *Amblyomma cajannense* engorged females and enzymatic characterisation of its chitinase. *Biocontrol Science and Technology, 23*(5), 584–594.

Carneiro, R., Gomes, M., Hidalgo-Diaz, L., Martins, I., Ayres, deS. SilvaK. F., Guimaraes, deS. M., & Tigano, M. S. (2011). Effect of nematophagous fungi on reproduction of *Meloidogyne enterolobii* on guava (*Psidium guajava*) plants. *Nematology, 13,* 721–728.

Cho, C. H., Kang, D. S., Kim, Y. J., & Whang, K. S. (2008). Morphological and Phylogenetic Characteristics of a Nematophagous Fungus, *Drechslerella brochopaga* Kan-23. The Korean. *Journal of Microbiology, 44,* 63–68.

Cooke, R. C. (1962). The ecology of nematode-trapping fungi in the soil. *Annals of Applied Biology, 50,* 507–513.

Cooke, R. C. (1963). Ecological characteristics of nematode-trapping Hyphomycetes I. Preliminary studies. *Annals of Applied Biology, 52,* 431–437.

Dackman, C., Olsson, S., Jansson, H. B., Lundgren, B., & Nordbring-Hertz, B. (1987). Quantification of predatory and endoparasitic nematophagous fungi in soil. *Microbial Ecology, 13,* 89–93.

Dackman, C., Jansson, H. B., & Nordbring-Hertz, B. (1992). Nematophagous fungi and their activities in soil. In G. Stotzky & B. J.M. (Eds.), *Soil biochemistry* (pp. 95–103). New York: Marcel Dekker.

De Boer, W., Verheggen, P., Klein Gunnewiek, P. J. A., George, A. K., & Johannes, A. V. (2003). Microbial community composition affects soil fungistasis. *Applied and Environmental Microbiology, 69,* 835–844.

De Almeida, G. L., Santurio, J. M., Filho, J. O., Zanette, R. A., Camillo, G., Flores, A. G., da Silva, J. H., & de, laR. M. L. (2012). Predatory activity of the fungus *Duddingtonia flagrans* in equine strongyle infective larvae on natural pasture in the southern region of Brazil. *Parasitological Research, 110,* 657–662.

Dijksterhuis, J., Sjollema, K. A., Veenhuis, M., & Harder, W. (1994). Competitive interactions between two nematophagous fungi during infection and digestion of the nematode *Panagrellus redivivus*. *Mycological Research, 98,* 1458–1462.

Dobbs, C. G., & Hinson, W. H. (1953). A widespread fungistasis in soil. *Nature, 172,* 197–199.

Dobbs, C. G., & Gash, M. J. (1965). Microbial and residual mycostasis in soils. *Nature, 207,* 1354–1356.

Drechsler, C. (1937). Some hyphomycetes that prey on free-living terricolous nematodes. *Mycologia, 29,* 447–552.

Duddington, C. L. (1951). The ecology of predaceous fungi I preliminary survey. *Transactions of the British Mycological Society, 34,* 322–331.

Duddington, C. L. (1954). Nematode-destroying fungi in agricultural soils. *Nature, 173,* 500–501.

Durand, D. T., Boshoff, H. M., Michael, L. M., & Krecek, R. C. (2005). Survey of nematophagous fungi in South Africa: The Onderstepoort. *Journal of Veterinary Research, 72,* 185–187.

El-Borai, F. E., Campos-Herrera, R., Stuart, R. J., & Duncan, L. W. (2011). Substrate modulation, group effects and the behavioral responses of entomopathogenic nematodes to nematophagous fungi. *Journal of Invertebrate Pathology, 106,* 347–356.

Ellis, R. J., Timms-Wilson, T. M., & Bailey, M. J. (2000). Identification of conserved traits in fluorescent pseudomonads with antifungal activity. *Environmental Microbiology, 2,* 274–284.

Falbo, M. K., Soccol, V. T., Sandini, I. E., Vicente, V. A., Robl, D., & Soccol, C. R. (2013). Isolation and characterization of the nematophagous fungus *Arthrobotrys conoides. Parasitology Research, 112,* 177–185.

Fernandez, A. S., Larsen, M., Wolstrup, J., Gronvold, J., Nansen, P., & Bjorn, H. (1999). Growth rate and trapping efficacy of nematode-trapping fungi under constant and fluctuating temperatures. *Parasitology Research, 85,* 661–668.

Gray, N. F. (1983a). Ecology of nematophagous fungi: Panagrellus redivivus as the target organism. *Plant and Soil, 73,* 293–297.

Gray, N. F. (1983b). Ecology of nematophagous fungi: Distribution and habitat. *Annals of Applied Biology, 102,* 501–509.

Gray, N. F. (1987). Nematophagous fungi with particular reference to their ecology. *Biological Review, 62,* 254–304.

Gray, N. F., & Bailey, F. (1985). Ecology of nematophagous fungi: Vertical distribution in a deciduous woodland. *Plant and Soil, 86,* 217–223.

Gronvold, J., Wolstrup, J., Nansen, P., Henriksen, S. A., Larsen, M., & Bresciani, J. (1993). Biological control of nematode parasites in cattle with nematode-trapping fungi: A survey of Danish studies. *Veterinary Parasitology, 48,* 311–325.

Hao, Y., Mo, M., Su, H., & Zhang, K. Q. (2005). Ecology of aquatic nematode-trapping hyphomycetes in southwestern China. *Aquatic Microbial Ecology, 40,* 175–181.

Hennebert, G. L. (Ed.). (1987). *Pleoanamorphy and its nomenclatural problem. Pleomorphic fungi: The diversity and its taxonomic implications.* Tokyo: Kodansha Ltd

Hidalgo-Diaz, L., Bourne, J. M., Kerry, B. R., & Rodriguez, M. G. (2000). Nematophagous *Verticillium* spp. in soils infested with *Meloidogyne* spp. in Cuba: Isolation and screening. *International Journal of Pest Management, 46,* 277–284.

Jaffee, B. A. (2000). Augmentation of soil with the nematophagous fungi *Hirsutella rhossiliensis* and *Arthrobotrys haptotyla. Phytopathology, 90,* 498–504.

Jaffee, B. A. (2003). Correlations between most probable number and activity of nematode-trapping fungi. *Phytopathology, 93,* 1599–1605.

Jaffee, B. A. (2004a). Do organic amendments enhance the nNematode-trapping fungi *Dactylellina haptotyla* and *Arthrobotrys oligospora*? *Journal of Nematology, 36,* 267–275.

Jaffee, B. A. (2004b). Wood, nematodes, and the nematode-trapping fungus *Arthrobotrys oligospora. Soil Biology & Biochemistry, 36,* 1171–1178.

Jaffee, B. A., Gaspard, J. T., & Ferris, H. (1989). Density-dependent parasitism of the soil-borne nematode *Criconemella xenoplax* by the nematophagous fungus *Hirsutella rhossiliensis. Microbial Ecology, 17,* 193–200.

Jaffee, B., Muldoon, A., Anderson, C., & Westerdahl, B. (1991). Detection of the nematophagous fungus *Hirsutella rhossiliensis* in California sugarbeet fields. *Biological Control, 1,* 63–67.

Jaffee, B., Phillips, R., Muldoon, A., & Mangel, M. (1992). Density-Dependent Host-Pathogen Dynamics in Soil Microcosms. *Ecology, 73,* 495–506.

Jaffee, B. A., Tedford, E. C., & Muldoon, A. E. (1993). Tests for density-dependent parasitism of nematodes by nematode-trapping and endo-parasitic fungi. *Biological Control, 3,* 329–336.

Jaffee, B. A., Strong, D. R., & Muldoon, A. E. (1996). Nematode-trapping fungi of a natural shrubland: Tests for food chain involvement. *Mycologia, 88,* 554–564.

Jaffee, B. A., Ferris, H., & Scow, K. M. (1998). Nematode-trapping fungi in organic and conventional cropping systems. *Phytopathology, 88,* 344–350.

Kelly, P., Good, B., Fitzpatrick, R., Hanrahan, J. P., & de Waal T. D. (2008). Development and application of a PCR diagnostic assay for the accurate and rapid identification of the nematophagous fungus Duddingtonia flagrans. *Mycological Research, 112,* 1026–1030.

Kelly, P., Good, B., Hanrahan, J. P., Fitzpatrick, R., & de Waal T. (2009). Screening for the presence of nematophagous fungi collected from Irish sheep pastures. *Veterinary Parasitology, 165,* 345–349.

Koppenhofer, A. M., Jaffee, B. A., Muldoon, A. E., Strong, D. R., & Kaya, H. K. (1996). Effect of nematode-trapping fungi on an entomopathogenic nematode originating from the same field site in California. *Journal of Invertebrate Pathology, 68,* 246–252.

Kumar, N., Singh, R. K., & Singh, K. P. (2011). Occurrence and colonization of nematophagous fungi in different substrates, agricultural soils and root galls. *Archives Of Phytopathology And Plant Protection, 44,* 1182–1195.

Li, T. F., Zhang, K. Q., & Liu, X. Z. (Eds.). (2000). *Taxonomy of nematophagous fungi*. Beijing: Chinese Scientific & Technological Publication.

Li, Y., Hyde, K. D., Jeewon, R., Cai, L., Vijaykrishna, D., & Zhang, K. Q. (2005). Phylogenetics and evolution of nematode-trapping fungi (Orbiliales) estimated from nuclear and protein coding genes. *Mycologia, 97,* 1034–1046.

Li, Z., Zou, C., He, Y., Mo, M., & Zhang, K. Q. (2008). Phylogenetic analysis on the bacteria producing non-volatile fungistatic substances. *The Journal of Microbiology, 46,* 250–256.

Li, L., Ma, M., Liu, Y., Zhou, J., Qu, Q., Lu, K., Fu, D., & Zhang, K. Q. (2011a). Induction of trap formation in nematode-trapping fungi by a bacterium. *FEMS Microbiology Letters, 322,* 157–165.

Li, Z. F., Xu, C. K., Zou, C. S., Qin, X. J., Qiu, H. Y., Qun, D. C., & Mo, M. H. (2011b). Fungistatic intensity of agricultural soil against fungal agents and phylogenetic analysis on the actinobacteria involved. *Current Microbiology, 62*(4), 1152–1159.

Linford, M. B., Yap, F., & Oliveira, J. M. (1938). Reduction of soil populations of the rootknot nematode during decomposition of organic matter. *Soil Science, 45,* 127–140.

Liu, X., & Zhang, K. Q. (1994). Nematode-trapping species of *Monacrosporium* with special reference to two new species. *Mycological Research, 98,* 862–868.

Liu, X., & Zhang, K. Q. (2003). *Dactylella shizishanna* sp. nov., from Shizi Mountain, China. *Fungal Diversity, 14,* 103–107.

Liu, X., Xiang, M., & Che, Y. (2009). The living strategy of nematophagous fungi. *Mycoscience, 50,* 20–25.

Lockwood, J. L. (1964). Soil fungistasis. *Annual Review of Phytopathology, 2,* 351–362.

Lockwood, J. L. (1977). Fungistasis in soils. *Biological Reviews, 52,* 1–43.

Mahoney, C. J., & Strongman, D. B. (1994). Nematophagous fungi from cattle manure in four states of decomposition at three sites in Nova Scotia, Canada. *Mycologia, 86,* 371–375.

Mankau, R. (1980). Biocontrol: Fungi as nematode control agents. *Journal of Nematology, 12,* 244–252.

Meyer, S. L. F., Carta, L. K., & Rehner, S. A. (2005). Morphological variability and molecular phylogeny of the nematophagous fungus *Monacrosporium drechsleri*. *Mycologia, 97,* 405–415.

Miao, Z., & Liu, X. (2003). Note on the seasonal fluctuation of the nematodetrapping fungi population in orchards at the vicinity of Beijing. *Mycosystema, 22,* 77–81.

Mo, M., Huang, X., Zhou, W., Huang, Y., Hao, Y., & Zhang, K. Q. (2005). *Arthrobotrys yunnanensis* sp. nov., the fourth anamorph of *Orbilia auricolor*. *Fungal Diversity, 18,* 107–115.

Mo, M. H., Chen, W. M., Su, H. Y., Zhang, K. Q., Duan, C. Q., & He, D. M. (2006). Heavy metal tolerance of nematode-trapping fungi in lead-polluted soils. *Applied Soil Ecology, 31,* 11–19.

Mo, M. H., Chen, W. M., Yang, H. R., & Zhang, K. Q. (2008). Diversity and metal tolerance of nematode-trapping fungi in Pb-polluted soils. *Journal of Microbiology, 46,* 16–22.

Mondal, S. N., & Hyakumachi, M. (1998). Carbon loss and germinability, viability, and virulence of chlamydospores of Fusarium solani f. sp. phaseoli after exposure to soil at different pH levels temperatures, and matric potentials. *Phytopathology, 88,* 148–155.

Morton, C. O., Mauchline, T. H., Kerry, B. R., & Hirsch, P. R. (2003). PCR-based DNA fingerprinting indicates host-related genetic variation in the nematophagous fungus Pochonia chlamydosporia. *Mycological Research, 107,* 198–205.

Nguyen, V. L., Bastow, J. L., Jaffee, B. A., & Strong, D. R. (2007). Response of nematode-trapping fungi to organic substrates in a coastal grassland soil. *Mycological Research, 111,* 856–862.

Niu, Q., Huang, X., Zhang, L., Xu, J., Yang, D., Wei, K., Niu, X., An, Z., Bennett, J. W., Zou, C., Yang, J., & Zhang, K. Q. (2010). A Trojan horse mechanism of bacterial pathogenesis against nematodes. *Proceedings of the National Academy of Sciences of the United States of America, 107,* 16631–16636.

Palma-Guerrero, J., Jansson, H. B., Salinas, J., & Lopez-Llorca, L. V. (2008). Effect of chitosan on hyphal growth and spore germination of plant pathogenic and biocontrol fungi. *Journal of Applied Microbiology, 104,* 541–553.

Paraud, C., Lorrain, R., Pors, I., & Chartier, C. (2012). Administration of the nematophagous fungus *Duddingtonia flagrans* to goats: An evaluation of the impact of this fungus on the degradation of faeces and on free-living soil nematodes. *Journal of Helminthology, 86,* 95–103.

Pathak, E., El-Borai, F. E., Campos-Herrera, R., Johnson, E. G., Stuart, R. J., Graham, J. H., & Duncan, L. W. (2012). Use of real-time PCR to discriminate parasitic and saprophagous behaviour by nematophagous fungi. *Fungal Biology, 116,* 563–573.

Persmark, L., & Jansson, H. B. (1997). Nematophagous fungi in the rhizosphere of agricultural crops. *FEMS Microbiology Ecology, 22,* 303–312.

Persmark, L., & Nordbring-Hertz, B. (1997). Conidial trap formation of nematode-trapping fungi in soil and soil extracts. *FEMS Microbiology Ecology, 22,* 313–323.

Persson, Y., Erland, S., & Jansson, H. B. (1996). Identification of the nematode-trapping fungi using RFLP analysis of the PCR-amplified ITS region of ribosomal DNA. *Mycological Research, 100,* 531–534.

Pfister, D. (1995). Two *Arthrobotrys* anamorphs from *Orbilia auricolor*. *Mycologia, 87,* 684–688.

Quinn, M. A. (1987). The influence of saprophytic competition on nematode predation by nematode-trapping fungi. *Journal of Invertebrate Pathology, 49,* 170–174.

Romine, M., & Baker, R. (1973). Soil fungistasis: Evidence for an inhibitory factor. *Phytopathology, 63,* 756–759.

Rosenzweig, W. D., & Pramer, D. (1980). Influence of cadmium, zinc, and lead on growth, trap formation, and collagenase activity of nematode-trapping fungi. *Applied and Environmental Microbiology, 40,* 694–696.

Rubner, A. (1994). Predaceous fungi from Ecuador. *Mycotaxon, 51,* 143–151.

Rubner, A. (1996). Revision of predacious hyphomycetes in the *Dactylella-Monacrosporium* complex. *Studies in Mycology, 39,* 1–134.

Sünder, A., & Lysek, G. (1988). Quantitative investigations on nematode-trapping hyphomycetes from woodland soils. *FEMS Microbiology Letters, 53,* 285–290.

Satchuthananthavale, V., & Cooke, R. C. (1967a). Carbohydrate nutrition of some nematode-trapping fungi. *Nature, 214,* 321–322.

Satchuthananthavale, V., & Cooke, R. C. (1967b). Vitamin requirements of some nematode-trapping fungi. *Transactions of the British Mycological Society, 50,* 221–228.

Saumell, C. A., & Padilha, T. (2000). Influence of weather and time of deposition on sheep faeces colonization by nematophagous fungi in the Mata region of Minas Gerais State, Brazil. *Applied Soil Ecology, 14,* 63–70.

Saxena, G., & Lysek, G. (1993). Observation of nematophagous fungi in natural soils by fluorescence microscopy and their correlation with isolation. *Mycological Research, 97,* 1005–1011.

Saxena, G., & Mukerji, K. G. (1991). Distribution of nematophagous fungi in Varanasi, India. *Nova Hedwigia, 52,* 487–495.

Saxena, G. (2008). Observations on the occurrence of nematophagous fungi in Scotland. Applied soil ecology: A section of. *Agriculture, Ecosystems & Environment, 39,* 352–357.

Scholler, M., Hagedorn, G., & Rubner, A. (1999). A reevaluation of predatory orbiliaceous fungi. II. A new generic concept. *Sydowia, 51,* 89–113.

Shenoy, B. D., Jeewon, R., & Hyde, K. D. (2007). Impact of DNA sequence-data on the taxonomy of anamorphic fungi. *Fungal Diversity, 26,* 1–54.

Shams Ghahfarokhi, M., Razzaghi Abyaneh, M., Ranjbar Bahadori, S., Eslami, A., Zare, R., & Ebrahimi, M. (2004). Screening of soil and sheep faecal samples for predacious fungi: Isolation and characterization of the nematode-trapping fungus *Arthrobotrys oligospora*. *Iranian Biomedical Journal, 8,* 135–142.

Singh, U. B., Sahu, A., Sahu, N., Singh, R., Prabha, R., Singh, D. P., Sarma, B., & Manna, M. (2012). Co-inoculation of *Dactylaria brochopaga* and *Monacrosporium eudermatum* affects disease dynamics and biochemical responses in tomato (*Lycopersicon esculentum* Mill.) to enhance bio-protection against *Meloidogyne incognita*. *Crop Protection, 35,* 102–109.

Singh, U. B., Sahu, A., Sahu, N., Singh, B. P., Singh, R., Singh, D. P., Jaiswal, R., Sarma, B., Singh, H., & Manna, M. (2013). Can endophytic *Arthrobotrys oligospora* modulate accumulation of defence related biomolecules and induced systemic resistance in tomato (*Lycopersicon esculentum* Mill.) against root knot disease caused by *Meloidogyne incognita*. *Applied Soil Ecology, 63,* 45–56.

Smith, M. E., & Jaffee, B. A. (2009). PCR primers with enhanced specificity for nematode-trapping fungi (Orbiliales). *Microbial Ecology, 58,* 117–128.

Soares, P. L. M., Nozaki, MdH., Barbosa, B. F. F., Santos, J. Md, & Barbosa, J. C. (2009). Growth and sporulation of two species of *Arthrobotrys* Corda in different culture media and two environments. *Bioscience Journal, 25,* 63–74.

Su, H., Hao, Y., Mo, M., & Zhang, K. Q. (2007). The ecology of nematode-trapping hyphomycetes in cattle dung from three plateau pastures. *Veterinary Parasitology, 144,* 293–298.

Su, H., Liu, S., Li, Y., Cao, Y., Chen, M., & Yang, X. (2011). *Arthrobotrys latispora*, a new nematode-trapping fungus from southwest China. *Mycotaxon, 117,* 29–36.

Swe, A. (2009). Biodiversity, systematics and ecology of nematode-trapping fungi from Hong Kong. PhD Thesis. University of Hong Kong, Hong Kong.

Swe, A., Jeewon, R., Pointing, S. B., & Hyde, K. D. (2008). Taxonomy and molecular phylogeny of *Arthrobotrys mangrovispora*, a new marine nematode-trapping fungal species. *Botanica Marina, 51,* 331–338.

Swe, A., Jeewon, R., Pointing, S. B., & Hyde, K. D. (2009). Diversity and abundance of nematode-trapping fungi from decaying litter in terrestrial, freshwater and mangrove habitats. *Biodiversity and Conservation, 18,* 1695–1714.

Vilela, V. L., Feitosa, T. F., Braga, F. R., de Araujo J. V., Souto, D. V., Santos, H. E., Silva, G. L., & Athayde, A. C. (2012). Biological control of goat gastrointestinal helminthiasis by *Duddingtonia flagrans* in a semi-arid region of the northeastern Brazil. *Veterinary Parasitology, 188,* 127–133.

Vilela, V. L. R., Feitosa, T. F., Braga, F. R., de Araújo J. V., de Lucena S. C., Dantas, E. S., Athayde, A. C. R., & Silva, W. W. (2013). Efficacy of *Monacrosporium thaumasium* in the control of goat gastrointestinal helminthiasis in a semi-arid region of Brazil. *Parasitology Research,* 112(2), 871–877.

Wolstrup, J., Nansen, P., Gronvold, J., Henriksen, A. S., & Larsen, M. (1996). Toward practical biological control of parasitic nematodes in domestic animals. *Journal of Nematodology, 28,* 129–132.

Xu, C., Mo, M., Zhang, L., & Zhang, K. Q. (2004). Soil volatile fungistasis and volatile fungistatic compounds. *Soil Biology and Biochemistry, 36,* 1997–2004.

Xu, L. L., Lai, Y. L., Wang, L., & Liu, X. Z. (2011). Effects of abscisic acid and nitric oxide on trap formation and trapping of nematodes by the fungus *Drechslerella stenobrocha* AS6.1. *Fungal Biology, 115,* 97–101.

Yang, Y., Yang, E., An, Z., & Liu, X. (2007). Evolution of nematode-trapping cells of predatory fungi of the Orbiliaceae based on evidence from rRNA-encoding DNA and multiprotein sequences. *Proceedings of the National Academy of Sciences of the United States of America, 104,* 8379–8384.

Yang, E., Xu, L., Yang, Y., Zhang, X., Xiang, M., Wang, C., An, Z., & Liu, X. (2012). Origin and evolution of carnivorism in the Ascomycota (fungi). *Proceedings of the National Academy of Sciences of the United States of America, 109,* 10960–10965.

Yeates, G. W., Skipp, R. A., Chen, L. Y., Waghorn, T. S., & Potter, J. F. (2007). Temporal and spatial aspects of the influence on soil nematodes of depositing artificial pats of sheep faeces containing a range of parasite management agents. *Applied Soil Ecology, 37,* 106–117.

Zhang, Y., Yu, Z. F., Xu, J., & Zhang, K. Q. (2011). Divergence and dispersal of the nematode-trapping fungus *Arthrobotrys oligospora* from China. *Environmental Microbiology Reports, 3,* 763–773.

Zhang, Y., Qiao, M., Xu, J., Cao, Y., Zhang, K. Q., & Yu, Z. F. (2013). Genetic diversity and recombination in natural populations of the nematode-trapping fungus *Arthrobotrys oligospora* from China. *Ecology and Evolution, 3,* 312–325.

Zhou, W., & Mo, M. H. (2002). Soil fungistasis on the spore germination of fungi. *Journal of Yunnan University (Natural Sciences), 24,* 312–315.

Zou, C. S., Mo, M. H., Gu, Y. Q., Zhou, J. P., & Zhang, K. Q. (2007). Possible contributions of volatile-producing bacteria to soil fungistasis. *Soil Biology and Biochemistry, 39,* 2371–2379.

Chapter 5
Biological Control of Plant-Parasitic Nematodes by Nematophagous Fungi

Jinkui Yang and Ke-Qin Zhang

Contents

Abstract Plant-parasitic nematodes cause severe damage to world agriculture each year. Environmental and health concerns over the use of chemical pesticides has increased the need for alternative measures to control plant-parasitic nematodes. Nematophagous fungi, a natural enemy of nematodes, have received most attention in biological control of plant-parasitic nematodes. This is due to their specific ability to capture and kill nematodes. Nematophagous fungi are divided into four groups according to their mode of action against nematodes, and several fungi such as *Pochonia chlamydosporia* and *Paecilomyces lilacinus* have been developed as commercial biological nematicides. In this chapter, important nematode parasitic and antagonistic fungi, and their taxonomy, biology and their mode of action are discussed. Progress in the study of highly virulent fungal strains for nematode biological control, and application of nematode-antagonistic agents are also discussed.

K.-Q. Zhang (✉) · J. Yang
Laboratory for Conservation and Utilization of Bio-Resources, and Key Laboratory of Microbial Diversity in Southwest China, Ministry of Education, Yunnan University, 650091 Kunming, Yunnan, China
e-mail: kqzhang1@ynu.edu.cn

K.-Q. Zhang, K. D. Hyde (eds.), *Nematode-Trapping Fungi,* Fungal Diversity Research Series 23, DOI 10.1007/978-94-017-8730-7_5, © Mushroom Research Foundation 2014

Keywords Biological control • Formulation • Genetic engineering • Infection; nematode • Nematophagous fungus • Trap device

Introduction

Nematodes are one of the most diverse groups of animals on earth. Nematode species are difficult to distinguish, over 28,000 have been described, of which over 16,000 are parasitic (Hugot et al. 2001). A handful of soil will contain thousands of these microscopic worms, many of which are parasites of insects, plants or animals. Plant-parasitic nematodes are the major group, feeding and reproducing on living plants and they are capable of active migration in the rhizosphere, on aerial plant parts, and inside the plant. They can cause significant plant damage ranging from negligible injury to total destruction of plant material. Although a few nematode species feed on above ground plant parts, such as leaves, stems, flowers, and seeds, most of these parasites feed on underground parts of plants, including roots, bulbs, and tubers, and as a consequence nematode damage to plants cannot always be readily diagnosed. Plant-parasitic nematodes represent a major biotic stress on world crops, causing over $100 billion in annual crop losses (Chitwood 2003). The majority of crop losses caused by plant-parasitic nematodes are inflicted by relatively few species belonging to two main groups, root-knot nematodes (*Meloidogyne* spp.) and cyst nematodes (*Heterodera* and *Globodera* spp.) (Molinari 2011). The high impact of these specific nematodes on world agriculture is a result of their wide distribution and ability to attack every species of cultivated plant (Sasser 1980).

Several control methods have been reported, including using cover crops, green manure, organic or inorganic soil amendments, resistant cultivars, hot water treatment, crop rotation, and fallow treatment (Barker and Koenning 1998). Nematicides have been widely used to control plant-parasitic nematodes, but these compounds are often associated with detrimental environmental effects leading to substantially reduced availability and use in recent years. For example, methyl bromide, one of the most important chemical fumigants used to control nematodes and other pests, affects a wide range of organisms, including beneficial organisms, and is a chemical that contributes to the depletion of the Earth's ozone layer (Carpenter et al. 2001). In recent decades, concerns about environmental hazards of using chemical nematicides, and limited alternative crops for rotation, have led to the development of biological control agents as a component of crop protection. Biological control is defined as suppress the population density or impact of a specific pest organism by using living organisms (Eilenberg et al. 2001). Biological control agents can regulate plant-parasitic nematode populations, and numerous organisms including fungi, bacteria, viruses, nematodes and other invertebrates have antagonistic activity against plant-parasitic nematodes (Siddiqui and Mahmood 1996; Dong and Zhang 2006; Tian et al. 2007). Nematophagous fungi have received most attention for biological control of plant-parasitic nematodes due to their specific ability to capture and kill nematodes. In this chapter, we will focus on the different nematophagous fungi, mechanisms of action, and future prospects.

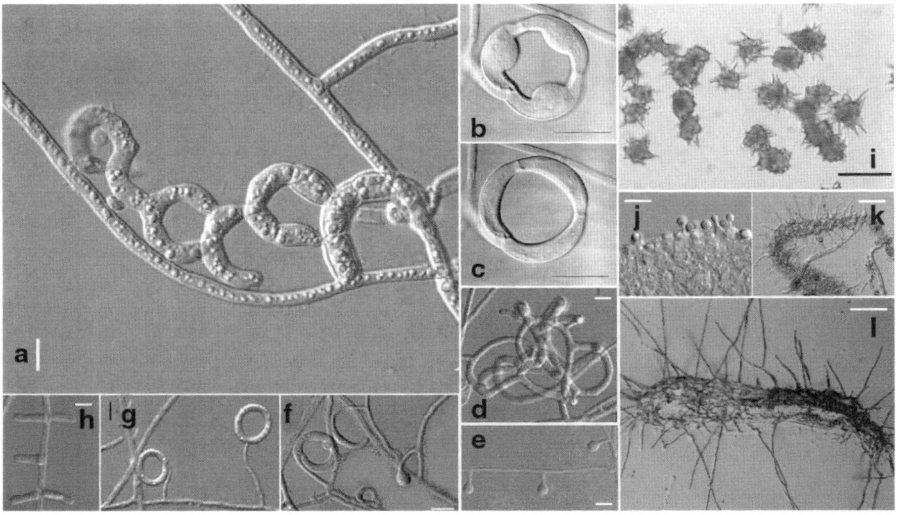

Fig. 5.1 Diversity of trapping structures and infection model of nematophagous fungi. **a** Adhesive network of *Dactylella dianchiensis*. Bar = 10 μm. **b** and **c** Constricting ring of *Arthrobotrys brochopaga*. Bar = 5 mm. **d** Adhesive network of *A. oligospora*. Bar = 10 μm. **e** Adhesive knobs of *Monacrosporium ellipososporum*. Bar = 10 μm. **f** adhesive knob and non-constricting rings of *M. candidum*. Bar = 10 μm. **g** Non-constricting rings of *M. candidum*. Bar = 10 μm. **h** Adhesive branches of *M. cionopagum*. Bar = 10 μm. **i** Acanthocytes of *Stropharia rugosoannulata*. Bar = 20 μm. **j** *Panagrellus redivivus* infected by *Plesiospora polyspora*. Bar = 10 μm. **k** *P. redivivus* infected by *Catenaria anguillulae*. Bar = 10 μm. **l** *Panagrellus redivivus* infected by *Lecanicillium psalliotae*. Bar = 100 μm. Figure reproduced from Yang et al. (2007a) with kind permission from Springer Science and Business Media

Nematophagous Fungi and Their Application in Biological Control of Plant-Parasitic Nematodes

Nematophagous fungi are those fungi with the capacity to capture, parasitise or paralyse nematodes at all stages of their life cycles. They play an important role as antagonists of plant-parasitic and animal-parasitic nematodes, therefore, there is great interest in using these fungi as model fungi in adaptative evolution researches and as biological control agents against plant-parasitic nematodes (Li et al. 2000; Nordbring-Hertz et al. 2006). More than 700 species of nematophagous fungi have been described (Zhang et al. 2011). Based on their infection mechanism, they are commonly subdivided into four main groups: (1) the nematode-trapping fungi (ca. 380 species) that capture free-living nematodes using specialized trapping devices (i.e., traps) (Fig. 5.1), (2) the endoparasitic fungi (ca. 120 species) that infect nematodes using adhesive spores (Fig. 5.1), (3) the toxin-producing fungi (ca. 270 species) that secrete a toxin which immobilizes nematodes before penetration of hyphae through the nematode cuticle, and 4) the egg- and cyst-parasitic fungi that infect these stages with their hyphal tips (Fig. 5.1) (Li et al. 2000; Nordbring-Hertz et al. 2006; Zhang et al. 2011).

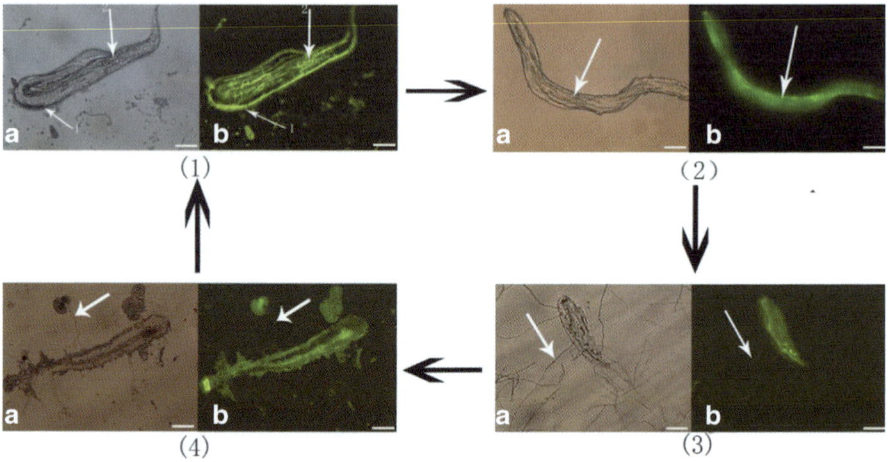

Fig. 5.2 The infection process of *Clonostachys rosea* against nematodes observed under light microscopy (**a**) and fluorescence microscopy (**b**). *1* Conidia adhered to the cuticle of nematodes (*arrow 1*) and bourgeoned (*arrow 2*). *2* Myceliaum (*arrow*) grows inside the body of nematodes. *3* Mycelium (*arrow*) growing from the body of a nematode. *4* Fungus produces conidiophores and abundant conidia (*arrow*). Scale bar =20 μm. Figure reproduced from Zhang et al. (2008) with kind permission from Springer Science and Business Media

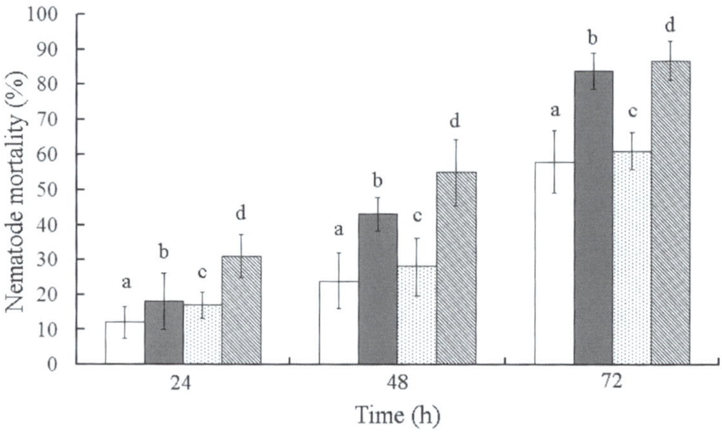

Fig. 5.3 The nematicidal activity of *Paecilomyces lilacinus* 112 and wild strain on corn meal agar plate at various time points. **a** Mortality of *Panagrellus redivivus* treated by wild strian. **b** Mortality of *P. redivivus* treated by *P. lilacinus* 112. **c** Mortality of *Caenorhabditis elegans* treated by wild strian. **b** Mortality of *C. elegans* treated by *P. lilacinus* 112. The bars represent the standard deviation values. Figure reproduced from Yang et al. (2011b) with kind permission from Springer Science and Business Media

Nematophagous fungi are found in most fungal taxa including Ascomycetes (asexual *Orbiliaceae* and *Clavicipitaceae*), Basidiomycetes (*Pleurotaceae*), Zygomycetes (*Zoopagales*), Chytridiomycetes and Oomycetes (Table 5.1) (Gams and Zare 2003). They are broadly distributed in terrestrial and aquatic ecosystems

Table 5.1 Taxonomy of nematophagous fungi and their infection modes[a]

Fungal group	Phyla	Fungi	Infection structures
Nematode-trapping fungi	Zygomycota	*Stylopage*	Adhesive hyphae
		Cystopage	Adhesive hyphae
	Ascomycota	*Arthrobotrys/Orbilia*	Adhesive networks
		Dactylellina/Orbilia	Adhesive knobs and/or nonconstricting rings
		Drechslerella/Orbilia	Constricting rings
		Gamsylella/Orbilia	Adhesive branches or unstalked knobs
	Basidiomycota	*Nematoctonus/Hohenbuehelia*	Adhesive "hour-glass" knobs
Endoparasitic fungi	Oomycota	*Myzocytiopsis*	Zoospores
	Chytridiomycota	*Haptoglossa*	"Gun cell", injection
	Blastocladiomycota	*Catenaria*	Zoospores
	Ascomycota	*Harposporium/Podocrella*	Ingested conidia
		Drechmeria	Adhesive conidia
		Haptocillium/Cordyceps	Adhesive conidia
		Hirsutella	Adhesive conidia
	Basidiomycota	*Nematoctonus/Hohenbuehelia*	Adhesive "hour-glass" knobs
Egg- and female-parasitic fungi	Oomycota	*Nematophthora*	Zoospores
	Ascomycota	*Pochonia/Metacordyceps*	Appressoria
		Paecilomyces/Cordyceps	Appressoria
		Lecanicillium/Cordyceps	Appressoria
Toxin-producing fungi	Basidiomycota	*Pleurotus*	Toxic droplets
		Coprinus	Toxin, "Spiny structures"

[a] Table modified from Moosavi and Zare (2012)

(Dackman et al. 1992; Hao et al. 2005; Kumar et al. 2011), and can reproduce in nearly all types of soil because of their few nutritional and vitamin requirements (Duddington 1962). It has been proposed that the nematode-trapping phenotype is an evolutionary response by cellulolytic or lignin-degrading fungi to nutrient deficiencies in nitrogen-limiting habitats (Barron 1992). Because nitrogen is essential to fungal growth and not freely available either in dead wood or in soil where carbon is abundant, direct capture of nitrogen compounds from other living life forms is an advantage (Barron 2003).

Most nematophagous fungi are facultative parasites, although some are obligate parasites of nematodes (Hallmann et al. 2009). The facultative parasites can infect nematodes through producing structures which trap migratory stages of nematodes, producing specialized adhesive spores (Table 5.1), or by means of developing appressoria on specialized hyphae that can penetrate the nematode cuticle or eggshell (Lòpez-Llorca et al. 2008). Obligate parasites initiate infection through their spores. The spores may be ingested, germinating in the nematode digestive system and breaching through its wall, or they may adhere to the nematode cuticle and penetrate directly (Barron 1977). The ability of nematophagous fungi to capture nematodes makes them an attractive candidate agent for controlling parasitic nematodes of plants and animals. Several commercial biological nematicides have been developed and applied in control of nematodes.

Nematode-Trapping Fungi

Nematode-trapping fungi are found in all regions of the world, from the tropics to Antarctica. They are commonly found in soils and decaying leaf litter, decaying wood, dung, compost and mosses (Li et al. 2000; Zhang et al. 2011). Nematode-trapping fungi grow vegetatively in soils as saprobes. Traps are initiated either spontaneously or in response to signals from the environment, including peptides and other compounds secreted by the host nematode (Dijksterhuis et al. 1994; Yang et al. 2011a). Different fungal species produce one or more types of trapping device varying from simple fungal hyphae covered with sticky secretions (*Stylopage* spp.) to much more complex structures (Moosavi and Zare 2012). Trapping devices provide an important function for obtaining nutrients and may confer competitive advantages over nonpredatory fungi (Rubner 1996; Scholler et al. 1999; Yang et al. 2007b). Three basic types of trapping devices are adhesive knobs, constricting rings and adhesive networks, and they can be subdivided further into seven types of trapping device (i.e., simple adhesive branches, unstalked adhesive knobs, stalked adhesive knobs, nonconstricting rings, constricting rings, two-dimensional networks and three-dimensional networks, Fig. 5.1) (Rubner 1996). The ultrastructure of nematode-trapping devices have been extensively studied (e.g., Nordbring-Hertz and Stalhammar-Carlemalm 1978; Dijksterhuis et al. 1994). Although there is variation in morphology, different types of adhesive traps (branches, nets and knobs) share some common features that clearly distinguish them from normal vegetative hyphae (Heintz and Pramer 1972; Dijksterhuis et al. 1994). One shared feature is the presence of numerous cytosolic organelles (dense bodies) within the trapping hyphal cells (Heintz and Pramer 1972; Nordbring-Hertz and Stalhammar-Carlemalm 1978). Another feature is the presence of extensive layers of extracellular polymers. These polymers have been considered important for the attachment of the traps to nematode surfaces (Tunlid et al. 1991a).

The majority of nematode-trapping fungi are asexaul taxa, mostly known as hyphomycetes; they are found in Zygomycota, Ascomycota and Basidiomycota (Table 5.1), the typical genera include *Cystopage* Drechsler, *Stylopage* Drechsler, *Zoophagus* Sommerst, *Triposporina* Höhnel, *Tridentaria* Preuss, *Nematoctonus* Drechsler, *Hyphoderma* Wallr., *Hohenbuehelia* Schulzer, *Arthrobotrys* Corda, *Drechslerella* Subram, *Orbilia* Fr., *Dactylellina* M. Morelet, *Dactylella* Grove, *Monacrosporium* Oudem (Zhang et al. 2011). MycoBank (Crous et al. 2004) reports 347 nematode-trapping species. Traditional taxonomy of nematode-trapping fungi relied on conidia and conidiophore morphology without taking into account the morphology of trapping devices. This has led to a situation where species with diverse types of trapping devices have been assigned to one genus, while other species with similar trapping devices are found in different genera (Glockling and Dick 1994; Liu and Zhang 1994, 2003; Zhang et al 1996). Recent studies with ITS and 18S rDNA sequences indicated that trapping devices are more informative than other morphological structures in delimiting genera (e.g., Liou and Tzean 1997; Ahrén et al. 1998; Scholler et al. 1999; Li et al. 2005). Ahrén et al. (1998) found that nematode-tapping fungi clustered into three lineages: species with constricting rings, species with various adhesive structures (net, hyphae, knobs and nonconstricting rings) and species have no trapping devices. Based on results obtained from morphological and molecular characters, Scholler et al (1999) classified nematode-trapping fungi in ascomycete into four genera: *Dactylellina* characterized by stalked adhesive knobs including species characterized by nonconstricting rings and stalked adhesive knobs; *Gamsylella* characterized by adhesive branches and unstalked knobs; *Arthrobotrys* characterized by adhesive networks; and *Drechslerella* characterized by constricting rings. This classification was questioned by Li et al. (2005) who suggested that the species in *Gamsylella* should be transferred to either *Arthrobotrys* or *Dactylellina* based on more and refined DNA sequencing. Li et al. (2005) hypothesised an evolutionary pathway of traps within the nematode-trapping *Orbiliales*. They suggested that two lines have evolved, both originating from adhesive knobs. In one line adhesive functionality was lost and they evolved to form constricting rings, whereas in the other evolutionary line adhesiveness was retained and became three-dimensional networks. Yang et al. (2007b) suggested an evolutionary pathway of five types of trapping devices based on comprehensive phylogenetic analysis. They suggested that the initial trapping structure evolved along two lineages with one developing into constricting rings and the other into adhesive traps. Among adhesive trapping devices, the adhesive network separated from the others early and evolved at a steady and gentle speed. The adhesive knob evolved through stalk elongation, with a final development of nonconstricting rings.

Trap Formation and Infection Mechanism

The formation of trapping devices by nematode-trapping fungi is an important indicator of their switch from saprobic to predacious lifestyles (Li et al. 2005; Yang et al. 2007b). Trapping involves a series of processes including attraction, adhesion, penetration, and immobilization of nematodes (Dijksterhuis et al. 1994; Nordbring-Hertz et al. 2006; Yang et al. 2007a). Detailed microscopic studies of net-forming species, in particular *Arthrobotrys oligospora*, showed that following physical contact between the trap cells and the nematodes, the nematodes become attached to the trap surface. In the electron microscope, the trap cells are seen to be surrounded by a layer of extracellular fibrils. After contact, these fibrils become directed perpendicularly to the nematode surface. Subsequently, the fungus pierces the cuticle by forming a penetration tube. This step probably involves both the activity of hydrolytic enzymes, which solubilize components of the cuticle and the activity of mechanical pressure (Tunlid et al. 1994; Yang et al. 2007a). Concomitant with penetration, the nematodes become paralyzed. Once inside the nematode, the penetration tube swells to form an infection bulb from which trophic hyphae develop and colonize the dead nematode. The infection bulb can be considered as an intermediate morphological structure, between the highly differentiated trap cells and the trophic hyphae which develop from the bulb. Upon maturation of the bulb the dense bodies (in trapping device) are degraded. At the same time, the endoplasmic reticulum in *A. oligospora* proliferates extensively. When the nematode cavity becomes invaded by the trophic hyphae, the internal tissues of the nematode are rapidly degraded. Some of the nematode content is converted to lipid droplets, which can be metabolized to support growth of new vegetative mycelium that develops outside the nematode. The infection process is usually completed in 48–60 h (Dijksterhuis et al. 1994).

Little is known about the molecular mechanism underlying trap formation or nematode-fungi interaction. The gene expression profiles in trap cells and vegetative hyphae of the nematode-trapping fungus *Monacrosporium haptotylum* were analyzed using microarray technology (Ahrén et al. 2005). In total, 23.3% (657 of 2822) of the putative genes were differentially expressed in knobs versus mycelium. Subsequently, the changes in the transcriptome of *M. haptotylum* during the adhesion, penetration and digestion of the nematode *Caenorhabditis elegans* were examined (Fekete et al. 2008), and dramatic shifts were found in the transcriptome of *M. haptotylum* during the different stages of the infection. The proteome of the mycelium of the knob-forming fungus *Monacrosporium lysipagum* has been analyzed using 2D gel electrophoresis and mass spectrometry. Out of the 250 proteins analyzed by 2D gel electrophoresis and mass spectrometry, 51 (20%) were identified by cross-species matches (Khan et al. 2008).

Although, the cDNA microarray experiments with *M. haptotylum* have shown that several hundreds of genes are regulated during infection, it is difficult to interpret this data due to the limited molecular background information on nematode-trapping fungi. The first genome sequence of a nematode-trapping fungus, *A. oligospora* was sequenced. Proteins differentially expressed in response to nematode extracts were identified using a proteomics approach and quantitative PCR (qPCR). Proteomics and qPCR analyses revealed that 90 genes were significantly up-regulated at the early stage of trap formation by nematode extracts and most of these genes were involved in translation, amino acid metabolism, carbohydrate metabolism, cell wall and membrane biogenesis (Yang et al. 2011a). Genomic and proteomic analyses of *A. oligospora* provided the first glimpse into the genome and proteome of this fascinating group of carnivorous fungi. The data should serve as a roadmap for further investigations into the interaction between nematode-trapping fungi and their host nematodes, providing broad foundations for research on the biocontrol of pathogenic nematodes.

Endoparasitic Fungi

Most endoparasitic fungi are obligate parasites or poor saprotrophic competitors in soil, but they usually have a broad nematode host range (Moosavi and Zare 2012). Obligate parasites live their whole vegetative life cycle inside the infected hosts (Li et al. 2000; Lòpez-Llorca et al. 2008). Endoparasitic fungi are found in Oomycota, Chytridiomycota, Blastocladiomycota, Ascomycota and Basidiomycota (Table 5.1), the typical genera include *Myzocytium* Schenk, *Drechmeria* W. Gams & H.-B. Jansson, *Nematoctonus* Drechsler, *Hirsutella* Pat. and *Harposporium* Lohde (Zare et al. 2000a; Zhang et al. 2011). So far, 122 nematode-endoparasitic species have been reported (Zhang et al. 2011). Endoparasitic fungi infect plant-parasitic nematodes through their spores (conidia or zoospores) which either penetrate the host cuticle or are injected (Table 5.1). Interestingly, *Drechmeria coniospora* forms large numbers of conidia in comparison to production of hyphal material. In a single infected nematode, *D. coniospora* may produce as many as 10,000 conidia while the endoparasite *Hirsutella rhossiliensis*, which sporulates singly, produces only 100-1000 conidia per infected nematode. Both fungi develop an adhesive bud on their conidia with which they infect the nematode (Jansson et al. 1987; Dijksterhuis et al. 1991; Timper et al. 1991; Tedford et al. 1993). The genus *Harposporium* contains fungi that produce spores with special shapes, which are ingested by the nematodes. Because of their shapes, the spores get stuck in the oesophagus and from there initiate infection of the nematodes (Hodge et al. 1997). *Catenaria anguillulae* infects nematodes through motile zoospores which encyst on and adhere to the nematode (Tunlid et al. 1991b; Jansson and Thiman 1992). Interestingly, in the genus *Haptoglossa* the spores form an infection "gun cell" which forcibly injects the infective principle into the nematode host (Robb and Barron 1982; Beakes and Glockling 1998).

The processes of *Drechmeria coniospora* conidiogenesis and penetration into nematode cuticle were illustrated by light- and electron-microscopy (Jansson et al. 1984; Dijksterhuis et al. 1991). *D. coniospora* secretes collagenase before and during penetration (Jansson et al. 1985a). The fungus occupies the pseudocoelum of the nematode without colonization of the internal organs. Nematode can ingest the conidia, but no germination is seen in the intestine (Jansson 1994). Thus, direct penetration of conidia through the cuticle is the only way of infection. Conidia of *D. coniospora* can adhere to the chemosensory organs of root-knot nematodes but do not penetrate and colonize the nematode (Jansson et al. 1985b). There are similar reports for insect-parasitic nematode species (*Neoaplectana* and *Heterorhabditis*) (Dijksterhuis et al. 1990) and *Acrobeloides* (Dijksterhuis et al. 1993) where conidial adhesion occurs without any penetration. Adhesive on the conidial surface of *D. coniospora* always keeps its fibrillar appearance (Jansson and Nordbring-Hertz 1988). The fibrillar layer is dissolved in pronase E. Infection was inhibited by chymostatin (a protease inhibitor), suggesting the involvement of chymotrypsin-like proteases in the infection process (Jansson and Friman 1999). After the binding of conidia of *D. coniospora* to nematode cuticle, an infection vesicle develops within the cuticle layers (Dijksterhuis et al. 1990; Sjollema et al. 1993).

Hirsutella rhossiliensis is a typical endoparasitic fungus of nematodes. It produces adhesive spores that attach to and penetrate the cuticle of passing nematodes (Jaffee and Zehr 1985). The conidia are infectious only if they are attached to the phialides (McInnis and Jaffee 1998), and one conidium is generally enough to infect a nematode. When the fungus penetrates its host, the nematode will be totally colonized, and within a few days new infectious conidia will be produced (Lackey et al. 1992). A neutral serine protease (Wang et al. 2009) and extracellular alkaline protease (Hasp) (Wang et al. 2007) were isolated from *H. rhossiliensis* and these may help the fungi penetrate the cuticle of nematodes.

Egg- and Female-Parasitic Fungi

Egg and cyst parasitizing fungi also named opportunistic fungi use appressoria or zoospores to infect their hosts (Lòpez-Llorca et al. 2008). This group of fungi are species of *Pochonia, Paecilomyces, Lecanicillium* and *Nematophthora.* The parasites of eggs and sedentary stages have attracted much attention because of their high potential in biological control of economically important nematodes. These fungi that can survive as saprobes in the rhizosphere, are relatively easy to mass-culture, and are more effective in infecting because their host life stage is sessile (eggs, developing juveniles and females). Among all nematophagous fungi, few have been considered as promising biocontrol agents as these fungi (Siddiqui and Mahmood 1996) and of these, the most frequently isolated are *Pochonia chlamydosporia* and *Paecilomyces lilacinus* (de Leij et al. 1991; Siddiqui and Mahmood 1996).

The eggshell of nematodes is composed of three distinct layers (an outer vitelline layer, a chitin layer and an inner lipoprotein layer) and mostly consists of protein and chitin organized in a microfibrillar and amorphous structure (Wharton 1980; Perry 2002). Penetration of the nematode eggshell occurs from an appressorium, a specialized penetration peg or lateral branches of mycelium (Lòpez-Llorca et al. 2008). Extracellular hydrolytic enzymes such as chitinases and proteases play an important role during eggshell penetration, and lead to disintegration of eggshell layers (Segers et al. 1996; Yang et al. 2007a). Egg- and female-parasitic fungi, which can produce more extracellular enzymes, are considered much more effective in infection of nematode eggs (Yang et al. 2007a). Fungi differ in their ability to degrade nematode eggshells, and the infection process can be affected by the nematode host (Segers et al. 1996, 1999). The infection of nematodes and their eggs by various nematophagous fungi follows a similar pattern (Lòpez-Llorca et al. 2008).

Pochonia chlamydosporia is a parasite of females and eggs of cyst and root-knot nematodes, and develops branched mycelial networks that form appressoria on the eggshell (Lòpez-Llorca and Duncan 1988; Viaene et al. 2006). A serine protease and chitinases of *P. chlamydosporia* are effective in degrading the eggshell (Tikhonov et al. 2002; Viaene et al. 2006). *P. chlamydosporia* secretes the VCP1 protease that can hydrolyze eggshell proteins of *Meloidogyne* species but not those of *Globodera* (Segers et al. 1996). Recently, an endochitinase gene (*pcchi44*) (Mi et al. 2010) and a new serine carboxypeptidase (SCP1) gene (Lòpez-Llorca et al. 2010) were isolated, identified and cloned from *P. chlamydosporia*. Comparison of the similarity of amino acid sequences between proteases from different nematophagous fungi showed a high level of conservation, with only minor insertions and deletions (Siezen and Leunissen 1997; Yang et al. 2007a). Minor variation in amino acid sequence may influence substrate utilization and host preference (Segers et al. 1995) that has been documented in VCP1 proteases from different isolates of *P. chlamydo-*

sporia (Morton et al. 2003a, b). Substitution of an alanine by a glycine in the S3 substrate-binding region of VCP1 confers enzymatic activity against eggshells of *Meloidogyne* (Morton et al. 2003a).

Clonostachys rosea (syn. *Gliocladium roseum*) is another widely distributed facultative saprobe in soil (Schroers et al. 1999). It can suppress sporulation of the plant pathogenic fungus *Botrytis cinerea* and can be used for the control of botrytis blight (Morandi et al. 2003). Dong et al. (2005) reported that certain chemicals isolated from *C. rosea* showed strong nematicidal activities against the nematodes, such as *Caenorhabditis elegans, Panagrellus redivivus*, and *Bursaphelenchus xylophilus*. Two extracellular serine proteases (Lmz1 and PrC) have been isolated from *C. rosea* and identified as important factors in fungal pathogenicity (Zhao et al. 2005; Li et al. 2006). To help understand the underlying mechanism, Zhang et al. (2008) constructed recombinant strains containing a plasmid with both the enhanced green fluorescent protein (GFP) gene *egfp* and the hygromycin resistance gene *hph*. Expression of the GFP was monitored using fluorescence microscopy. The transformant strain of *C. rosea* began to attack the nematodes after co-incubation for 1 day on a potato dextrose agar plate (Fig. 5.2). The conidia of the fungus secreted a glutinous substance, leading to the conglutination of the nematode. It was difficult for the conglutinated nematodes to escape from the mucous liquid containing abundant conidia. When the nematodes were exhausted from their escape attempts and stopped moving, conidia could germinate and penetrate the body of the nematodes, and grow by digesting the tissue of nematodes (Zhang et al. 2008).

Toxin-Producing Fungi

Toxin-producing fungi secrete a toxin that immobilizes nematodes before penetration of hyphae through the nematode cuticle (Lòpez-Llorca et al. 2008). Most of these fungi are Basidiomycota (e.g., *Pleurotus, Coprinus*) although other species (eg, *Lecanicillium, Paecilomyces, Pochonia*) also produce nematocidal compounds. *Paecilomyces lilacinus* secretes acetic acid that paralyzes juvenile nematodes (Djian et al. 1991), and some bioactive compounds have been isolated from in vitro cultures of *Pochonia chlamydosporia* (Kerry and Hominick 2002; Niu et al. 2010). So far, more than 270 fungi have been reported to immobilize nematodes by a toxin, and 230 nematocidal compounds have been identified from fungi (Li et al. 2007; Zhang et al. 2011). These compounds include alkaloids, peptide compounds, terpenoids, macrolide compounds, oxygen heterocycle and benzo compounds, aliphatic compounds, quinones, simple aromatic compounds and sterols (Li et al. 2007). Recently, three novel oligosporons, (arthrobotrisins A-C (1–3)), were isolated from *Arthrobotrys oligospora* and identified by spectroscopic analysis in combination with X-ray diffraction (Wei et al. 2011).

A new nematicidal mechanism was reported in *Coprinus comatus*, which can immobilize *Panagrellus redivivus* by two strategies (Luo et al. 2007). Firstly, with the spiny ball structures, *Coprinus comatus* can damage the cuticle of the nematode with the mechanical force provided by its sharp projections. Secondly, this fungus can also secret potent toxins to immobilize and kill nematodes within hours. The combined mechanisms are different from the typical patterns of attacking nematodes with sticking materials and killing them with potent toxins.

Screening and Genetic Improvement of High Virulence Fungal Strains for Nematode Biological Control

Although many microbial antagonists of nematodes have been found and tested for their activity against nematodes, they have not led to the development of commercial products as cost-effective as chemical nematicides (Oka et al. 2000). The most common strategy to control plant-parasitic nematodes by the use of nematode-trapping fungi has been to mass-produce the fungi that infects the nematodes on solid substrates, followed by the application of these fungi to the soil.. Based on this has emerged the demand for new strategies in order to control parasitic nematodes, one of these being the generation of improved fungal strains (Casas-Flores and Herrera-Estrella 2007).

One way to improve fungal pathogenicity is to increase the copy number of virulence genes or to introduce exogenous virulence genes. This strategy has been successfully used in entomopathogenic fungi and other pathogenic fungi (Åhman et al. 2002; St Leger and Wang2010; Wang et al. 2010). Wang and St Leger (2007) found that high-level expression of an insect-specific neurotoxin from the scorpion *Androctonus australis* in hemolymph by *M. anisopliae* increased fungal toxicity 22-fold against tobacco hornworm (*Manduca sexta*) caterpillars and ninefold against adult yellow fever mosquitoes (*Aedes aegypti*) without compromising host specificity.The engineered *M. anisopliae* was also dramatically more virulent against coffee berry borer (CBB), increasing mortality by by almost one-third. The medial lethal concentration (LC50) was reduced 15.7-fold and the average survival time was reduced by 20 % (Pava-Ripoll et al. 2008). Qin et al. (2010) constructed a transgenic *Beauveria bassiana* strain by expressing an insecticidal protein (Vip3A) from *Bacillus thuringiensis*. The engineered *B. bassiana* strain showed enhanced fungal virulence to *Spodoptera litura* larvae.

There are few successful examples of improving the virulence of a nematode-trapping fungus using genetic engineering (Åhman et al. 2002; Yang et al. 2011b). A cuticle-degrading protease PII has been isolated from *A. oligospora* (Tunlid et al. 1994), and the encoding gene cloned (Åhman et al. 1996). Subsequently, the transformation system of *A. oligospora* was constructed using the hygromycin-B phosphotransferase gene from *Escherichia coli* as screen marker (Tunlid et al. 1999). Åhman et al. (2002) disrupted the PII gene by homologous recombination and constructed mutants containing additional copies of the PII gene by restriction enzyme-mediated integration (REMI) transformation. The pathogenicity of two PII mutants (TJÅDPII.2 and −3) and two overexpressing mutants (TJÅMPII.1 and TATMPII.D17) were examined using a bioassay with the nematode *P. redivivus*. TDisruption of PII had a limited effect on the capture of nematodes, as significant differences between the PII mutants and the wild type in the capture of nematodes were only observed after 10 h of infection. The over-expression mutants had a significantly higher number of traps than the wild-type strain and had an increased speed of capturing nematodes than the wild type. The effect of the gene-disrupted and the over-expressing mutants on the immobilization of captured nematodes was also investigated. No significant difference in the immobilization of nematodes was observed between the two PII mutants and the wild-type strain. However, the percentage of immobilized nematodes was higher in the over-expressing strains than in the wild type; thus, the over-expressing mutants showed a more rapid killing of nematodes than the wild-type strain (Åhman et al. 2002).

Paecilomyces lilacinus is an opportunistic fungal pathogen that can infect both nematodes and insects, and it has been widely studied and successfully implemented for controlling plant-parasitic nematodes (e.g., Basualdo et al. 2000; Brand et al. 2004; Kalele et al. 2007). This fungus can infect eggs and cyst nematodes using through secreted hydrolyzing enzymes (Khan et al. 2003, 2004). A basic serine protease pSP-3 was identified from *P. lilacinus* culture filtrate by affinity chromatography (Bonants et al. 1995), which could be inhibited by PMSF, and this enzyme shared a high degree of sequence similarity to subtilisin-like serine proteases. The fungus *Lecanicillium psalliotae* (syn. *Verticillium psalliotae*) is an effective parasite of plant-parasitic nematodes (Zare et al. 2000b), and a cuticle-degrading protease (Ver112) has been purified and cloned from it (Yang et al. 2005a, b). This enzyme showed a strong activity to hydrolyze the proteins in nematode cuticles. Recently, the gene *ver112* was introduced into the commercial biocontrol fungal agent *P. lilacinus* by REMI transformation (Yang et al. 2011b). Compared to the wild strain, the transformant *P. lilacinus*112 showed significantly greater protease activity, with nematicidal activities increased at the second day by 79–96% against *P. redivivus* and *C. elegans*. The crude protein extract isolated from the culture filtrate of *P. lilacinus*112 also showed 20–25% higher nematicidal activity than that of the wild type strain (Fig. 5.3). Reverse transcription PCR results showed that the expression of gene *ver112* in *P. lilacinus*112 was correlated to protease activity of the culture filtrate (Yang et al. 2011b). This demonstrated the first successful transfer of a virulence gene from one nematophagous fungus to another nematophagous fungus, and improved the pathogenicity of the recipient fungus against pest nematodes.

The complete genome of *A. oligospora* has been sequenced and it contains 24 genes encoding putative subtilases, which can be categorized into four subtilisin families (Yang et al. 2011a). Twenty of them belong to the proteinase K-like family, which can be further classified into five subfamilies (SF1-SF5). The four proteases (PII, P186, P233, P12) are clustered into one of the five sub-families (SF4). Among them, *P12* shares a high nucleotide sequence identity (77.9%) with the cuticle-degrading protease encoding gene *spr1* of another nematode-trapping fungus *Monacrosporium megalosporum* (Kanda et al. 2008). Interestingly, when *A. oligospora* was induced by nematode extracts for 10 h, the transcription levels of *P12* and *P186* increased by 5.9-fold and 23.4-fold respectively, whereas those of *PII* and

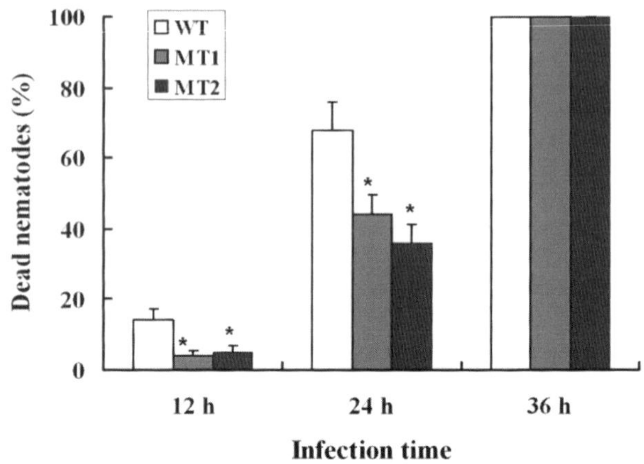

Fig. 5.4 Disruption of the gene *p186* and nematocidal activity analysis. WT, wild-type strain. MT1 and MT2, mutant strains of *Arthrobotrys oligospora*. Test nematodes are *Caenorhabditis elegans*. The bars represent the standard deviation values. *$p < 0.05$ versus wild strain. Figure reproduced from Yang et al. (2011a)

P233 did not change significantly. In order to identify the role of P186 in infection against nematodes, the gene *P186* was disrupted by homologous recombination. No obvious differences in phenotypic properties such as growth rates, spore number, trap number and morphology, were found between the mutants and the wild strain. However, disruption of *P186* attenuated the pathogenicity of *A. oligospora*. The fatality rate of nematodes infected by P186-deletion mutants decreased by 24–32 % (Fig. 5.4) at 24 h after infection. The results suggested that *P186* likely plays major roles during nematode infection by *A. oligospora*. In contrast, the lack of a significant change in PII expression level upon NE induction is consistent with an earlier finding that PII gene disruption had only a limited effect on the pathogenicity of *A. oligospora* (Åhman et al. 2002).

Other virulence genes including chitinase, adhesion and recognition-related genes may be alternative candidates for future fungal engineering with over-expression. Recently, the crystal structures of serine proteases and chitinases from nematophagous fungi were determined (Liang et al. 2010, 2011; Yang et al. 2010), thus providing a solid basis for improving the pathogenicity of this group of fungi against nematodes by genetic and protein engineering.

Production and Formulation of Nematode-Antigonistic Agents

Fermentation and formulation of biocontrol microorganisms are often limiting and deciding factors in the success of biocontrol measures (Burges 1998). This is particularly true for the use of nematophagous fungi to control plant-parasitic nematodes. Production and formulation of biocontrol fungi are the key procedures for the application of nematode-antigonistic agents. Solid and liquid fermentation are the common methods for mass production of fungi. Solid culturing is the most widely used method, and it is especially suited for those fungi that can not produce spores in liquid. However, it has several drawbacks, including long producing period, high waste and low yield. Liquid culturing of fungi for mass production of spores and mycelium has often been considered best for biological control (Papavizas et al. 1984). Moreover, solid combined with liquid fermentation is perhaps an effective way for mass production of fungi. Mass mycelia can be produced using liquid fermentation, and conidia can then be produced using solid fermentation (Feng et al. 1994). Cornmeal and potato dextrose media have been used for mass culturing of endoparasitic fungi such as *Drechmeria coniospora, Verticillium balanoides* and *Harposporium anguillulae* (Lohmann and Sikora 1989). Production of large amounts of these fungi has an advantage as a biocontrol agent (BCA) because they do not rely on an energy source, as do egg-parasitic and nematode-trapping fungi.

A good formulation is the key to the commercial success of BCAs. Formulation of BCAs includes powder, wettable powder, emulsion, oil solution, granular formulation, blending agent and microcapsule (Liu et al. 2004). Formulation of BCAs can be used to stabilize the organisms during production, distribution and storage; aid in the handling and application of the product; protect the agent from harmful environmental factors; and enhance the activity of the organisms (Jones and Burges 1998). From a technical efficacy standpoint, an effective formulation requires a thorough knowledge of the biocontrol organism, pathogen, environment, and interactions with other organisms (e.g. host plant, soil microflora). From an end-user standpoint, there is a need to understand common application practices and equipment, as well as the desires of customers for formulation handling (Leggett et al. 2011). It is important to start with an understanding of the biology of the BCA and the target, and to develop a clear vision of the requirements of a given product before formulation research is started. Once this is established, a comprehensive research programme needs to be developed. Potential adjuvants must be safe and acceptable to regulatory agencies in all areas where the product will be used. The experiences of a broad spectrum of specialists should be consulted to identify formulation ingredients and aid in the development of an appropriate formulation (Leggett et al. 2011).

Microencapsulation is a promising step towards commercialization for biocontrol fungi. Patel et al. (2011) investigated the effect of fermentation conditions to suppress pellet formation of *Hirsutella rhossiliensis* in shake flask culture, and to develop a novel capsule system with optimised nutrient content. When incubated in 2 % (w/w) glucose and 0.5 % (w/w) yeast extract medium in a 1-L Erlenmeyer flask without baffles, heavy pellet formation was observed. Only 40 % of the mycelium had a size less than 500 μm. When a flask with three baffles was used, the portion of mycelium <500 μm rose to 95 %. In the next step, the influence of aeration rate and stirrer speed on production of finely dispersed mycelium in a stirred tank reactor was investigated. The best fermentation results were obtained at 0.4 vvm and 400 rpm stirrer speed with 90 % mycelium <500 μm and 5 g/L biomass. Then, mycelium was microencapsulated in hollow beads based on sulfoethylcellulose. Experiments on the capsule nutrient reservoir showed that 15 % (w/w) corn gluten and 0.5 % (w/w) yeast extract could be replaced with 3 % (w/w) autoclaved baker's yeast which had not been previously used as a capsule additive. Radial growth of mycelium out of dried hollow beads containing 1 % (w/w) biomass and 3 % (w/w) baker's yeast was faster than for alginate beads containing equivalent amounts of biomass and yeast indicating a higher bio-control potential. Recently, Jin and Custis (2011) developed a method for microencapsulation of *Trichoderma* conidia with sugar through spray drying. Microencapsulation with sugars, such as sucrose, molasses or glycerol, significantly ($p<0.05$) increased the survival percentages of conidia after drying. Microencapsulation of conidia with 2 % sucrose solution resulted in the highest survival percentage when compared with other sucrose concentrations and had about 7.5×10^{10} cfu/g of dried conidia, and 3.4 mg of sucrose added to each gram of dried conidia. The optimal inlet/outlet temperature setting was 60/31 °C for spray drying and microencapsulation. The particle size of microencapsulated conidia balls ranged from 10 to 25 μm. The spray dried biomass of *Trichoderma harzianum* was a flow-able powder with over 99 % conidia, which could be used in a variety of formulation developments from seed coatings to sprayable formulations.

There are different techniques of cell immobilization such as pellet, granular and liquid formulations that have been evaluated and employed to develop carriers for the field application of BCAs. Encapsulation of liquid suspension of BCAs by an appropriate gelant (gel forming material), especially natural polymers (alginate, carragenan, cellulose, agar, agarose, egg white, gelatin, etc.) as well as synthetic polymers (polyacrylamie, photo cross-linkable resins, etc.) is one of the important approaches to develop an efficient delivery system (D'Souza and Melo 1991). However, the technique using gelant to encapsulate the BCA has failed commercially due to difficulty in handling the formulation, low viability of the organism and shorter shelf life (Khan and Gupta 1998). Powder or fine granular carrier systems for BCAs are more useful than gelant or liquid formulations, and compatible with existing farm machinery (Khan 2007). The powder formulations available are less expensive as they are developed from low cost agriculture/industrial wastes or by-products and their handling is also easy. Numerous solid materials and their combinations such as wheat bran-sand mixture, sawdust-sand-molasses mixture, corn cob-sand-molasses mixture, bagasse-sand-molasses mixture, organic cakes, cowdung-sand mixture, compost/farm manure, inert charcoal, diatomaceous earth and fly ash have been tested to prepare granular/powder formulations of BCAs (Kerry 1988; Khan and Khan 1994; Khan et al. 2005).

Recently, a novel process was described to produce biopesticides of *Trichoderma harzianum* Rifai, *Pochonia chlamydosporia* Zare and Gams, *Bacillus subtilis* Cohn and *Pseudomonas fluorescens* (Threvesan) Migula by taking 1 part of stock culture (sawdust:soil:5 % molasses, 15:5:1) of the BCAs and 20 parts carrier (flyash:soil:5 % molasses mixture, 5:3:1) (w/w) (Khan et al. 2011). Greatest colony forming unit (CFU) counts of the microorganisms were recorded at 25 °C or room temperature during 2–12 weeks of 32 weeks long shelf life test. Seed treatment with the biopesticides 5 g/kg seeds carried $10^{3\text{-}6}$ CFU/g seed of chickpea and pigeonpea. The treatments with *Trichoderma harzianum* and *Pochonia chlamydosporia* effectively controlled the wilt fungi (*Fusarium oxysporum* f. sp. *ciceri* Padwick, *F. udum* Butler) and root knot nematode (*Meloidogyne incognita*) on chickpea (*Cicer arietinum* L.) and pigeonpea (*Cajanus cajan* L.) and greatly reduced the soil population of these pathogens. The BCAs established in the soil and their CFU increased

Table 5.2 Partial studies evaluating the efficacy of nematophagous fungi as a biological control of nematodes in grazing livestock

Nematophagous fungi	Animal	Countries	References
Duddingtonia flagrans	Sheep	Australia	Knox and Faedo 2001; Waller et al. 2001
Arthrobotrys robusta, Monacrosporium thaumasium	Cattle, dog	Brazil	Araujo et al. 1998; Saumell et al. 1999; Carvalho et al. 2009; Braga et al. 2009
Duddingtonia flagrans	Sheep, cattle	Denmark	Faedo et al. 2002; Githigia et al. 1997
Duddingtonia flagrans	Goat	France	Paraud and Chartier 2003; Chartier and Pors 2003
Duddingtonia flagrans	Sheep	India	Sanyal 2001
Duddingtonia flagrans	Cattle	Lithuania	Sarkunas et al. 2000
Duddingtonia flagrans, Dactylaria sp., *Arthrobotrys oligospora*	Sheep	Mexico	Flores-Crespo et al. 2003
Duddingtonia flagrans	Sheep	Malaysia	Chandrawathani et al. 2002, 2004
Duddingtonia flagrans	Goats, sheep	New Zealand	Waghorn et al. 2003; Wright et al. 2003
Duddingtonia flagrans	Cattle, sheep	Sweden	Dimander et al. 2003; Yeates et al. 2002; Waller et al. 2006
Duddingtonia flagrans	Horses, sheep, goat	United States	Baudena et al. 2000; Fontenot et al. 2003; Terrill et al. 2004

significantly ($p \leq 0.05$), being greater in pathogen infested soils ($p \leq 0.05$) than non infested soil during a 4-month period. In summary, formulation of biocontrol microorganisms are still the important factors in the success of BCAs, while more and more effective formulations of BCAs will be developed in the future.

Use of Nematophagous Fungi to Control Animal Nematodes

Nematophagous fungi are potential BCAs for controlling of animal nematodes, and several fungi have been applied successfully in controlling of nematodes parasitic to animals such as cattle, horses, sheep and pigs. The net-trapping predacious fungus *Duddingtonia flagrans* is the typical fungus for controlling the animal nematodes, it produces thick walled resting spores, chlamydospores, which are able to survive passage through the gastrointestinal tract of cattle, horses, sheep and pigs (Soder and Holden 2005). This fungus forms sticky traps that catch developing larval stages of parasitic nematodes in the fecal environment. When chlamydospores of this fungus are fed daily to grazing animals for a period of time, the pasture infectivity and thus, the worm burden of grazing animals are lowered, especially in young lambs (Larsen 2006). Work with *D. flagrans* in France, Australia, USA, and Mexico has confirmed the strong biological control potential of this fungus (Table 5.2) (Chandrawathani et al. 2002, 2004; Soder and Holden 2005; Larsen 2006).

Field studies were conducted on two government managed small ruminant research farms, located in different geo-climatic regions and approximately 300 km separate from each other, on Peninsula Malaysia (Chandrawathani et al. 2004). The Infoternak trial (48 weeks) and the Chalok trial (43 weeks) each compared nematode parasite control in separately managed groups of young sheep, either short-term rotationally grazed around a suite of 10 paddocks in addition to receiving a daily supplement of *Duddingtonia flagrans* spores (Fungus Group); or similar groups of sheep being rotationally grazed alone (Control Group). The prevailing weather conditions at Infoternak farm were of below average rainfall conditions for the most of the trial. As a consequence, only very low worm infections (almost exclusively *Haemonchus contortus*) were acquired by the 17 sets of tracer lambs that grazed sequentially with the experimental lambs. However on all except two occasions in the early part of the trial, the mean tracer worm burdens were significantly lower ($p<0.05$) and the experimental lambs grew significantly better ($p=0.054$) in the Fungus Group. Rainfall at Chalok farm during the course of the trial was also below average. As a consequence infectivity of pastures was assumed to be relatively low based on faecal egg counts (epg) of the experimental sheep, which following an anthelmintic treatment prior to allocation, remained very low in both treatment groups. Faecal egg counts of undosed replacement lambs in the latter half of the Chalok study, showed a progressive increase in the Control Group to levels exceeding 3000 epg, whereas the Fungus Group remained static at approximately 500 epg. These results show that the deployment of the nematophagous fungus, *D. flagrans*, can improve the level of parasite control of sheep in the tropics above that which can be achieved by the short-term rotational grazing strategy alone.

The viability of a fungal formulation using the fungus *Duddingtonia flagrans* was assessed for the biological control of horse cyathostomin (Braga et al. 2009). Two groups (fungus-treated and control without fungus treatment), consisting of eight crossbred mares (3–18 years of age) were fed on *Cynodon* sp. pasture naturally infected with equine cyathostome larvae. Each animal of the treated group received oral doses of sodium alginate mycelial pellets (1 g/(10 kg live weight week)), during 6 months. Significant reduction ($p<0.01$) in the number of eggs per gram of feces and coprocultures was found for animals of the fungus-treated group compared with the control group. There was difference ($p<0.01$) of 78.5 % reduction in herbage samples collected up to (0–20 cm) between the fungus-treated group and the control group, during the experimental period (May–October). Difference of 82.5 % ($p<0.01$) was found between the fungus-treated group and the control group in the sampling distance (20–40 cm) from fecal pats. During the last three months of the experimental period (August, September and October), fungus treated mares had significant weight gain ($p<0.01$) compared with the control group, an increment of 38 kg. The treatment with sodium alginate pellets containing the nematode-trapping fungus *D. flagrans*reduced cyathostomin in tropical southeastern Brazil and could be an effective tool for biological control of this parasitic nematode in horses.

Pochonia chlamydosporia is also an effective biocontrol fungus for controlling the animal nematodes (Braga et al. 2010; Ferreira et al. 2011). The *in vitro* ovicidal effect of four isolates of the nematophagous fungi *P. chlamydosporia* (VC1 and

VC4), *Duddingtonia flagrans* (AC001) and *Monacrosporium thaumasium* (NF34) was evaluated on egg capsules of *Dipylidium caninum*, a cestode parasite of dogs, cats and humans (Araujo et al. 2009). One thousand egg capsules of *D. caninum* were plated on 2% water-agar with the grown isolates and control without fungus. The ovicidal activity of these fungi was evaluated 5, 10 and 15 days after incubation. Only *Pochonia chlamydosporia* showed ovicidal activity ($p<0.05$) on *Dipylidium caninum* egg capsules, of 19.6% (VC1) and 20% (VC4) on the 5th day; 44.2% (VC1) and 31.5% (VC4) on the 10th day; and 49.2% (VC1) and 41.9% (VC4) on the 15th day. *Duddingtonia flagrans* and *Monacrosporium thaumasium* caused nomorphological damage to egg capsules. The results demonstrated that *Pochonia chlamydosporia* was *in vitro* effective against capsules and eggs of *Dipylidium caninum*, and can be considered as a potential BCA for this helminth.

Recently, the *in vitro* effect of four isolates of the nematophagous fungi *Duddingtonia flagrans* (AC001), *Monacrosporium thaumasium* (NF34a) and *Pochonia chlamydosporia* (VC1 and VC4) on the eggs of *Trichuris vulpis* was evaluated (Silva et al. 2010). One thousand eggs of *T. vulpis* were plated on Petri-dishes with 2% water–agar with the fungal isolates grown and without fungus as control. After 7, 14 and 21 days 100 eggs were removed from each plate and classified according to the following parameters: type 1, lytic effect without morphological damage to eggshell; type 2, lytic effect with morphological alteration of embryo and eggshell; and type 3, lytic effect with morphological alteration of embryo and eggshell, besides hyphal penetration and internal egg colonization. *Pochonia chlamydosporia* demonstrated ovicidal activity ($p<0.05$) on the eggs of *T. vulpis* in the studied intervals presenting type 3 effect of 29.5% (VC1) and 36.5% (VC4), 59.5% (VC1) and 2.5% (VC4), 94.8% (VC1) and 2.95% (VC4) at 7, 14 and 21 days, respectively. The other fungi showed no type 3 effect. *Pochonia chlamydosporia* should be a potential biological control agent of *Trichuris vulpis* eggs. An assessment was made of the ovicidal activity of egg-parasitizing fungi *Pochonia chlamydosporia* (isolates VC1 and VC4) and *Paecilomyces lilacinus* on *Toxocara canis eggs* in vitro. The result showed that *Pochonia chlamydosporia* and *Paecilomyces lilacinus* were effective in vitro on *Toxocara canis* eggs and can be considered a potential candidate to biological controller of those nematodes (Carvalho et al. 2010).

The predatory capacity of the nematophagous fungus *Pochonia chlamydosporia* (isolate VC4) after passage through the gastrointestinal tract of dogs was assessed in vivo against *Toxocara canis* eggs. Twelve dogs previously wormed were divided into two groups of six animals and caged. The treatments consisted of a fungus-treated group (VC4) and a control group without fungus. Each dog of the fungus-treated group received a single 4 g dose of mycelial mass of *P. chlamydosporia* (VC4). Fecal samples from animals of both groups (treated and control) were collected at five different times (6, 12, 24, 36, and 48 h) after fungal administration, and placed in Petri dishes. Each Petri dish of both groups for each studied time interval received approximately 1000 *Toxocara canis* eggs. Thirty days after the fecal samples were collected, approximately one hundred eggs were removed from each Petri dish of each studied time interval and evaluated by light microscopy (LM) and scanning electron microscopy (SEM). Microscopy examination of plates inoculated with the

Table 5.3 Commercial biological nematicides based on nematophagous fungi

Products	Fungi	Modes of action	Nematode targets	References
Biocon	*Paecilomyces lilacinus*	Egg and female parasitism	Unspecified	Poinar and Georgis 1994
DiTera	*Myrothecium* sp.	Toxin production	*Globodera rostochiensis, Globodera pallid, Heterodera glycines, Meloidogyne incognita, Radopholus* spp.	Twomey et al. 2000; Fernández et al. 2001
Miexianning	*P. lilacinus*	Egg and female parasitism	Root-knot nematodes parasitizing tobacco	Zhu et al. 2001; Sun et al. 2002; Zhou and Mo 2002
Nemout	Unspecified NTF[a]	Trapping	*Meloidogyne javanica*	Al-Hazmi et al. 1982; Ibrahim 1994
Paecil/Bioact	*P. lilacinus*	Egg and female parasitism	Unspecified	Holland et al. 1999
Royal 350	*Arthrobotrys irregularis*	Trapping	*Meloidogyne* spp.	Cayrol et al. 1978
Royal 300	*Arthrobotrys robusta*	Trapping	Unspecified	Cayrol 1983
Xianchongbike	*Pochonia chlamydosporia*	Egg and female parasitism	Root-knot nematodes parasitizing tobacco, peanut, soybean and watermelon	Zhu et al. 2001; Zhou and Mo 2002

[a] Nematode-trapping fungus. Table modified from Dong and Zhang (2006)

fungus showed that the isolate VC4 was able to destroy the *T. canis* eggs with destruction percentages of 28.6% (6 h), 29.1% (12 h), 32.0% (24 h), 31.7% (36 h), and 37.2% (48 h). These results suggest that *Pochonia chlamydosporia* can be used as a tool for the biological control of *Toxocara canis* eggs in feces of contaminated dogs.

From the above, nematophagous fungi are potential BCAs of animal nematodes, several fungi, such as *Duddingtonia flagrans* and *Pochonia chlamydosporia* have been applied widely in biological control of animal nematodes. We can anticipate that more BCAs would be developed based on nematophagous fungi in the future, and these BCAs would be applied widely in grazing livestock.

Use of Nematophagous Fungi to Control Plant Nematodes

Nematophagous fungi are potential BCAs for controlling of plant nematodes, and several commercial biological nematicides (Table 5.3) based on nematophagous fungi have been developed (Dong and Zhang 2006). These biological nematicides have been shown capable of promoting plant growth and reducing the damage caused by nematodes.

Since the nematode-trapping fungi *Arthrobotrys irregularis* (Cayrol et al. 1978) and *A. robusta* (Cayrol 1983) were developed as commercial BCAs, more nematophagous fungi were used for biological control of nematodes. *Paecilomyces lilacinus* and *Pochonia chlamydosporia* have been extensively studied as BCAs for plant-parasitic nematodes. *Paecilomyces lilacinus* is a biocontrol fungus with a potential range of activity to control the worldwide most important plant parasitic nematodes (Kiewnick and Sikora 2003; Mendoza et al. 2004). Brand et al. (2004) produced a bionematicide with *P. lilacinus* by solid-state fermentation, and coffee husks, cassava bagasse, and defatted soybean cake were utilized as substrates, and sugarcane bagasse was used as support. The products were evaluated for their nematicide activity in pot experiments containing one seedling of the plant Coleus inoculated with the nematode *Meloidogyne incognita*. The plants were evaluated two months after inoculation. Fermented products showed a reduction in the number of nematodes. The best results were obtained with defatted soybean cake, which showed almost 100 % reduction in the number of nematodes; the reduction with coffee husk was 80 % and with cassava bagasse was about 60 % (Brand et al. 2004).

Subsequently, greenhouse experiments were conducted with the root-knot nematodes *Meloidogyne incognita* and *M. hapla* on tomato (Kiewnick and Sikora 2004). *Paecilomyces lilacinus*, formulated as WG (Bioact WG), was incorporated into soil inoculated with root-knot nematode eggs prior to transplanting the susceptible tomato cultivar. Furthermore, soil treatments were combined with seedling treatments 24 hours before transplanting and a soil drench two weeks after planting, respectively. Seedling and post planting treatment was also combined with a soil treatment at planting. All single or combination treatments tested decreased the gall index and the number of egg masses compared to the untreated control 12 weeks after planting. The results showed that the above mentioned combination of pre-planting application plus the seedling and one post plant drench gave the best control and resulted in a significant fruit yield increase in concurrence with a decrease in number of galls per root.

Paecilomyces lilacinus and *Monacrosporium lysipagum* were assayed for their ability to reduce the populations of three economically important plant-parasitic nematodes in pot trials (Khan et al. 2006). The fungi were tested individually and in combination against the root-knot nematode *Meloidogyne javanica*, cereal cyst nematode *Heterodera avenae*, or burrowing nematode *Radopholus similis* on tomato, barley and tissue cultured banana plants, respectively. In all cases, nematode populations were controlled substantially by both individual and combined applications of the fungi. Combined application of *Paecilomyces lilacinus* and *Monacrosporium lysipagum* reduced 62 % of galls and 94 % of *Meloidogyne javanica* juveniles on tomato when compared to the experiment with no fungi added. Sixty-five percent of *Heterodera avenae* cysts were reduced on barley by combined application of fungi. Control of *Radopholus similis* on banana, both in the roots and in the soil, was greatest when *Monacrosporium lysipagum* was applied alone (86 %) or in combination with *Paecilomyces lilacinus* (96 %), using a strategy where the fungi were inoculated twice in 18 weeks growth period. Overall, combined application of *P. lilacinus* and *Monacrosporium lysipagum* was the most effective treatment in controlling nematode populations, although in some cases *M. lysipagum* alone was as effective as the combined application of fungi, particularly against *Meloidogyne javanica*.

The efficacies of three nematophagous fungi, *Paecilomyces lilacinus, Plectosphaerella cucumerina* and *Pochonia chlamydosporia*, for controlling potato cyst nematodes (PCN) as part of an Integrated Pest Management (IPM) regime were studied. The compatibility of the nematophagous fungi with commonly used chemical pesticides and their ability to compete with the soil fungi *Rhizoctonia solani, Chaetomium globosum, Fusarium oxysporum, Penicillium bilaii* and *Trichoderma harzianum* were tested *in vitro*. *Paecilomyces lilacinus* was the most successful competitor when the ability to grow and inhibit growth of an opposing colony at both 10 and 20 °C was considered. *Paecilomyces lilacinus* also showed potential for control of the soil-borne fungal pathogen *Rhizoctonia solani*, releasing a diffusable substance *in vitro* which inhibited its growth and caused morphological abnormalities in its hyphae. *Pochonia chlamydosporia* was least susceptible to growth inhibition by other fungi at 20 °C *in vitro*, but the isolate tested did not grow at 10 °C. *Plectosphaerella cucumerina* was a poor saprophytic competitor (Jacobs et al. 2003). In combination other experiments, *P. lilacinus* showed the greatest potential for use in combination with selected fungicides and nematicides as part of an IPM programme for the control of PCN, but further work is required to confirm whether it is effective against PCN in soil.

Recently, more nematophagous fungi were tested as potential BCAs for plant nematodes. For example, the nematode trapping and mycoparasitic potential of *Arthrobotrys oligospora* was tested in vitro against *Meloidogyne graminicola* and *Rhizoctonia solani* (Singh et al. 2012), respectively. Five isolates of *Arthrobotrys oligospora* were isolated from different locations of India. Diversity of the trapping structures is large and highly dependent on the environmental condition and nature of the fungus. In *A. oligospora*, a three-dimensional adhesive net (in response to nematode) and hyphal coils developed around the hyphae of *Rhizoctonia solani*. In vitro trap formation and predacity were tested against second-stage juveniles of *Meloidogyne graminicola* (J2) and the interactions between *Arthrobotrys oligospora* and *Rhizoctonia solani* were recorded. All the isolates of *Arthrobotrys oligospora* parasitized and killed *Meloidogyne graminicola* and *Rhizoctonia solani*. Application of *Arthrobotrys oligospora*, isolate VNS-1, in soil infested with *Meloidogyne graminicola* and *Rhizoctonia solani* reduced the number of root knot by 57.58–62.02 %, sheath blight incidence by 55.68–59.32 % and lesion length by 54.91–66.66 % under green house and miniplot (field) conditions. Applications of *Arthrobotrys oligospora* to the soil increased plant growth: shoot length by 56.4–68.8 %, root length by 44–54.55 %, fresh weight of shoot and root by 62.91–65.4 % and 38.9–44.19 %, respectively, as compared to the plants grown in nematode infested soil. From the above reports, we can anticipate that more BCAs would be developed based on nematophagous fungi in the future, and these BCAs would be applied widely in agriculture and forestry production.

Conclusion

Nematophagous fungi are ubiquitous organisms with the capacity to attack, infect and digest living nematodes at all stages, adults, juveniles and eggs. They may use trapping organs, spores and appressoria to initiate infection of their nematode hosts. The nematophagous fungi may not only infect nematodes, but may also infect other fungi as mycoparasites, and colonize plant roots endophytically. These various capabilities of nematophagous fungi, the latter in particular, may render them good candidates for biological control of plant root diseases. Some nematophagous fungi have the ability to colonize plant roots as a probable survival strategy (Lòpez-Llorca et al. 2008). The egg parasite fungi, like *Pochonia* spp., that grow as endophytic fungi may have higher chance to parasitize eggs of economically important endoparasitic nematodes (like cyst and root-knot species) inside the roots and to decrease succeeding spread and roots infection by the second generation of juveniles. Some structures similar to trapping devices were seen in epidermal cells colonized by *A. oligospora*, which can use to entrap newly hatched juveniles escaping the roots. The ability to colonize plant roots by nematophagous fungi is a novel area of research that deserves in-depth investigations. Strains and formulation are the two important aspects for developing a commercial BCA. Screening and genetic improvement of high virulence fungal strains for nematode biological control would be a potential development direction. Meanwhile, developing a suitable formulation is an effective method to enhance the bio-control potential of BCA.

References

Ahrén, D., Ursing, B. M., & Tunlid, A. (1998). Phylogeny of nematode-trapping fungi based on 18 S rDNA sequences. *FEMS Microbiology Letters, 158,* 179–184.

Ahrén, D., Tholander, M., Fekete, C., Rajashekar, B., Friman, E., Johansson, T., & Tunlid, A. (2005). Comparison of gene expression in trap cells and vegetative hyphae of the nematophagous fungus *Monacrosporium haptotylum. Microbiology (Reading, England), 151,* 789–803.

Åhman, J., Ek, B., Rask, L., & Tunlid, A. (1996). Sequence analysis and regulation of a gene encoding a cuticle-degrading serine protease from the nematophagous fungus *Arthrobotrys oligospora. Microbiology (Reading, England), 142,* 1605–1616.

Åhman, J., Johansson, T., Olsson, M., Punt, P. J., van den Hondel, C. A., & Tunlid, A. (2002). Improving the pathogenicity of a nematode-trapping fungus by genetic engineering of a subtilisin with nematotoxic activity. *Applied and Environmental Microbiology, 68,* 3408–3415.

Al-Hazmi, A. S., Schmitt, D. P., & Sasser, J. N. (1982). The effect of *Arthrobotrys conoides* on *Meloidogyne incognita* population densities in corn as influenced by temperature, fungus inoculum density, and time of fungus introduction in the soil. *Journal of Nematology, 14,* 168–173.

Araujo, J. V., Gomes, A. P. S., & Guimares, M. P. (1998). Biological control of bovine gastrointestinal parasites in south Brazil by the nematode-trapping fungus *Arthrobotrys robusta. Revista Brasileira de Parasitologia Veterinária, 7,* 117–122

Araujo, J. M., de Araújo, J. V., Braga, F. R., Carvalho, R. O., & Ferreira, S. R. (2009). Activity of the nematophagous fungi *Pochonia chlamydosporia, Duddingtonia flagrans* and *Monacrosporium thaumasium* on egg capsules of *Dipylidium caninum. Veterinary Parasitology, 66,* 86–89.

Barker, K. R., & Koenning, S. R. (1998). Developing sustainable systems for nematode management. *Annual Review of Phytopathology, 36,* 165–205.

Barron, G. L. (1977). *The nematode-destroying fungi*. Guelph: Canadian Biological Publications Ltd.

Barron, G. L. (1992). Lignolytic and cellulolytic fungi as predators and parasites. In G. C. Carroll & D. T. Wicklow (Eds.), *The fungal community, its organization and role in the ecosystems*. New York: Marcel Dekker.

Barron, G. L. (2003). Predatory fungi, wood decay, and the carbon cycle. *Biodiversity, 4,* 3–9.

Basualdo, J., Ciarmela, M., Sarmiento, P., & Minvielle, M. (2000). Biological activity of *Paecilomyces* genus against *Toxocara canis* eggs. *Parasitology Research, 86,* 854–859.

Baudena, M. A., Chapman, M. R., Larsen, M., & Klei, T. R. (2000). Efficacy of the nematophagous fungus *Duddingtonia flagrans* in reducing equine cyathostome larvae on pasture in south Louisiana. *Veterinary Parasitology, 89,* 219–230.

Beakes, G. W., & Glockling, S. L. (1998). Injection tube differentiation in gun cells of a *Haptoglossa* species which infects nematodes. *Fungal Genetics and Biology, 24,* 45–68.

Bonants, P. J. M., Fitters, P. F. L., Thijs, H., den Belder, E., Waalwijk, C., & Henfling, J. W. D. M. (1995). A basic serine protease from *Paecilomyces lilacinus* with biological activity against *Meloidogyne hapla* eggs. *Microbiology (Reading, England), 141,* 775–784.

Braga, F. R., Araújo, J. V., Silva, A. R., Araujo, J. M., Carvalho, R. O., Tavela, A. O., Campos, A. K., Carvalho, G. R. (2009). Biological control of horse cyathostomin (Nematoda: Cyathostominae) using the nematophagous fungus *Duddingtonia flagrans* in tropical southeastern Brazil. *Veterinary Parasitology, 163,* 335–340.

Braga, F. R., Ferreira, S. R., Araújo, J. V., Araujo, J. M., Silva, A. R., Carvalho, R. O., Campos, A. K., Freitas, L. G. (2010). Predatory activity of *Pochonia chlamydosporia* fungus on *Toxocara* (syn. *Neoascaris*) *vitulorum* eggs. *Tropical Animal Health and Production, 42,* 309–314.

Brand, D., Roussos, S., Pandey, A., Zilioli, P., Pohl, J., & Soccol, C. (2004). Development of a bionematicide with *Paecilomyces lilacinus* to control *Meloidogyne incognita. Applied Biochemistry and Biotechnology, 118,* 81–88.

Burges, H. D. (1998). *Formulation of microbial pesticides*. Dordrecht: Kluwer.

Carvalho, R. O., Araújo, J. V., Braga, F. R., Ferreira, S. R., Araujo, J. M., Silva, A. R., Frassy, L.N., Alves, C. D. (2009). Biological control of Ancylostomosis in dogs using the nematode-trapping fungus *Monacrosporium thaumasium* in southeastern Brazil. *Veterinary Parasitology, 165,* 179–183.

Carvalho, R. O., Araújo, J. V., Braga, F. R., Araujo, J. M., & Alves, C. D. (2010). Ovicidal activity of *Pochonia chlamydosporia* and *Paecilomyces lilacinus* on *Toxocara canis* eggs. *Veterinary Parasitology, 169,* 123–127.

Casas-Flores, S., & Herrera-Estrella, A. (2007). Antagonism of plant parasitic nematodes by fungi. In C. P. Kubicek & I. S. Druzhinina (Eds.), *The mycota VI: Environmental and microbial relationships* (2nd ed.). Berlin: Springer.

Carpenter, J., Lynch, L., & Trout, T. (2001). Township limits on 1,3-D will impact adjustment to methyl bromide phase-out. *California Agriculture, 55,* 12–18.

Cayrol, J. C., Frankowski, J. P., Laniece, A., d'Hardemare, G., & Talon, J. P. (1978). Contre les nématodes en champignonniere. Mise an point d'une méthode de lute biologique a l'aide d'un hyphomycete predateur: *Arthrobotrys robustus* souche antipolis' (Royal 300). *Rev Hortie, 184,* 23–30.

Cayrol, J. C. (1983). Lutte biologique contre les *Meloidogyne* au moyen d'*Arthrobotrys irregularis* (in French). *Revue de Nématologie, 6,* 265–273.

Chandrawathani, P., Jamnah, O., Waller, P. J., Höglund, J., Larsen, M., & Zahari, W. M. (2002). Nematophagous fungi as a biological control agent for nematode parasites of small ruminants in Malaysia: A special emphasis on *Duddingtonia flagrans. Veterinary Research, 33,* 685–696.

Chandrawathani, P., Jamnah, O., Adnan, M., Waller, P. J., Larsen, M., & Gillespie, A. T. (2004). Field studies on the biological control of nematode parasites of sheep in the tropics, using the microfungus *Duddingtonia flagrans. Veterinary Parasitology, 120,* 77–87.

Chartier, C., & Pors, I. (2003). Effect of the nematophagous fungus, *Duddingtonia flagrans*, on the larval development of goat parasitic nematodes: A plot study. *Veterinary Research, 34,* 221–230.

Chitwood, D. J. (2003). Research on plant-parasitic nematode biology conducted by the United States Department of Agriculture-Agricultural Research Service. *Pest Management Science, 59,* 748–753.

Crous, P. W., Gams, W., Stalpers, J. A., Robert, V., & Stegehuis, G. (2004). MycoBank: An online initiative to launch mycology into the 21st century. *Studies in Mycology, 50,* 19–22.

Dackman, C., Jansson, H. B., & Nordbring-Hertz, B. (1992). Nematophagous fungi and their activities in soil. In G. Stotsky & J. M. Bollag (Eds.), *Soil biochemistry* (Vol. 7). New York: Marcel Dekker.

de Leij, F. A. A. M., & Kerry, B. R. (1991). The nematophagous fungus *Verticillium chlamydosporium* as a potential biological control agent for *Meloidogyne arenaria. Revue de Nématologie, 14,* 157–164.

Djian, C., Pijarowski, L., Ponchet, M., Arpin, N., & Favrebonvin, J. (1991). Acetic-acid a selective nematocidal metabolite from culture filtrates of *Paecilomyces lilacinus* (Thom) Samson and *Trichoderma longibrachiatum* Rifai. *Nematologica, 37,* 101–112.

Dijksterhuis, J., Veenhuis, M., & Harder, W. (1990). Ultrastructural study of adhesion and initial stages of infection of nematodes by conidia of *Drechmeria coniospora. Mycological Research, 94,* 1–8.

Dijksterhuis, J., Harder, W., Wyss, U., & Veenhuis, M. (1991). Colonization and digestion of nematodes by the endoparasitic nematophagous fungus *Drechmeria coniospora. Mycological Research, 95,* 873–878.

Dijksterhuis, J., Veenhuis, M., & Harder, W. (1993). Conidia of the nematophagous fungus *Drechmeria coniospora* adhere to but barely infect *Acrobeloides buetschlii. FEMS Microbiology Letters, 113,* 183–188.

Dijksterhuis, J., Veenhuis, M., Harder, W., & Nordbring-Hertz, B. (1994). Nematophagous fungi: Physiological aspects and structure-function relationships. *Advances in Microbial Physiology, 36,* 111–143.

Dimander, S. O., Hogland, J., & Waller, P. J. (2003). Seasonal translation of infective larvae of gastrointestinal nematodes of cattle and the effect of *Duddingtonia flagrans*; a 3-year plot study. *Veterinary Parasitology, 117,* 99–116.

Dong, J. Y., He, H. P., Shen, Y. M., & Zhang, K. Q. (2005). Nematicidal epipolysulfanyldioxopiperazines from *Gliocladium roseum. Journal of Natural Products, 68,* 1510–1513.

Dong, L. Q., & Zhang, K. Q. (2006). Microbial control of plant-parasitic nematodes a five-party interaction. *Plant and Soil, 288,* 31–34.

D'Souza, S. F., & Melo, J. S. (1991). A method for the preparation of coimmobilizates by adhesion using polyethylenimine. *Enzyme and Microbial Technology, 13,* 508–511.

Duddington, C. L. (1962). Predaceous fungi and the control of eelworms. In C. L. Duddington & J. D. Carthy (Eds.), *Viewpoints in biology* (Vol. 1). London: Butterworths.

Eilenberg, J., Hajek, A., & Lomer, C. (2001). Suggestions for unifying the terminology in biological control. *BioControl, 46,* 387–400.

Faedo, M., Larsen, M., & Grønvold, J. (2002). Predacious activity of *Duddingtonia flagrans* within the cattle faecal pat. *Journal of Helminthology, 76,* 295–302.

Fekete, C., Tholander, M., Rajashekar, B., Ahrén, D., Friman, E., Johansson, T., & Tunlid, A. (2008). Paralysis of nematodes: Shifts in the transcriptome of the nematode-trapping fungus *Monacrosporium haptotylum* during infection of *Caenorhabditis elegans. Environmental Microbiology, 10,* 364–375.

Feng, M. G., Poprawski, T. J., & Khachatourians, G. G. (1994). Production, formulation and application of the entomopathogenic fungus *Beauveria bassiana* for insect control: Current status. *Biocontrol Science and Technology, 4,* 3–34.

Fernández, C., Rodríguez-Kábana, R., Warrior, P., & Kloepper, J. W. (2001). Induced soil suppressiveness to a root-knot nematode species by a nematicide. *Biological Control, 22,* 103–114.

Ferreira, S. R., de Araújo, J. V., Braga, F. R., Araujo, J. M., Frassy, L. N., & Ferreira, A. S. (2011). Biological control of *Ascaris suum* eggs by *Pochonia chlamydosporia* fungus. *Veterinary Research Communications, 35,* 553–558.
Flores-Crespo, J., Herrera-Rodriguez, D., Mendoza-de Gives, P., Liebano-Hernandez, E., Bazquez, V. M., & Lopez-Arellano, M. E. (2003). The predatory capability of three nematophagous fungi in the control of *Haemonchus contortus* infective larvae in ovine faeces. *Journal of Helminthology, 77,* 297–303.
Fontenot, M. E., Miller, J. E., Pea, M. T., Larsen, M., & Gillespie, A. (2003). Efficiency of feeding *Duddingtonia flagrans* chlamydospores to grazing ewes on reducing availability of parasitic nematode larvae on pasture. *Veterinary Parasitology, 118,* 203–213.
Gams, W., & Zare, R. (2003). A taxonomic review of the clavicipitaceous anamorphs parasitizing nematodes and other microinvertebrates. In J. F. White Jr., C. W. Bacon, N. L. Hywel-Jones & J. W. Spatafora (Eds.), *Clavicipitalean fungi: Evolutionary biology, chemistry, and cultural impacts*. New York: Marcel Dekker.
Githigia, S. M., Thamsborg, S. M., Larsen, M., Kyvsgaard, N. C., & Nansen, P. (1997). The preventative effect of the fungus *Duddingtonia flagrans* on trichostrongyle infections of lambs on pasture. *International Journal for Parasitology, 27,* 931–939.
Glockling, S. L., & Dick, M. W. (1994). *Dactylella megalobrocha*, a new species of nematophagous fungus with constricting ring. *Mycological Research, 98,* 845–853.
Hallmann, J., Davies, K. G., & Sikora, R. A. (2009). Biological control using microbial pathogens, endophytes and antagonists. In R. N. Perry, M. Moens, & J. L. Starr (Eds.), *Root-knot nematodes*. Wallingford: CABI.
Hao, Y. E., Mo, M. H., Su, H. Y., & Zhang, K. Q. (2005). Ecology of aquatic nematode-trapping hyphomycetes in southwestern China. *Aquatic Microbial Ecology, 40,* 175–181.
Heintz, C. E., & Pramer, D. (1972). Ultrastructure of nematode-trapping fungi. *Journal of Bacteriology, 110,* 1163–1170.
Hodge, K. T., Viaene, N. M., & Gams, W. (1997). Two *Harposporium* species with *Hirsutella* synanamorphs. *Mycological Research, 101,* 1377–1382.
Holland, R. J., Williams, K. L., & Khan, A. (1999). Infection of *Meloidogyne javanica* by *Paecilomyces lilacinus. Nematology, 1,* 131–139.
Hugot, J. P., Baujard, P., & Morand, S. (2001). Biodiversity in helminths and nematodes as a field of study: An overview. *Nematology, 3,* 199–208.
Ibrahim, A. A. M. (1994). Effect of cadusafos, *Paecilomyces lilacinus* and Nemout on reproduction and damage potential of *Meloidogyne javanica. Pakistan Journal of Nematology, 12,* 141–147.
Jacobs, H., Gray, S. N., & Crump, D. H. (2003). Interactions between nematophagous fungi and consequences for their potential as biological agents for the control of potato cyst nematodes. *Mycological Research, 107,* 47–56.
Jaffee, B. A., & Zehr, E. I. (1985). Parasitic and saprophytic abilities of the nematode-attacking fungus *Hirsutella rhossiliensis. Journal of Nematology, 17,* 341–345.
Jansson, H. B. (1994). Adhesion of conidia of *Drechmeria coniospora* to *Caenorhabditis elegans* wild type and mutants. *Journal of Nematology, 26,* 430–435.
Jansson, H. B., & Friman, E. (1999). Infection-related surface proteins on conidia of the nematophagous fungus *Drechmeria coniospora. Mycological Research, 103,* 249–256.
Jansson, H. B., & Nordbring-Hertz, B. (1988). Infection events in the fungus-nematode system. In G. O. Poinar & H. B. Jansson (Eds.), *Diseases of nematodes*. Boca Raton: CRC Press.
Jansson, H. B., & Thiman, L. (1992). A preliminary study of chemotaxis of zoospores of the nematodeparasitic fungus *Catenaria anguillulae. Mycologia, 84,* 109–112.
Jansson, H. B., Hofsten, A. V., & Mecklenburg, C. V. (1984). Life cycle of the endoparasitic nematophagous fungus *Meria coniospora*: A light and electron microscopic study. *Antonie van Leeuwenhoek, 50,* 321–327.
Jansson, H. B., Jeyaprakash, A., & Zuckerman, B. M. (1985a). Differential adhesion and infection of nematodes by the endoparasitic fungus *Meria coniospora* (Deuteromycetes). *Applied and Environmental Microbiology, 49,* 552–555.

Jansson, H. B., Jeyaprakash, A., & Zuckerman, B. M. (1985b). Control of root knot nematodes on tomato by the endoparasitic fungus *Meria coniospora*. *Journal of Nematology, 17,* 327–329.
Jansson, H. B., Dackman, C., & Zuckerman, B. M. (1987). Adhesion and infection of plant parasitic nematodes by the fungus *Drechmeria coniospora*. *Nematologica, 33,* 480–487.
Jin, X. X., & Custis, D. (2011). Microencapsulating aerial conidia of *Trichoderma harzianum* through spray drying at elevated temperatures. *Biological Control, 56,* 202–208.
Jones, K. A., & Burges, H. D. (1998). Technology of formulation and application. In H. D. Burges (Ed.), *Formulation of microbial pesticides-beneficial microorganisms, nematodes and seed treatments* (pp. 7–30). Dordrecht: Kluwer Academic.
Kalele, D. N., Affokpon, A., & Coosemans, J. (2007). Efficacy of *Paecilomyces lilacinus* strain 251 against root knot nematodes in tomato under greenhouse conditions. *Communications in Agricultural and Applied Biological Sciences, 72,* 209–213.
Kanda, S., Aimi, T., Kano, S., Ishihara, S., Kitamoto, Y., & Morinaga, T. (2008). Ambient pH signaling regulates expression of the serine protease gene (spr1) in pine wilt nematode-trapping fungus, *Monacrosporium megalosporum*. *Microbiological Research, 163,* 63–72.
Kerry, B. R. (1988). Two microorganisms for the biological control of plant parasitic nematodes. In *Proceedings of the Brighton Crop Protection Conference, 2,* 603–607.
Kerry, B. R., & Hominick, W. M. (2002). Biological control. In D. L. Lee (Ed.), *The biology of nematodes*. London: Taylor & Francis.
Khan, M. R. (2007). Prospects of microbial control of root-knot nematodes infecting vegetable crops. In N. Sharma, & H. B. Singh (Eds.), *Biotechnology: Plant heatlh management* (pp. 643–665). Lucknow: International Book Distributing Company.
Khan, M. R., & Gupta, J. (1998). Antagonistic effects of *Trichoderma* species against *Macrophomina phaseolina* on eggplant. *Journal of Plant Diseases and Protection, 105,* 387–393.
Khan, M. R., & Khan, M. W. (1994). Single and interactive effects of root-knot nematode and coal smoke on okra. *New Phytologist, 126,* 337–342.
Khan, A., Williams, K., Molloy, M. P., & Nevalainen, H. (2003). Purification and characterization of a serine protease and chitinases from *Paecilomyces lilacinus* and detection of chitinase activity on 2D gels. *Protein Expression and Purification, 32,* 210–220.
Khan, A., Willianms, K., & Nevalainen, H. (2004). Effects of *Paecilomyces lilacinus* protease and chitinase on the eggeshell structures and hatching of *Meloidogyne javanica* juveniles. *Biological Control, 31,* 346–352.
Khan, M. R., Khan, S. M., & Mohiddin, F. A. (2005). Root-knot problem of some winter ornamental plants and its biomanagement. *Journal of Nematology, 37,* 198–206.
Khan, A., Williams, K. L., & Nevalainen, H. K. M. (2006). Control of plant-parasitic nematodes by *Paecilomyces lilacinus* and *Monacrosporium lysipagum* in pot trials. *Biological Control, 51,* 643–658
Khan, A., Williams, K. L., Soon, J., & Nevalainen, H. K. (2008). Proteomic analysis of the knob-producing nematode-trapping fungus *Monacrosporium lysipagum*. *Mycological Research, 112,* 1447–1452.
Khan, M. R., Majid, S., Mohidin, F. A., & Khan, N. (2011). A new bioprocess to produce low cost powder formulations of biocontrol bacteria and fungi to control fusarial wilt and root-knot nematode of pulses. *Biological Control, 59,* 130–140.
Kiewnick, S., & Sikora, R. A. (2003). Efficacy of *Paecilomyces lilacinus* (strain 251) for the control of root-knot nematodes. *Communications in Agricultural and Applied Biological Sciences, 68,* 123–128.
Kiewnick, S., & Sikora, R. A. (2004). Optimizing the efficacy of *Paecilomyces lilacinus* (strain 251) for the control of root-knot nematodes. *Communications in Agricultural and Applied Biological Sciences, 69,* 373–380.
Knox, M. R., & Faedo, M. (2001). Biological control of field infections of nematode parasites of young sheep with *Duddingtonia flagrans* and effects of spore intake on efficacy. *Veterinary Parasitology, 101,* 155–160.

Kumar, N., Singh, R. K., & Singh, K. P. (2011). Occurrence and colonization of nematophagous fungi in different substrates, agricultural soils and root galls. *Archives of Phytopathology and Plant Protection, 44*, 1182–1195.

Lackey, B. A., Jaffee, B. A., & Muldoon, A. E. (1992). Sporulation of the nematophagous fungus *Hirsutella rhossiliensis* from hyphae produced in vitro and added to soil. *Phytopathology, 82,* 415–420.

Larsen, M. (2006). Biological control of nematode parasites in sheep. *Journal of Animal Science, 84,* E133–E139.

Leggett, M., Leland, J., Kellar, K., & Epp, B. (2011). Formulation of microbial biocontrol agents—an industrial perspective. *Canadian Journal of Plant Pathology, 33,* 101–107.

Li, T. F., Zhang, K. Q., & Liu, X. Z. (2000). *Taxonomy of nematophagous fungi (Chinese)*. Beijing: Science Press.

Li, Y., Hyde, K. D., Jeewon, R., Cai, L., Vijaykrishna, D., & Zhang, K. Q. (2005). Phylogenetics and evolution of nematode-trapping fungi (*Orbiliales*) estimated from nuclear and protein coding genes. *Mycologia, 97,* 1034–1046.

Li, J., Yang, J. K., Huang, X. W., & Zhang, K. Q. (2006). Purification and characterization of an extracellular serine protease from *Clonostachys rosea* and its potential as a pathogenic factor. *Process Biochemistry, 41,* 925–929.

Li, G. H., Zhang, K. Q., Xu, J. P., Dong, J. Y., & Liu, Y. J. (2007). Nematicidal substances from fungi. *Recent Patents on Biotechnology, 1,* 212–233.

Liang, L. M., Meng, Z. H., Ye, F. P., Yang, J. K., Liu, S. Q., Sun, Y., Guo, Y., Mi, Q. L., Huang, X. W., Zou, C. G., Rao, Z. H., Lou, Z.Y., Zhang, K. Q. (2010). The crystal structures of two cuticle-degrading proteases from nematophagous fungi and their contribution to infection against nematodes. *FASEB Journal, 24,* 1391–1400.

Liang, L. M., Liu, S. Q., Yang, J. K., Meng, Z. H., Lei, L. P., & Zhang, K. Q. (2011). Comparison of homology models and crystal structures of cuticle-degrading proteases from nematophagous fungi: Structural basis of nematicidal activity. *FASEB Journal, 25,* 1894–1902.

Liou, G. Y., & Tzean, S. S. (1997). Phylogeny of the genus *Arthrobotrys* and allied nematode-trapping fungi based on rDNA sequences. *Mycologia, 89,* 876–884.

Liu, X. Z., & Zhang, K. Q. (1994). Nematode-trapping species of *Monacrosporium* with special reference to two new species. *Mycological Research, 98,* 862–868.

Liu, X. F., & Zhang, K. Q. (2003). *Dactylella shizishanna* sp. nov., from Shizi Mountain, China. *Fungal Diversity, 14,* 103–107.

Liu, X. Z., Zhang, K. Q., & Li, T. F. (2004). *Biological control of plant-parasitic nematodes*. Beijing: China Science and Technology Press.

Lohmann, U., & Sikora, R. A. (1989). Mass production of the endoparasitic fungi *Drechmeria coniospora, Verticollium balanoides* and *Harposporium anguillulae* in liquid culture. *Nematologica, 35,* 97–104.

Lòpez-Llorca, L. V., & Duncan, G. H. (1988). A scanning electron microscopy study of fungal endoparasitism of cereal cyst nematode (*Heterodera avenae*). *Canadian Journal of Microbiology, 34,* 613–619.

Lòpez-Llorca, L. V., Macia-Vicente, J. G., & Jansson, H. B. (2008). Mode of action and interactions of nematophagous fungi. In A. Ciancio & K. G. Mukerji (Eds.), *Integrated management and biocontrol of vegetable and grain crops nematodes* (pp. 51–76). Dordrecht: Springer.

Lòpez-Llorca, L. V., Gomez-Vidal, S., Monfort, E., Larriba, E., Casado-Vela, J., Elortza, F., Jansson, H. B., Salinas, J., Martin-Nieto, J. (2010). Expression of serine proteases in egg-parasitic nematophagous fungi during barley root colonization. *Fungal Genetics and Biology, 47,* 342–351.

Luo, H., Liu, Y. J., Fang, L., Li, X., Tang, N., & Zhang, K. Q. (2007). *Coprinus comatus* damages nematode cuticles mechanically with spiny balls and produces potent toxins to immobilize nematodes. *Applied and Environmental Microbiology, 73,* 3916–3923.

McInnis, T. M., & Jaffee, B. A. (1998). An assay for *Hirsutella rhossiliensis* spores and the importance of phialides for nematode inoculation. *Journal of Nematology, 21,* 229–234.

Mendoza, A., Sikora, R. A., & Kiewnick, S. (2004). Efficacy of *Paecilomyces lilacinus* (strain 251) for the control of *Radopholus similis* in banana. *Communications in Agricultural and Applied Biological Sciences, 69,* 365–372.

Mi, Q. L., Yang, J. K., Ye, F. P., Gan, Z. W., Wu, C. W., Niu, X. M., Zou, C. G., Zhang, K. Q. (2010). Cloning and overexpression of *Pochonia chlamydosporia* chitinase gene *pcchi44*, a potential virulence factor in infection against nematodes. *Process Biochemistry, 45,* 810–814.

Molinari, S. (2011). Natural genetic and induced plant resistance, as a control strategy to plant-parasitic nematodes alternative to pesticides. *Plant Cell Reports, 30,* 311–323.

Moosavi, M. R., & Zare, R. (2012). Fungi as biological control agents of plant-parasitic nematodes. J. M Mérillon & K. G. Ramawat (Eds.), Plant defence: Biological control, progress in biological control (Vol. 12, Part 2, pp. 67–107). Berlin, Heidelberg: Springer Science and Business Media.

Morton, C. O., Hirsch, P. R., Peberdy, J. P., & Kerry, B. R. (2003a). Cloning of and genetic variation in protease VCP1 from the nematophagous fungus *Pochonia chlamydosporia. Mycological Research, 107,* 38–46.

Morton, C. O., Hirsch, P. R., Peberdy, J. P., & Kerry, B. R. (2003b). PCR-based DNA fingerprinting indicates host-related genetic variation in the nematophagous fungus *Pochonia chlamydosporia. Mycological Research, 107,* 198–205.

Morandi, M. A. B., Maffia, L. A., Mizubuti, E. S. G., Alfenas, A. C., & Barbosa, J. G. (2003). Suppression of *Botrytis cinerea* sporulation by *Clonostachys rosea* on rose debris: A valuable component in *Botrytis* blight management in commercial greenhouses. *Biological Control, 26,* 311–317.

Nordbring-Hertz, B., & Stalhammar-Carlemalm, M. (1978). Capture of nematodes by *Arthrobotrys oligospora*, an electron microscope study. *Canadian Journal of Botany, 56,* 1297–1307.

Nordbring-Hertz, B., Jansson, H. B., & Tunlid, A. (2006). *Nematophagous fungi. In Encyclopedia of life sciences*. Chichester: John Wiley & Sons.

Niu, X. M., Wang, Y. L., Chu, Y. S., Xue, H. X., Li, N., Wei, L. X., Mo, M. H., Zhang, K. Q. (2010). Nematodetoxic aurovertin-type metabolites from a root-knot nematode parasitic fungus *Pochonia chlamydosporia. Journal of Agricultural and Food Chemistry, 58,* 828–834.

Oka, Y., Koltai, H., Bar-Eyal, M., Mor, M., Sharon, E., Chet, I., & Spiegel, Y. (2000). New strategies for the control of plant-parasitic nematodes. *Pest Management Science, 56,* 983–988.

Patel, A. V., Jakobs-Schönwandt, D., Rose, T., & Vorlop, K. D. (2011). Fermentation and micro-encapsulation of the nematophagous fungus *Hirsutella rhossiliensis* in a novel type of hollow beads. *Applied and Environmental Microbiology, 89,* 1751–1760.

Papavizas, G. C., Dunn, M. T., Lewis, J. A., & Beagle-Ristaino, J. (1984). Liquid fermentation technology for experimental production of biocontrol fungi. *Phytopathology, 74,* 1171–1175.

Paraud, C., & Chartier, C. (2003). Biological control of infective larvae of a gastro-intestinal nematode (*Teladorsagia circumcincta*) and a small lungworm (*Muellerius capillaries*) by *Duddingtonia flagrans* in goat faeces. *Parasitology Research, 89,* 102–106.

Pava-Ripoll, M., Posada, F. J., Momen, B., Wang, C., St Leger R (2008). Increased pathogenicity against coffee berry borer, *Hypothenemus hampei* (Coleoptera: Curculionidae) by *Metarhizium anisopliae* expressing the scorpion toxin (AaIT) gene. *Journal of Invertebrate Pathology, 99,* 220–226.

Perry, R. N. (2002). Hatching. In D. L. Lee (Ed.), *The biology of nematodes*. London: Taylor & Francis.

Poinar, G. O. Jr., & Georgis, R. (1994). Biological control for plant-parasitic nematodes attacking turf and ornamentals. In A. R. Leslie (Eds.), *Handbook of integrated pest management for turf and ornamentals* (pp. 491–501). Boca Raton, FL: CRC Press.

Qin, Y., Ying, S. H., Chen, Y., Shen, Z. C., & Feng, M. G. (2010). Integration of insecticidal protein Vip3Aa1 into *Beauveria bassiana* enhances fungal virulence to *Spodoptera litura* larvae by cuticle and *per os* infection. *Applied and Environmental Microbiology, 76,* 4611–4618.

Robb, E. J., & Barron, G. L. (1982). Nature's ballistic missile. *Science, 218,* 1221–1222.

Rubner, A. (1996). Revision of predacious hyphomycetes in the *Dactylella-Monacrosporium* complex. *Studies in Mycology, 39,* 1–134.

Sanyal, P. K. (2001). Biological control of *Haemonchus contortus* in sheep by nematophagous fungus, *Duddingtonia flagrans. Indian Journa of Veterinary, 78,* 1091–1095.
Sarkunas, M., Larsen, M., Nansen, P., & Hansen, J. W. (2000). Biological control of trichostrongylid infections in calves on pasture in Lithuania using *Duddingtonia flagrans*, a nematode-trapping fungus. *Journal of Helminthology, 74,* 355–359.
Sasser, J. N. (1980). Root-knot nematodes: A global menace to crop production. *Plant Disease, 64,* 36–41.
Saumell, C. A., Padiha, T., Santos, C. D. P., & Roque, M. V. C. (1999). Nematophagous fungi in fresh feces of cattle in the Mata region of Mina Gerais state, Brazil. *Veterinary Parasitology, 82,* 217–220.
Scholler, M., Hagedorn, G., & Rubner, A. (1999). A reevaluation of predatory orbiliaceous fungi. II. A new generic concept. *Sydowia, 51,* 89–113.
Schroers, H. J., Samuels, G. J., Seifert, K. A., & Gams, W. (1999). Classification of the mycoparasite *Gliocladium roseum* in *Clonostachys* as *G. rosea*, its relationship to *Bionectria ochroleuca*, and notes on other Gliocladium-like fungi. *Mycologia, 91,* 365–385.
Segers, R., Butt, T. M., Keen, J. N., Kerry, B. R., & Peberdy, J. F. (1995). The subtilisins of the invertebrate mycopathogens *Verticillium chlamydosporium* and *Metarhizium anisopliae* are serologically and functionally related. *FEMS Microbiology Letters, 126,* 227–231.
Segers, R., Butt, T. M., Kerry, B. R., Beckett, A., & Peberdy, J. F. (1996). The role of the proteinase VCP1 produced by the nematophagous *Verticillium chlamydosporium* in the infection process of nematode eggs. *Mycological Research, 100,* 421–428.
Segers, R., Butt, T. M., Carder, J. H., Keen, J. N., Kerry, B. R., & Peberdy, J. F. (1999). The subtilisins of fungal pathogens of insects, nematodes and plants: Distribution and variation. *Mycological Research, 103,* 395–402.
Siddiqui, Z. A., & Mahmood, I. (1996). Biological control of plant parasitic nematodes by fungi: A review. *Bioresource Technology, 58,* 229–239.
Siezen, R. J., & Leunissen, J. A. M. (1997). Subtilases: The superfamily of subtilisin-like serine proteases. *Protein Science, 6,* 501–523.
Silva, A. R., Araújo, J. V., Braga, F. R., Alves, C. D., & Frassy, L. N. (2010). In vitro ovicidal activity of the nematophagous fungi *Duddingtonia flagrans, Monacrosporium thaumasium* and *Pochonia chlamydosporia* on *Trichuris vulpis* eggs. *Veterinary Parasitologyol, 172,* 76–79.
Singh, U. B., Sahu, A., Singh, R. K., Singh, D. P., Meena, K. K., Srivastava, J. S., Renu, Manna, M. C. (2012). Evaluation of biocontrol potential of *Arthrobotrys oligospora* against *Meloidogyne graminicola* and *Rhizoctonia solani* in Rice (*Oryza sativa* L.). *Biological Control, 60,* 262–270.
Sjollema, K., Dijksterhuis, J., & Veenhuis, M. H. W. (1993). An electron microscopical study of the infection of the nematode *Panagrellus redivivus* by the endoparasitic fungus *Verticillium balanoides. Mycological Research, 97,* 479–484.
Soder, K. J., & Holden, L. A. (2005). Use of nematode-trapping fungi as a biological control in grazing livestock. *Professional Animal Scientist, 21,* 30–37.
St Leger, R. J., Wang, C. S. (2010). Genetic engineering of fungal biocontrol agents to achieve greater efficacy against insect pests. *Applied and Environmental Microbiology, 85,* 901–907.
Sun, S. H., Zhou, L., Xia, Z. Y., Zhao, Y. H., & Duan, L. C. (2002). Study of microorganic nematicide to control tobacco root-knot nematodes (in Chinese, with English abstract). *Acta Tabacaria Sinica, 8,* 30–33.
Tedford, E. C., Jaffee, B. A., Muldoon, A. E., Anderson, C. E., & Westerdahl, B. B. (1993). Parasitism of *Heterodera schachtii* and *Meloidogyne javanica* by *Hirsutella rhossiliensis* in microplots over two growing seasons. *Journal of Nematology, 25,* 427–433.
Terrill, T. H., Larsen, M., Samples, O., Husted, S., Miller, J. E., Kaplan, R. M., & Gelaye, S. (2004). Capability of the nematodetrapping fungus *Duddingtonia flagrans* to reduce infective larvae of gastrointestinal nematodes in goat feces in the southeastern United States: Dose titration and dose time interval studies. *Veterinary Parasitology, 120,* 285–296.

Tian, B. Y., Yang, J. K., & Zhang, K. Q. (2007). Bacteria used in the biological control of plant-parasitic nematodes: Populations, mechanisms of action, and future prospects. *FEMS Microbiology Ecology, 61,* 197–213.

Tikhonov, V. E., Lopez-Llorca, L. V., Salinas, J., & Jansson, H. B. (2002). Purification and characterization of chitinases from the nematophagous fungi *Verticillium chlamydosporium* and *V. suchlasporium. Fungal Genetics and Biology, 35,* 67–78.

Timper, P., Kaya, H. K., & Jaffee, B. A. (1991). Survival of entomogenous nematodes in soil infested with the nematode-parasitic fungus *Hirsutella rhossiliensis* (Deuteromycotina: Hyphomycetes). *Biological Control, 1,* 42–50.

Tunlid, A., Johansson, T., & Nordbring-Hertz, B. (1991a). Surface polymers of the nematode-trapping fungus *Arthrobotrys oligospora. Journal of General Microbiology, 137,* 1231–1240.

Tunlid, A., Nivens, D. E., Jansson, H. B., & White, D. C. (1991b). Infrared monitoring of the adhesion of *Catenaria anguillulae* zoospores to solid surfaces. *Experimental Mycology, 15,* 206–214.

Tunlid, A., Rosén, S., Ek, B., & Rask, L. (1994). Purification and characterization of an extracellular serine protease from the nematode-trapping fungus *Arthrobotrys oligospora. Microbiology (Reading, England), 140,* 1687–1695.

Tunlid, A., Ahman, J., & Oliver, R. P. (1999). Transformation of the nematode-trapping fungus *Arthrobotrys oligospora. FEMS Microbiology Letters, 173,* 111–116.

Twomey, U., Warrior, P., Kerry, B. R., & Perry, R. N. (2000). Effects of the biological nematicide, DiTera, on hatching of *Globodera rostochiensis* and *G. pallida. Nematology, 2,* 355–362.

Viaene, N., Coyne, D. L., & Kerry, B. R. (2006). Biological and cultural management. In R. N. Perry, M. Moens (Eds.) *Plant nematology*. Wallingford: CABI.

Waghorn, T. S., Leathwick, D. M., Chen, L. Y., & Skipp, R. A. (2003). Efficacy of the nematode-trapping fungus *Duddingtonia flagrans* against three species of gastro-intestinal nematodes in laboratory faecal cultures from sheep and goats. *Veterinary Parasitology, 118,* 227–234.

Waller, P. J., Knox, M. R., & Faedo, M. (2001). The potential of nematophagous fungi to control the free-living stages of nematode parasites of sheep: Feeding and block studies with *Duddingtonia flagrans. Veterinary Parasitology, 102,* 321–330.

Waller, P. J., Ljungström, B. L., Schwan, O., Martin, L. R., Morrison, D. A., & Rydzik, A. (2006). Biological control of sheep parasites using *Duddingtonia flagrans*: Trials on commercial farms in Sweden. *Acta Veterinaria Scandinavica, 47,* 23–32.

Wang, C. S., St Leger, R. J. (2007). A scorpion neurotoxin increases the potency of a fungal insecticide. *Nature Biotechnology, 25,* 1455–1456.

Wang, B., Wu, W. P., & Liu, X. Z. (2007). Purification and characterization of a neutral serine protease with nematicidal activity from *Hirsutella rhossiliensis. Mycopathologia, 163,* 169–176.

Wang, B., Liu, X., Wu, W., Liu, X., & Li, S. D. (2009). Purification, characterization, and gene cloning of an alkaline serine protease from a highly virulent strain of the nematode-endoparasitic fungus *Hirsutella rhossiliensis. Microbiological Research, 164,* 665–673.

Wang, J., Wang, J., Liu, F., & Pan, C. (2010). Enhancing the virulence of *Paecilomyces lilacinus* against *Meloidogyne incognita* eggs by overexpression of a serine protease. *Biotechnology Letters, 32,* 1159–1166.

Wei, L. X., Zhang, H. X., Tan, J. L., Chu, Y. S., Li, N., Xue, H. X., Wang, Y. L., Niu, X. M., Zhang, Y., Zhang, K. Q. (2011). Arthrobotrisins A-C, oligosporons from the nematode-trapping fungus *Arthrobotrys oligospora. Journal of Natural Products, 74,* 1526–1530.

Wharton, D. A. (1980). Nematode egg-shells. *Parasitology, 81,* 447–463.

Wright, D. A., McAnulty, R. W., Noonan, M. J., & Stankiewicz, M. (2003). The effect of *Duddingtonia flagrans* on trichostrongyle infections of Saanen goats on pasture. *Veterinary Parasitology, 118,* 61–69.

Yang, J. K., Huang, X. W., Tian, B. Y., Wang, M., Nui, Q. H., & Zhang, K. Q. (2005a). Isolation and characterization of a serine protease from the nematophagous fungus, *Lecanicillium psalliotae*, displaying nematicidal activity. *Biotechnology Letters, 27,* 1123–1128.

Yang, J. K., Huang, X. W., Tian, B. Y., Sun, H., Duan, J. X., Wu, W. P., & Zhang, K. Q. (2005b). Characterization of an extracellular serine protease gene from the nematophagous fungus *Lecanicillium psalliotae*. *Biotechnology Letters, 27,* 1329–1334.

Yang, J. K., Tian, B. Y., Liang, L. M., & Zhang, K. Q. (2007a). Extracellular enzymes and the pathogenesis of nematophagous fungi. *Applied and Environmental Microbiology, 75,* 21–31.

Yang, Y., Yang, E., An, Z., & Liu, X. (2007b). Evolution of nematode-trapping cells of predatory fungi of the *Orbiliaceae* based on evidence from rRNA-encoding DNA and multiprotein sequences. *Proceedings of the National academy of Sciences of the United States of America, 104,* 8379–8384.

Yang, J. K., Gan, Z. W., Lou, Z. Y., Tao, N., Mi, Q. L., Liang, L. M., Sun, Y., Guo, Y., Huang, X. W., Zou, C. G., Rao, Z. H., Meng, Z.H., Zhang, K. Q. (2010). Crystal structure and mutagenesis analysis of chitinase CrChi1 from the nematophagous fungus *Clonostachys rosea* in complex with the inhibitor caffeine. *Microbiology (Reading, England), 156,* 3566–3574.

Yang, J. K., Wang, L., Ji, X. L., Feng, Y., Li, X. M., Zou, C. G., Xu JP., Ren, Y., Mi, Q. L., Wu, J. L., Liu, S. Q., Liu, Y., Huang, X. W., Wang, H. Y., Niu, X. M., Li, J., Liang, L. M., Luo, Y. L., Ji, K. F., Zhou, W., Yu, Z. F., Li, G. H., Liu, Y. J., Li, L., Qiao, M., Feng, L., Zhang, K. Q. (2011a). Genomic and proteomic analyses of the fungus *Arthrobotrys oligospora* provide insights into nematode-trap formation. *PLoS Pathogens, 7,* e1002179.

Yang, J. K., Zhao, X. N., Liang, L. M., Xia, Z. Y., Lei, L. P., Niu, X. M., Zou, C. G., Zhang, K. Q. (2011b). Overexpression of a cuticle-degrading protease Ver112 increases the nematicidal activity of *Paecilomyces lilacinus*. *Applied and Environmental Microbiology, 89,* 1895–1903.

Yeates, G. W., Dimander, S. O., Waller, P. J., & Hoglund, J. (2002). Environmental impact on soil nematodes following the use of the Ivermectin sustained-release bolus or the nematophagous fungus *Duddingtonia flagrans* to control nematode parasites of cattle in Sweden. *Acta Agriculturae Scandinavica. Section A- Animal Science, 52,* 233–242.

Zare, R., Gam, W., & Evans, H. C. (2000a). A revision of *Verticillium* section Prostrata. V. The genus *Pochonia*, with notes on *Rotiferophthora*. *Nova Hedwigia, 73,* 51–86.

Zare, R., Gams, W., & Culham, A. (2000b). A revision of *Verticillium* section prostrate. I. Phylogenetic studies using ITS sequences. *Nova Hedwigia, 71,* 465–480.

Zhang, K. Q., Liu, X. Z., & Cao, L. (1996). Nematophagous species of *Monacrosporium* from China. *Mycological Research, 100,* 274–276.

Zhang, L., Yang, J. K., Niu, Q. H., Zhao, X. N., Ye, F. P., Liang, L. M., & Zhang, K. Q. (2008). Investigation on the infection mechanism of the fungus *Clonostachys rosea* against nematodes using the green fluorescent protein. *Applied and Environmental Microbiology, 78,* 983–890.

Zhang, Y., Li, G. H., & Zhang, K. Q. (2011). A review on the research of nematophagous fungal species (In Chinese). *Mycosystema, 30,* 836–845.

Zhao, M. L., Huang, J. S., Mo, M. H., & Zhang, K. Q. (2005). A potential virulence factor involved in fungal pathogenicity: Serine-like protease activity of nematophagous fungus *Clonostachys rosea*. *Fungal Diversity, 19,* 217–234.

Zhou, W., & Mo, M. H. (2002). Soil fungistasis on the spore germination of fungi (in Chinese, with English abstract). *Journal of Yunnan University (Natural Science Edition), 24,* 312–315.

Zhu, M. L., Zhang, K. Q., Li, T. F., Xia, Z. Y., & Yang, S. J. (2001). Study of microorganic nematicide for controlling tobacco knot nematodiasis (in Chinese, with English abstract). *Journal of Mountain Agriculture and Biology, 20*, 28–31.K.-Q. Zhang and J. Yang

Chapter 6
Molecular Mechanism of Nematophagous Fungi Infection of Nematodes

Jinkui Yang, Lianming Liang, Chenggang Zou and Ke-Qin Zhang

Contents

K.-Q. Zhang (✉) · J. Yang · L. Liang · C. Zou
Laboratory for Conservation and Utilization of Bio-Resources, and Key Laboratory of Microbial Diversity in Southwest China, Ministry of Education, Yunnan University, 650091 Kunming, Yunnan, China
e-mail: kqzhang1@ynu.edu.cn

K.-Q. Zhang, K. D. Hyde (eds.), *Nematode-Trapping Fungi,* Fungal Diversity Research Series 23, DOI 10.1007/978-94-017-8730-7_6,

Abstract Nematophagous fungi are an important group of soil microorganisms that can suppress the populations of plant-parasitic nematodes. At present, the detailed molecular pathogenic mechanisms against nematodes by nematophagous fungi have not yet been fully elucidated. However, increasing evidence show that extracellular hydrolytic enzymes including proteases, collagenase, and chitinase may be involved in nematode-cuticle penetration and host-cell digestion. Recently, the crystal structures of proteases (Ver112 and PL646) and chitinase CrChi1 from nematophagous fungi were resolved, which can help us to identify the active site residues and to elucidate the catalytic mechanism of these enzymes involved in infection against hosts. The expression and regulation of protease PrC from *Clonostachys rosea* by different environmental conditions has also been reported. The genome of *Arthrobotrys oligospora* has been sequenced, and a model of nematode trap formation in *A. oligospora* suggested; thus the genome data may serve as a roadmap for further investigations into the interaction between nematode-trapping fungi and their host nematodes, providing broad foundations for research on the biocontrol of pathogenic nematodes. In this chapter, we describe the characterization of extracellular enzymes from nematophagous fungi, the expression and regulation of serine protease prC in *Clonostachys rosea*, and the genome and proteomic analyses of the nematode-trapping fungus *Arthrobotrys oligospora*.

Keywords *Arthrobotrys oligospora* • Chitinase • Crystal structure • Expression • Extracellular enzyme • Genome • Nematophagous fungi • Proteome • Regulation • Serine protease • Trap formation • Virulence factor

Introduction

Nematophagous fungi include a diverse group that can antagonize nematodes. They can be grouped into three categories: nematode-trapping fungi, parasitic fungi and toxic fungi based on to their different pathogenic mechanisms (Siddiqui and Mahmood 1996; Li et al. 2000; Nordbring-Hertz et al. 2006). Extracellular enzymes can break the physical and physiological integrity of the cuticles of nematodes resulting in penetration and colonization, and several extracellular enzymes have been identified as important virulence factors in nematophagous fungi. Since the first cuticle-degrading protease P32 was identified from *Pochonia rubescens* (syn. *Verticillium suchlasporia*) in 1990 (Lopez-Llorca 1990), more than ten proteases have been identified from different nematophagous fungi (Yang et al. 2007a). Moreover, chitinase and collagenase are also reported to involved in nematode-cuticle penetration and host-cell digestion, and several chitinases have been cloned and identified (Gan et al. 2007a, b; Dong et al. 2007; Mi et al. 2010). The crystal structures of serine proteases and chitinases of nematophagous fungi have also recently been resolved (Liang et al. 2010, 2011a; Yang et al. 2010); these data provide a basis to improve the nematacidal activity of these enzymes.

The molecular background behind the infection of nematodes by nematophagous fungi however, is not fully elucidated. Recently, the entire genome of *Arthrobotrys oligospora* was sequenced (Yang et al. 2011a) and is the first sequence of a nematophagous fungus. The genome data will serve as a roadmap for further investigations into the interactions between nematode-trapping fungi and their host nematodes. This may provide foundations for research on the biocontrol of pathogenic nematodes.

The purpose of this chapter is to characterize extracellular enzymes produced by nematophagous fungi, the exoression and regulation of serine protease prC in *Clonostachys rosea*, and relay genome and proteomic analyses of the nematode-trapping fungus *Arthrobotrys oligospora*.

Extracellular Enzymes Involved in the Infection of Nematophagous Fungi

Extracellular Enzymes Produced by Nematophagous Fungi

Many types of enzymes are produced by microbes during growth, and several extracellular enzymes can be secreted from the cell and aid the pathogens in attacking their hosts. The chemical composition of the host surface is of particular important for hydrolytic enzymes involved in infection (Tunlid and Jansson 1991). Extracellular enzymes corresponding to the main chemical constituents of the insect cuticle (protein, chitin, and lipids) have been detected in various entomopathogenic fungi (St Leger et al. 1986; Bidochka and Khachatourians 1987).

The nematode cuticle consists mainly of proteins, including collagens (Cox et al. 1981). Nematode-trapping fungi possess the unique ability to capture and infect nematodes, and the infection by nematode-trapping fungi involves a series of processes including adhesion, penetration, and immobilization of nematodes (Nordbring-Hertz 2004, 2006; Yang et al. 2007a). During the infection process, the cuticle is penetrated, the nematode is immobilized, and the prey is finally invaded and digested by the fungus (Tunlid and Jansson 1991). *Arthrobotrys oligospora* produces extracellular proteases when grown in a liquid culture, however the extracellular protease activity was inhibited by phenylmethylsulfonyl fluoride (PMSF) and other serine protease inhibitors (Tunlid and Jansson 1991). Substrate gel electrophoresis showed that the *Arthrobotrys oligospora* produced several different proteases, including multiple serine proteases. Other studies showed that if trap-bearing mycelium are incubated with a serine protease inhibitor, PMSF, antipain, or chymostatin, or the metalloprotease inhibitor phenanthroline this resulted in significant decreases in the immobilization of nematodes captured by the fungus. Cysteine or aspartic protease inhibitors did not affect the immobilization of captured nematodes (Tunlid and Jansson 1991). These results indicate that serine proteases are involved in the degradation of nematode cuticles. Serine proteases have subsequently been isolated and identified from several different nematophagous fungi (Lopez-Llorca 1990; Tunlid et al. 1994; Zhao et al. 2004; Yang et al. 2007a). Other hydrolytic enzymes, such as chitinases and collagenases are also produced by nematophagous fungi and have been shown to be involved in infection of nematodes (Yang et al. 2007a).

Characterization of Extracellular Enzymes from Nematophagous Fungi

Serine Proteases

Serine proteases are a family of enzymes that utilize a uniquely activated serine residue in the substrate-binding pocket to catalytically hydrolyze peptide bonds (Schultz and Liebman 1997; Siezen and Leunissen 1997). Serine proteases carry out a diverse array of physiological functions (Yousef et al. 2003) and have been reported as pathogenic factors found in bacterial or fungal pathogens against insects, nematodes and even humans (e.g., Tunlid et al. 1994; Joshi et al. 1995; Yang et al. 2007a). For example, serine proteases are the major extracellular enzymes produced in large amounts by virulent isolates of *Verticillium lecanii* (Jackson et al. 1985) and *Metarhizium anisopliae* (St Leger et al. 1992).

The first pathogenicity-related serine protease P32 was identified from *Pochonia rubescens* (Lopez-Llorca 1990). Subsequently, similar proteases were also found in other nematophagous fungi. Recently, serine proteases have been identified from the nematophagous taxa, such as *Paecilomyces lilacinus* (Bonants et al. 1995), *Arthrobotrys oligospora* (Zhao et al. 2004), *Lecanicillium psalliotae* (Yang et al. 2005a), *Monacrosporium microscaphoides* (Wang et al. 2006a), *Clonostachys rosea* (syn. *Gliocladium rosea*) (Li et al. 2006) and *Dactylella shizishanna* (Wang et al. 2006b). So far, more than 20 serine proteases have been isolated or cloned from different nematophagous fungi, the partial biochemical properties of these serine proteases are summarized in Table 6.1.

These proteases (Table 6.1) from nematophagous fungi are highly sensitive to inhibitor phenylmethylsulfonyl fluoride (PMSF) indicating that they belong to the subtilisin-like serine protease family (Siezen and Leunissen 1997). They have similar molecular weights, ranging from 32 to 39 kDa, and share a broad range of protein substrates including casein, gelatin, nematode cuticle and eggshells. The biochemical properties of proteases PII (from *Arthrobotrys oligospora*), Aoz1 (*A. oligospora*), Mlx (*Monacrosporium microscaphoides*) and Ds1 (*Dactylella shizishanna*) (Table 6.1) are similar and share a lower isoelectric point (pI); all being isolated from nematode-trapping fungi. Interestingly, the biochemical properties of proteases P32 (from *Pochonia rubescens*), VCP1 (*Pochonia chlamysporia*), pSP-3 (*Paecilomyces lilacinus*), Ver112 (*Lecanicillium psalliotae*) and PrC (*Clonostachys rosea*) (Table 6.1) are also similar to each other except that they share a higher pI; these enzymes were isolated from parasitic fungi.

Table 6.1 Partial properties of serine proteases isolated from nematophagous fungi

Nematophagous fungi	Protease	MW (kDa)	Inhibitors	pI	Optim-um pH	References
Arthrobotrys oligospora	PII	35	PMSF, Chymostatin	4.6	7–9	Tunlid et al. (1994)
Arthrobotrys oligospora	Aoz1	38	PMSF, SSI	4.9	6–8	Zhao et al. (2004)
Monacrosporium microscaphoides	Mlx	39	PMSF	6.8	9	Wang et al. (2006a)
Dactylella shizishanna	Ds1	32	PMSF	–[a]	10	Wang et al. (2006b)
Arthrobotrys conoides	Ac1	35	PMSF	–	7.0	Yang et al. (2007b)
Dactylellina varietas	Dv1	30	PMSF	–	8.0	Yang et al. (2007c)
Monacrosporium cystosporium	Mc1	38	PMSF	–	7–9	Yang et al. (2008)
Monacrosporium megalosporum	Spr1	–	–	–	–	Kanda et al. (2008)
Dactylellina haptotyla	Dha1	–	–	–	–	Fekete et al. (2008)
Dactylellina cionopaga	Dc1	–	–	–	–	AEP02888[a]
Pochonia rubescens	P32	32	PMSF, pCMB	–	8.5	Lopez-Llorca (1990); Larriba et al. (2012)
Pochonia chlamydosporia	VCP1	33	PMSF	10.2	–	Segers et al. (1994); Esteves et al. (2009); Ward et al. (2012)
Paecilomyces lilacinus	pSP-3	33.5	PMSF	10.2	10.3	Bonants et al. (1995)
Lecanicillium psalliotae	Ver112	32	PMSF	–	10	Yang et al. (2005b)
Clonostachys rosea	PrC	33	PMSF	10	9–10	Li et al. (2006; Liang et al. (2011b)
Cordyceps sinensis	Csp1	36	PMSF	–	7.0	Zhang et al. (2008)
Cordyceps sinensis	Csp2	35	PMSF	–	7.0	Zhang et al. (2008)
Hirsutella rhossiliensis	Hnsp	32	PMSF	–	7.0	Wang et al. (2007)
Hirsutella rhossiliensis	Hasp	33	PMSF	–	9.0	Wang et al. (2009)
Hirsutella minnesotensis	Hm1	–	–	–	–	ABQ51089[a]
Duddingtonia flagrans	Df1	38	PMSF	–	8.0	Braga et al. (2012)
Monacrosporium thaumasium	Mt1	40	PMSF	–	7–8	Soares et al. (2012)
Trichoderma pseudokoningii	SprT	31	–	–	8.5	Chen et al. (2009)

– No data in references, *pI* isoelectric point

[a] Only reported in GenBank

PII_CAA63841	100%																						
Mlx_AAW21809	84.7%	100%																					
Moe1_AAX54902	84.9%	94.8%	100%																				
Mc1_AAX54901	84.9%	91.9%	92.8%	100%																			
Am1_ABL74282	89.5%	84.7%	84.9%	85.9%	100%																		
Ay1_ABL74283	82.6%	81.7%	81.0%	82.0%	83.1%	100%																	
Mp1_ABL74284	83.5%	84.1%	81.9%	82.7%	83.3%	84.4%	100%																
Ac1_AAX54903	88.2%	81.0%	81.5%	81.5%	88.0%	81.2%	79.8%	100%															
Mco1_ABL74285	80.7%	80.9%	82.1%	80.9%	79.3%	78.8%	80.5%	81.6%	100%														
Ds1_ABL74286	69.7%	69.7%	69.7%	70.8%	70.7%	69.0%	68.1%	68.5%	68.1%	100%													
Dv1_ABF72182	65.8%	67.3%	67.1%	66.6%	64.2%	63.1%	65.2%	64.2%	65.1%	67.7%	100%												
Dha1_ABV46589	56.2%	57.2%	57.0%	58.8%	57.3%	57.8%	57.3%	55.8%	57.0%	56.2%	53.5%	100%											
Dc1_AEP02888	54.4%	55.6%	56.1%	56.3%	56.3%	55.3%	53.9%	55.5%	55.2%	53.2%	54.3%	55.7%	100%										
Spr1_BAD44716	52.5%	54.6%	54.3%	53.8%	51.8%	54.1%	53.5%	52.6%	53.0%	51.0%	51.9%	55.5%	57.1%	100%									
PrC_ACS66684	43.1%	44.5%	45.8%	44.5%	43.0%	42.6%	43.0%	42.2%	42.3%	43.6%	41.6%	42.2%	38.5%	37.1%	100%								
Csp1_ABX79998	41.6%	43.3%	44.1%	43.0%	41.8%	42.1%	43.4%	42.3%	42.1%	40.9%	43.1%	41.6%	36.6%	40.9%	51.7%	100%							
Csp2_ABX79999	40.5%	43.5%	43.5%	43.8%	41.2%	41.6%	43.7%	42.3%	41.6%	41.7%	41.8%	41.7%	37.0%	36.6%	50.4%	69.2%	100%						
P32_ADK74313*	51.8%	53.1%	53.9%	52.4%	50.7%	51.3%	49.5%	51.1%	52.0%	52.2%	53.3%	51.3%	46.4%	46.7%	60.0%	71.0%	69.9%	100%					
VCP1_CAD20578	44.4%	45.8%	45.6%	45.6%	44.0%	45.7%	46.8%	44.3%	43.6%	43.4%	43.2%	43.0%	42.7%	41.0%	52.6%	63.6%	64.2%	75.5%	100%				
Ver112_AAU01968	40.4%	42.9%	42.7%	43.5%	41.4%	42.3%	42.8%	42.5%	41.2%	40.8%	42.7%	40.5%	37.8%	39.4%	48.3%	58.8%	62.7%	72.1%	58.4%	100%			
Hm1_ABQ51089	44.0%	45.4%	45.7%	45.4%	43.9%	43.5%	45.5%	43.6%	42.9%	43.3%	42.8%	40.8%	36.9%	37.5%	52.0%	63.8%	75.6%	69.5%	61.3%	64.6%	100%		
Hnsp_ABD96101	41.1%	43.8%	43.3%	43.5%	41.5%	41.3%	43.4%	41.8%	40.8%	41.4%	43.1%	41.4%	36.5%	37.1%	51.2%	72.5%	91.3%	70.6%	62.9%	63.5%	76.1%	100%	
pSP-3_ABO32256	40.8%	44.4%	44.9%	43.8%	41.8%	41.3%	42.8%	41.5%	40.0%	41.9%	41.2%	40.3%	35.6%	38.1%	55.1%	73.1%	76.1%	75.6%	66.8%	65.4%	71.0%	78.4%	100%

Fig. 6.1 Homologous analysis of serine proteases from different nematophagous fungi. Proteases highlighted in *grey* are identified from the nematode-trapping fungi *Arthrobotrys* spp. and *Monacrosporium* spp. Non highlighted proteases are identified from nematode-trapping fungi reported in GenBank. The proteases highlighted in *red* are identified from *Dactylella* spp., and the proteases highlighted in *yellow* are identified from parasitic fungi. *the sequence of P32 is incomplete, thus the predicted identity is inaccurate

Serine proteases from nematophagous fungi share a high degree of similary (Fig. 6.1). Protease PII (from *Arthrobotrys oligospora*) is most similar to proteases from nematode-trapping fungi than proteases from opportunistic fungi. Protease PII shared 80.7–89.5 % similarity to proteases (e.g., Mlx, Moe1, Mc1, Mco1, Am1, Ac1 and Mp1) from species of *Arthrobotrys* and *Monacrosporium*, while PII shared 65.8 and 69.7 % similarity to proteases Dv1 and Ds1, respectively. PII shared 40.5–44.4 % similarity to proteases from opportunistic fungi (Fig. 6.1). Interestingly, serine proteases from opportunistic fungi were only 48.3–55.1 % similar to each other.

When serine proteases from nematophagous fungi are aligned, the mature peptides are more conversed than signal peptide and propeptides, and they all shared conserved Asp-His-Ser catalytic triads (Fig. 6.2). Five disulfide bonds were found in the serine proteases from opportunistic fungi (alkaline proteases), while only one was found in the serine proteases from nematode-trapping fungi (neutral proteases) (Fig. 6.2). The alkaline proteases (e.g., Ver112) can form two disulfide bonds, Cys36–C125 and C180–C251 (Ver112 numbering), whereas the neutral protease (e.g., PII) contains no disulfide bond; and Ver112 displays higher thermal stability and stronger nematicidal/catalytic activity than PII (Liang et al. 2011a). Therefore, the presence of disulfide bonds not only enhances the local, but also the global stability of the protease, thus explaining the higher thermal stability of the alkaline protease Ver112 as compared to that of the neutral protease PII.

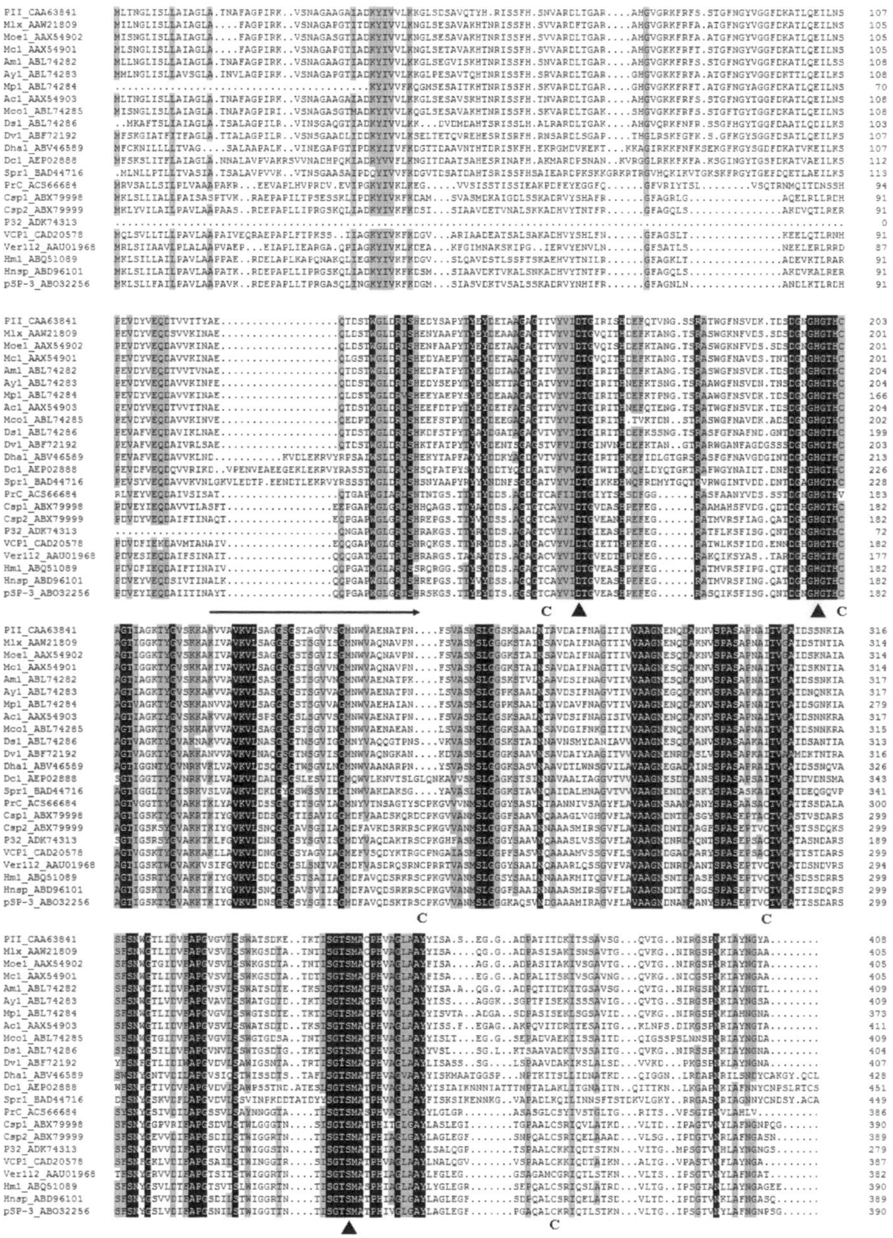

Fig. 6.2 Alignment of deduced peptide sequences from different fungi. Areas highlighted in *black* are conserved regions (100 % similarity), areas highlighted in *grey* have high degrees of homology (more than 75 % similarity) and non highlighted areas are regions of variability between these proteases. The predicted N-terminal sequences of mature peptides are marked with arrows. ▲ indicates the Asp-His-Ser catalytic triad. *C* indicates cysteine

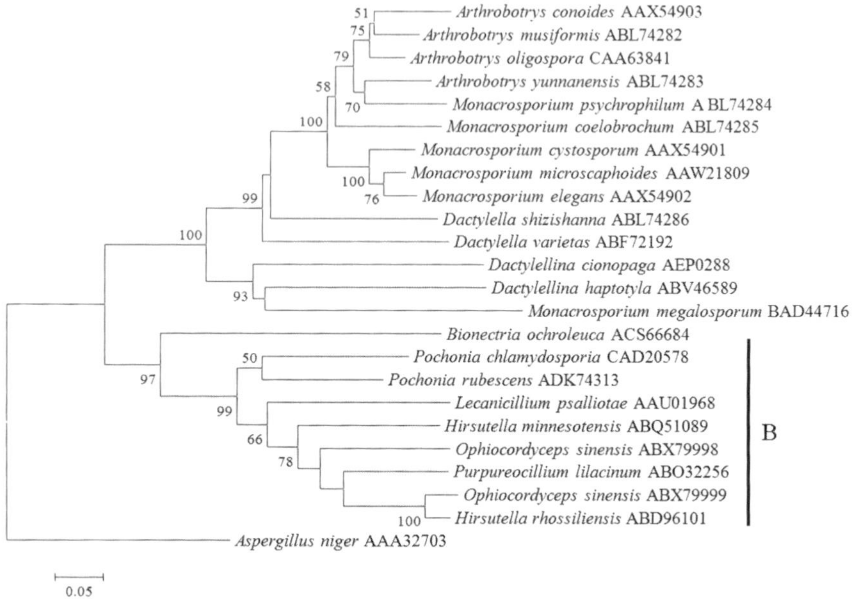

Fig. 6.3 Phylogenetic analyses based on the deduced peptide sequences of serine protease from different fungi. The GenBank accession number of proteases was shown in Fig. 6.3. *Aspergillus niger* (accession number: AAA32703) is used as outgroup. The phylogenetic tree was obtained by the neighbor-joining method using the Mega 4.0 software package. The numbers *below* the branches indicate the percentage at which a given branch was supported in 1,000 bootstrap replications. (Reproduced from Yang et al. (2013) with kind permission from Springer Science and Business Media)

The disulfide bonds also increase the flexibility of substrate-binding pockets located relatively far from disulfide bonds, thus explaining why alkaline proteases have higher substrate affinity and catalytic activity than neutral proteases (Liu et al. 2011; Tao et al. 2010).

Serine proteases from nematophagous fungi cluster in clades A and B (Fig. 6.3), with those from nematode-trapping fungi clustering in clade A, and those from parasitic fungi clustering in clade B. Therefore, pathogenicity-related proteases from nematophagous fungi can be divided into two categories according to the differences of their biochemical properties and phylogenetic analyses. Class I comprising proteases from nematode-trapping fungi and class II consists of proteases from parasitic fungi (Yang et al. 2005b, 2007a).

Previous results indicated that the higher pI value was important for the hydrolytic activity and binding of the enzyme to fragments to insect cuticles (St Leger et al. 1986). Serine proteases also showed different nematicidal activity against different nematodes (Wang et al. 2006a), suggesting that substrate recognition sites of these proteases might differ among fungi with different host preferences. Recently, two cuticle-degrading proteases were resolved with X-ray crystallography, experimental studies showed that the shape, size and hydrophilicity or hydrophobicity of the substrate-binding pockets influence the affinity and specifity of proteases to substrates (Liang et al. 2011a).

Increasing evidence has shown that increasing the copy number of virulence genes can increase the pathogenicity of pathogenic fungi. This strategy has been successfully used in nematophagous and entomopathogenic fungi (Åhman et al. 2002; St Leger and Wang 2011). PII is an important pathogenicity factor in *Arthrobotrys oligospora*, the transcript of *PII* was not detected during the early stages of infection (adhesion and penetration), but high levels were expressed concurrent with the killing and colonization of the nematode. Disruption of the *PII* gene by homologous recombination had a limited effect on the pathogenicity of the fungus. However, mutants containing additional copies of the *PII* gene developed a higher number of infection structures and had an increased speed of capturing and killing nematodes as compared to the wild type (Åhman et al. 2002). Recently, the encoding gene of cuticle-degrading proteases Ver112 (from *Lecanicillium psalliotae*) was transformed into the commercial biocontrol fungal agent *Paecilomyces lilacinus* by the restriction enzyme-mediated integration transformation. Compared to the wild strain, the transformant *P. lilacinus* 112 showed significantly greater protease activity, with nematicidal activities increasing by 79 and 96% to *Panagrellus redivivus* and *Caenorhabditis elegans* at the second day, respectively. The crude protein extract isolated from the culture filtrate of *Paecilomyces lilacinus* 112 also showed 20–25% higher nematicidal activity than that of the wild-type strain (Yang et al. 2011b).

Serine proteases are the important virulence factors in nematophagous fungi, which can help fungi to penetrate the cuticle of hosts. In this part, the biochemical of serine proteases from different nematophagous fungi are compared, the amino acid sequences are aligned, and the phylogenetic relationships were analyzed. Recently, increasing serine proteases have been identified from different nematophagous fungi, they share similar biochemical properties and a high degree of similarities in amino acid sequences, suggesting they play similar roles in infection of nematodes. However, the expression and regulation of serine proteases during the infection is unknown, which may be the key for improving the nematocidal activity of nematophagous fungi.

Chitinases

Chitin is an important structural polymer found in the cell walls of fungi and in the exoskeletons of invertebrates. It is an important component of the middle layer of nematode eggshells (Wharton 1980; Bird and Self 1995). Egg-parasitic fungi, such as *Pochonia rubescens* and *P. chlamydosporia*, must penetrate the nematode eggshell first to cause infection (Lýsek and Krajcí 1987). There is extensive evidence for the production of chitinases by fungal parasites during infection based on ultrastructural studies (e.g., Lopez-Llorca and Robertson 1992; Tikhonov et al. 2002; Khan et al. 2004).

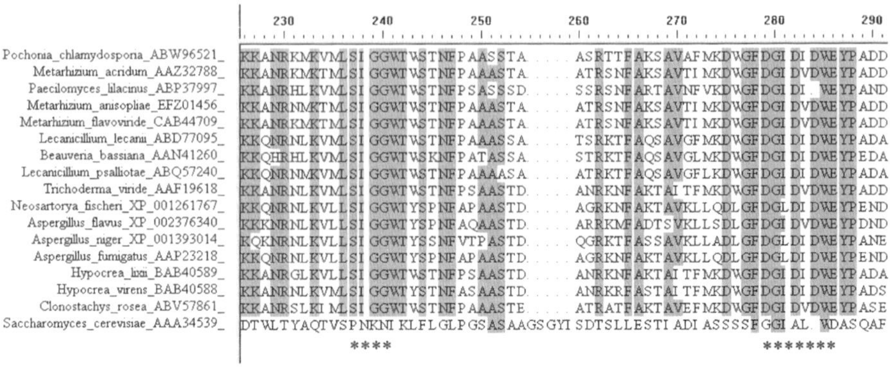

Fig. 6.4 Alignment of the amino acid sequences of chitinases from different fungi. Areas shaded in *grey* are high degree homology (more than 75 % identify) and unshaded areas are regions of variability between the chitinases. *** indicated the conserved substrate binding and catalytic domains (SXGG and DXXDXDXE) of glycosyl hydrolase family 18

An extracellular chitinase CHI43 was identified from *Pochonia chlamydosporia* and *P. rubescens*, and it was found to serve as a nematicidal factor in infecting nematode eggs (Tikhonov et al. 2002). Chitinase activity was also observed in the culture supernatant of the nematophagous fungus *Paecilomyces lilacinus* strain 251 in a minimal medium containing chitin, and after separation by isoelectric focusing, six proteins were detected that showed chitinolytic activity (Khan et al. 2003). Chitinase activity was further confirmed on nondenaturing one-dimensional and two-dimensional gels using a sandwich assay with glycol chitin as a substrate. Studies on extracellular enzymes of *P. lilacinus* revealed that the application of the protease and chitinase drastically altered the eggshell structures when applied individually or in combination (Khan et al. 2004).

Recently, four chitinases were cloned and identified from *Paecilomyces lilacinus* (Dong et al. 2007), *Lecanicillium psalliotae* (Gan et al. 2007a), *Clonostachys rosea* (Gan et al. 2007b) and *Pochonia chlamydosporia* (Mi et al. 2010), they shared similar molecular weight and structure, they were synthesized as a preproenzyme, which consist of signal peptide, propeptide and mature chitinase, all belong the glycosyl hydrolase 18 (GH18) family. The alignment of chitinases from nematophagous fungi with other reported chitinases from a large range of organisms indicated that the binding site S-X-G-G and catalytic domain D-X-D-X-E are highly conserved (Fig. 6.4). The two motifs might be the chitinase catalytic centre (Henrissat et al. 1993). Among these residues, the locations of active sites Asp and Glu were the most conserved residues. Glu acts as a general acid catalyst which donates a proton to the glycosidic oxygen, and contributes to the lowering of the energy barrier of the reaction by stabilizing the transient carbonium ion intermediate electrostatically (Phillips 1967). Chemical modification and site-directed mutagenesis confirmed that Asp/Glu residues were essential for chitinase activity (Mileweski et al. 1992; Watanabe et al. 1993).

The gene families encoding the chitinases are large and the genes from the same as well as different families have undergone considerable divergence. However, the chitinases from pathogenic fungi are very conserved, the chitinase PCCHI44 from *P. chlamydosporia* share a high degree of similarity (range from 65.2–81.1 %) to chitinases from mycoparasitic, nematophagous and entomopathogenic fungi, and the PCCHI44 also share a moderate degree of similarity (about 50 %) to chitinases from member of genus *Aspergillus*, such as *A. flavus* (48.6 %), *A. niger* (51.7 %) and *A. fumigatus* (53.7 %), while they shared a low similarity (12.5 %) with *S. cerevisiae*. Chitinases from *Metarhizium* are very highly conserved, the chitinase from *M. acridum* shares 95.3 % identity with *M. anisopliae* and 99.8 % identity with *M. flavoviride*. Chitinases identified from parasitic *Beauveria bassiana* (Bbchit1) and *Hypocrea lixii* (CHIT36) showed low similarity (16.7–21.4 %) to other members of glycosyl hydrolase family 18. Despite these differences, they all share conserved substrate binding and catalytic domains (SXGG and DXXDXDXE) (Fig. 6.4).

The phylogenetic tree (Fig. 6.5) revealed that chitinases from different fungi evolved in two branches, chitinases from mycoparasitic, nematophagous and entomopathogenic fungi clustering in clade (A), suggesting that these chitinases play an important role in infection against hosts. However, these fungi did not cluster according to their hosts, suggesting that they may infect fungi, nematode and insects (Li et al. 2010). More than one type of chitinases have been identified from the same fungal species, for example, 18 putative chitinases were identified from the *Hypocrea jecorina* (anamorph: *Trichoderma reesei*) by genome-wide analysis, all of them belonging to GH18 family (Seidl et al. 2005). The genomes of the entomopathogenic fungi *Metarhizium anisopliae* and *M. acridum* have now been sequenced (Gao et al. 2011), and 30 and 21 putative chitinases respectively, were found in *M. anisopliae* and *M. acridum*. In the mycoparasitic *Trichoderma*, multiple chitinases also have been identified, however, only ech42 (Lorito et al. 1998) and chit33 (Limon et al. 1999) were found capable for biological activity enhancement. The multiplicity of the chitinase genes within the same species may reflect their functional differences between related proteins (Orikoshi et al. 2005), and these chitinases may play different roles during the fungal growth and differentiation (St Leger et al. 1993).

The function of chitinase LPCHI1 from *Lecanicillium psalliotae* in infection against nematode eggs was identified by expression in *Pichia pastoris*. Incubation of eggs of the root-knot nematode (*Meloidogyne incognita*) in the presence of purified chitinase for 3 days significantly inhibited egg hatching *in vitro* (Gan et al. 2007a). The immature eggs of *M. incognita* were degraded after treatment with the purified chitinase; enzyme-treated eggs were swollen and the eggshells lost their original structural features as indicated by light microscopy (Fig. 6.6). There were large vacuoles in the chitin layer of nematode eggs after treatment with chitinase (Fig. 6.4b), the eggshell was partially degraded (Fig. 6.4c), and some eggs were deformed (Fig. 6.4d). The immature eggs did not hatch after treatment with chitinase (Fig. 6.4f). The chitinase PCCHI44 from *Pochonia chlamydosporia* has been expressed in *Eschericola coli* BL21, and the recombinant chitinase PCCH144 was

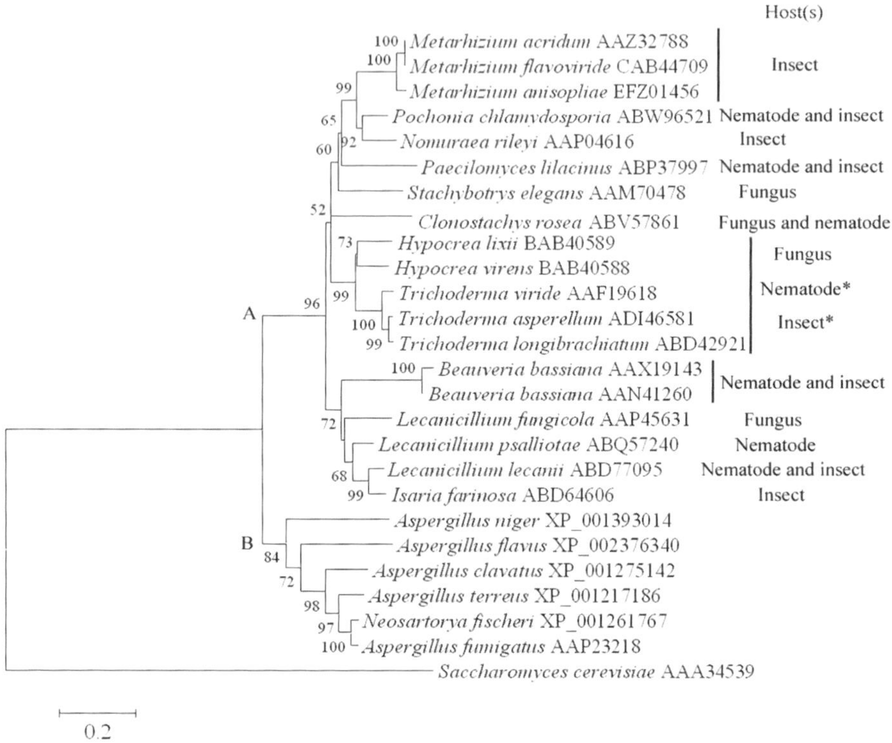

Fig. 6.5 Phylogenetic tree showing the relationship of chitinases from different fungi. The tree was constructed with the Mega version 4.0 program packages. The chitinase (AAA34539) of *Saccharomyces cerevisiae* was used as an outgroup in order to root the tree. * indicated that partial fungi were reported to infect against nematode or insect. (Reproduced from Yang et al. (2013) with kind permission from Springer Science and Business Media)

shown to be harmful to eggs of the nematode *Meloidogyne incognita*, and also to the insect *Bombyx mori* (Mi et al. 2010).

Chitinases have been isolated and identified in opportunistic nematophagous fungi, however, no chitinase has been identified from nematode-trapping fungi. The genome of the nematode-trapping fungus *Arthrobotrys oligospora* has been sequenced (Yang et al. 2011a), and 16 putative chitinases were predicted, which provided a basis for studying the roles of chitinase in nematode-trapping fungi. Similar to serine proteases, an increase the copy number of chitinase in pathogenic fungi can enhance the biocontrol potential of pathogens (St Leger and Wang 2011). The chitinase gene, *Bbchit1*, was transformed into the fungus *Beauveria bassiana*, and insect bioassays revealed that overproduction of Bbchit1 enhanced the virulence of *B. bassiana* for aphids, as indicated by significantly lower 50% lethal concentrations and 50% lethal times of the transformants compared to the values for the wild-type strain (Fang et al. 2005).

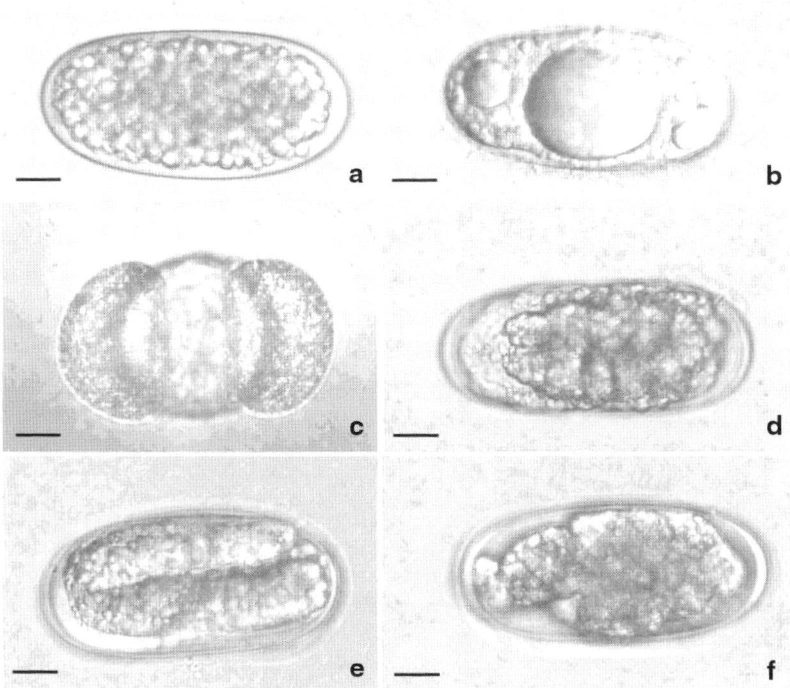

Fig. 6.6 Eggs of the root-knot nematode *Meloidogyne incognita* after treatment with the purified chitinase. **a** and **e** the control; **b** large vacuole was observed in eggs; **c** the eggshell was partially degraded; **d** some eggs deformed; **e** immature eggs developed into eggs containing a juvenile after treating by control. **f** Immature eggs did not hatched after treating by chitinase. Photographs were taken on days 3. Scale bar, 10 ìm. (Reproduced from Gan et al. (2007a) with kind permission from Springer Science and Business Media)

Collagenases and Other Hydrolytic Enzymes

Collagenases are enzymes that can catalyze the hydrolysis of collagen and gelatin rather than other proteinaceous substrates (Maclennan et al. 1953). Collagen is the main constitutive component of the nematode cuticle (Blaxter and Robertson 1998). Therefore, collagenase from nematophagous fungi may play an important role in infection against nematodes. However, reports on collagenase production by nematopahgous fungi are relatively rare (Schenck et al. 1980; Tosi et al. 2001). Collagenases have been identified in *Arthrobotrys* spp. (Tosi et al. 2001) and all of the *Arthrobotrys* species examined produced collagenase when they were grown in liquid medium free of proteosepeptone (proteosepeptone induces collagenase production). This result shows that collagenase is a constitutive enzyme in these fungi. A putative gene encoding for collagenase was found in the genome of *A. oligospora* (Yang et al. 2011a).

Other hydrolytic enzymes are also reported to be involved in the infection of nematodes by fungi. For example, eggs of *Heterodera schachtii* infected by fungi appeared to have their inner lipid layers degraded in comparison to uninfected eggs, and this was attributed to the lipolytic activity of the fungus (Perry and Trett 1986). In the future, the catalytic mechanism of these enzymes may be elucidated in detail.

Crystal Structure Analyses of Serine Proteases and Chitinase

Crystal Structures and Catalytic Properties of Cuticle-Degrading Proteases

Cuticle-degrading proteases are extracellular subtilisin-like serine proteases that are secreted by pathogenic fungi. These proteases can hydrolyze the host cuticle during invasion and serve as a group of important virulence factors during the infection of nematodes by nematophagous fungi (Yang et al. 2007a). Increasing evidence have shown that cuticle-degrading proteases from nematophagous fungi are involved nematode infection, however, little is known about the catalytic mechanism of these proteases due to lack of available tertiary structure. We have purified two cuticle-degrading proteases, Ver112 from *Lecanicillium psalliotae*. and PL646 from *Paecilomyces lilacinus*, the Ver112 protein and the complex between PL646 and the substrate-like tetrapeptide inhibitor methoxysuccinyl-Ala-Ala-Pro-Val-chloromethyl ketone were crystallized using the hanging-drop vapour-diffusion method, the crystals were analyzed by X-ray diffraction to resolutions of 1.65 and 2.2 Å (Ye et al. 2009), respectively.

Alignment of five cuticle-degrading proteases showed that they share a high degree of sequence identity, the sequence identities of PL646 with Pr1A, VCP1, Ver112, and PRK are 76.2, 69.0, 72.6, and 62.5 %, respectively. As expected, the amino acid residues within the catalytic triad, substrate binding sites, and Ca^{2+}-binding site are highly conserved (Fig. 6.7). Comparison of crystal structures of PL646 and Ver112 showed that they are very similar to each other, and similar to that of proteinase K (PRK) from the fungus *Tritirachium album*. Differences between the structures were found among residues of the substrate binding sites (S1 and S4) (Fig. 6.7) (Liang et al. 2010).

Three substrates (PAPF, NAPF and AAPD) were designed according to the reported substrate, Suc-AAPF-pNA (AAPF), and the catalytic properties of PRK, PL646 and Ver112 were compared. Kinetics analyses showed that three proteases had different Km values to synthesized substrates. For the PRK, the strongest substrate affinity occurred toward AAPD due to the lowest *K*m value, as shown in Table 6.2. However, the cuticle-degrading enzyme PL646 showed relatively weak affinity toward AAPD, and this substrate may not bind to Ver112, as the kinetic parameters could not be determined. We speculate that it is the Asp165 in the S1

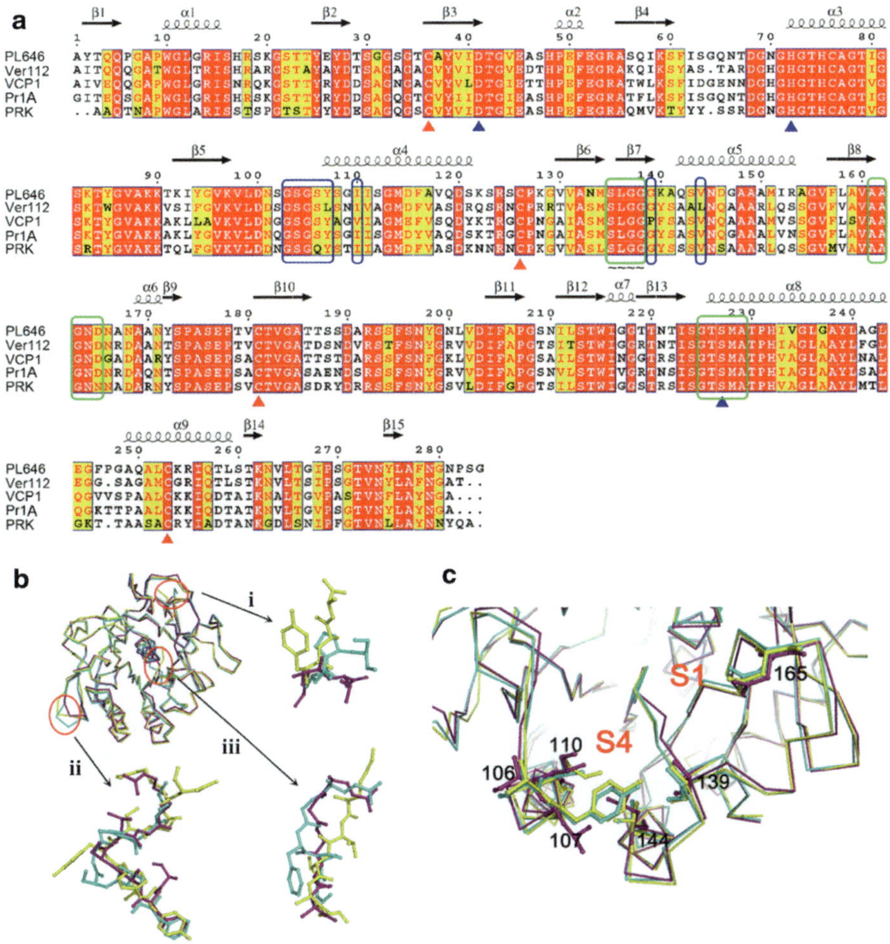

Fig. 6.7 Structural and sequence comparisons of cuticle-degrading proteases. **a** Multiple sequence alignment of PL646 (ABO32256), Ver112 (AAU01968), VCP1 (CAD20580), Pr1A (P29138), and PRK (P06873). Secondary structure elements are marked on top of the alignment; α-helices and β-strands are presented as curves and arrows, respectively. Residues belonging to the S1 site are circled by *green* frames; residues belonging to the S4 site are circled by *blue* frames; ~ indicates residues involved in both the S1 and S4 sites. *Blue* triangles showed the catalytic triad (Asp, His and Ser). *Red* triangles showed the cysteines involved in the disulfide bridge. **b** Superposition of the structures of Ver112 (*magenta*), PL646 (*cyan*), and PRK (*yellow*). Structures of the three enzymes are highly identical, except for some surface-exposed loops such as the N and C termini (*i*), and those located between β4 and α3 (*ii*), and between α6 and α7 (*iii*). **c** Differences in residues in the S1 and S4 pockets between Ver112 (*magenta*), PL646 (*cyan*), and PRK (*yellow*). Positions involved in amino acid variations are labelled (PL646 numbering); corresponding residues are shown as sticks. (Reproduced from Liang et al. (2010))

Table 6.2 Kinetics analysis of Ver112, PL646 and PRK against synthesized substrates. (Reproduced from Liang et al. (2010))

Enzymes	AAPF[a]		AAPD[a]		NAPF[a]		PAPF[a]	
	Km	Kcat/Km	Km	Kcat/Km	Km	Kcat/Km	Km	Kcat/Km
Ver112	0.145	27354	N/A[b]	N/A[b]	0.164	2067	0.226	5206
PL646	1.42	17957	1.21	166	0.695	453	0.472	1576
PRK	0.31	33142	0.093	1229	N/A[b]	N/A[b]	0.111	3558

[a] AAPF, N-succinyl-Ala-Ala-Pro-Phe-p-nitroanilide; AAPD, N-succinyl-Ala-Ala-Pro-Asp-p-nitroanilide; NAPF, N-succinyl-Asn-Ala-Pro-Phe-p-nitroanilide; PAPF, N-succinyl-Pro-Ala-Pro-Phe-p-nitroanilide

[b] not available (kinetics parameters cannot be determined)

pocket that hinders (in PL646) or abolishes (in Ver112) AAPD binding because of electrostatic repulsion. The kinetic parameters of PRK towards NAPF have also not determined, and likely caused by the relatively small S4 pocket of PRK when compared to that of Ver112 and PL646. Ver112 and PL646 both show high affinity towards substrates NAPF and PAPF, reflecting their capability of accommodating peptide substrate with large P1 and P4 residues. Interestingly, Ver112 binds with slightly higher affinity than does PL646 to NAPF and PAPF, which is probably caused by a larger S4 pocket in Ver112 than in PL646. In addition, the relatively smaller S4 pocket in PL646 may cause difficulty in stereochemical rearrangement when interacting with P1 Ala of AAPF and AAPD, thereby leading to PL646's relatively weaker affinity toward these two substrates than Ver112 (Liang et al. 2010).

The crystal structures of PRK, Ver112, and PL646, in conjunction with the multiple sequence alignment of the cuticle-degrading proteases (Fig. 6.7), allowed us to infer the substrate specificities of these proteases at the amino acid level. The substrate-binding pockets within both enzymes are large and in the case of S1, hydrophobic. Therefore, the P1 substrate residues with bulky hydrophobic side chains (such as phenylalanine) are favoured by the S1 pocket. Although the S4 pocket is smaller than the S1 pocket, it is still large enough to accommodate large P4 substrate residues, a fact that is reflected by the lower *K*m values of PL646 towards PAPF and NAPF, than toward AAPD and AAPF. The possible reasons for this include that greater number of bonding contacts (such as hydrogen bonding, hydrophobic and van der Waals interactions) that can be formed between large P4 residues and S4 pocket than between small P4 residues and S4 pocket. Cuticle-degrading proteases investigated here have a tyrosine (Tyr107) in the S4 pocket with the exception of Ver112. The side chain of the Tyr, which is a large phenoxyl ring, is located in the external region of the S4 pocket and acts as a lid to this pocket, thereby restricting to a certain extent the binding of the peptide substrate with large P4 side chains. The results of our kinetics analyses, *i.e.*, the reduced affinities of PL646 toward NAPF and PAPF in comparison with those of Ver112, support this hypothesis.

Crystal Structure and Mutagenesis Analysis of Chitinase CrChi1 from Clonostachys rosea

CrChi1 is a chitinase from the nematophagous fungus *Clonostachys rosea* that plays a role in the infection against nematodes. The Crchi1 gene from *C. rosea* was cloned and expressed in *E. coli* and the recombinant CrChi1 protein was purified, the molecular weight of the protein was found to be 44 kDa. The recombinant CrChi1 was crystallized at 291 K using PEG 3350 and ammonium dihydrogen phosphate as precipitant and a 1.8 Å resolution X-ray data set was collected from a single flash-cooled crystal (100 K). The crystals belonged to space group P21, with unit-cell parameters $a=44.1$, $b=71.7$, $c=59.1$ Å, $\alpha=\gamma=90$, $\beta=91.3°$ (Gan et al. 2009).

The structure of apo CrChi1 consists of two parts: eight-strands of parallel β-barrels surrounded by eight α-helices to form the core domain. The core domain, which was named as a $(\beta/\alpha)_8$ TIM-barrel, have been observed in other family 18 chitinases from *Aspergillus fumigatus* (afChi) (Rao et al. 2005a), *Coccidioides immitis* (cmChi) (Hollis et al. 2000), human chitotriosidase (huChi) (Fusetti et al. 2002), and chitinases A (chiA) (Papanikolau et al. 2003) and B (chiB) (Houston et al. 2002) from *Serratia marcescens*. An additional α/β domain, composed of five antiparallel β-strands and two α-helices, is inserted in the loop between strand β9 and helix α8, which might give the active site a groove character. Like all other family 18 chitinases, the CrChi1 has the DXDXE motif at the end of strand β5 with Glu174 being the catalytic residue in the middle of the open end of the $(\beta/\alpha)_8$ barrel (Fig. 6.8). The chitinase CrChi1 showed sequence similarities with five other chitinases (afChi, cmChi, huChi, ChiA and ChiB) ranging from 24.2 to 54.8 %, but the amino acid residues around substrate binding sites and catalytic centre are very conserved. The main differences occurred in the N and C-terminal domains of chitinases (Fig. 6.8). These chitinases from different organisms shared low sequence similarity, but their structures were highly conserved, especially for the amino acid residues corresponding to the substrate binding domain and catalytic domain (Fig. 6.8). Our analyses suggested that these chitinases might share a common catalytic mechanism.

In general, family 18 chitinases bind to their substrates in an extended recognition site. By convention, the sugars on the non-reducing end of the substrate are given negative numbers, and those on the reducing side giving positive numbers. Soaking experiment of CrChi1 with caffeine was carried out, and the structure of the CrChi1-caffeine complex was solved and refined against 1.6 Å resolution of the X-ray diffraction data (Yang et al. 2010). The structure of CrChi1 in complex with caffeine defined the active centre of CrChi1 (Fig. 6.9). In this complex, the scissile glycosidic bond lies between the −1 and +1 subsites. Previous structural studies of afChi, cmChi, huChi, ChiA and ChiB or substrate-inhibitor bound to them defined the GlcNAc subsites in the enzyme, which formed a deep groove on the surface of the protein, were also observed in the CrChi1 structure (Fig. 6.9a). The groove consists of six sugar-binding subsites, numbered from −4 (non-reducing end) to +2 (reducing end), with hydrolysis taking place on the glycosidic bond between the −1 and +1 subsites (van Aalten et al. 2001; Fusetti et al. 2002). The subsites

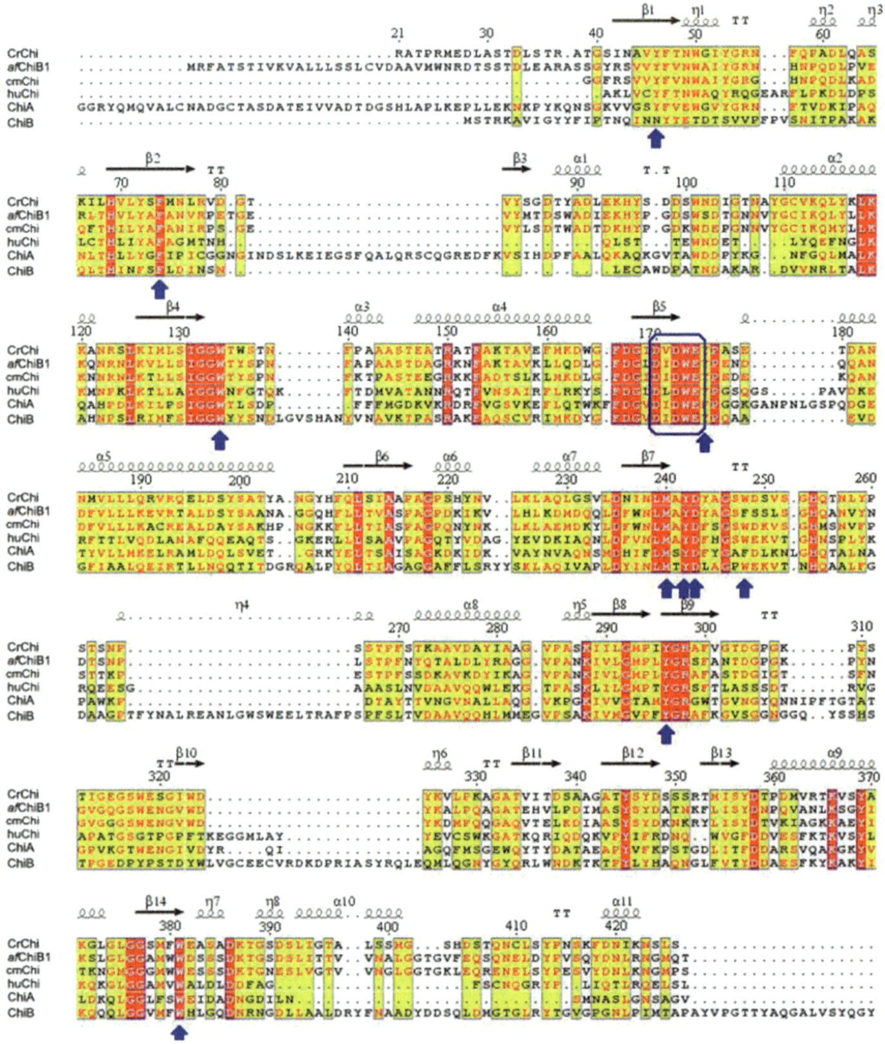

Fig. 6.8 Alignment of chitinase CrChi1 and homologues from different organisms. *Background-red* residues indicate those that are conserved; *background-yellow* denotes residues identified to be more than 80 % conserved. The key DXDXE motif and residues, which are important for inhibitor binding are labelled out by using *blue* frame and arrows. (Reproduced from Yang et al. (2010))

−2 to +2 formed a deep groove, lined by side chains that were highly conserved in family 18 chitinases, including the conserved DXDXE motif. In the −1 subsite, Asp172/Glu174 in CrChi1 are equivalent to Asp155/Glu157 in ScCTS1 (*Saccharomyces cerevisiae*) (Hurtado-Guerrero and van Aalten 2007) and Asp125/Glu127 in hevamine (Terwisscha van Scheltinga et al. 1995), respectively, and they are located at the end of β5 and form the conserved DXDXE motif in family 18 chitinase. Like

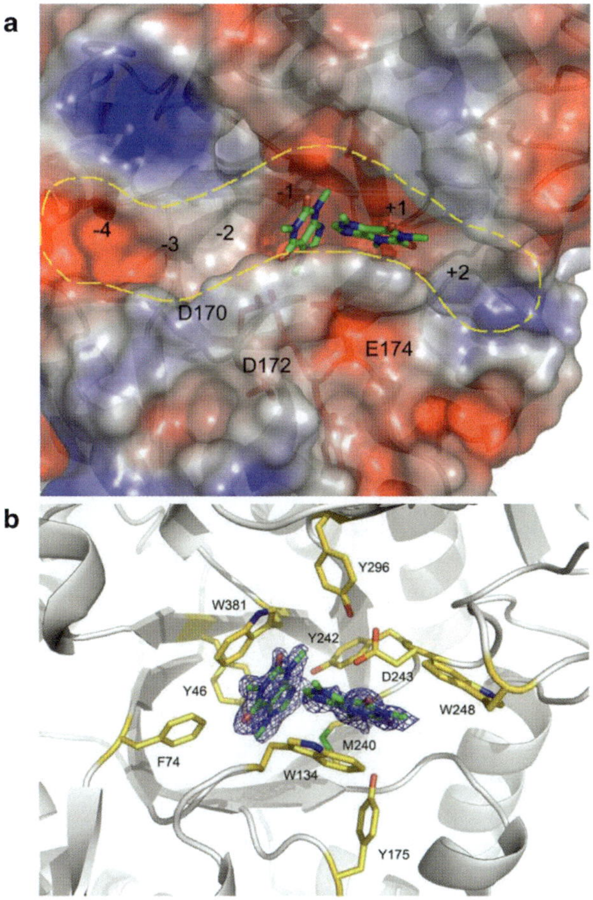

Fig. 6.9 **a** The crystal structure of caffeine in complex with CrChi1 is shown as a transparent surface. The substrate-binding cleft is represented by *yellow circle*. Caffeines are shown in *blue* stick representation. Three important residues (Asp170, Asp172 and Glu174) from the family 18 chitinase DXDXE motif are shown. **b** Caffeine bound to active site of CrChi1. The two bound caffeine molecules were shown as *green* sticks and covered with omit density map at 1.1σ. The residues, which are important to form the binding pocket and caffeine binding, were shown as gold sticks. (Reproduced from Yang et al. (2010))

other family 18 chitinases, Glu174 is the catalytic residue, with Asp172 stabilizing the oxazolinium ring of the reaction intermediate. Trp381 is the most conserved residues in all family 18 chitinases, forming the bottom of −1 subsite, with some conserved residues forming the sidewall.

The interactions between the caffeine moiety and the enzyme are similar to those observed in the *Af*ChiB1-caffeine complex and the *Af*ChiB1-C2-dicaffeine complex (Rao et al. 2005b; Schüttelkopf et al. 2006). The active site Asp172 points down into the catalytic core. The electron density map shows that the primary caffeine moiety is sandwiched between Trp134 and Trp381, and accepts two hydrogen bonds, one from the backbone amide of Trp134 and one from the hydroxyl of Tyr242 (Fig. 6.9b).

Residues Tyr46, Trp134 and Tyr242 in CrChi1 play important roles in the formation of hydrogen bonds between the enzyme and caffeine (Fig. 6.9b). To establish the role of these residues in catalytic reaction, a mutagenesis study of CrChi1 was conducted and biochemical properties were determined. These amino acid substitutions

Table 6.3 Mutagenesis and enzymology analysis of Crchi1[a]. (Reproduced from Yang et al. (2010))

Mutant sites	Protein	*Km*	*Vmax*	*kcat*	*kcat/Km*
		uM	$uM\ s^{-1}$	s^{-1}	$uM^{-1}s^{-1}$
Control	Soluble	9.5±0.3	0.005±0.00007	2.59±0.03	0.27
Y46F	Insoluble	11.5±1.0	0.001±0.00003	0.0062±0.0002	0.0005
W134G	Soluble	ND	ND	ND	ND
E174Q	Soluble	ND	ND	ND	ND
Y242F	Soluble	10.8±2.4	0.001±0.00003	0.0054±0.0002	0.0005
Y242G	Soluble	6. 9±1.0	0.0009±0.0001	0.005±0.0005	0.0008

ND no significant signal
[a] All experiments were performed in triplicate

caused a decrease of over 1000-fold in the enzymatic activity (Table 6.3). Two mutations, E174Q and W134G, completely inactivated the enzyme, confirming the essential roles of these two residues in the catalytic process. The Glu174 residue was considered to be a proton donor in its catalysis. Trp134 is located at the substrate-binding domain and forms hydrogen bond with caffeine (Fig. 6.9b). The benzene ring of Trp134 may be important for the formation and maintenance of the substrate-binding sites. From the sequence comparison (Fig. 6.8), the Glu174 and Trp134 were found conserved in all family 18 chitinases. The loss of function of E174Q mutation in CrChi1 is also consistent with the importance of the acid residue at that position, similar to a previous finding for the E144Q mutation of ChiB from *S. marcescens* (Houston et al. 2002). The W134G mutant abrogated the detectable fluorescence signal and showed no detectable binding in the tryptophan fluorescence binding experiments, similar to that reported previously (Rao et al. 2005a). Tyr242 was also conserved in family 18 chitinases and site-directed mutagenesis of Tyr242 in CrChi1 impaired the enzymatic activity, but did not completely eliminate the activity. Residue Tyr242 forms hydrogen bonds with caffeine.Changing residue Tyr242 to Gly and Phe reduced *kcat* by three orders of magnitude, while no large effects on *Km* were observed (Table 6.3). This result suggests that this residue is crucial for catalysis rather than substrate binding, mostly in agreement with mutational studies of the residue in other family 18 chitinases (Bortone et al. 2002; Rao et al. 2005a; van Aalten et al. 2001). Tyr46 is also high conserved in family 18 chitinase, the enzymatic activity was impaired when Tyr46 in CrChi1 was mutated. Tyr46 is located near the substrate binding domain (Fig. 6.9b), which suggests that it may influence substrate binding. Interestingly, when Tyr46 was changed to phenylalanine, a highly conserved residue among family 18 chitinases, the protein was expressed as inclusion bodies in *E. coli*, suggesting that this residue may be required for proper structure formation. The *kcat* reduced by three orders of magnitude after renaturation and refolding of CrChi1, while no obvious effects on *Km* were observed (Table 6.3), suggesting that mutation of Tyr46 affected the catalytic activity and resulted in *kcat* decrease, while the substrate binding was not affected.

Serine proteases and chitinases play important roles during the nematophagous fungi infection against nematodes (Åhman et al. 2002; Gan et al. 2007a; Yang et al. 2007a). The crystal structures of serine proteases and chitinases from the nematophagous fungi were resolved, which can help us to identify the active site residues and to elucidate the catalytic mechanism of these enzymes involved in infection against hosts, and exploit more effective pesticides, fungicides, and antimalarial drugs.

Regulation of Subtilisin-Like Protease prC Expression in the Nematophagous Fungus Clonostachys rosea

PrC Functions as a Virulence Factor of Clonostachys rosea

Nematophagous fungi have been used as biological control agents against nematodes parasitic to plants and animals. These fungi can secret subtilisin-like extracellular serine proteases during the infection of nematodes. Previous reports indicate that these proteases play an important role during the penetration of nematodes (Åhman et al. 2002; Ahrén et al. 2005; Yang et al. 2007a; Fekete et al. 2008) and over 10 subtilisin-like serine proteases have been isolated from nematophagous fungi (Yang et al. 2007a; Li et al. 2010).

As a mycoparasite, *Clonostachys rosea* has been tested successfully as a biological control agent against divergent fungal plant pathogens (Sutton et al. 1997; Xue 2003). The mycoparasitic activity has been attributed to the secretion of its cell-wall degrading enzymes, including serine proteases and chitinases. We purified and cloned a subtilisin-like extracellular protease (designated PrC) from the nematophagous fungus *C. rosea* (Li et al. 2006; Liang et al. 2011b). The purified PrC could immobilize nematodes and hydrolyse proteins of the nematode cuticle. To investigate the potential role of PrC, the encoding gene was knocked-out by homologous recombination (Zou et al. 2010a). The effect of the gene *prC* on pathogenicity in *C. rosea* was examined using a bioassay with the nematode *Panagrellus redivivus*. The bioassay was initiated by adding nematodes to each YNB agar plate innoculated with 10^4 of conidia of *Clonostachys rosea*. The conidia could adhere to passing nematodes and kill them. The percentage of dead nematodes was significantly lower in the mutant strains than in the wild-type stains (Fig. 6.10). These results suggested that the protease PrC functions as a virulence factor in the infection of *C. rosea* against nematodes.

The Promoter Features and Transcription Regulation of the Gene prC

To investigate the mechanism underlying the regulation of *prC* expression by various conditions in *Clonostachys rosea*, the 5′-upstream DNA sequence of *prC* was amplified by DNA walking strategy. The sequence data have been submitted to the

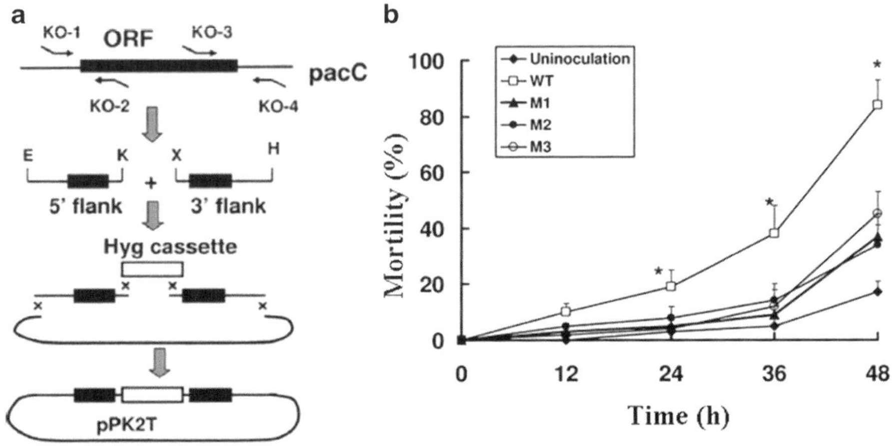

Fig. 6.10 Deletion of the gene *pacC* and analysis of nematicidal activity. **a** The sketch of deletion of the gene *pacC*. Two truncated *pacC* fragments joining the hygromycin resistance cassette in the vector pPK2 were amplified by primers KO-1/KO-2 and KO-3/KO-4 from *Clonostachys rosea* genomic DNA to produce the gene replacement vector pPK2T. **b** Nematicidal activities of the wild and mutant strains. *WT*: wild-type strain; *M1*, *M2* and *M3*: *pacC* mutant strains; *Uninoculation*, without inoculation of the fungus *C. rosea*. Nematodes were added to plates for 48 h. At time intervals, the number of killed (not moving) nematodes was counted by using a microscope. (Reproduced from Zou et al. (2010a))

GenBank databases under accession no. FJ966246. The promoter region contains putative transcription control sites for nitrogen regulation (5′-GATA), carbon regulation (5′-SYGGRG), pH regulation (5′-GCCARG), and stress response element (STRE) (5′-AGGGG), suggesting expression of prC may be regulated by nitrogen sources, environmental pH and other stress conditions (Zou et al. 2010a, b, c).

The Expression of PrC Is Regulated by Environmental pH

It is now widely accepted that extracellular pH functions as a critical signal for virulence in fungal pathogens of humans, plants and insects (St Leger et al. 1999; Davis et al. 2000; Yakoby et al. 2000; Eshel et al. 2002). PrC, like many other subtilisin-like proteases from nematophagous fungi, had the highest activity under alkaline conditions. Thus, environmental pH might be a determining factor for the pathogenesis of nematophagous fungi against nematodes. Regulation of gene expression by ambient pH in fungi is mediated via a PacC/Rim101 pathway, including seven genes, *palA*, *palB*, *palC*, *palF*, *palH*, *palI* and *pacC* (Peñalva et al. 2008). PacC is the terminal element of the pH signalling pathway and the transcription factor for regulation of pH-dependent gene expression (Rollins and Dickman 2001). PacC has a Cys_2His_2 zinc finger DNA-binding domain for the cis-transcript element 5′-GCCARG-3′.

The *pacC* gene of *Clonostachys rosea* was amplified by PCR using degenerate primers designed from conserved region of aligned *pacC* homologues. Then the 5′- and 3′-flanking sequences were obtained by DNA walking strategy (Zou et al. 2010a). The *C. rosea* PacC protein has considerable amino acid identities to other PacC homologues. Quantitative real-time RT-PCR was performed to investigate the expression of pacC from mycelia of *C. rosea* grown in defined medium at different pH values. As a result, pacC mRNA levels were much higher under alkaline conditions than under acidic conditions, independent of the carbon source used (including glucose, glycerol and acetate). The levels of *pacC* transcript at extreme acidic conditions (pH 3) were five fold lower than those at pH 4.

To investigate the role of *pacC* in *Clonostachys rosea*, the gene was knocked-out by homologous recombination. The wild-type strain of *C. rosea* grew better than the *pacC* mutant strain on potato dextrose agar (PDA) medium while the colony growth of *pacC* mutant strain was similar to that of the wild-type strain on minimum medium at pH 4, and severely impaired on minimum medium at pH 8. These phenotypes are similar to other *pacC* mutants, such as *Aspergillus nidulans* and *Fusarium oxysporum* (Tilburn et al. 1995; Caracuel et al. 2003).

In the sequence between −699 and −705 bp upstream of the start codon of C. rosea *PrC* there is one copy of the inverted PacC consensus binding site 5′-CCTG-GC-3′. To determine whether PacC influence the expression of *PrC*, we examined the transcript level of *PrC* in mycelia from the wild-type and mutant strains of *pacC* in minimum medium using glucose as the carbon source and proline as the nitrogen source. The *pacC* mutant strain showed significantly lower mRNA levels of *PrC* than the wild-type strain (Fig. 6.11a). Additionally, the mRNA levels of *PrC* were significantly higher at pH 6 and 8 than that at pH 4 in the wild-type strain (Fig. 6.11b). These results suggest that the expression of *PrC* gene is regulated by the pH in a PacC-dependent manner.

To determine whether the pH regulate the expression of *PrC* under the transcriptional level, we constructed two *PrC* promoter-GoxA reporter vectors: pGX-B, contained 890 bp of the 5′ promoter region with the inverted PacC consensus binding site 5′-CCTGGC-3′; pGX-B0 does not contain 5′-CCTGGC-3′. When these experiments were performed using pGX-B, the promoter activities at pH 4 were only 32 % of those at pH 8 (Fig. 6.11c). In contrast, when using pGX-B0, we observed approximately 80 % reduction in the promoter activities, even at pH 8. These results suggested that 5′-CCTGGC-3′ box in the promoter region was probably required for the activity of the gene promoter at pH 8. An electrophoretic mobility shift assay (EMSA) was then performed (Fig. 6.11d). Protein extract from the strain grown at pH 8 contained significantly higher amounts of slow-migrating and fast-migrating PacC-DNA complexes than the strain grown at pH 8. This indicated that the PacC DNA binding activities were higher at pH 8 than those at pH 4.

Besides of the pH regulation by the *pacC*, we also found that the expression of *PrC* was not stimulated by the addition of nematode cuticles in the *pacC* mutant strain. Furthermore, the inducible expression of *PrC* by nematode cuticles was lost in the wild-type strain at pH 4.

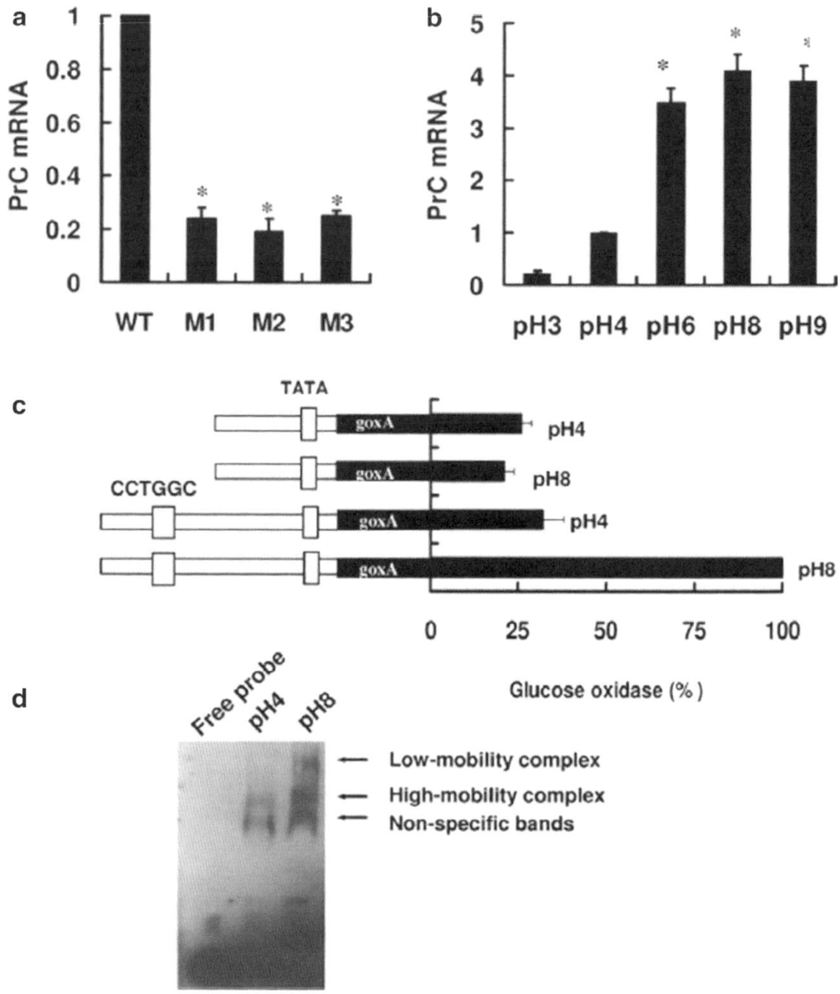

Fig. 6.11 PacC is required for regulating the expression of a subtilisin-like protease PrC. **a** Mycelia of *C. rosea* wild-type (WT) and *pacC* mutant strains (M1, M2 and M3) were grown on PDA. **b** Mycelia of *Clonostachys rosea* wild-type strains were grown in minimum medium containing 1 % (w/v) glucose as the carbon source and 0.1 % proline as the nitrogen source for 12 h at different pH values, then total RNA was extracted. The mRNA levels were detected by real-time PCR. **c** *PrC* promoter-GoxA reporter vectors pGX-B and pGX-B0 were transformed with *C. rosea*, respectively. The strains were grown in minimum medium containing 1 % (w/v) glucose as the carbon source and 0.1 % proline as the nitrogen source for 12 h at pH 4 and pH 8, respectively. **d** EMSA using the *C. rosea PrC* promoter fragment as a probe with protein extracts from *C. rosea* wild-type grown in minimum medium at pH 4 and pH 8 respectively. (Reproduced from Zou et al. (2010a))

The Expression of prC Is Regulated by Nitrogen Sources

The expression of prC regulated by nitrogen sources was investigated. Firstly, the basal expression of *prC* were tested from mycelia of *C. rosea* grown in YNB medium with glucose as the carbon source and with and without one of the preferred nitrogen sources, glutamine or ammonia. As a result of real-time PCR test, the *prC* mRNA levels were down regulated approximately three fold by glutamine. The inhibitory effect of glutamine on the basal expression of *prC* was dose-dependent, and 1 % was sufficient to achieve a maximal effect. Similar results were obtained when using ammonia as the preferred nitrogen source. However, pre-treatment using rapamycin (a specific inhibitor of TOR) can suppress the inhibitory effect of *prC* expression by glutamine or ammonia.

The basal level of PrC was found essential for its inducible expression by nematode cuticle. When adding nematode cuticle extracted from the nematode, *Panagrellus redivivus*, there was increased by threefold in the expression of *prC* after 12 h. However, when pre-treated with PMSF (1 mM), the inducible expression of *prC* by nematode cuticle as well as the protease activity were significantly inhibited. Thus, it seems that the cuticle degrading product by the protease prC may be mediated the up regulation of *prC* gene. The effects of the mixture of degrading products by purified prC were investigated. As a result, only materials with molecule weights less than 3 kD were capable of inducing prC expression.

To investigate the transcriptional regulation of *prC* by nitrogen sources or nematode cuticle, three *prC*-promoter glucose oxidase (*goxA*) reporter vectors were constructed: pGX-A1, A2 and A3, which include three, two and one consensus binding site 5′-GATA-3′, respectively. Transformants were grown in YNB medium using glucose as the carbon source. The promoter activity assays indicated that the region −404 to −202, containing one GATA element, was involved in the basal expression of *prC* in the absence of preferred nitrogen sources. Glutamine (1 %) significantly inhibited promoter activities. In contrast, the addition of nematode cuticle (1 mg/ml) did not influence the activities of the gene promoter.

An EMSA was performed with nuclear extracts from *Clonostachys rosea* mycelia by using three *prC* promoter fragments as probes. Each probe contained one GATA element in the *prC* promoter. As a result, only the probe P2 containing the second GATA element located in the −246 to −242, had protein DNA binding activities. Protein extracts from the strain grown in YNB medium contained significantly higher amounts of protein-DNA complexes than the strain grown in the YNB-glutamine medium. These results indicated that the GATA protein DNA binding activities were suppressed in the presence of glutamine. In contrast, nematode cuticle did not influence the DNA binding activities, suggesting that glutamine, rather than nematode cuticle, regulated *prC* expression at the transcript level (Zou et al. 2010b).

Based on above data, we propose the scheme given in Fig. 6.12 by which nematode cuticle may induce the expression of subtilisin-like extracellular protease *prC*.

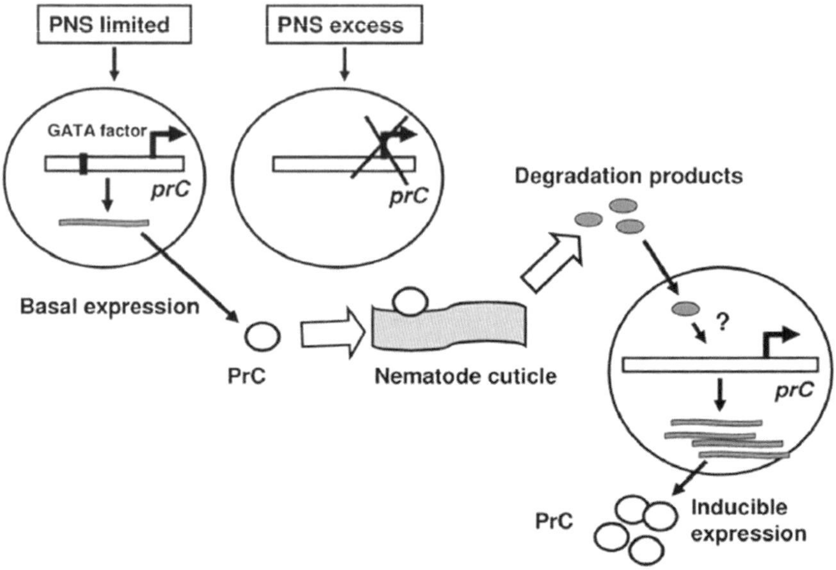

Fig. 6.12 A proposed mechanism of nematode cuticle-induced *prC* expression in *C. rosea*. The extracellular subtilisin-like protease *prC* is expressed at a basal level mediated by the GATA factors in the absence of the preferred nitrogen sources (PNS). The enzyme is secreted into the medium and digests nematode cuticle to produce the degradation products. The degradation products enter into cells, which in turn cause a significantly higher inducible expression of *prC*. The preferred nitrogen sources inhibit the basal expression of *prC*. (Reproduced from Zou et al. (2010b))

In the absence of preferred nitrogen sources, *Clonostachys rosea prC* is expressed at a high basal level through a mechanism involving in GATA factors. The enzyme is released into the medium. After the addition of nematode cuticle, PrC digests a small amount of nematode cuticle to produce the degradation products. The low molecule weight materials of the degradation products enter into *C. rosea* cells, which in turn cause a significantly higher inducible expression of *prC*. The preferred nitrogen sources inhibit the basal expression of *prC*, eventually resulting in reduced inducible expression of this enzyme.

The Expression of PrC Is Regulated by Oxidative Stress

When growing in the soil, fungi likely encounter many stressful conditions such as osmotic shock, high temperature, heavy metals, and UV irradiations. To survive and grow in such conditions, fungi have evolved multiple defence systems with some relying on enzymatic actions, among which, one system is characterized by a rapid increase in transcriptional expression of stress response element (STRE) gene upon stress (Gasch et al. 2000; Bose et al. 2005). Given that the promoter of *Clonostachys rosea prC* contains three conserved STRE sites, investigation of the effect of environmental stress on the expression of prC were carried out.

Several stressful conditions were tested for prC expression, as a result, we found that prC expression (in the mRNA level) increased by oxidant treatments, and this effect is dose-dependent. The expression of prC was also rapidly induced after shifting cells from 28 °C to higher temperature (33–37 °C). In contrast, hyperosmotic shock (0.5–1.5 M NaCl) and treatment with cadmium (0.01–0.1 mM) and lead (0.02–0.2 mM) did not affect the expression of prC. The activity of the protease secreted into the media was found to be consistent with the transcript level (Zou et al. 2010c).

To determine whether the up-regulation of prC expression by oxidants and heat shock were the result of transcriptional activation, two reporter vectors were constructed. One contains 844 bp of the 5′ promoter region, including the three consensus STRE sites and the other one contains 422 bp of the 5′ promoter with no STRE site. When treated with oxidate or heat shock, only transformants containing the first vector have significant increase of reporter gene expression, indicating that the region containing three STRE sites was involved in the regulation of prC expression upon environmental stress.

The oxidate stress can markedly inhibited conidial germination (Fig. 6.13a), but when treated with oxidants (H_2O_2 or manadione) in the presence of nematode cuticle, the conidial germination rate of wide type was significantly higher than a *prC* mutant (Fig. 6.13b). The germination rates of conidia in the prC mutant strain were similar to those in the wild-type strain in the absence of nematode cuticle at 37 °C. In contrast, 56 % of conidia in the wild-type strain germinated 15 h after the addition of nematode cuticle at 37 °C, whereas only 32 % of conidia in the prC mutant strain germinated under the same conditions (Fig. 6.13c). When applied with a serine protease inhibitor PMSF, the protective effect of nematode cuticle against oxidants-induced inhibition of conidial germination in the wild type strain was attenuated (Fig. 6.13d).

The reactive oxygen species (ROS) level in the conidia were examined using a fluorescence dye 2′, 7′-dichlorodihydrofluorescein diacetate (H_2DCFDA). The ROS production induced by H_2O_2, menadione or heat shock was found markedly reduced by the addition of nematode cuticle in the wild type strain, but not in the prC mutant strain. In addition, the degrading products of nematode cuticle by prC can directly antagonize the inhibitory action on conidial germination by oxidation and heat shock, and this effect was not affected by PMSF. In conclusion, the subtilisin-like extracellular protease prC was regulated by environmental stress, such as oxidants and heat shock. The increase expression of prC acts as another role besides of degrading nematode, but participation in the detoxification of ROS. The above results showed protease PrC is not only involved in degradation of nematode cuticle, but play a role in fungi to adapt to environmental stress.

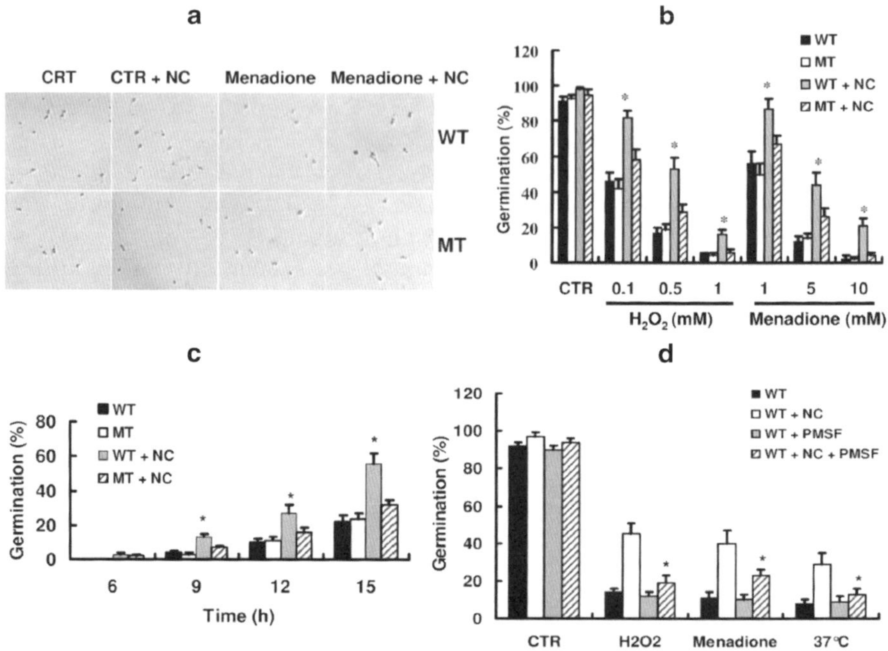

Fig. 6.13 The *prC* mutant is more sensitive to oxidants and heat shock. **a** Microscopic examination (X100) of germinated conidia after 9 h at 28 °C in YNB agar in the presence or absence of menadione (5 mM) or nematode cuticle (NC) (1 mg/ml). **b** Germination frequency of the wild-type (WT) and the *prC* mutant (MT) strains under oxidative stress conditions. **c** Germination frequency of the wild-type and the *prC* mutant strains under heat shock conditions. **d** PMSF inhibited the protective effect of nematode cuticle (1 mg/ml) on conidial germination in the wild-type strain. (Reproduced from Zou et al. (2010c))

Genomic and Proteomic Analyses of the Nematode-Trapping Fungus Arthrobotrys oligospora

The Life Cycle of Arthrobotrys oligospora

Nematode-trapping fungi are a heterogeneous group of organisms broadly distributed in terrestrial and aquatic ecosystems (Nordbring-Hertz et al. 2006). These fungi are capable of developing specific trapping devices such as adhesive networks, adhesive knobs, and constricting rings to capture nematodes and then extract nutrients from their nematode prey (Nordbring-Hertz et al. 2006; Schmidt et al. 2007; Yang et al. 2007d). The ultrastructure of the nematode-trapping structures has been examined extensively. These studies have shown that despite the large variation in morphology, the adhesive types of trap (branches, nets and knobs) have a unique ultrastructure that clearly distinguishes them from vegetative hyphae (Dijksterhuis et al. 1994; Nordbring-Hertz 2004). First, which is common to all of these traps, is

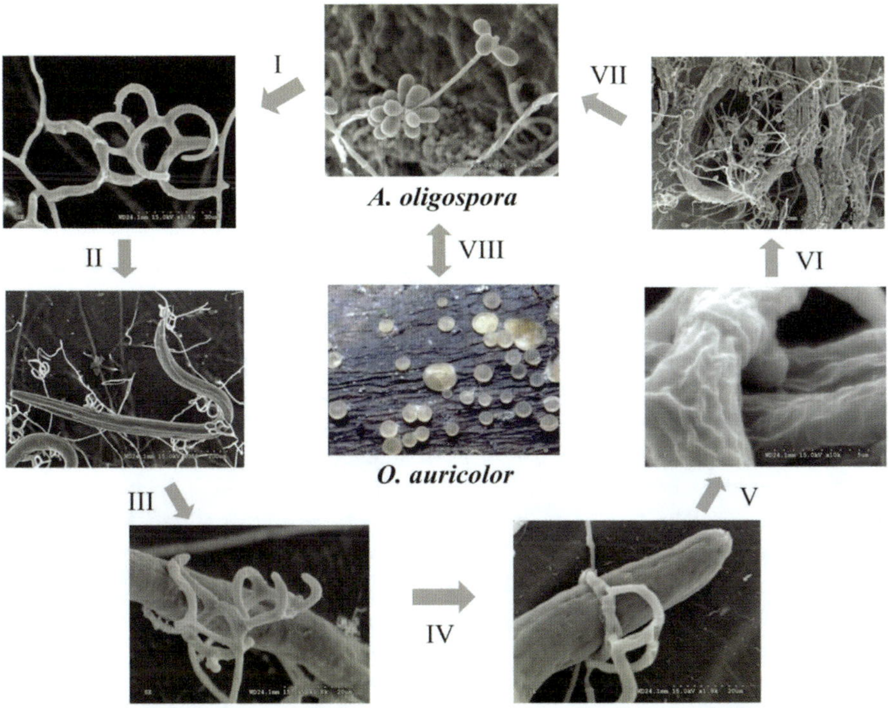

Fig. 6.14 The lifestyle of nematode-trapping fungus *Arthrobotrys oligospora*. The life cycle of *A. oligospora* includes three stages: saprophytic stage, transitional stage, and parasitic stage. The parasitic stage can be divided into the following six steps: I. Trap formation; II. Attraction; III. Adhesion; IV. Capturing; V, Penetration and immobilization; and VI. Digestion and assimilation. The saprobic stage (VII) and sexual stage (VIII) are also shown. (Reproduced from Yang et al. (2011a))

the presence of numerous cytosolic organelles, the so-called dense bodies. Although the function of these organelles is not yet clear, the fact that they exhibit catalase and D-amino acid oxidase activity indicates that the dense bodies are peroxisomal in nature (Dijksterhuis et al. 1994). Another feature common to the trap cells is the presence of extensive layers of extracellular polymers, which are thought to be important for attachment of the traps to the surface of the nematode (Tunlid et al. 1991).

Arthrobotrys oligospora (teleomorph *Orbilia auricolor*) is one of the best-studied nematode-trapping fungi (Nordbring-Hertz 2004, 2006). *A. oligospora* have been found in diverse soil environments including heavy metal-polluted soils and decaying wood (Pfister and Liftik 1995; Mo et al. 2006) where they live mainly as saprobes. In the presence of nematodes, *A. oligospora* enters the parasitic stage by forming complex three-dimensional networks to trap nematodes (Fig. 6.14). The trapping initiates a series of processes including adhesion, penetration, and immobilization of nematodes (Nordbring-Hertz 2004; Nordbring-Hertz et al. 2006). The ability to trap nematodes makes it an attractive candidate agent for controlling parasitic nematodes of plants and animals.

Traps may be formed directly on germination of conidia (spores) to form the so-called conidial traps. This developmental pattern occurs in practically all trap-forming species when conidia are allowed to germinate in natural substrates, such as cow dung or rhizosphere soil (Persmark and Nordbring-Hertz 1997). The conidial trap of *Arthrobotrys oligospora* (ATCC 24927) was first detected during a study where the efficiency of *A. oligospora* as a biological control agent against animal-parasitic nematodes in cow faeces was evaluated. The production of conidial traps may indicate an increased potential of these fungi as antagonists to nematodes (Nordbring-Hertz 2004). The conidial trap also contains numerous electron-dense bodies characteristic of normal hyphal network traps (Nordbring-Hertz et al. 1995). Conidial traps are fully functional in trapping nematodes. They adhere to a passing nematode and may be carried away and spread by the nematode in a way similar to adhesive conidia of endoparasitic nematophagous fungi.

Apart from attacking nematodes, nematophagous fungi also have the capacity to infect other fungi. *Arthrobotrys oligospora* attack their host fungi by coiling of the hyphae around the host hyphae, which results in disintegration of the host cell cytoplasm without penetration of the host. It was shown that nutrient transfer took place between the *A. oligospora* and its host *Rhizoctonia solani* using radioactive phosphorous tracing (Olsson and Persson, 1994). Moreover, *Arthrobotrys oligospora* and other nematophagous fungi, such as *Pochonia chlamydosporia* have the capacity to colonize plant roots (Bordallo et al. 2002). The growth of the nematophagous fungi in plant roots is endophytic, and endophytic growth of *P. chlamydosporia* in barley and wheat roots appeared to increase plant growth and reduce growth of the plant parasitic take-all fungus *Gaeumannomyces graminis* var. *tritici* (Monfort et al. 2005). Mycoparasitism and plant endophytism may be important issues for extension of the biological control potential of the nematophagous fungi.

Genome Sequencing and Assembly

High molecular weight DNA was used to construct plasmid libraries with insert size of 4–5 kb and fosmid libraries with insert size of 35 kb. In total, 78 Mb high quality paired-end data was generated by AB 3730 Sanger sequencing platform from plasmid and fosmid libraries. In addition, 1009 MB pyrosequencing data with average read length of 345 bp and 375 Mb pyrosequencing data with average read length of 242 bp were obtained using the Roche 454 Genome Sequencer Titanium/FLX platforms. Pyrosequencing reads and Sanger sequencing reads were assembled by gsAssembler (Margulies et al. 2005) with default parameter settings. Resequencing of low quality regions and closing of gaps were performed by walking on plasmid and fosmid clones and by PCR using custom primers designed by Consed (Gordon et al. 2001).

Ab initio gene prediction was performed on the genome assembly by Augustus (Stanke and Waack 2003), Glimmer HMM (Majoros et al. 2004), and SNAP (Korf 2004) trained with transcript sequences of *Arthrobotrys oligospora* from this study,

Table 6.4 Properties of the *Arthrobotrys oligospora* genome. (Reproduced from Yang et al. (2011a))

General features	Value
Size of assembled genome including gaps (bp)	40,072,829
Number of scaffolds larger than 2 kb	215
Number of scaffolds	323
Length of gaps within scaffolds (bp)	106,613
N50 scaffold size (bp)	2,037,373
N50 contig size (bp)	575,766
GC content (%)	44.45
Repeat regions (%)	0.47
Coding regions (%)	48.47
Number of predicted gene models	11,479
Number of single-exon genes	2616
Average gene length (bp)	1690
Average number of introns per multi-exon gene	2.8
Average intron size (bp)	89
Average exon size (bp)	473
Number of tRNA genes	145

and by GeneMark (Ter-Hovhannisyan et al. 2008) formulated for fungal genomes. A final set of gene models were selected by EvidenceModeler (Haas et al. 2008) with weight for evidence from each gene prediction software determined by comparing *ab initio* predicted genes with transcript sequences of *A. oligospora* using Exonerate (Slater and Birney 2005), and with transcript alignment supports from using Exonerate to align transcript sequences of *A. oligospora* and other fungal species downloaded from NCBI and TGI (Quackenbush et al. 2001). Predicted genes were annotated by BLASTP (Altschul et al. 1997) searches against protein databases with E-value $1e^{-10}$: NR (www.ncbi.nlm.nih.gov), KOGs and COGs (Tatusov et al. 2003), KEGG (Kanehisa et al. 2004), UniRef100 (Jensen et al. 2009), STRING (Chang et al. 2009), and by InterProScan (Quevillon et al. 2005) searches against protein domain databases with default parameter settings. Pathway mapping was conducted by associating EC assignment and KO assignment with KEGG metabolic pathways based on BLASTP search result. Multiple gene families were constructed by searching each annotated gene against all other genes using BLASTP. The matches with $E \leq 1e^{-5}$ and at least 30% sequence identity over 60% of both gene lengths were used for clustering genes based on single linkage transitive closure (Galagan et al. 2003).

The genome of *Arthrobotrys oligospora* strain ATCC24927 was sequenced to 36.6-fold coverage through a Sanger/pyrosequencing hybrid shotgun approach from multiple clone types. The 40.07 Mb genome is similar in size to that of the model ascomycete fungus *Neurospora crassa* (Galagan et al. 2003). The *A. oligospora* genome was assembled into 215 scaffolds, containing a long-range continuity as reflected by N50 scaffold size of 2037 Kb and N50 contig size of 575.8 Kb (Table 6.4). The assembly represents 99% of the coding regions of the genome, as assessed by mapping 50,121 assembled transcript sequences (8.9 Mb) to the genome assembly. Almost all of the transcripts (99.6%) were mapped onto the genome.

A total of 11,479 protein-coding genes were predicted: 23.6 % belonged to multigene families and 44.6 % were mapped in the KOG/COG database. The average gene density was one gene per 3.50 kb, with an average gene length of 1.69 kb, similar to that of *Neurospora crassa* (1.67 kb) (Galagan et al. 2003). Compared to *N. crassa*, genes in *Arthrobotrys oligospora* have more introns (2.8 VS. 1.7) but with a shorter average intron length (90 bp VS. 134 bp) (Galagan et al. 2003). As is typical of ascomycete fungal genomes, approximately one-third (34.6 %) of the predicted *A. oligospora* genes lacked significant homologies to known proteins from public databases (Table 6.4).

Comparative Analysis

Predicted proteins in *Arthrobotrys oligospora* were compared with the predicted proteins of ten sequenced fungal genomes. All proteins were searched against all other proteins in these genomes using BLASTP. The matches with $E \leq 1e^{-5}$ and at least 30 % sequence identity (Rost 1999) over 60 % of both protein lengths (Galagan et al. 2003) were taken as homologous sequences. The orthologous genes between *A. oligospora* and other 10 fungal genomes were identified based on bidirectional best hits (BBHs) using BLAST (Overbeek et al. 1999). The other 10 genomes used for comparison were divided into non-pathogen (*Aspergillus nidulans*, *Neurospora crassa* and *Saccharomyces cerevisiae*) and pathogen (*Fusarium graminearum*, *Magnaporthe oryzae*, *Verticillium dahliae*, *Aspergillus fumigatus*, *Coccidioides immitis*, *Histoplasma capsulatum* and *Chaetomium globosum*) groups. A total of 529 orthologous proteins were found in all the 11 analyzed fungal genomes. These orthologous sequences were concatenated to infer phylogenomic relationships among these fungi using Neighbor-joining, Maximum parsimony and Maximum likelihood methods (Tamura et al. 2007). The same tree topology was found by all the three phylogenetic methods. From the phylogenomic trees, Eurotiomycete fungi (*Aspergillus nidulans*, *A. fumigatus*, *Coccidioides immitis* and *Histoplasma capsulatum*) clustered into one clade, Sordariomycete fungi (*Fusarium graminearum*, *Verticillium dahliae*, *Chaetomium globosum*, *Neurospora crassa* and *Magnaporthe grisea*) clustered into a different clade, while *Arthrobotrys oligospora* formed a separate branch.

Based on orthology analysis, the genes in *Arthrobotrys oligospora* can be classified into four categories: ao (only found in *A. oligospora*, 6157 genes), ao/pathogen (found in *A. oligospora* and pathogen genomes, 961 genes), ao/pathogen/non-pathogen (found in all genomes, 4249 genes), and ao/non-pathogen (found in *A. oligospora* and non-pathogen genomes, 112 genes) (Fig. 6.15a). The results showed that *A. oligospora* shared many more genes with pathogenic fungi than with non-pathogenic fungi. The genes shared between *A. oligospora* and other pathogenic fungi may be functionally related to pathogenicity in these fungi. The whole genome blast analysis against the pathogen-host interaction (PHI) gene database (Winnenburg et al. 2006) identified 398 putative PHI genes in *A. oligospora*, 294 of

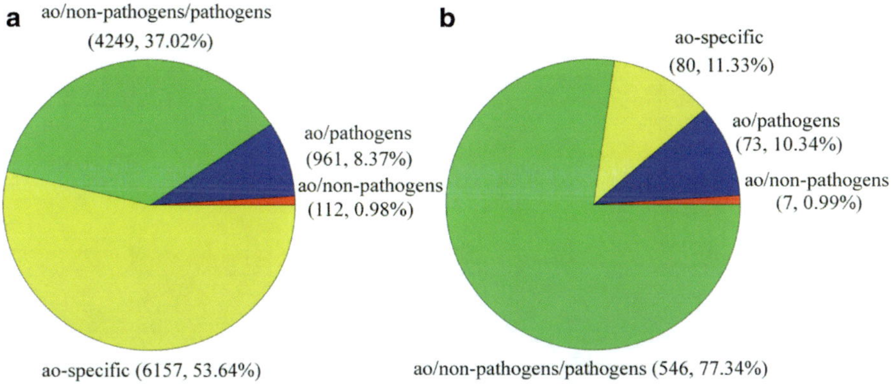

Fig. 6.15 The distributions of *Arthrobotrys oligospora* genes and PHI putative genes in different categories. The other 10 fungal genomes were divided into non-pathogen and pathogen genomes. Based on orthology analysis, the genes in *A. oligospora* were classified into four categories: ao-specific (only found in *A. oligospora*), ao/pathogen (also found in pathogenic fungi), ao/pathogen/non-pathogen (found in all genomes) and ao/non-pathogen (also found in non-pathogenic fungi). **a** The distribution of all *A. oligospora* genes in different categories. **b** The distribution of PHI putative genes in different categories. (Reproduced from Yang et al. (2011a))

which belong to 86 multigene families. The gene number expanded to 706 if single linkage transitive closure was applied. The distribution analysis of the othologous sequences showed that the putative PHI genes included 80 ao-specific genes and 73 genes only found in pathogen genomes, more than those in the ao/non-pathogen category (Fig. 6.15b). It should be noted that many putative PHI genes were not included in the pathogenicity-related gene families, suggesting that there are potentially more pathogenicity-related genes that remain to be experimentally confirmed.

Multigene families were constructed from the homologous sequences based on single linkage transitive closure (Galagan et al. 2003). A total of 2882 of the 11479 genes clustered into 789 multigene families. Moreover, the number of genes in several gene families related to fungal pathogenicity was found expanded in the *Arthrobotrys oligospora* genome (Fig. 6.16). For example, subtilisins, a group of proteases essential for infection (Åhman et al. 2002; Yang et al. 2007a), were found in a greater number in *A. oligospora* (24) than in several model ascomycetes such as the animal pathogens *Aspergillus fumigatus* (4), *Coccidioides immitis* (15) and *Histoplasma capsulatum* (3), plant pathogens *Fusarium graminearum* (15), *Verticillium dahliae* (16) and *Magnaporthe grisea* (22), and non-pathogens *Aspergillus nidulans* (3), *Neurospora crassa* (5). Besides subtilisin genes, the *Arthrobotrys oligospora* genome also contains a larger number of genes within enzyme families of cellulases (58), pectinesterases (18) and cellobiohydrolases (17) than other sequenced model fungi (Fig. 6.16). As a comparison, the *M. grisea* genome contains the highest number of genes within the cytochrome P450 family (P450; 112); the *Aspergillus fumigatus* genome contains the highest numbers of genes within enzyme families of non-ribosomal peptide synthases (NRPS; 21), chitinases (18), and polygalacturonases (16); the *Verticillium dahliae* genome contains the highest

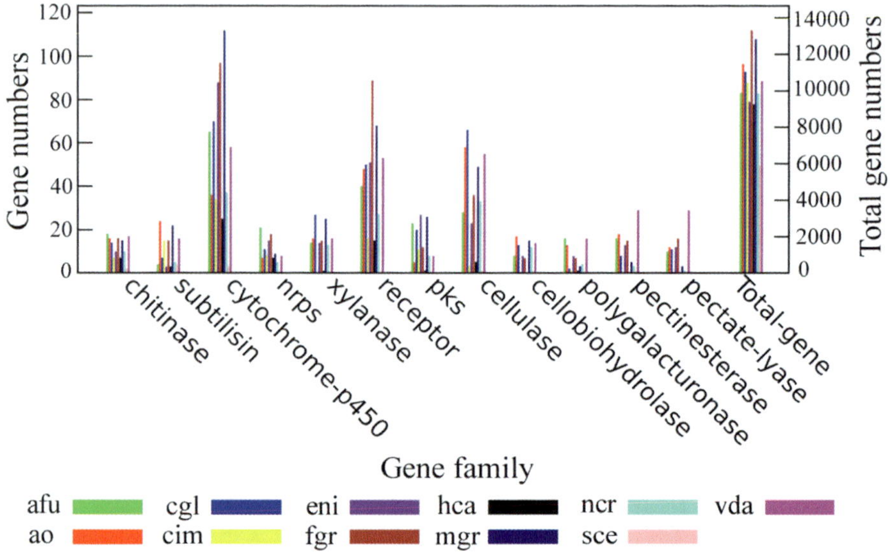

Fig. 6.16 Comparison of pathogenicity-related gene families between *Arthrobotrys oligospora* and other sequenced fungi. Several pathogenicity-related enzyme families were compared between *A. oligospora* and other 10 sequenced fungi including *Aspergillus fumigatus* (*afu*), *Chaetomium globosum* (*cgl*), *Coccidioides immitis* (*cim*), *Aspergillus nidulans* (*eni*), *Fusarium graminearum* (*fgr*), *Histoplasma capsulatum* (*hca*), *Magnaporthe grisea* (*mgr*), *Neurospora crassa* (*ncr*), *Saccharomyces cerevisiae* (*sce*) and *Verticillium dahliae* (*vda*). (Reproduced from Yang et al. (2011a))

numbers of genes within enzyme families of polygalacturonases (16), pectinesterases (29), and pectate lyases (29); and the *Chaetomium globosum* genome contains the highest numbers of genes within enzyme families of cellulases (66) and xylanases (27).

Proteomic Analysis Under Induced by Nematode Extraction

The formation of traps likely involves other genes and is a prerequisite for infection, and serving as the indicator for lifestyle switch from saprobic to predacious stages in nematode-trapping fungi (Nordbring-Hertz et al. 2006; Yang et al. 2007d; Schmidt et al. 2007). At present, little is known about the molecular mechanism of trap formation. To identify the proteins involved in trap formation, a proteomic study was performed and the profiles of intracellular proteins from *Arthrobotrys oligospora* cells at two developmental stages representing the early nematode extract (NE) induction stage (10 h after treatment with NE) and the late stage of trap formation (48 h after treatment with NE) were compared. *A. oligospora* was cultured in the liquid PL-4 medium for 6 days. The vegetative hyphae were harvested by filtration through sterile cheesecloth, washed with sterile deionized water for three times, and then induced using the diluted *Caenorhabditis elegans* extract (10 %, v/v) in 88-mm petri dishes at 26 °C. Crude *C. elegans* extract was obtained by ultrasonication for 15 min and centrifugation at 10,000 × g for 30 min at 4 °C. About 30 ml extract was obtained from 10 g *C. elegans* (wet weight). The extract was sterilized by passing through a 0.22 μm membrane. In each Petri dish, about 200 mg hyphae were incubated in the dilute *C. elegans* extract (40 ml). Hyphae incubated with an equal volume of sterile deionized water at the same condition were considered as a negative control. After induction, hyphae were collected by filtration through sterile cheesecloth, washed with sterile deionized water three times, and frozen immediately in liquid nitrogen and maintained at −80 °C prior to use.

To extract cytosolic proteins, approximately 2 g mycelia were ground to a fine powder in liquid nitrogen using a cooled mortar. Mycelium powder was resuspended in 10 % (w/v) trichloroacetic acid in acetone, and proteins were allowed to precipitate for 2.5 h at −40 °C. The protein pellet was recovered by centrifugation at 20,200 × g for 30 min, rinsed twice with 90 % acetone, air dried, and solubilized in 8 M urea, 2 M thiourea, 4 % CHAPS (w/v), 1 % dithiothreitol (DTT) (w/v) and 0.8 % Bio-lyte 3/10 ampholyte (Bio-Rad Laboratories, Hercules, California) by ultrasonication for 15 min using a UP200S sonicator (Fernández-Acero et al. 2007) (Hielscher, Teltow, Germany). Insoluble material was removed by centrifugation at 20,200 × g for 30 min. The supernatant was collected and stored at −80 °C prior to use. Protein concentration was determined by the method of Bradford (1976) with bovine serum albumin as the standard.

Approximately 1 mg of protein samples were applied on 17 cm IPG strips, pH 4-7 (Bio-Rad Laboratories) or pH 6-11 (GE Healthcare Life Sciences, Piscataway, New Jersey), using the anode Ettan IPGphor Cup Loading Manifold (GE Healthcare Life Sciences). The first dimension (IEF) was performed in the IPGphor Isoelectric Focusing System (GE Healthcare Life Sciences) by stepwise increase of the voltage as follows: 0V-500V for 2 h, 500V for 5 h, 500V-3500V for 3 h and finally 3500V continuing until the total volt-hours reached 54 kVh. After completion of IEF, IPG strips were incubated for 15 min in equilibration solution I (50 mM Tris pH6.8, 6 M urea, 30 % glycerol, 2 % SDS, 2 % DTT, trace bromophenol blue) and 15 min in equilibration solution II (solution I with 2.5 % iodoacetamide instead of DTT). The second dimension was performed on 12 % SDS-PAGE gels using Bio-Rad Protean Plus Dodeca Cell, at 15 mA per gel for first 15 min and then at a constant voltage of 250V, until the dye front reached the bottom of the gel. Low-molecular-weight markers were applied next to the acidic end of the IPG strips. Proteins were visualized by Coomassie Brilliant Blue G-250 (BBI Research, Madison, Wisconsin).

The gel images were acquired by scanning with a UMAX Powerlook 2100XL scanner. Analysis of profiles and statistical analysis of protein spot data were performed with the PDQuest 7.3.0 software (Bio-Rad Laboratories). Gels were normalized based on the total spot integrated volume (area × density) in each gel of the match set. The value assigned to a protein spot was calculated as a percentage of the sum of volumes of all spots detected and multiply a factor of 1,000,000.

Protein spots were manually excised from Coomassie Brilliant Blue G-250 stained gels. Excised gel spots were first washed with pure water for 15 min and then with 50 % (v/v) acetonitrile containing 25 mM NH_4HCO_3 three times for

15 min, then again with pure water for 15 min. The gel spots were dehydrated by in 50 μl 100% ACN for 20 min at room temperature and digested in 25 mM NH_4HCO_3 containing 5 μg/ml modified trypsin (Promega, Madison, Wisconsin) for 16–20 h at 37 °C to generate peptides.

Peptides were spotted on a MALDI sample plate with the same volume of matrix (10 mg/ml α-Cyano-4-hydroxycinnamic acid in 50% acetonitrile, 0.1% trifluoroacetic acid), and analyzed on a 4700 series Proteomics Analyzer (Applied Biosystems). Peptide mass spectra were obtained by averaging 1000 aquired spectra in the positive ion reflector mode with a m/z range of 700–4000, and internally calibrated with a mass standard kit for the 4700 Proteomics Analyzer. The 4000 series explorer software 3.0 (Applied Biosystems) was used for peak list generating (S/N > 10). Proteins were identified by automated peptide mass fingerprinting using the mascot algorithm of the GPS Explorer 3.5 software (Applied Biosystems) against an in-house sequence database of *Arthrobotrys oligospora*. Positive identifications were accepted up to 95% of confidence level. The following criteria were used for the database searches: at least four matching peptide masses; maximum one missed cleavage per peptide; mass tolerance of 0.1 Da, and the acceptation of carbamidomethylation for cysteine and oxidation for methionine.

The expressions of 91 and 16 proteins were found up-regulated ($P < 0.05$), while 24 and 95 were down-regulated ($P < 0.05$) at 10 h and 48 h, respectively. Most of the proteins up-regulated at 10 h were involved in translation, posttranslational modification, amino acid metabolism, carbohydrate metabolism, energy conversion, cell wall and membrane biogenesis (Table 6.5). The results suggest very active growth and metabolism during the transition from vegetative hyphae to trap cells. In contrast, compared to those at the saprobic stage, the expressions of most proteins up-regulated at 10 h were found either down-regulated or unchanged at 48 h when the traps were already formed, consistent with the hypothesis that the proteins up-regulated at 10 h were likely involved in trap formation of *Arthrobotrys oligospora*.

Real Time PCR Analysis

Total RNA was extracted with Trizol Reagent (Invitrogen, Carlsbad, California), purified with a RNAse Mini Kit (Qiagen, Valencia, Califonia), and then reverse transcribed with PrimeScript RT reagent Kit (Takara Bio Inc., Shiga, Japan) by following the manufacturer's instructions. Quantitative real-time PCR was conducted with 2 μl reverse transcribed product in a 7300 Real-Time PCR system (Applied Biosystems, California, USA) using Power SYBR Green PCR Master Mix (Applied Biosystems). 18S rDNA gene was used as the internal control. Fold changes were calculated using the formula $2^{-(\Delta\Delta Ct)}$, where ΔΔCt is ΔCt (treatment)-ΔCt (control), ΔCt is Ct (target gene)—Ct (18S), and Ct is the threshold cycle (User's Manual for ABI 7300 Real-Time PCR System).

Fifty-three genes were selected for analyzing the change of transcriptional level during the formation of traps (treated with NE for 10 h) in comparison to veg-

Table 6.5. Differentially expressed genes in *Arthrobotrys oligospora* during the formation of traps (treated with NE for 10 h) in comparison to vegetative mycelia as revealed by proteomics analysis. Proteins with changes greater than 1.5 folds between the two conditions are listed

COD/KOG classification	Induced by10h		Induced by 48h	
	Up-regulated	Down-regulated	Up-regulated	Down-regulated
Signal transduction mechanisms	5	0	0	4
Translation, ribosomal structure and biogenesis	8	0	2	10
Posttranslational modification, protein turnover, chaperones	10	2	2	8
Replication, recombination and repair	2	0	0	1
RNA processing and modification	0	1	0	2
Nucleotide transport and metabolism	3	0	0	1
Proteases	0	1	1	0
Amino acid transport and metabolism	6	1	6	15
Carbohydrate transport and metabolism	10	2	3	8
Lipid transport and metabolism	3	2	2	5
Energy production and conversion	15	1	3	12
Cell wall/membrane/envelope biogenesis	3	0	1	0
Cytoskeleton	5	0	0	3
Cell cycle control, cell division	2	1	0	1
Coenzyme transport and metabolism	1	0	0	5
Secondary metabolites biosynthesis, transport and catabolism	2	0	1	0
Inorganic ion transport and metabolism	1	0	1	2
Intracellular trafficking, secretion, and vesicular transport	3	0	0	3
General function prediction only	3	1	1	10
Others	9	4	1	5
Total	91	16	24	95

etative mycelia by Real time PCR. These genes involved in different biological processes, for example, TCA and glyoxylate cycle, cell division and biosynthesis of adhesive proteins, cell wall, peroxisomal proteins and glycerol (Table 6.6). The transcriptional level of 40% genes (21/53) was up-regulated during the formation of traps, especially, the expression level of several genes (e.g., adhesive proteins *AOL_s00043g50* and *AOL_s00210g231*, Glycerol-3-phosphate dehydrogenase *AOL_s00054g748*, and serine proteases *AOL_s00170g103* and *AOL_s00215g702*) increased by more than 5-fold (Table 6.6), suggesting that they involved in trap formation.

The Mechanism of Trap Formation

According to the proteomic and real time PCR analysis, the expression levels of 112 genes/proteins were up-regulated during the trap formation, these genes involved

Table 6.6 Differentially expressed genes in *Arthrobotrys oligospora* during the formation of traps (treated with NE for 10 h) in comparison to vegetative mycelia as revealed by Real time PCR analysis

Gene ID	Gene annotation	Real time PCR results		
		1	2	3
Adhesive proteins				
AOL_s00043g50	Cell adhesion complex protein (bystin)	4.36	7.71	4.19
AOL_s00083g511	Jacalin-like lectin domain-containing protein	0.99	2.86	1.47
AOL_s00007g5	Putative cell agglutination protein	6.76	2.45	3.48
AOL_s00080g55	Cell membrane glycoprotein	2.0	1.26	1.29
AOL_s00004g121	Adhesin (aidA-I)	0.83	0.99	1.06
AOL_s00076g567	Adhesin protein Mad1 [*Metarhizium anisopliae*]	2.04	1.65	1.43
AOL_s00081g40	Fasciclin and related adhesion glycoproteins	0.68	0.65	0.78
AOL_s00076g207	Collagen adhesion protein	1.71	1.40	1.74
AOL_s00043g531	Lectin, (mannose-binding 1)	1.65	1.41	1.15
AOL_s00043g494	Lectin, mannose-binding 2	0.95	1.21	0.81
AOL_s00097g409	Fasciclin and related adhesion glycoproteins	0.91	1.52	1.76
AOL_s00210g231	Carcinoembryonic cell adhesion molecule	11.28	9.32	44.58
AOL_s00054g942	Extracellular matrix glycoprotein	1.08	4.12	0.51
AOL_s00215g185	GPI-anchored cell surface glycoprotein	0.57	0.92	1.02
AOL_s00054g276	Fucose-specific lectin [*Aspergillus fumigatus*]	1.09	0.82	0.69
Cell wall biosynthesis				
AOL_s00079g31	Chitin synthase/hyaluronan synthase	0.79	0.81	0.79
AOL_s00210g143	Chitin synthase/hyaluronan synthase	1.16	0.65	2.29
AOL_s00004g638	Glucosamine-phosphate N-acetyltransferase	1.58	1.26	1.31
AOL_s00215g548	Protein involved in beta-1,3-glucan synthesis	1.06	1.41	1.27
AOL_s00210g38	Chitin synthase/hyaluronan synthase	1.39	1.54	1.31
AOL_s00210g37	Chitin synthase/hyaluronan synthase	2.19	2.27	2.46
AOL_s00075g119	Chitin synthase/hyaluronan synthase	1.83	0.75	1.76
AOL_s00075g153	Chitin synthase/hyaluronan synthase	0.46	0.40	0.76
AOL_s00054g465	Phosphoglucomutase/phosphomannomutase	0.43	0.35	0.36
AOL_s00054g491	1,3-Beta-glucan synthase	2.12	1.71	1.76
AOL_s00215g806	Alpha-1,3/alpha-1,6-mannosyltransferase	1.15	1.32	0.92
AOL_s00110g288	1,3-Beta-glucan synthase	2.01	1.45	1.49
AOL_s00076g99	Glucosamine 6-phosphate synthetases	5.44	4.73	3.28
AOL_s00078g76	Chitin synthase/hyaluronan synthase	1.98	2.91	2.19
TCA and Glyoxylate cycle				
AOL_s00004g654	Fumarase	2.00	0.97	2.58
AOL_s00210g140	NAD-dependent malate dehydrogenase	3.01	0.78	2.98
AOL_s00215g565	NAD-dependent malate dehydrogenase	1.56	2.03	1.57
Peroxisomal proteins				
AOL_s00081g131	Peroxisomal long-chain acyl-CoA transporter,	1.31	1.10	0.51
AOL_s00004g606	ABC superfamily	1.92	1.42	1.51
AOL_s00007g540	Peroxisomal biogenesis protein (peroxin 16)	1.50	2.19	1.64
AOL_s00083g431	Peroxisomal biogenesis protein (peroxin 11)	2.46	1.46	2.88
AOL_s00043g697	Peroxisome assembly factor 2 (peroxin 6)	1.38	1.26	1.13
AOL_s00080g243	Peroxisomal biogenesis factor 2 (peroxin 2)	1.35	1.25	0.92
AOL_s00081g263	Peroxisomal membrane anchor protein (Pex14p)	0.77	1.35	0.93
AOL_s00215g610	Peroxisomal assembly protein (PEX3)	1.49	1.36	1.07
AOL_s00215g40	Peroxisomal biogenesis protein (peroxin 11)	1.22	1.85	1.32

Table 6.6 (continued)

Gene ID	Gene annotation	Real time PCR results		
		1	2	3
AOL_s00054g525	Peroxisomal biogenesis protein (peroxin 13)	1.56	1.49	2.46
Cell division				
AOL_s00054g368	Protein kinase of the CDC7 subfamily	1.09	1.06	0.90
AOL_s00083g388	DNA polymerase delta catalytic subunitCdc2	1.28	1.84	0.88
AOL_s00188g11	Cell division control protein 7	0.83	0.98	0.84
AOL_s00188g9	Cell division control protein7	0.55	0.71	0.82
Serine proteases				
AOL_s00075g8	Serine protease [*Dactylella varietas*]	1.486	1.29	1.03
AOL_s00170g103	Serine protease [*Monacrosporium megalosporum*]	3.16	5.22	9.27
AOL_s00076g4	Serine protease [*Arthrobotrys oligospora*]	0.98	0.94	0.56
AOL_s00215g702	Serine protease [*Dactylella varietas*]	31.93	24.08	14.06
Glycerol biosynthesis				
AOL_s00112g97	Glycerol 3-phosphatase 1	1.47	2.27	0.67
AOL_s00054g748	Glycerol-3-phosphate dehydrogenase (NAD+)	4.48	1.09	6.84

in different biological processes in *Arthrobotrys oligospora*, such as signaling pathways and energy synthesis. For example, genes encoding the components of common fungal signal transduction pathways are all found in the *A. oligospora* genome. Specifically, glycosylphosphatidylinositol-specific phospholipase C (AOL_s00109g54), mitogen-activated protein kinase (MAPK, AOL_s00173g235), serine/threonine protein phosphatase 2A (regulatory subunit, AOL_s00007g146), calcyclin binding protein (AOL_s00054g214), and Ca^{2+}/calmodulin-dependent protein kinase (AOL_s00078g95) were all up-regulated during the formation of traps. This result is consistent with the importance of signal sensing and transduction in the shift from saprobic to carnivorous lifestyles in *A. oligospora*. Moreover, in *A. oligospora*, the tricarboxylic acid (TCA) cycle was up-regulated in response to NE, as indicated by the enhanced expression of genes in the TCA cycle, such as citrate synthase (AOL_s00079g361), aconitase (AOL_s00110g24), isocitrate dehydrogenase (AOL_s00075g141), succinyl-CoA synthetase (AOL_s00043g45), and malate dehydrogenase (AOL_s00210g140 and AOL_s00215g565). In addition, malate synthase (AOL_s00112g112) and isocitrate lyase (AOL_s00075g130), two key enzymes in the glyoxylate cycle, were up-regulated at 10 h, indicating that this pathway is also important for trap formation.

A model of nematode trap formation in *Arthrobotrys oligospora* (Fig. 6.17) was suggested by combined the genomic, proteomic, and qPCR data. In this model, *A. oligospora* recognizes nematode signals and activates a diversity of downstream cellular processes (Yang et al. 2011a). These processes include (i) increased cell proliferation and septum formation with enhanced cell wall biosynthesis and cytoskeleton assembling; (ii) enhanced glycerol synthesis and accumulation leading to increased intracellular turgor pressure in preparation for host colonization and penetration within the nascent trapping cells; (iii) formation of dense bodies possibly involving the peroxisome biogenesis; and (iv) synthesis of adhesive proteins

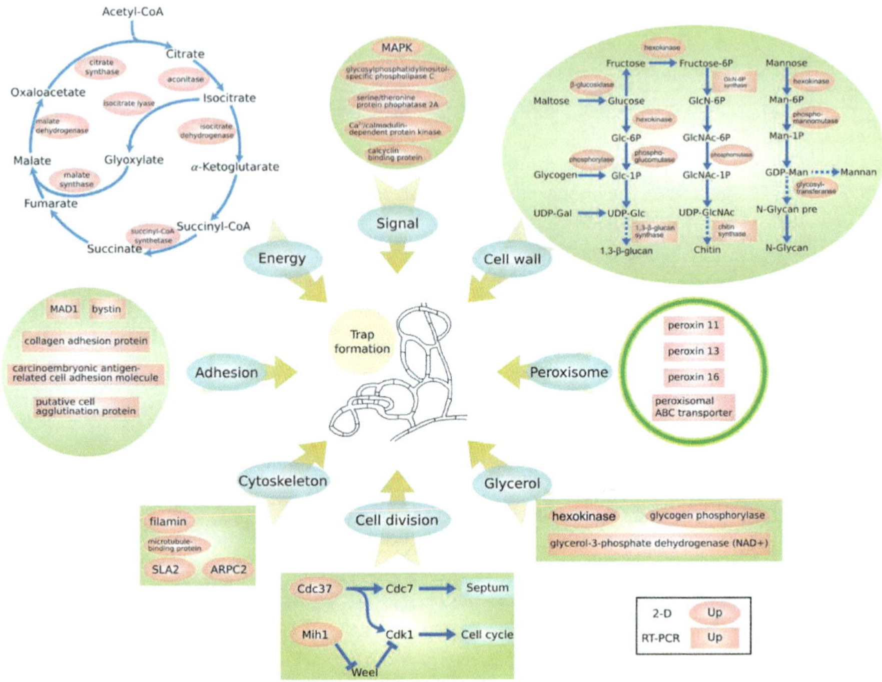

Fig. 6.17 A proposed model for trap formation in *Arthrobotrys oligospora*. In this model, multiple fungal signal transduction pathways are activated by its nematode prey to further regulate downstream genes associated with diverse cellular processes such as energy metabolism, biosynthesis of the cell wall and adhesive proteins, cell division, glycerol accumulation and peroxisome biogenesis. *Black* letters in *red* background represented the up-regulated proteins. Oval represented the 2-D results and quadrate represented the qPCR results. (Reproduced from Yang et al. (2011a))

on the surface of the trapping cells to enhance nematode capture. These complex and coordinated processes are fueled by energy and building blocks supplied from the glyoxylate and TCA cycles (Yang et al. 2011a).

Arthrobotrys oligospora is the first orbiliomycete fungus to have its whole genome sequenced. Compared with other pathogenic and non-pathogenic fungi, *A. oligospora* contained abundant orphan genes not found in other sequenced fungi (53.64%) (Fig. 6.15a), a result consistent with the phylogenomic analysis that *A. oligospora* is phylogenetically very distant from other sequenced Ascomycota. The *A. oligospora* specific genes may be related to its complex life-styles. Specifically, *A. oligospora* is not only a saprobe, but also a nematode pathogen, a pathogen of other fungi, and a endophyte colonizer of plant roots (Bordallo et al. 2002). The genes shared between *A. oligospora* and ten other pathogenic and non-pathogenic ascomycetous fungi (37.02%, Fig. 6.15a) likely represent housekeeping genes for this group of organisms. Interestingly, *A. oligospora* share many more genes with pathogenic fungi than with non-pathogenic fungi. Those shared between *A. oligospora* and pathogens are likely to contribute to fungal pathogencity in general.

Fungi that are pathogenic to invertebrates, especially those targeting nematodes and insects, are of great importance for maintaining ecological balance in natural environments and for improving agricultural production (Nordbring-Hertz et al. 2006; St Leger and Wang 2011). Here the first genome of this group of fungi was reported and the key features related to their pathogenicity were described. The information presented here on a nematode-trapping fungus should facilitate our understanding of these fascinating carnivorous fungi, and provide a basis to analyze the similarities and differences between nematophagous and other pathogenic fungi. These data also have important ramification for the practical development of improved biological control agents.

Conclusion

Nematopahous fungi are a heterogeneous group of organisms that are ubiquitous in terrestrial and aquatic ecosystems (Nordbring-Hertz et al. 2006). They play important roles in maintaining nematode population dynamics in natural environments. Most nematophagous fungi can live as saprophytic and parasitic life. Their ability to live both saprophytically and parasitically makes them great value in ecology and agricultural economy, such as for controlling parasitic nematodes of plants and animals. In this chapter, we discussed the extracellular enzymes, especially serine proteases and chitinases, which involved in the infection of fungi against nematodes. Recent year, increasing extracellular enzymes were identified from different fungi, especially, the structures of proteases and chitinases were resolved, which provided a basis for elucidating the catalytic mechanism of these enzymes. Meanwhile, we discussed the regulation of serine protease PrC in *Clonostachys rosea*, the expression of PrC was regulated by environmental pH, nitrogen sources and oxidative stress, which enhance our knowledge about the expression and regulation of a virulence factor in nematophagous fungi. Finally, we discussed the molecular mechanism of nematode trap formation by genome sequencing, proteomic and real time PCR analyses. These studies help us to partialy elucidate the molecular mechanism of trap formation and infection. We have a better understanding of nematophagous fungi with the application of biochemical and molecular biology technology, especially developing sequencing technology, which opens ways to find the secrets of these special organsisms.

References

Åhman, J., Johanson, T., Olsson, M., Punt, P. J., van den Hondel, C. A., & Tunlid, A. (2002). Improving the pathogenicity of a nematode-trapping fungus by genetic engineering of a subtilisin with nematotoxic activity. *Applied and Environmental Microbiology, 68,* 3408–3415.

Ahrén, D., Tholander, M., Fekete, C., Rajashekar, B., Friman, E., Johansson, T., & Tunlid, A. (2005). Comparison of gene expression in trap cells and vegetative hyphae of the nematophagous fungus *Monacrosporium haptotylum*. *Microbiology, 151,* 789–803.

Altschul, S. F., Madden, T. L., Schäffer, A. A., Zhang, J., Zhang, Z., Miller, W., & Lipman, D. J. (1997). Gapped BLAST and PSI-BLAST: A new generation of protein database search programs. *Nucleic Acids Research, 25,* 3389–3402.

Bidochka, M. J., & Khachatourians, G. G. (1987). Purification and properties of an extracellular protease produced by the entomopathogenic fungus *Beauveria bassiana*. *Applied and Environmental Microbiology, 53,* 1679–1684.

Bird, A. F., & Self, P. G. (1995). Chitin in *Meloidogyne javanica*. *Fundamental and Applied Nematology, 18,* 235–239.

Blaxter, M. L., & Robertson, W. M. (1998). The cuticle. In R. N. Perry & D. J. Wright (Eds.), *The physiology and biochemistry of free-living and plant-parasitic nematodes* (pp. 25–48). Wallingford: CABI Publishing.

Bonants, P. J. M., Fitters, P. F. L., Thijs, H., den Belder, E., Waalwijk, C., & Henfling, J. W. D. M. (1995). A basic serine protease from *Paecilomyces lilacinus* with biological activity against *Meloidogyne hapla* eggs. *Microbiology, 141,* 775–784.

Bordallo, J. J., Lopez-Llorca, L. V., Jansson, H. B., Salinas, J., Persmark, L., & Asensio, L. (2002). Colonization of plant roots by egg-parasitic and nematode-trapping fungi. *New Phytologist, 154,* 491–499.

Bortone, K., Monzingo, A. F., Ernst, S., & Robertus, J. D. (2002). The structure of an allosamidin complex with the *Coccidiodes immitis* defines a role for second acid residue in substrate assisted mechanism. *Journal of Molecular Biology, 320,* 293–302.

Bose, S., Dutko, J. A., & Zitomer, R. S. (2005). Genetic factors that regulate the attenuation of the general stress response of yeast. *Genetics, 169,* 1215–1226.

Bradford, M. (1976). A rapid and sensitive method for the quantitation of microgram quantities of protein utilizing the principle of protein-dye binding. *Analytical Biochemistry, 72,* 248–254.

Braga, F. R., Araújo, J. V., Soares, F. E. F., Genièr, H. L. A., & Queiroz, J. H. (2012). An extracellular serine protease of an isolate of *Duddingtonia flagrans* nematophagous fungus. *Biocontrol Science and Technology, 22,* 1131–1142.

Caracuel, Z., Roncero, M. I., Espeso, E. A., Gonzalez-Verdejo, C. I., Garcia-Maceira, F. I., Di Pietro, A. (2003). The pH signalling transcription factor PacC controls virulence in the plant pathogen *Fusarium oxysporum*. *Molecular Microbiology, 48,* 765–779.

Chang, A., Scheer, M., Grote, A., Schomburg, I., & Schomburg, D. (2009). BRENDA, AMENDA and FRENDA the enzyme information system: New content and tools in 2009. *Nucleic Acids Research, 37,* D588–D592.

Chen, L. L., Liu, L. J., Shi, M., Song, X. Y., Zheng, C. Y., Chen, X. L., & Zhang, Y. Z. (2009). Characterization and gene cloning of a novel serine protease with nematicidal activity from *Trichoderma pseudokoningii* SMF2. *FEMS Microbiology Letters, 299,* 135–142.

Cox, G. N., Kusch, M., & Edgar, R. S. (1981). Cuticle of *Caenorhabditis elegans*: Its isolation and partial characterization. *Journal of Cell Biology, 90,* 7–17.

Davis, D., Edwards, J. E., Jr Mitchell, A. P., & Ibrahim, A. S. (2000). *Candida albicans* RIM101 pH response pathway is required for host-pathogen interactions. *Infection and Immunity, 68,* 5953–5959.

Dijksterhuis, J., Veenhuis, M., Harder, W., & Nordbring-Hertz, B. (1994). Nematophagous fungi: Physiological aspects and structurefunction relationships. *Advances in Microbial Physiology, 36,* 111–143.

Dong, L. Q., Yang, J. K., & Zhang, K. Q. (2007). Cloning and phylogenetic analysis of the chitinase gene from the facultative pathogen *Paecilomyces lilacinus*. *Journal of Applied Microbiology, 103,* 2476–2488.

Eshel, D., Miyara, I., Ailing, T., Dinoor, A., & Prusky, D. (2002). pH regulates endoglucanase expression and virulence of *Alternaria alternata* in persimmon fruit. *Molecular Plant-microbe Interactions 15*, 774–779.

Esteves, I., Peteira, B., Atkins, S. D., Magan, N., & Kerry, B. (2009). Production of extracellular enzymes by different isolates of *Pochonia chlamydosporia*. *Mycological Research, 113,* 867–876.

Fang, W., Leng, B., Xiao, Y., Jin, K., Ma, J., Fan, Y., Feng, J., Yang, X., Zhang, Y., & Pei, Y. (2005). Cloning of *Beauveria bassiana* chitinase gene *Bbchit1* and its application to improve fungal strain virulence. *Applied and Environmental Microbiology, 71,* 363–370.

Fekete, C., Tholander, M., Rajashekar, B., Ahrén, D., Friman, E., Johansson, T., & Tunlid, A. (2008). Paralysis of nematodes: Shifts in the transcriptome of the nematode-trapping fungus *Monacrosporium haptotylum* during infection of *Caenorhabditis elegans*. *Environmental Microbiology, 10,* 364–375.

Fernández-Acero, F. J., Jorge, I., Calvo, E., Vallejo, I., Carbú, M., Camafeita, E., Garrido, C., López, J. A., Jorrin, J., & Cantoral, J. M. (2007). Proteomic analysis of phytopathogenic fungus *Botrytis cinerea* as a potential tool for identifying pathogenicity factors, therapeutic targets and for basic research. *Archives of Microbiology, 187,* 207–215.

Fusetti, F., von Moeller, H., Houston, D., Rozeboom, H. J., Dijkstra, B. W., Boot, R. G., Aerts, J. M., & van Aalten, D. M. F. (2002). Structure of human chitotriosidase. Implications for specific inhibitor design and function of mammalian chitinase-like lectins. *Journal of Biological Chemistry, 277,* 25537–25544.

Galagan, J. E., Calvo, S. E., Borkovich, K. A., Selker, E. U., Read, N. D., Jaffe, D., FitzHugh, W., Ma, L. J., Smirnov, S., Purcell, S., Rehman, B., Elkins, T., Engels, R., Wang, S., Nielsen, C. B., Butler, J., Endrizzi, M., Qui, D., Ianakiev, P., Bell-Pedersen, D., Nelson, M. A., Werner-Washburne, M., Selitrennikoff, C. P., Kinsey, J. A., Braun, E. L., Zelter, A., Schulte, U., Kothe, G. O., Jedd, G., Mewes, W., Staben, C., Marcotte, E., Greenberg, D., Roy, A., Foley, K., Naylor, J., Stange-Thomann, N., Barrett, R., Gnerre, S., Kamal, M., Kamvysselis, M., Mauceli, E., Bielke, C., Rudd, S., Frishman, D., Krystofova, S., Rasmussen, C., Metzenberg, R. L., Perkins, D. D., Kroken, S., Cogoni, C., Macino, G., Catcheside, D., Li, W., Pratt, R. J., Osmani, S. A., DeSouza, C. P., Glass, L., Orbach, M. J., Berglund, J. A., Voelker, R., Yarden, O., Plamann, M., Seiler, S., Dunlap, J., Radford, A., Aramayo, R., Natvig, D. O., Alex, L. A., Mannhaupt, G., Ebbole, D. J., Freitag, M., Paulsen, I., Sachs, M. S., Lander, E. S., Nusbaum, C., & Birren, B. (2003). The genome sequence of the filamentous fungus *Neurospora crassa*. *Nature, 422,* 859–868.

Gan, Z. W., Yang, J. K., Tao, N., Liang, L. M., Mi, Q. L., Li, J., & Zhang, K. Q. (2007a). Cloning of the gene *Lecanicillium psalliotae* chitinase Lpchi1 and identification of its potential role in the biocontrol of root-knot nematode *Meloidogyne incognita*. *Applied Microbiology and Biotechnology, 76,* 1309–1317.

Gan, Z. W., Yang, J. K., Tao, N., Yu, Z. F., & Zhang, K. Q. (2007b). Cloning and expression analysis of a chitinase gene *Crchi1* from the mycoparasitic fungus *Clonostachys rosea* (syn. *Gliocladium roseum*). *Journal of Microbiology, 45,* 422–430.

Gan, Z. W., Yang, J. K., Tao, N., Lou, Z. Y., Mi, Q. L., Meng, Z. H., & Zhang, K. Q. (2009). Crystallization and preliminary crystallographic analysis of a chitinase from *Clonostachys rosea*. *Acta Crystallographica Section F: Structural Biology and Crystallization Communications, 65,* 386–388.

Gao, Q., Jin, K., Ying, S. H., Zhang, Y., Xiao, G., Shang, Y., Duan, Z., Hu, X., Xie, X. Q., Zhou, G., Peng, G., Luo, Z., Huang, W., Wang, B., Fang, W., Wang, S., Zhong, Y., Ma, L. J., St Leger, R. J., Zhao, G. P., Pei, Y., Feng, M. G., Xia, Y. X., & Wang, C. S. (2011). Genome sequencing and comparative transcriptomics of the model entomopathogenic fungi *Metarhizium anisopliae* and *M. acridum*. *PLoS Genetics, 7,* e100126.

Gasch, A. P., Spellman, P. T., Kao, C. M., Carmel-Harel, O., Eisen, M. B., Storz, G., Botstein, D., & Brown, P. O. (2000). Genomic expression programs in the response of yeast cells to environmental changes. *Molecular Biology of the Cell, 11,* 4241–4257.

Gordon, D., Desmarais, C., & Green, P. (2001). Automated finishing with autofinish. *Genome Research, 11,* 614–625.

Haas, B. J., Salzberg, S. L., Zhu, W., Pertea, M., Allen, J. E., Orvis, J., White, O., Buell, C. R., & Wortman, J. R. (2008). Automated eukaryotic gene structure annotation using EVidenceModeler and the program to assemble spliced alignments. *Genome Biology, 9,* R7.

Henrissat, B., & Bairoch, A. (1993). New families in the classification of glycosyl hydrolases based on amino acid sequence similarities. *Biochemical Journal, 293,* 781–788.

Hollis, T., Monzingo, A. F., Bortone, K., Ernst, S., Cox, R., & Robertus, J. D. (2000). The X-ray structure of a chitinase from the pathogenic fungus *Coccidioides immitis*. *Protein Science, 9,* 544–551.

Houston, D. R., Eggleston, I., Synstad, B., Eijsink, V. G., & van Aalten, D. M. F. (2002). The cyclic dipeptide C1-4[cyclo-(L-Arg-D-Pro)] inhibits family 18 chitinases by structural mimicry of a reaction intermediate. *Biochemical Journal, 368,* 23–27.

Hurtado-Guerrero, R., & van Aalten, D. M. F. (2007). Structure of *Saccharomyces cerevisiae* chitinase 1 and screening-based discovery of potent inhibitors. *Chemistry & Biology, 14,* 589–599.

Jackson, C. W., Heale, J. B., & Hall, R. A. (1985). Traits associated with virulence to the aphid *Macrosiphoniella sanbornii* in eighteen isolates of *Verticillium lecanii*. *Annals of Applied Biology, 106,* 39–48.

Jensen, L. J., Kuhn, M., Stark, M., Chaffron, S., Creevey, C., Muller, J., Doerks, T., Julien, P., Roth, A., Simonovic, M., Bork, P., & von Mering, C. (2009). STRING 8-a global view on proteins and their functional interactions in 630 organisms. *Nucleic Acids Research, 37,* D412–D416.

Joshi, L., St Leger, R. J., & Bidochka, M. J. (1995). Cloning of a cuticle-degrading protease from the entomopathogenic fungus, *Beauveria bassiana*. *FEMS Microbiology Letters, 125,* 211–217.

Kanda, S., Aimi, T., Kano, S., Ishihara, S., Kitamoto, Y., & Morinaga, T. (2008). Ambient pH signaling regulates expression of the serine protease gene (*spr1*) in pine wilt nematode-trapping fungus, *Monacrosporium megalosporum*. *Microbiological Research, 163,* 63–72.

Kanehisa, M., Goto, S., Kawashima, S., Okuno, Y., & Hattori, M. (2004). The KEGG resources for deciphering the genome. *Nucleic Acids Research, 32,* D277–D280.

Khan, A., Williams, K. L., Molloy, M. P., & Nevalainen, H. (2003). Purification and characterization of a serine protease and chitinases from *Paecilomyces lilacinus* and detection of chitinase activity on 2D gels. *Protein Expression and Purification, 32,* 210–220.

Khan, A., Williams, K. L., & Nevalainen, H. K. M. (2004). Effects of *Paecilomyces lilacinus* protease and chitinase on the eggshell structures and hatching of *Meloidogyne javanica* juveniles. *Biological Control, 31,* 346–352.

Korf, I. (2004). Gene finding in novel genomes. *BMC Bioinformatics, 5,* 59.

Larriba, E., Martín-Nieto, J., & Lopez-Llorca, L. V. (2012). Gene cloning, molecular modeling, and phylogenetics of serine protease P32 and serine carboxypeptidase SCP1 from nematophagous fungi *Pochonia rubescens* and *Pochonia chlamydosporia*. *Canadian Journal of Microbiology, 58,* 815–827.

Li, T. F., Zhang, K. Q., & Liu, X. Z. (2000). *Taxonomy of nematophagous fungi (Chinese)*. Beijing: Science Press.

Li, J., Yang, J. K., Huang, X. W., & Zhang, K. Q. (2006). Purification and characterization of an extracellular serine protease from *Clonostachys rosea* and its potential as a pathogenic factor. *Process Biochemistry, 41,* 925–929.

Li, J., Yu, L., Yang, J. K., Dong, L. Q., Tian, B. Y., Yu, Z. F., Liang, L. M., Zhang, Y., Wang, X., & Zhang, K. Q. (2010). New insights into the evolution of subtilisin-like serine protease genes in Pezizomycotina. *BMC Evolutionary Biology, 10,* 68.

Liang, L. M., Meng, Z. H., Ye, F. P., Yang, J. K., Liu, S. Q., Sun, Y., Guo, Y., Mi, Q. L., Huang, X. W., Zou, C. G., Rao, Z. H., Lou, Z. Y., & Zhang, K. Q. (2010). The crystal structures of two cuticle-degrading proteases from nematophagous fungi and their contribution to infection against nematodes. *FASEB Journal, 24,* 1391–1400.

Liang, L. M., Liu, S. Q., Yang, J. K., Meng, Z. H., Lei, L. P., & Zhang, K. Q. (2011a). Comparison of homology models and crystal structures of cuticle-degrading proteases from nematophagous fungi: Structural basis of nematicidal activity. *FASEB Journal, 25,* 1894–1902.

Liang, L. M., Yang, J. K., Li, J., Mo, Y. Y., Li, L., Zhao, X. Y., & Zhang, K. Q. (2011b). Cloning and homology modeling of a serine protease gene (*PrC*) from the nematophagous fungus *Clonostachys rosea*. *Annals of Microbiology, 61,* 511–516.

Limon, M. C., Pintor-Toro, J. A., & Benitez, T. (1999). Increased anti-fungal activity of *Trichoderma harzianum* transformants that overexpress a 33-kDa chitinase. *Phytopathology, 89,* 254–261.

Liu, S. Q., Tao, Y., Meng, Z. H., Fu, Y. X., & Zhang, K. Q. (2011). The effect of calciums on the molecular motions of proteinase K. *Journal of Molecular Modeling, 17,* 289–300.

Lorito, M. S., Woo, L., Garcia, I., Colucci, G., Harman, G. E., Pintor-Toro, J. A., Filippone, E., Muccifora, S., Lawrence, C. B., Zoina, A., Tuzun, S., & Scala, F. (1998). Genes from mycoparasitic fungi as a source for improving plant resistance to fungi pathogens. *Proceedings of the National academy of Sciences of the United States of America, 95,* 7860–7865.

Lopez-Llorca, L. V. (1990). Purification and properties of extracellular proteases produced by the nematophagous fungus *Verticillium suchlasporium*. *Canadian Journal of Microbiology, 36,* 530–537.

Lopez-Llorca, L. V., & Robertson, W. M. (1992). Immumocytochemical locolization of a 32-kDa protease from the nematophagous fungus *Verticillium suchlasporium* in infected nematode eggs. *Experimental Mycology, 16,* 261–267.

Lýsek, H., & Krajcí, D. (1987). Penetration of ovicidal fungus *Verticillium chlamydosporium* through the Ascaris lumbricoides egg-shells. *Folia Parasitologica, 34,* 57–60.

Maclennan, J. D., Mandl, I., & Howes, E. L. (1953). Bacterial digestion of collagen. *Journal of Clinical Investigation, 32,* 1317–1322.

Majoros, W. H., Pertea, M., & Salzberg, S. L. (2004). TigrScan and GlimmerHMM: Two open source ab initio eukaryotic gene-finders. *Bioinformatics, 20,* 2878–2879.

Margulies, M., Egholm, M., Altman, W. E., Attiya, S., Bader, J. S., Bemben, L. A., Berka, J., Braverman, M. S., Chen, Y. J., Chen, Z., Dewell, S. B., Du, L., Fierro, J. M., Gomes, X. V., Godwin, B. C., He, W., Helgesen, S., Ho, C. H., Irzyk, G. P., Jando, S. C., Alenquer, M. L., Jarvie, T. P., Jirage, K. B., Kim, J. B., Knight, J. R., Lanza, J. R., Leamon, J. H., Lefkowitz, S. M., Lei, M., Li, J., Lohman, K. L., Lu, H., Makhijani, V. B., McDade, K. E., McKenna, M. P., Myers, E. W., Nickerson, E., Nobile, J. R., Plant, R., Puc, B. P., Ronan, M. T., Roth, G. T., Sarkis, G. J., Simons, J. F., Simpson, J. W., Srinivasan, M., Tartaro, K. R., Tomasz, A., Vogt, K. A., Volkmer, G. A., Wang, S. H., Wang, Y., Weiner, M. P., Yu, P., Begley, R. F., & Rothberg, J. M. (2005). Genome sequencing in microfabricated high-density picolitre reactors. *Nature, 437,* 376–380.

Mi, Q. L., Yang, J. K., Ye, F. P., Gan, Z. W., Wu, C. W., Niu, X. M., Zou, C. G., & Zhang, K. Q. (2010). Cloning and overexpression of *Pochonia chlamydosporia* chitinase gene *pcchi44*, a potential virulence factor in infection against nematodes. *Process Biochemistry, 45,* 810–814.

Mileweski, S., O'Donnell, R. W., & Gooday, G. W. (1992). Chemical modification studies of the active center of *Candida albicans* chitinase and its inhibition by allosamidin. *Journal of General Microbiology, 138,* 2545–5501.

Mo, M. H., Chen, W. M., Su, H. Y., Zhang, K. Q., Duan, C. Q., & He, D. M. (2006). Heavy metal tolerance of nematode-trapping fungi in lead-polluted soils. *Applied Soil Ecology, 31,* 11–19.

Monfort, E., Lopez-Llorca, L. V., & Jansson, H.-B. (2005). Colonisation of seminal roots of wheat and barley by egg-parasitic nematophagous fungi and their effects on *Gaeumannomyces graminis* var. *tritici* and development of root-rot. *Soil Biology and Biochemistry, 37,* 1229–1235.

Nordbring-Hertz, B. (2004). Morphogenesis in the nematode-trapping fungus *Arthrobotrys oligospora*-an extensive plasticity of infection structures. *Mycologist, 18,* 125–133.

Nordbring-Hertz, B., Neumeister, H., Sjollema, K., & Veenhuis, M. (1995). A conidial trap - forming mutant of *Arthrobotrys oligospora*. *Mycological Research, 99,* 1395–1398.

Nordbring-Hertz, B., Jansson, H. B., & Tunlid, A. (2006). Nematophagous fungi. *Encyclopedia of Life Sciences*. doi:10.1002/9780470015902.a0000374.pub3. (John Wiley & Sons, Ltd: Chichester).

Olsson, S., & Persson, Y. (1994). Transfer of phosphorus from *Rhizoctonia solani* to the mycoparasite *Arthrobotrys oligospora*. *Mycological Research, 98,* 1065–1068.

Orikoshi, H., Nakayama, S., Miyamoto, K., Hanato, C., Yasuda, M., Inamori, Y., & Tsujibo, H. (2005). Roles of four chitinases (chia, chib, chic, and chid) in the chitin degradation system of marine bacterium *Alteromonas* sp. strain O-7. *Applied and Environmental Microbiology, 71,* 1811–1815.

Overbeek, R., Fonstein, M., D'Souza, M., Pusch, G. D., & Maltsev, N. (1999). The use of gene clusters to infer functional coupling. *Proceedings of the National academy of Sciences of the United States of America, 96,* 2896–2901.

Papanikolau, Y., Tavlas, G., Vorgias, C. E., & Petratos, K. (2003). De novo purification scheme and crystallization conditions yield highresolution structures of chitinase A and its complex with the inhibitor allosamidin. *Acta Crystallographica Section D: Biological Crystallography, 59,* 400–403.

Peñalva, M. A., Tilburn, J., Bignell, E., & Arst, H. N. Jr. (2008). Ambient pH gene regulation in fungi: Making connections. *Trends in Microbiology, 16,* 291–300.

Perry, R. N., & Trett, M. W. (1986). Ultrastructure of the egg shell of *Heterodera schachtii* and *H. glycines* (Nematoda: Tylenchida). *Revue de Nématologie, 9,* 399–403.

Persmark, L., & Nordbring-Hertz, B. (1997). Conidial trap formation of nematode-trapping fungi in soil and soil extracts. *FEMS Microbiology Ecology, 22,* 313–323.

Pfister, D. H., & Liftik, M. E. (1995). Two *Arthrobotrys* anamorphs from *Orbilia auricolor*. *Mycologia, 87,* 684–688.

Phillips, D. C. (1967). The hen-egg-white lysozyme molecule. *Proceedings of the National academy of Sciences of the United States of America, 57,* 484–495.

Quackenbush, J., Cho, J., Lee, D., Liang, F., Holt, I., Karamycheva, S., Parvizi, B., Pertea, G., Sultana, R., & White, J. (2001). The TIGR Gene Indices: Analysis of gene transcript sequences in highly sampled eukaryotic species. *Nucleic Acids Research, 29,* 159–164.

Quevillon, E., Silventoinen, V., Pillai, S., Harte, N., Mulder, N., Apweiler, R., & Lopez, R. (2005). InterProScan: Protein domains identifier. *Nucleic Acids Research, 33,* W116–W120.

Rao, F. V., Andersen, O. A., Vora, K. A., Demartino, J. A., & van Aalten, D. M. (2005a). Methylxanthine drugs are chitinase inhibitors: Investigation of inhibition and binding modes. *Chemistry & Biology, 12,* 973–980.

Rao, F. V., Houston, D. R., Boot, R. G., Aerts, J. M., Hodkinson, M., Adams, D. J., Shiomi, K., Omura, S., & van Aalten, D. M. (2005b). Specificity and affinity of natural product cyclopentapeptide inhibitors against *A. fumigatus*, human, and bacterial chitinases. *Chemistry & Biology, 12,* 65–76.

Rollins, J. A., & Dickman, M. B. (2001). pH signaling in Sclerotinia sclerotiorum: Identification of a *pacC*/RIM1 homolog. *Applied and Environmental Microbiology, 67,* 75–81.

Rost, B. (1999). Twilight zone of protein sequence alignments. *Protein Engineering, 12,* 85–94.

Schenck, S., Chase, T. J., Rosenzweig, W. D., & Pramer, D. (1980). Collagenase production by nematode-trapping fungi. *Applied and Environmental Microbiology, 40,* 567–570.

Schmidt, A. R., Dörfelt, H., & Perrichot, V. (2007). Carnivorous fungi from Cretaceous Amber. *Science, 318,* 1743.

Schultz, R. M., & Liebman, M. N. (1997). Structure-function relationship in protein families. In T. M. Devlin (Ed.), *Textbook of biochemistry with clinical correlations* (4th ed., pp. 1–2). New York: Wiley-Liss.

Schüttelkopf, A. W., Andersen, O. A., Rao, F. V., Allwood, M., Lloyd, C., Eggleston, I. M., & van Aalten, D. M. (2006). Screening-based discovery and structural dissection of a novel family 18 chitinase inhibitor. *Journal of Biological Chemistry, 281,* 27278–27285.

Segers, R., Butt, T. M., Kerry, B. R., & Peberdy, J. F. (1994). The nematophagous fungus *Verticillium chlamydosporium* produces a chymoelastase-like protease which hydrolyses host nematode proteins in situ. *Microbiology, 140,* 2715–2723.

Seidl, V., Huemer, B., Seiboth, B., & Kubicek, C. P. (2005). A complete survey of *Trichoderma* chitinases reveals three distinct subgroups of family 18 chitinases. *FEBS Journal, 272,* 5923–5939.

Siddiqui, Z. A., & Mahmood, I. (1996). Biological control of plant parasitic nematodes by fungi: A review. *Bioresource Technology, 58,* 229–239.

Siezen, R. J., & Leunissen, J. A. M. (1997). Subtilases: The superfamily of subtilisin like serine protease. *Protein Science, 6,* 501–523.

Slater, G. S. C., & Birney, E. (2005). Automated generation of heuristics for biological sequence comparison. *BMC Bioinformatics, 6,* 31.

Soares, F. E., Braga, F. R., Araújo, J. V., dos Santos Lima, W., Mozer, L. R., & Queiróz, J. H. (2012). In vitro activity of a serine protease from *Monacrosporium thaumasium* fungus against first-stage larvae of *Angiostrongylus vasorum*. *Parasitology Research, 110,* 2423–2427.

Stanke, M., & Waack, S. (2003). Gene prediction with a hidden Markov model and a new intron submodel. *Bioinformatics, 19,* ii215–ii225.

St Leger, R. J., & Wang, C. (2011). Genetic engineering of fungal biocontrol agents to achieve greater efficacy against insect pests. *Applied Microbiology and Biotechnology, 85,* 901–907.

St Leger, R. J., Charnley, A. K., & Cooper, R. M. (1986). Cuticle-degrading enzymes of entomopathogenic fungi: Synthesis in culture on cuticle. *Journal of Invertebrate Pathology, 48,* 85–95.

St Leger, R. J., Frank, D. C., Roberts, D. W., & Staples, R. C. (1992). Molecular cloning and regulatory analysis of the cuticle-degrading protease structural gene from the entomopathogenic fungus *Metarhizium anisopliae*. *European Journal of Biochemistry, 204,* 991–1001.

St Leger, R. J., Staples, R. C., & Roberts, D. W. (1993). Entomopathogenic isolates of *Metarhizium anisopliae*, *Beauvaria bassiana*, and *Aspergillus flavus* produce multiple extracellular chitinase isozymes. *Journal of Invertebrate Pathology, 61,* 81–84.

St Leger, R. J., Nelson, J. O., & Screen, S. E. (1999). The entomopathogenic fungus *Metarhizium anisopliae* alters ambient pH, allowing extracellular protease production and activity. *Microbiology, 145,* 2691–2699.

Sutton, J. C., Li, D. W., Peng, G., Yu, H., Zhang, P., & Valdebenito-Sanhueza, R. M. (1997). Gliocladium roseum a versatile adversary of a Botrytis cinerea in crops. *Plant Disease, 81,* 316–328.

Tamura, K., Dudley, J., Nei, M., & Kumar, S. (2007). MEGA4: Molecular evolutionary genetics analysis (MEGA) software version 4.0. *Molecular Biology and Evolution, 24,* 1596–1599.

Tao, Y., Rao, Z. H., & Liu, S. Q. (2010). Insight derived from molecular dynamics simulation into substrate-induced changes in protein motions of proteinase K. *Journal of Biomolecular Structure and Dynamics, 28,* 143–157.

Tatusov, R. L., Fedorova, N. D., Jackson, J. D., Jacobs, A. R., Kiryutin, B., Koonin, E. V., Krylov, D. M., Mazumder, R., Mekhedov, S. L., Nikolskaya, A. N., Rao, B. S., Smirnov, S., Sverdlov, A. V., Vasudevan, S., Wolf, Y. I., Yin, J. J., & Natale, D. A. (2003). The COG database: An updated version includes eukaryotes. *BMC Bioinformatics, 4,* 41.

Ter-Hovhannisyan, V., Lomsadze, A., Chernoff, Y. O., & Borodovsky, M. (2008). Gene prediction in novel fungal genomes using an ab initio algorithm with unsupervised training. *Genome Research, 18,* 1979–1990.

Terwisscha van Scheltinga, A. C., Armand, S., Kalk, K. H., Isogai, A., Henrissat, B., & Dijkstra, B. W. (1995). Stereochemistry of chitin hydrolysis by a plant chitinase/lysozyme and X-ray structure of a complex with allosamidin: Evidence for substract assisted catalysis. *Biochemistry, 34,* 15619–15623.

Tikhonov, V. E., Lopez-Llorca, L. V., Salinas, J., & Jansson, H. B. (2002). Purification and characterization of chitinases from the nematophagous fungi *Verticillium chlamydosporium* and *V. suchlasporium*. *Fungal Genetics and Biology, 35,* 67–78.

Tilburn, J., Sarkar, S., Widdick, D. A., Espeso, E. A., Orejas, M., Mungroo, J., Peñalva, M. A., & Arst, H. N. Jr. (1995). The *Aspergillus* PacC zinc finger transcription factor mediates regulation of both acidand alkaline-expressed genes by ambient pH. *EMBO Journal, 14,* 779–790.

Tosi, S., Annovazzi, L., Tosi, I., Iadrola, P., & Caretta, G. (2001). Collagenase production in an antarctic strain of *Arthrobotrys tortor* Jarowaja. *Mycopathologia, 153,* 157–162.

Tunlid, A., & Jansson, S. (1991). Proteases and their involvement in the infection and immobilization of nematodes by the nematophagous fungus *Arthrobotrys oligospora. Applied and Environmental Microbiology, 57,* 2868–2872.
Tunlid, A., Rosen, S., Ek, B., & Rask, L. (1994). Purification and characterization of an extracellular serine protease from the nematode-trapping fungus *Arthrobotrys oligospora. Microbiology, 140,* 1687–1695.
van Aalten, D. M., Komander, D., Synstad, B., Gåseidnes, S., Peter, M. G., & Eijsink, V. G. (2001). Structural insights into the catalytic mechanism of a family 18 exo-chitinase. *Proceedings of the National academy of Sciences of the United States of America, 98,* 8979–8984.
Wang, B., Wu, W. P., & Liu, X. Z. (2007). Purification and characterization of a neutral serine protease with nematicidal activity from *Hirsutella rhossiliensis. Mycopathologia, 163,* 169–176.
Wang, M., Yang, J. K., & Zhang, K. Q. (2006a). Characterization of an extracellular protease and its cDNA from the nematode-trapping fungus *Monacrosporium microscaphoides. Canadian Journal of Microbiology, 52,* 130–139.
Wang, R. B., Yang, J. K., Lin, C., & Zhang, K. Q. (2006b). Purification and characterization of an extracellular serine protease from the nematode-trapping fungus *Dactylella shizishanna. Letters in Applied Microbiology, 42,* 589–594.
Wang, B., Liu, X., Wu, W. P., Liu, X., & Li, S. (2009). Purification, characterization, and gene cloning of an alkaline serine protease from a highly virulent strain of the nematode-endoparasitic fungus *Hirsutella rhossiliensis. Microbiological Research, 164,* 665–673.
Ward, E., Kerry, B. R., Manzanilla-López, R. H., Mutua, G., Devonshire, J., Kimenju, J., & Hirsch, P. R. (2012). The *Pochonia chlamydosporia* serine protease gene vcp1 is subject to regulation by carbon, nitrogen and pH: Implications for nematode biocontrol. *PLoS One, 7,* e35657.
Watanabe, T., Kobori, K., Miyashita, K., Fujii, T., Sakai, H., Uchida, M., & Tanaka, H. (1993). Identification of glutamic acid 204 and aspartic acid 200 in chitinase A1 of *Bacillus circulans* WL212 as essential residues for chitinase activity. *Journal of Biological Chemistry, 268,* 18567–18572.
Wharton, D. A. (1980). Nematode egg-shells. *Parasitology, 81,* 447–463.
Winnenburg, R., Baldwin, T. K., Urban, M., Rawlings, C., Köhler, J., & Hammond-Kosack, K. E. (2006). PHI-base: A new database for pathogen host interactions. *Nucleic Acids Research, 34,* D459–D464.
Xue, A. G. (2003). Biological control of pathogens causing root rot complex in field pea using *Clonostachys rosea* strain ACM941. *Phytopathology, 93,* 329–335.
Yakoby, N., Kobiler, I., Dinoor, A., & Prusky, D. (2000). pH regulation of pectate lyase secretion modulates the attack of *Colletotrichum gloeosporioides* on avocado fruits. *Applied and Environmental Microbiology, 66,* 1026–1030.
Yang, J. K., Huang, X. W., Tian, B. Y., Wang, M., Niu, Q. H., & Zhang, K. Q. (2005a). Isolation and characterization of a serine protease from the nematophagous fungus, *Lecanicillium psalliotae*, displaying nematicidal activity. *Biotechnology Letters, 27,* 1123–1128.
Yang, J. K., Huang, X. W., Tian, B. Y., Sun, H., Duan, J. X., Wu, W. P., & Zhang, K. Q. (2005b). Characterization of an extracellular serine protease gene from the nematophagous fungus *Lecanicillium psalliotae. Biotechnology Letters, 27,* 1329–1334.
Yang, J. K., Tian, B. Y., Liang, L. M., & Zhang, K. Q. (2007a). Extracellular enzymes and the pathogenesis of nematophagous fungi. *Applied Microbiology and Biotechnology, 75,* 21–31.
Yang, J. K., Li, J., Liang, L. M., Tian, B. Y., Zhang, Y., Chen, C. M., & Zhang, K. Q. (2007b). Cloning and characterization of an extracellular serine protease from the nematode-trapping fungus *Arthrobotrys conoides. Archives of Microbiology, 188,* 167–174.
Yang, J. K., Liang, L. M., Zhang, Y., Li, J., Zhang, L., Ye, F. P., Gan, Z. W., & Zhang, K. Q. (2007c). Purification and cloning of a novel serine protease from the nematode-trapping fungus *Dactylellina varietas* and its potential roles in infection against nematodes. *Applied Microbiology and Biotechnology, 75,* 557–565.
Yang, Y., Yang, E. C., An, Z. Q., & Liu, X. Z. (2007d). Evolution of nematode-trapping cells of predatory fungi of the Orbiliaceae based on evidence from rRNA-encoding DNA and

multiprotein sequences. *Proceedings of the National academy of Sciences of the United States of America, 104,* 8379–8384.

Yang, J. K., Ye, F. P., Mi, Q. L., Tang, S. Q., Li, J., & Zhang, K. Q. (2008). Purification and cloning of an extracellular serine protease from the nematode-trapping fungus *Monacrosporium cystosporium*. *Journal of Microbiology and Biotechnology, 18,* 852–858.

Yang, J. K., Gan, Z. W., Lou, Z. Y., Tao, N., Mi, Q. L., Liang, L. M., Sun, Y., Guo, Y., Huang, X. W., Zou, C. G., Rao, Z. H., Meng, Z. H., & Zhang, K. Q. (2010). Crystal structure and mutagenesis analysis of chitinase CrChi1 from the nematophagous fungus *Clonostachys rosea* in complex with the inhibitor caffeine. *Microbiology, 156,* 3566–3574.

Yang, J. K., Wang, L., Ji, X. L., Feng, Y., Li, X. M., Zou, C. G., Xu, J. P., Ren, Y., Mi, Q. L., Wu, J. L., Liu, S. Q., Liu, Y., Huang, X. W., Wang, H. Y., Niu, X. M., Li, J., Liang, L. M., Luo, Y. L., Ji, K. F., Zhou, W., Yu, Z. F., Li, G. H., Liu, Y. J., Li, L., Qiao, M., Feng, L., & Zhang, K. Q. (2011a). Genomic and proteomic analyses of the fungus *Arthrobotrys oligospora* provide insights into nematode-trap formation. *PLoS Pathogens, 7,* e1002179.

Yang, J. K., Zhao, X. N., Liang, L. M., Xia, Z. Y., Lei, L. P., Niu, X. M., Zou, C. G., & Zhang, K. Q. (2011b). Overexpression of a cuticle-degrading protease Ver112 increases the nematicidal activity of *Paecilomyces lilacinus*. *Applied Microbiology and Biotechnology, 89,* 1895–1903.

Yang, J. K., Liang, L. M., Li, J., & Zhang, K. Q. (2013). Nematicidal enzymes from microorganisms and their applications. *Applied Microbiology and Biotechnology, 97,* 7081–7095.

Ye, F. P., Liang, L. M., Mi, Q. L., Yang, J. K., Lou, Z. Y., Sun, Y. N., Guo, Y., Meng, Z. H., & Zhang, K. Q. (2009). Preliminary crystallographic study of two cuticle-degrading proteases from the nematophagous fungi *Lecanicillium psalliotae* and *Paecilomyces lilacinus*. *Acta Crystallographica Section F: Structural Biology and Crystallization Communications, 65,* 271–274.

Yousef, G. M., Kopolovic, A. D., Elliott, M. B., & Diamandis, E. P. (2003). Genomic overview of serine proteases. *Biochemical and Biophysical Research Communications, 305,* 28–36.

Zhang, Y. J., Liu, X. Z., & Wang, M. (2008). Cloning, expression, and characterization of two novel cuticle-degrading serine proteases from the entomopathogenic fungus *Cordyceps sinensis*. *Research in Microbiology, 159,* 462–469.

Zhao, M. L., Mo, M. H., & Zhang, K. Q. (2004). Characterization of a neutral serine protease and its full-length cDNA from the nematode-trapping fungus *Arthrobotrys oligospora*. *Mycologia, 96,* 16–22.

Zou, C. G., Tu, H. H., Liu, X. Y., Tao, N., & Zhang, K. Q. (2010a). PacC in the nematophagous fungus *Clonostachys rosea* controls virulence to nematodes. *Environmental Microbiology, 12,* 1868–1877.

Zou, C. G., Tao, N., Liu, W. J., Yang, J. K., Huang, X. W., Liu, X. Y., Tu, H. H., Gan, Z. W., & Zhang, K. Q. (2010b). Regulation of subtilisin-like protease prC expression by nematode cuticle in the nematophagous fungus *Clonostachys rosea*. *Environmental Microbiology, 12,* 3243–3252.

Zou, C. G., Xu, Y. F., Liu, W. J., Zhou, W., Tao, N., Tu, H. H., Huang, X. W., Yang, J. K., & Zhang, K. Q. (2010c). Expression of a serine protease gene prC is up-regulated by oxidative stress in the fungus *Clonostachys rosea*: Implications for fungal survival. *PLoS One, 5,* e13386.

Chapter 7
Nematode-Toxic Fungi and their Nematicidal Metabolites

Guo-Hong Li and Ke-Qin Zhang

Contents

K.-Q. Zhang (✉) · G.-H. Li
Laboratory for Conservation and Utilization of Bio-Resources, and Key Laboratory of Microbial Diversity in Southwest China, Ministry of Education, Yunnan University, 650091 Kunming, Yunnan, China
e-mail: kqzhang1@ynu.edu.cn

K.-Q. Zhang, K. D. Hyde (eds.), *Nematode-Trapping Fungi,* Fungal Diversity Research Series 23, DOI 10.1007/978-94-017-8730-7_7,

Abstract This chapter summarizes more than 200 compounds from fungi that have been shown to possess nematicidal activities. These compounds belong to diverse chemical groups including alkaloid, quinone, isoepoxydon, pyran, furan, peptide, macrolide, terpenoid, fatty acid, diketopiperazine, aphthalene and simple aromatics. They have mainly been isolated from a variety of ascomycetous and basidiomycetous fungal taxa. Their nematicidal activities are described and their potential roles in the biocontrol of nematodes are discussed.

Keywords Fungi • Nematode • Nematicidal compounds • Biocontrol

Introduction

Nematodes can attack, infect and consume a wide variety of organisms, including animals, microorganisms and plants. In recent years, the use of synthetic chemical compounds has been the most common strategy for controlling parasitic nematodes (Haydock et al. 2006). While these have been effective in certain circumstances, the widespread use of man-made chemical nematicides has caused significant problems to both the environment and human health. Consequently, their use for pest control in agriculture and forestry has been reduced significantly. The reduced use of synthetic chemical nematicides has generated significant demands for environmentally friendly alternative strategies (Meira et al. 2006). Biological control is a potentially effective alternative for the management of nematode pests. Biological agents include both live organisms as well as their metabolic products (Willis and Thomas 1998). Fungi are known to possess a huge diversity of metabolic pathways and they have provided several large classes of commercial compounds, including many antibiotics used in medicine (Harvey 2000). Therefore, secondary metabolites in fungi could have much potential in their novel structures and nematicidal activities.

Many nematicidal compounds had been discovered from nematode-toxic fungi, a group of nematophagous fungi. The aim of this chapter is to review the nematode-toxic fungi, the different types, their structure and the nematicidal activity of compounds isolated from nematode-toxic fungi.

Nematode-Toxic Fungi and Their Nematicidal Metabolites

About 280 fungal species in 150 genera of Ascomycota and Basidiomycota have been reported to possess nematicidal activity as they produce toxic compounds which are active against nematodes. More than 200 of these compounds with nematicidal activities are summarized in this chapter.

Nematode-Toxic Ascomycetes and Their Nematicidal Metabolites

About 80 genera comprising more than 120 species have been reported to produce nematicidal active components. *Lachnum papyraceum* (*Hyaloscyphaceae*) is one of the most prolific producers of nematicidal secondary metabolites and a number of nematicidal metabolites have been isolated from this taxa, culturing under different conditions. The structures of 30 compounds isolated from *L. papyraceum* have been elucidated to be isoepoxydon, isocoumarin, mycorrhizin and furan, and

24 compounds including 15 new isolates having nematicidal activities (Stadler and Anke 1993a, b; Stadler et al. 1995a, b, c, d, e; Shan et al. 1996).

The production of nematicidal compounds by three species of *Nematoctonus* have been demonstrated by Giuma and Cooke (1971) and Giuma et al. (1973), and these compounds were termed nematoctoxins. The result indicated that *Nematoctonus* species quickly retard and kill their nematode hosts by the production of toxin, but the toxin had not been identified. *Nematoctonus robustus* was shown to have nematode-immobilizing activity in culture filtrate (Kennedy and Tampion 1978). The nematicidal effect of filtrates from 15 asexual ascomyctes were tested against *Meloidogyne incognita*, and *Acremonium strictum*, *Alternaria alternata*, *Curvularia pallescens*, *Nigrospora sphaerica*, *Paecilomyces lilacinus*, *Penicillium spinulosu*, *Trichoderma harzianum*, were most effective hatch inhibitors of root-knot nematodes and the nematicidal action of culture filtrates against nematode might be attributed to the production of certain toxic metabolites (Khan and Kgan 1992).

Besides producing special structures that trap nematodes, some trapping fungi can produce nematicidal compounds at the same time as the process of trapping nematodes. The nematicidal compound linoleic acid was isolated from the nematophagous *Arthrobotrys brochopaga*, *A. conoides*, *A. dactyloides*, *A. oligospora*, and oligosporon, with 4',5'-dihydro-oligosporon and arthrobotrisin A being obtained from *A. oligospora* (Stadler et al. 1993; Anke et al. 1995; Anderson et al. 1995; Wei et al. 2011). These compounds have nematicidal activity and thus the process by which these fungi overcome and capture the nematode is complex.

Beauveria bassiana is an important insect pathogenic fungus which produces the bioactive substance beauvericin which has nematicidal activity against *Meloidogyne incognita* (Hamill et al. 1969; Mayer 1995), *Caenorhabditis elegans* and *Bursaphelenchus xylophilus* (Shimada et al. 2010). The compound is also produced by *Fusarium* sp. (Mayer 1995) and *Paecilomyces fumoso-roseus* (Bernardini et al. 1975).

Isolates of 130 freshwater fungal taxa have been assayed for nematicidal activity against *B. xylophilus* and 22 filtrates and 13 water-soluble extracts of broken fungal mycelia were found to be active against the nematode. The mobility of over 90 % of nematodes were inhibited by filtrates from *Annulatascus* sp., *Caryospora callicarpa*, *Massarina thalassioidea*, *Ophioceras commune*, *Pseudohalonectria adversaria*, *Pseudohalonectria lignicola*, and mycelia extracts from *Helicomyces roseus*, *Phomatospora berkeleyi* and *P. lignicola* (Dong et al. 2003). Several new nematicidal compounds were obtained from these freshwater fungi (Dong et al. 2007, 2008, 2010). A novel class of potent nematicidal thermolides was isolated from a thermophilic fungus *Talaromyces thermophilus*. Thermolides A and B showed the strongest activities against nematodes with similar activity of avermectins (Guo et al. 2012). These and other nematode-toxic ascomycetes (and anamorphs), and their nematicidal compounds are lised in Table 7.1.

Table 7.1 Nematode-toxic ascomycetes and their nematicidal compounds

Species	Test nematodes	Nematicidal compounds	References
Acremonium sp.	*Meloidogyne incognita*	–	Yan et al. (2010)
Acremonium sp.	*B. xylophilus*	–	Meng et al. (2012)
A. strictum	*Tylenchulus Semipenetrans*	–	Verdejo-Lucas et al. (2009)
A. strictum	*M. incognita*	–	Khan and Kgan (1992)
Acrophiolophora funsispore	*M. incognita*	–	Khan and Kgan (1992)
Alternaria alternata	*H. cortortus* *M. incognita*	Helmidiol	Khan and Kgan (1992); Kind et al. (1996)
A. carthami	*A. aceti*	Brefeldin A	BačÍkovÁ et al. (1965); Vurro et al. (1998)
A. zinnia	*A. aceti*	Brefeldin A	BačÍkovÁ et al. (1965); Vurro et al. (1998)
Aniptodera sp.	*B. xylophilus*	–	Dong et al. (2003)
Annulatascus sp.	*B. xylophilus*	–	Dong et al. (2003)
A. triseptata	*B. xylophilus*	–	Dong et al. (2003)
Apiocrea chrysosperma	Nematicidal agent	Chrysospermins A, B, C, and D	Metzger et al. (1994); Dornberger et al. (1995)
Arthrobotrys brochopaga	*C. elegans* *M. incognita*	Linoleic acid	Anke et al. (1995)
A. conoides	*C. elegans* *M. incognita*	Linoleic acid	Anke et al. (1995)
A. dactyloides	*C. elegans* *M. incognita*	Linoleic acid	Anke et al. (1995)
A. oligospora	*H. contortus* *C. elegans* *M. incognita* *P. redivivus*	Oligosporon, Linoleic acid, 4',5'-dihydro-oligosporon, Arthrobotrisin A	Stadler et al. (1993); Anke et al. (1995); Anderson et al. (1995); Wei et al. (2011)
Ascochyta imperfecta	*A. aceti*	Brefeldin A	BačÍkovÁ et al. (1965); Suzuki et al. (1970)

Table 7.1 (continued)

Species	Test nematodes	Nematicidal compounds	References
Aspergillus clavatus	*A. aceti*	Brefeldin A	BačÍkovÁ et al. (1965); Wang et al. (2002)
A. flavus	*M. incognita*	–	Khan and Kgan (1992)
A. fumigatus	*A. aceti* *P. penetrans* *B. xylophilus*	Fumagillin, Gliotoxin, Fumiquinones A and B, Spinulosin, LL-S490, Pseurotin A	Tarbell et al. (1960); BačÍkovÁ et al. (1965); Beecham et al. (1966); Hayashi et al. (2007)
A. glaucus	*M. incognita*	Emodin	Anke et al. (1980a); Anke et al. (1980b); Mayer (1995)
A. melleus	*P. penetrans*	Aspyrone	Kimura et al. (1996)
A. niger	*A. suum* *H. cortortus*	Nafuredin, Nafuredin-γ	Ui et al. (2001); Ōmura et al. (2001); Shiomi et al. (2005)
A. niger	*M. incognita*	–	Khan and Kgan (1992)
Aspergillus sp.	*B. xylophilus* *C. elegans*	5-hydroxymethyl-2-furoic acid	Kimura et al. (2007)
Aspergillus sp.	*H. glycines*	–	Meyer et al. (2004)
Aspergillus sp.	*M. incognita* *A. aceti*	Patulin, Penicillic acid	BačÍkovÁ et al. (1965); Mayer (1995)
Aspergillus sp.	*P. penetranas*	*βγ*-dehydrocurvularin, *αβ*-dehydrocurvularin, 7-oxo-curvularin, 8-*β*-hydroxy-7-oxocurvularin	Kusano et al. (2003)
Aspergillus sp.	*T. colubriformis* *H. contortus*	Aspergillimide, 16-keto aspergillimide, VM54159, SB203105, SB200437	Banks et al. (1997)
A. ustus	*C. elegans*	Ophiobolin K, 6-epiophiobolin K	Singh et al. (1991)
Beauveria bassiana	*M. incognita* *C. elegans* *B. xylophilus*	Beauvericin	Hamill et al. (1969); Mayer (1995); Shimada et al. (2010)
Bulgaria inquinans	*C. elegans*	Bulgarialactone A, B	Stadler et al. (1995)
Byssochlamys nivea	*C. elegans*	–	Park et al. (2001)

Table 7.1 (continued)

Species	Test nematodes	Nematicidal compounds	References
Caryospora callicarpa	*B. xylophilus*	Caryospomycins A, B and C, 4,8-dihydroxy-3,4-dihydronaphthalen-1(2*H*)-one, 4,6-dihydroxy—3,4-dihydronaphthalen-1(2*H*)-one, 4,6,8-trihydroxy-3,4-dihydronaphthalen-1(2*H*)-one, 3,4,6,8-tetrahydroxy-3,4-dihydronaphthalen-1(2*H*)-one (*cis*-4-hydroxyscytalone)	Dong et al. (2007); Zhu et al. (2008)
Chaetomium globosum	*M. incognita* *H. glycines*	Flavipin	Nitao et al. (2002)
C. robustum	*T. semipenetrans*	–	Verdejo-Lucas et al. (2009)
Chaetomium sp.	*M. incognita*	–	Yan et al. (2010)
Chlorosplenium sp.	*C. elegans* *M. incognita*	Linoleic acid	Anke et al. (1995)
Cladobotryum rubrobrunnescens	*M. incognita*	Cladobotrin I	Wagner et al. (1998)
Cladosporium sp.	*B. xylophilus*	–	Zhao et al. (2004)
Clonostachys cylindrospora	*H. cortortus*	Clonostachydiol	Grabley et al. (1993); Rao et al. (1995)
Cochliobolus heterostrophus	*C. elegans*	Ophiobolin M, 6-epiophiobolin M, Ophiobolin C, 6-epiophiobolin C	Tsipouras et al. (1996)
C. miyabeanus	*C. elegans*	Cochlioquinone A	Barrow and Murphy (1972); Schaeffer et al. (1990)
Coelomycetes sp.	*P. penetranas*		
B. xylophilus	Preussomerins C, D, E, (4*RS*)4,8-dihydroxy-3,4-dihydronaphthalen-1(2H)-one, 4,6,8-trihydroxy-3,4-dihydronaphthalen-1(2H)-one	Zhou et al. (2009)	

Table 7.1 (continued)

Species	Test nematodes	Nematicidal compounds	References
Cordyceps ophioglossoides	*C. elegans*	–	Stadler and Sterner (1998)
Coronophora gregaria	*C. elegans*	MK7924	Kumazawa et al. (2003)
Curvularia pallescens	*M. incognita*	–	Khan and Khan (1992)
Cylindrocarpon olidum	*Heterorhabditis nematode*	Cannabiorcichromenic acid, 8-chloro-cannabiorci-chromenic acid	Quaghebeur et al. (1994)
Dactylella candida	*C. elegans* *M. incognita*	Linoleic acid	Anke et al. (1995)
Daldinia concentrica	*C. elegans* *M. incognita*	1-methoxy-8-hydroxynaphthalene, 1,8-dimethoxynaphthalene	Dasenbrock (1994); Anke et al. (1995)
Diaporthe sp.	*B. xylophilus*	–	Dong et al. (2003)
Emericella rugulosa	*T. semipenetrans*	–	Verdejo-Lucas et al. (2009)
Emericellopsis poonensis	Against helminths	Antiamoebin I	Thirumalachur (1968); Pandey et al. (1977)
E. synnematicola	Against helminths	Antiamoebin I	Thirumalachur (1968); Pandey et al. (1977)
Epicoccum nigrum	*M. incognita* *H. glycines*	Favipin	Burge et al. (1976); Nitao et al. (2002)
E. purpurascens	*M. incognita* *H. glycines*	Favipin	Brown et al. (1987); Nitao et al. (2002)
Fusarium bulbicola	*M. incognita* *C. elegans* *B. xylophilus*	Beauvericin	Mayer (1995); Shimada et al. (2010)
F. compactum	*H. glycines* *M. incognita*	–	Meyer et al. (2004)
F. equiseti	*M. incognita*	–	Meyer et al. (2004)
F. oxysporum	*M. incognita* *Radopholus similis* *Pratylenchus goodeyi* *Helicotylenchus multicinctus*	–	Khan and Kgan (1992); Van Dessel et al. (2011)

Table 7.1 (continued)

Species	Test nematodes	Nematicidal compounds	References
F. redolens	*M. incognita* *C. elegans* *B. xylophilus*	Beauvericin	Mayer (1995); Xu et al. (2010); Shimada et al. (2010)
F. roseum	*A. aceti*	Trichothecolone	Freeman et al. (1959); BačÍkovÁ et al. (1965)
Fusarium sp.	*M. incognita*	Enniatin A, Enniatin B	Mayer (1995); Tomoda et al. (1992)
Fusarium sp.	*M. incognita* *C. elegans* *B. xylophilus*	Beauvericin	Mayer (1995); Bernardini et al. (1975); Shimada et al. (2010)
Fusarium sp.	*M. incognita*	–	Ruanpanun et al. (2010)
Geotrichum sp.	*P. redivivus, B. xylophilus*	1-((2*R**,4*S**,5*S**)-2-chloro-4-methyl-1,3-oxazinan-5-yl)ethanone, 1-((2*R**,4*S**,5*R**)-2-chloro-4-methyl-1,3-oxazinan-5-yl)ethanone, 2',4'-dihydroxyacetophenone	Li et al. (2007)
Gliocladium deliquescens	*M. javanica* *M. incognita*	–	Sankaranarayanan et al. (1997)
G. fimbriatum	*A. aceti*	Gliotoxin	BačÍkovÁ et al. (1965); Beecham et al. (1966)
G. roseum	*C. elegans* *P. redivivus* *B. xylophilus*	Gliocladine A, B, C, D and F, verticillin A, 11'-deoxyverticillin A, Sch52900, Sch52901, Glioclasine	Dong et al. (2005); Dong et al. (2006)
G. virens	*A. aceti*	Viridin	BačÍkovÁ et al. (1965); Blight et al. (1968)
Gymnoascus reesii	*M. incognita*	(3E,5E)-2,5-dihydroxy-2,7-dihydrooxepine-3-carboxylic anhydride	Liu et al. (2011)
Helicomyces roseus	*B. xylophilus*	–	Dong et al. (2003)
Helminthosporium leersii	*C. elegans*	Cochlioquinone A	Barrow and Murphy (1972); Schaeffer et al. (1990)
H. sativum	*C. elegans*	Cochlioquinone A	Barrow and Murphy (1972); Schaeffer et al. (1990)

Table 7.1 (continued)

Species	Test nematodes	Nematicidal compounds	References
Hemicarpenteles paradoxus	*A. aceti*	Brefeldin A	BačÍkovÁ et al. (1965); Anke et al. (1995)
Hirsutella thompsonii var. *synnematosa*	*M. incognita*	Phomalactone	Krasnoff and Gupta (1994); Khambay et al. (2000)
Hymenoscyphus sp.	*C. elegans*	–	Anke et al. (1995)
Hypomyces sp.	*B. xylophilus*	Hypocrellin A, Elsinochrome A	Dong et al. (2001)
Imperfect fungus, D1084	*B. xylophilus*	Bursaphelocides A, B	Kawazu et al. (1993)
Imperfect fungus, PF1022	*A. galli*	Cyclodepsipeptide PF1022A	Sasaki et al. (1992); Samson-Himmelstjerna et al. (2005)
Isaria cicadae	*B. xylophilus*	–	Zhao (2004)
Lachnum papyraceum	*C. elegans* *M. incognita*	Lachnumon, Lachnumon A, Lachnumon B1, Lachnumon B2, Mycorrhizin A, Chloromycorrhizin A, (1'-*E*)-dechloromycorrhizin A, 6-hydroxymellein, 4-chloro-6-hydroxymellein, 4-bromo-6-hydroxymellein, 6-methoxymellein, 4-chloro-6-methoxymellein, 4-chloro-6,7-dihydroxymellein, (1'-Z)-dechloromycorrhizin, Papyracon A, Papyracon B, Papyracon C, Mycorrhizin B1, Mycorrhizin B2, Papyracon, 6-*O*-methylpapyracon B, 6-*O*-methylpapyracon C, Lachnumfuran A, Lachnumlactone A	Stadler and Anke (1993a, b); Stadler et al. (1995a, b, c, d, e); Shan et al. (1996)
Leptosphaeria sp.	*B. xylophilus*	–	Dong et al. (2003)
Massarina bipolaris	*B. xylophilus*	–	Dong et al. (2003)
M. thalassioidea	*B. xylophilus*	–	Dong et al. (2003)
Melanconium betulinum	*M. incognita* *C. elegans*	3-hydroxypropionic acid	Schwarz et al. (2004)
Mollisia sp.	*C. elegans*	–	Anke et al. (1995)
Monacrosporium doedycoides	*C. elegans* *M. incognita*	Linoleic acid	Anke et al. (1995)

Table 7.1 (continued)

Species	Test nematodes	Nematicidal compounds	References
Muscodor albus	*M. chitwoodi* *M. hapla* *P. allius* *P. penetrans*	–	Ekaterini et al. (2008)
Mycosphaerella lethalis	*C. elegans*	Lethaloxin	Stadler (1993); Arnone et al. (1993)
Myrothecium verrucaria	*T. semipenetrans*	–	Verdejo-Lucas et al. (2009)
Nectria radicola	soil nematode	Radicicol, Radicicol B, Radicicol C	Mirrington et al. (1964); Stadler (1993)
Nectria sp.	*B. xylophilus* *C. elegans*	–	Anke et al. (1995); Dong et al. (2003)
Nematoctonus concurrens	*P. redivivus*	–	Giuma and Cooke (1971); Giuma et al. (1973)
N. haptocladus	*P. redivivus*	–	Giuma and Cooke (1971); Giuma et al. (1973)
N. robustus	*P. redivivus*	–	Kennedy and Tampion (1978)
N. tripolitanus	*P. redivivus*	–	Giuma and Cooke (1971); Giuma et al. (1973)
Neobulgaria pura	*C. elegans*	14-epicochlioquinone B	Lorenzen et al. (1994); Anke et al. (1995)
Nigrospora sphaerica	*M. incognita*	Phomalactone	Khambay et al. (2000); Kim et al. (2001)
Oidiodendron sp.	*P. penetrans* *B. xylophilus*	4-hydroxyphenylacetic acid (4-HPA) Oidiolactone D	Ohtani et al. (2011)
Ophioceras commune	*B. xylophilus*	–	Dong et al. (2003)
O. dolichostomum	*B. xylophilus*	Isoamericanoic acid A, Caffeic acid	Dong et al. (2010)
Paecilomyces cateniannulatus	*P. redivivus*	–	Yao et al. (2006)
P. fumoso-roseus	*M. incognita* *C. elegans* *B. xylophilus*	Beauvericin	Bernardini et al. (1975); Mayer (1995); Shimada et al. (2010)

Table 7.1 (continued)

Species	Test nematodes	Nematicidal compounds	References
P. lilacinus	*C. elegans* *M. incognita* *M. javanica* *T. semipenetrans*	Acetic acid	Djian et al. (1991); Park et al. (2004) Verdejo-Lucas et al. (2009)
Paecilomyces sp.	*R. pseudoelongata*	Paeciloxazine	Kanai et al. (2004)
Paecilomyces sp.	*B. xylophilus*	Cerebroside A and B	Zhang et al. (2010)
Paecilomyces sp.	*M. incognita* *B. xylophilus*	4-(4'-carboxy-2'-ethyl-hydroxypenty)-5,6-dihydro-6-methyl-cyclobuta[*b*]pyridine-3,6-dicarboxylic acid	Liu et al. (2009)
Paecilomyces sp.	*A. aceti*	Brefeldin A	BačÍkovÁ et al. (1965); Wang et al. (2002)
Paecilomyces sp.	*M. incognita*	–	Ruanpanun et al. (2010)
P. varioti	*M. incognita*	–	Khan and Kgan (1992)
Paraniesslia sp.	*B. xylophilus*	3,5-dicarboxyaldehyde-4-hydroxy-acetophenone	Dong (2005)
Paraniesslia sp.	*B. xylophilus*	(2*S*,2'*R*,3*R*,3'*E*,4*E*,8*E*)-1-O-(β-D-glucopyranosyl)-3-hydroxyl-2-[N-2'-hydroxyl-3'-eicosadecenoyl] amino-9-methyl-4,8-octadecadiene, (2*S*,2'*R*-,3*R*,3'*E*,4*E*,8*E*)-1-O-(β-D-glucopyranosyl)-3-hydroxyl-2-[N-2'-hydroxyl-3'-octadecenoyl] amino-9-methyl-4,8-octadecadiene	Dong et al. (2005)
Penicillium bilaiae	*P. penetrans*	Penipratynolene, 6-methoxy-carbonylpicolinic acid, 2,6-pyridinedicarboxylic acid	Alfaro et al. (2003); Kimura et al. (1981); Mori et al. (1982); Nakahara et al. (2004)
P. brefeldianum	*A. aceti*	Brefeldin A	BačÍkovÁ et al. (1965); Kim and Kochevar (1995)
P. camemberti	*A. aceti*	Brefeldin A	BačÍkovÁ et al. (1965); Abraham and Arfmann (1992)
P. charlesii	*C. elegans* *T. columbriformis* *H. contortus*	Paraherquamide, Paraherquamides B, C, D, E, F and G	Ondeyka et al. (1990); Liesch and Wichmann (1990); Blanchflower et al. (1991)
P. decumbens	*A. aceti*	Brefeldin A	Singleton et al. (1958); BačÍkovÁ et al. (1965)

Table 7.1 (continued)

Species	Test nematodes	Nematicidal compounds	References
P. nigricans	*A. aceti*	Fumagillin	Tarbell et al. (1960)
P. paraherquei	*C. elegans* *T. columbriformis* *H. contortus*	Paraherquamide	Yamazaki et al. (1981); Blanchflower et al. (1991)
P. roqueforti	*plant parasitic nematode*	Marcfortine A, B, and C	Polondky et al. (1980); Prangé et al. (1981)
P. simplicissimum	*P. penetrans* *C. elegans*	Peniprequinolone, Penigequinolones A and B, 3-methoxy-4,6-dihydroxy-4-(4'-methoxyphenyl) quinolinone	Kusano et al. (2000)
Penicillium sp.	*B. xylophilus*	–	Zhao (2004)
Penicillium sp.	*T. columbriformis* *H. contortus*	Paraherquamide, VM55594, VM54158, VM54159, VM55595, VM55596, VM55597, VM55599	Blanchflower et al. (1991); Blanchflower et al. (1993)
Penicillium sp.	*M. incognita* *A. aceti*	Patulin, Penicillic acid	BačÍkovÁ et al. (1965); Mayer (1995)
Penicillium sp.	*A. aceti*	Gliotoxin	BačÍkovÁ et al. (1965); Beecham et al. (1966)
Penicillium sp.	*M. incognita*	–	Ruanpanun et al. (2010)
P. spinulosum	*M. incognita*	–	Khan and Kgan (1992)
Rhizopus stolonifer	*M. incognita*	–	Khan and Kgan (1992)
Phoma multirostrata	*M. incognita*	–	Khan and Kgan (1992)
Phoma sp.	*B. xylophilus*	–	Dong et al. (2003)
Phomatospora berkeleyi	*B. xylophilus*	–	Dong et al. (2003)
Phomopsis phaseoli	*M. incognita* *C. elegans*	3-hydroxypropionic acid	Schwarz et al. (2004)
Phyllosticta sp.	*M. incognita*	–	Yan et al. (2010)
Pochonia chlamydosporia	*P. redivivus*	Aurovertins F and D	Niu et al. (2010)
Pseudohalonectria adversaria	*B. xylophilus*	Pseudohalonectrin A and B	Dong et al. (2006)

Table 7.1 (continued)

Species	Test nematodes	Nematicidal compounds	References
P. lignicola	*B. xylophilus*	–	Dong et al. (2003)
Selenosporella sp.	*M. incognita*	–	Reyes-Estebanez et al. (2011)
Syncephalastrum racemosum	*B. xylophilus*	–	Sun (1997); Zhang (2004); Sun et al. (2008)
Talaromyces cyanescens	*Tylenchulus semipenetrans*	–	Verdejo-Lucas et al. (2009)
T. thermophilus	*B. xylophilus* *M. incognita* *P. redivivus*	Thermolides A, B, C and D	Guo et al. (2012)
Theilavia terricola	*M. incognita*	–	Khan and Kgan (1992)
Trichoderma album	*M. incognita*	–	Radwan et al. (2012)
Trichoderma compactus	*P. redivivus*	–	Yang (2008)
T. harzianum	*M. javanica* *M. incognita*	–	Khan and Khan (1992); Sankaranarayanan et al. (1997); Khan and Haque (2011); Radwan et al. (2012)
T. koningii	*M. javanica* *M. incognita* *R. reniformis*	–	Robertson et al. (2002); Fardos (2009)
T. longibrachiatum	*C. elegans*	Acetic acid	Djian et al. (1991)
T. piluliferum	*C. elegans*	–	Yang (2008)
T. pseudokoningii	*M. javanica* *M. incognita*	–	Sun (1997)
T. reesei	*P. redivivus, C. elegans*	Trichodermin	Yang et al. (2010); Watts et al. (1988)
Trichoderma sp.	*B. xylophilus*	–	Zhao (2004)
Trichoderma sp.	*B. xylophilus, P. redivivus, C. elegans*	6-pentyl-2H-pyran-2-one	Yang et al. (2012)
Trichoderma sp.	*P. redivivus, C. elegans*	Trichodermin	Yang et al. (2010)

Table 7.1 (continued)

Species	Test nematodes	Nematicidal compounds	References
T. viride	*M. javanica* *M. incognita*	–	Sun (1997)
Trichothecium roseum	*A. aceti*	Trichothecolone	Freeman et al. (1959); BačÍkovÁ et al. (1965); Konishi et al. (2003)
unidentified ascomycete of *Dermateaceae*	*C. elegans* *A. besseyi* *M. incognita*	5-pentyl-2-furaldehyde	Anke et al. (1995)
unidentified freshwater fungus YMF 1.01029	*B. xylophilus*	Ymf 1029 A, B, C, D and E, Preussomerin C and D, (4*RS*)- 4,8-dihydroxy-3,4-dihydronaphthalen-1(2*H*)-one, 4,6,8-trihydroxy-3,4-dihydronaphthalen-1(2*H*)-one	Dong et al. (2008)
unidentified ascomycete A111-95	*C. elegans* *M. incognita*	5-(2*E*)-2-buten-1-ylidene-3-(1*E*)-1-propen-1-yl-2(5*H*)-furanone, Pregaliellalactone, 5(R)-(1E)-1,3-butadien-1-yl-3-(1E)-1-propen-1-yl-2(5H)-furanone, 5(R)-(3-buten-1-yl) dihydro-3-vinyldihy-2(3H)-furanone	Köpcke et al. (2002)
Verticillium chlamydosporium	*M. incognita*	Phomalactone	Khambay et al. (2000)
V. leptobactrum	*M. incognita*	–	Regaieg et al. (2010)
Verticillium sp.	*B. xylophilus*	–	Zhao (2004)
Xylaria sp.	*C. elegans* *B. xylophilus*	–	Anke et al. (1995); Yuan et al. (2010)

No nematicidal compounds were reported from the taxon

Nematicidal Metabolites from Alternaria, Ascochyta, Aspergillus, Hemicarpenteles, Paecilomyces and Penicillium Species

Brefeldin A (**1**) is identical to two known chemicals ascotoxin and decumbin. It was first obtained from *Penicillium decumbens* (Singleton et al. 1958), and subsequently isolated from several other fungal species including *P. brefeldianis* (Kima and Kochevar 1995), *P. camemberti* (Abraham and Arfmann 1992), *Hemicarpenteles paradoxus* (Anke et al. 1995), *Alternaria carthami*, *A. zinniae* (Vurro et al. 1998), *Paecilomyces* sp., *Aspergillus clavatus* (Wang et al. 2002) and *Ascochyta imperfecta* (Suzuki et al. 1970). Screening against the nematode *A. aceti* with brefeldin A (**1**) resulted in significant nematicidal activity (BačÍkovÁ et al. 1965). A symmetric 16-membered macrodiolide helmidiol (**2**) was produced by *Alternaria alternata* (Kind et al. 1996). The compound had activity against *Haemunchus cortortus* (Kind et al. 1996) and *Meloidogyne incognita* (Khan and Kgan 1992).

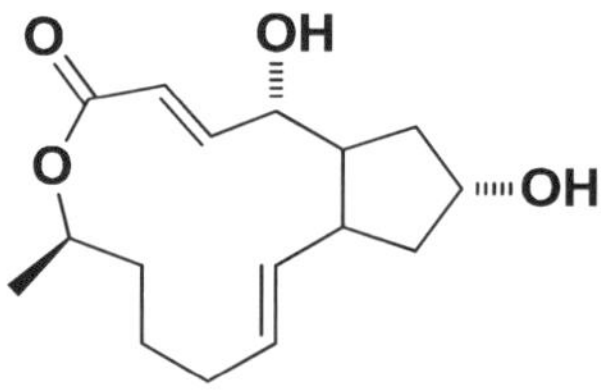

brefeldin A (1)

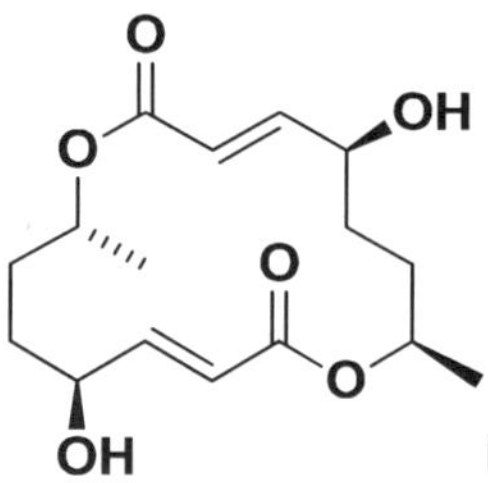

helmidiol (2)

Nematicidal Metabolites from Apiocrea Chrysosperma

Four linear lipophilic peptides chrysospermins A (**3**), B (**4**), C (**5**) and D (6) were isolated from the mycelium of *Apiocrea chrysosperma* Ap101 (Dornberger et al. 1995). These compounds have been patented as nematicidal and anthelminthic agents (Metzger et al. 1994). Each of these four peptides contains 19 amino acids and possesses a C-terminal Trpol and one labile Aib-Pro bond (Table 7.2, Bodo et al. 1985).

Nematicidal Metabolites from Nematode-Trapping Fungi Arthrobotrys, Chlorosplenium, Dactylella and Monacrosporium

An aliphatic compound linoleic acid (**7**) was detected in the mycelial extracts of the nematode-trapping fungi *Arthrobotrys conoides*, *A. brochopaga*, *A. dactyloides*, *A. oligospora*, *Dactylella candida* and *Monacrosporium doedycoides*. The LD_{50} of the compound towards *Caenorhabditis elegans* was 10 μg mL^{-1} and the LD_{30} to *Meloidogyne incognita* was 100 μg mL^{-1} (Stadler et al. 1994c; Anke et al. 1995). Besides linoleic acid (**7**), two isoepoxydon compounds oligosporon (**8**) and its dihydro-derivative 4',5'-dihydro-oligosporon (**9**) were isolated from *Arthrobotrys oligospora*. These two compounds were active against *Haemunchus cortortus* with LD_{50} values of 25 and 50–100 μg mL^{-1}. However, they were inactive against the nematode

Table 7.2 The structures of chrysospermins A, B, C and D

	1	2	3	4	5	6	7	8	9	10	11	12	13	14	15	16	17	18	19
A	AcPhe-	Aib-	Ser-	Aib-	Aib-	Leu-	Gln-	Gly-	Aib-	Aib-	Ala-	Ala-	Aib-	Pro-	Aib-	Aib-	Aib-	Gln-	Trpol
B	AcPhe-	Aib-	Ser-	Aib-	Aib-	Leu-	Gln-	Gly-	Aib-	Aib-	Ala-	Ala-	Aib-	Pro-	Iva-	Aib-	Aib-	Gln-	Trpol
C	AcPhe-	Aib-	Ser-	Aib-	Iva-	Leu-	Gln-	Gly-	Aib-	Aib-	Ala-	Ala-	Aib-	Pro-	Aib-	Aib-	Aib-	Gln-	Trpol
D	AcPhe-	Aib-	Ser-	Aib-	Iva-	Leu-	Gln-	Gly-	Aib-	Aib-	Ala-	Ala-	Aib-	Pro-	Iva-	Aib-	Aib-	Gln-	Trpol

Caenorhabditis elegans at concentrations up to 100 μg mL^{-1} (Anderson et al. 1995; Stadler et al. 1993). Recently, three novel oligosporons, named arthrobotrisins A-C (**10–12**) were isolated from *A. oligospora*, but only arthrobotrisin A (**10**) had nematicidal activity against *Panagrellus redivivus* (Wei et al. 2011). Linoleic acid (**7**) was also found in *Chlorosplenium* sp. (Anke et al. 1995), and the basidiomycete *Hericium coralloides* (Xiang and Feng 2001) and *Pleurotus pulmonarius* (Stadler et al. 1994c).

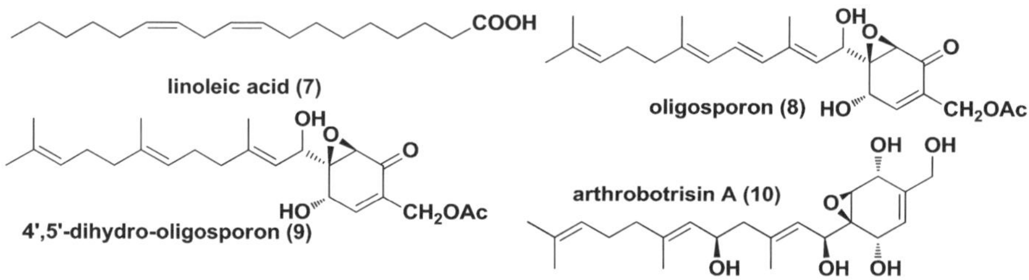

Nematicidal Metabolites from Aspergillus Fumigatus, Penicillium Nigricans, Gliocladium Fimbriatum and Penicillium Sp.

A terpenoid, fumagillin (**13**), isolated from both *Aspergillus fumigatus* and *Penicillium nigricans*, was reported to be moderately active against nematode *Anguillula aceti* (Tarbell et al. 1960; BačÍkovÁ et al. 1965; Beecham et al. 1966). Gliotoxin (**14**), a known antibiotic, was also weakly active against *A. aceti* (BačÍkovÁ et al. 1965). Gliotoxin (**14**) had been isolated from *Gliocladium fimbriatum*, *Aspergillus fumigatus*, *Penicillium* sp. and other fungi (Tarbell et al. 1960; Beecham et al. 1966). Besides the two compounds, five compounds including two new active fumiquinones A (**15**), B (**16**) and three known spinulosin (**17**), LL-S490 (**18**) and pseurotin A (**19**) were also isolated from *A. fumigatus*. Fumiquinone A (**15**) showed effective nematicidal activities against *Pratylenchus penetrans* and *Bursaphelenchus xylophilus*, but fumiquinone B (**16**) and the three known compounds only showed activity against *B. xylophilus*. All of these five compounds had no nematicidal activities against *Caenorhabditis elegans* (Hayashi et al. 2007).

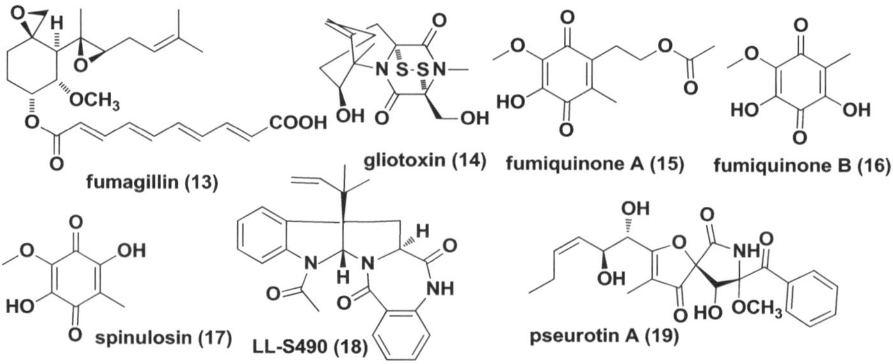

Nematicidal Metabolites from Aspergillus Glaucus, A. Melleus and A. Niger

A widely distributed anthraquinone in plants, emodin (**20**) was also obtained from *Aspergillus glaucus* (Anke et al. 1980a, b). Emodin (**20**) has shown activity against *Meloidogyne incognita* (Mayer 1995). Aspyrone (**21**) was isolated from *Aspergillus melleus* and showed a nematicidal activity against *Pratylenchus penetrans* with killing rates of 39 % and 80.8 % at concentrations of 100 mg L^{-1} and 300 mg L^{-1}, respectively (Kimura et al. 1996). Nafuredin (**22**) was isolated as an inhibitor of an anaerobic electron transporter from the culture broth of *Aspergillus niger* FT-0554 (Ui et al. 2001; Ōmura et al. 2001). *In vivo* trials with sheep indicated that nafuredin (**22**)had significant nematicidal activity against *Haemunchus cortortus*. Nafuredin (**22**) could be easily converted to nafuredin-γ (**23**) by weak alkaline treatment. The latter also showed an inhibitory activity similar to nafuredin (**22**) (Shiomi et al. 2005). The IC_{50} values of nafuredin (**22**) and nafuredin-γ (**23**) were 9.7 nM and 6.4 nM respectively in their inhibition against NADH-fumarate reductase (NFRD) of *Ascaris suum* (Shiomi et al. 2005). Two new nematicidal ophiobolins, ophiobolin K (**24**) and 6-epiophiobolin K (**25**) were obtained from *Aspergillus ustus* (Singh et al. 1991). The two compounds were also isolated from *Cochliobolus heterostrophus* (Rosegay et al. 1996).

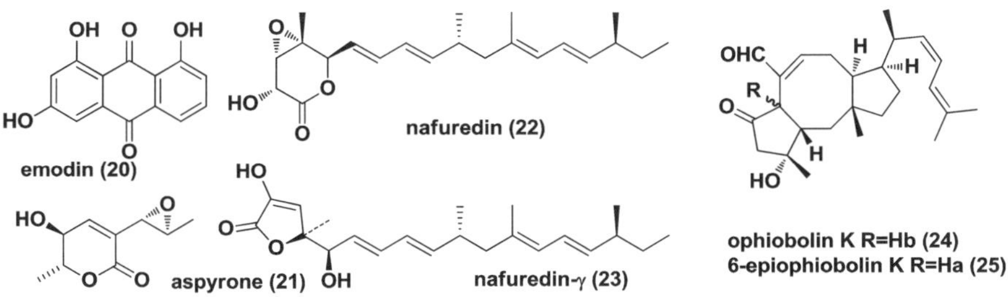

Nematicidal Metabolites from Aspergillus Spp.

Four macrolides including a new compound*ββ,γ*-dehydrocurvularin (**26**) and three known ones *αβ*-dehydrocurvularin (**27**), 8-*β*-hydroxy-7-oxocurvularin (**28**) and 7-oxocurvularin (**29**) were obtained from *Aspergillus* sp. These four macrolides have shown nematicidal activities against the root-lesion nematode *Pratylenchus penetrans* (Kusano et al. 2003). However, none of the four compounds had any observable effects on *Caenorhabditis elegans* at the tested concentrations (1–1000 mg L^{-1}). The three known compounds are produced by many species in the genera *Alternaria, Cochliobolus, Curvularia* and *Penicillium* (Munro et al. 1967; Hyeon et al. 1976; Robeson and Strobel 1981, 1985; Kobayashi et al. 1988; Arai et al. 1989; Lai et al. 1989, 1990; Ghisalberti and Rowland 1997). Two members of a new class of anthelmintics, aspergillimide (VM55598) (**30**) and 16-keto aspergillimide (SB202327) (**31**), were isolated from *Aspergillus* sp. IMI 337664 (Banks et al. 1997). In addition, three new paraherquamides SB203105 (**32**), SB200437 (**33**) and VM54159 (**34**) were also isolated from this strain. This study was the first to report paraherquamides from an organism outside the fungal genus *Penicillium*. These compounds had activity against *Trichjostrongylus columbriformis*. Tests showed that the 16-keto analogue of aspergillimide (**31**) was active against *Haemunchus contortus* L_3 larvae *in vitro* but not *in vivo* (Banks et al. 1997). A new nematicide 5-hydroxymethyl-2-furoic acid (**35**) was obtained from the cultures of an *Aspergillus* sp. The compound showed effective nematicidal activities against *Bursaphelenchus xylophilus* and *Caenorhabditis elegans* (Kimura et al. 2007). A pyran compound patulin (**36**) was proven to be active against *Meloidogyne incognita* with the LD_{30} dose at 100 µg mL^{-1} and a oxygen heterocycle penicillic acid (**37**) was found to possess weak activities against *Anguillula aceti* (BačÍkovÁ et al. 1965; Mayer 1995). Patulin (**36**) was found in several fungi including *Aspergillus* spp. (Lopez-Diaz and Flannigan 1997), *Penicillium* spp. (Adams et al. 1976; Alfaro et al. 2003; Dombrink-Kurtzman and Blackburn 2005), and *Byssochlamys* spp. (Moulé and Hatey 1977). Penicillic acid (**37**) has been isolated from several fungal species and strains belonging to *Aspergillus* (He et al. 2004; Kang and Kim 2004), *Penicillium* (Wirth et al. 1956; Reimerdes et al. 1975), and *Malbranchea aurantiaca* (Martínez-Luis et al. 2005).

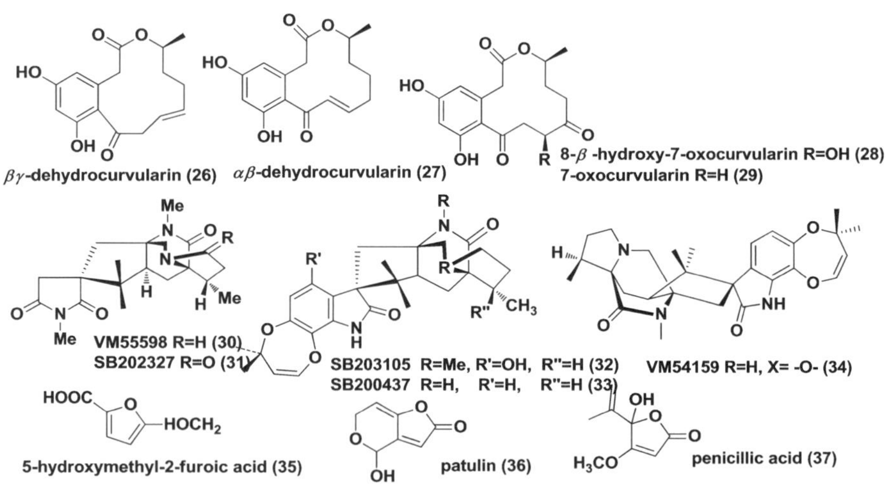

Nematicidal Metabolites from Beauveria Bassiana, Bulgaria Inquinans and Paecilomyces Fumoso-Roseus

The cyclic depsipeptide beauvericin (**38**) was weakly active against *Meloidogyne incognita* (Mayer 1995), and then was proved to be active against *Caenorhabdi-*

tis elegans and *Bursaphelenchus xylophilus* (Shimada et al. 2010). This peptide was isolated from *Beauveria bassiana* (Hamill et al. 1969), *Paecilomyces fumosoroseus* (Bernardini et al. 1975), *Fusarium* spp. (Bernardini et al. 1975), *Beauveria* sp. FKI-1366 (Fukuda et al. 2004), *Fusarium bulbicola* (Shimada et al. 2010) and *F. redolens* (Xu et al. 2010), and the basidiomycete *Polyporus sulphureus* (Deol et al. 1978). Two new azaphilones, bulgarialactone A (**39**) and B (**40**) were isolated from both the mycelia and fruit bodies of the *Bulgaria inquinans*. The LD_{50} value of bulgarialactone A (**39**) and B (**40**) against the nematode *Caenorhabditis elegans* was 5 μg mL^{-1} and 10–25 μg mL^{-1} respectively (Stadler et al. 1995).

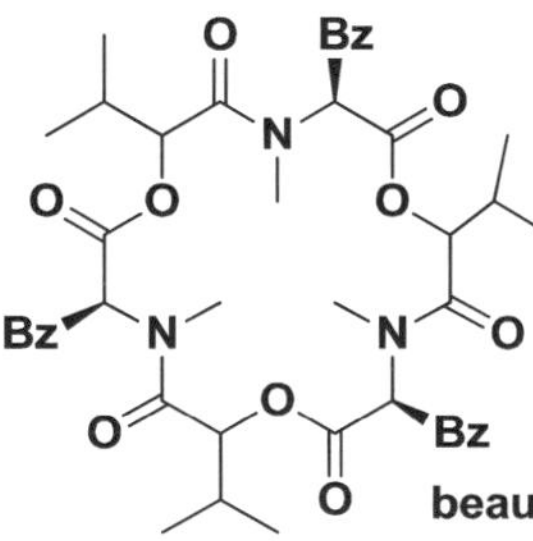

beauvericin (38)

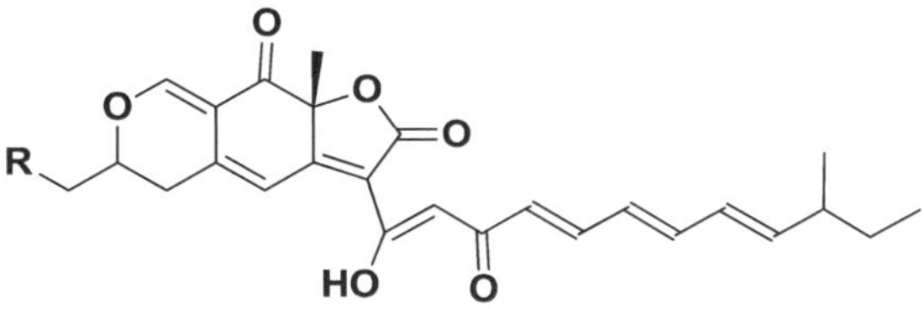

bulgarialactone A=H (39)
bulgarialactone B=OH (40)

Nematicidal Metabolites from Caryospora Callicarpa

Three new macrolide caryospomycins A (**41**), B (**42**) and C (**43**), and four known compounds (**44**–**47**) were isolated from the mycelium and fermentation broth of freshwater taxon *C. callicarpa* YMF1.01026 (Dong et al. 2007; Zhu et al. 2008). These compounds were active against *Bursaphelenchus xylophilus* (Dong et al. 2007; Zhu et al. 2008).

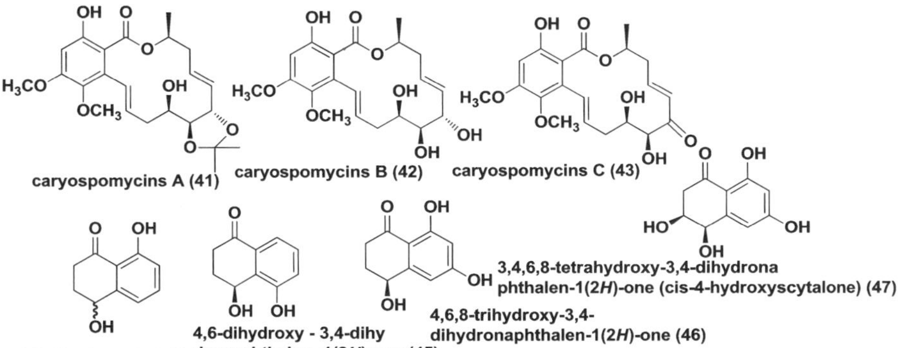

caryospomycins A (41)
caryospomycins B (42)
caryospomycins C (43)
3,4,6,8-tetrahydroxy-3,4-dihydronaphthalen-1(2*H*)-one (cis-4-hydroxyscytalone) (47)
4,6,8-trihydroxy-3,4-dihydronaphthalen-1(2*H*)-one (46)
4,6-dihydroxy - 3,4-dihydronaphthalen- 1(2*H*)-one (45)
4,8-dihydroxy-3,4-dihydronaphthalen-1(2*H*)-one (44)

Nematicidal Metabolites from Chaetomium Globosum, Cladobotryum Rubrobrunnescens, Clonostachys Cylindrospora, Coelomycetes Sp., Epicoccum Nigrum and E. Purpurascens

A simple aromatic flavipin (**48**), produced by *Chaetomium globosum*, could inhibit *in vitro* egg hatch and juvenile mobility of *Meloidogyne incognita*, and could also inhibit the hatch of *Heterodera glycines* (Nitao et al. 2002). Flavipin (**48**) was also found in *Epicoccum nigrum* (Burge et al. 1976) and *E. purpurascens* (Brown et al. 1987). Cladobotrin I (**49**) exhibited nematicidal activity towards *Meloidogyne incognita* with an $LD_{50\ at}$ 100 μg mL^{-1}, and it was isolated from *Cladobotryum rubrobrunnescens* (Wagner et al. 1998). A 14-membered macrodiolide clonostachydiol (**50**) was isolated from the *Clonostachys cylindrospora* (Grabley et al. 1993). Its synthesis *in vitro* has been achieved (Rao et al. 1995). A dose of 2.5 mg kg^{-1} subcutaneously administered to sheep artificially infected with the nematode *Haemunchus cortortus* caused 80 to 90 % reduction of *nematode* (Grabley et al. 1993). Five nematicidal compounds preussomerin C (**51**), preussomerin D (**52**), preussomerin E (**53**), (4RS)4,8-dihydroxy-3,4-dihydronaphthalen-1(2H)-one (**54**) and 4,6,8-trihydroxy-3,4-dihydronaphthalen-1(2H)-one (**55**) were isolated from an aquatic fungus *Coelomycetes* sp. YMF 1.01029 (Zhou et al. 2009).

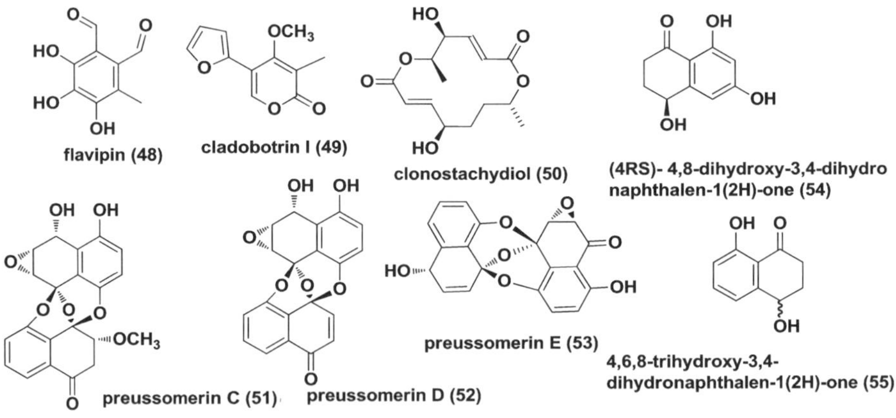

Nematicidal Metabolites from Cochliobolus Heterostrophus, C. Miyabeanus, Helminthosporium Leersii and H. Sativum

Four ophiobolane-type sesterterpenes including ophiobolin M (**56**), 6-epiophiobolin M (**57**), ophiobolin C (**58**) and 6-epiophiobolin C (**59**) have been isolated from *Cochliobolus heterostrophus* (Tsipouras et al. 1996). Ophiobolin C (**58**) was first obtained from *Helminthosporium* species (Cutler et al. 1984) and it was the most active compound among these compounds with an LD_{50} value of 5 μM against *Caenorhabditis* *elegans*. These compounds are non-competitive inhibitors of ivermectin binding to membranes prepared from *C. elegans* (Tsipouras et al. 1996). A quinone cochlioquinone A (**60**) has been isolated from *Cochliobolus miyabeanus*, *Helminthosporium leersii* and *H. sativum* (Barrow and Murphy 1972; Schaeffer et al. 1990). The ED_{50} of cochlioquinone A (**60**) against *Caenorhabditis elegans* was 135 μM (Snook et al. 1998). Cochlioquinone A (**60**) may have a similar mode of action as that of the widely used avermectin because the compound is a competitive inhibitor of [^{3}H] ivermectin and both can bind to the cell membrane of *Caenorhabditis elegans*.

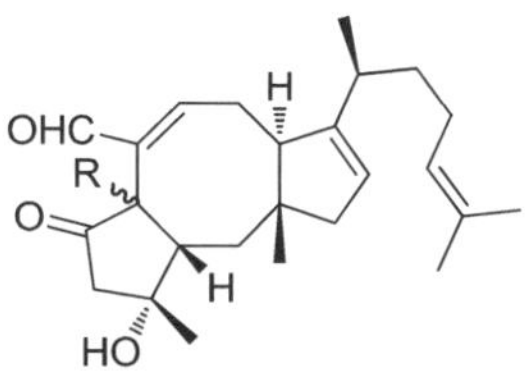

ophiobolin M R=H$_b$ (56)
6-epiophiobolin M R=H$_a$ (57)

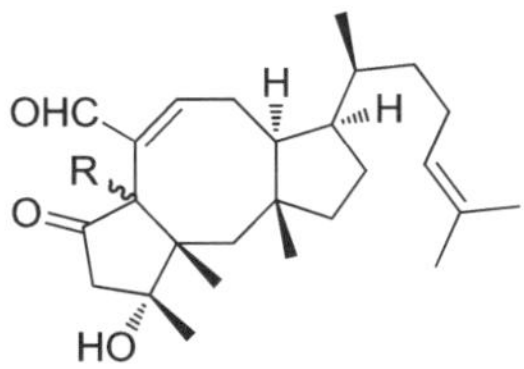

ophiobolin C R=H$_b$ (58)
6-epiophiobolin C R=H$_a$ (59)

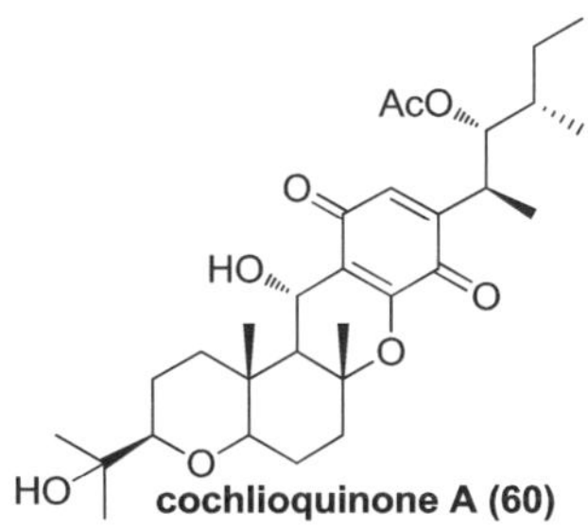

cochlioquinone A (60)

Nematicidal Metabolites from Coronophora Gregaria and Cylindrocarpon Olidum

A highly methylated polyketide MK7924 (**61**) was isolated from the culture broth of *Coronophora gregaria* L2495 and the compound exhibited significant nematicidal activity against *Caenorhabditis elegans* at 100 μg mL^{-1} (Kumazawa et al. 2003). There were significant structural differences between MK7924 (**61**) and other known anthelmintic agents. Therefore, MK7924 (**61**) could be developed as a promising new type of anthelmintic. Two nematicidal compounds cannabiorcichromenic acid (**62**) and its 8-chloro derivative (**63**) were isolated from *Cylindrocarpon olidum* (Quaghebeur et al. 1994). The mixture of these two compounds could kill 50% of *Heterorhabditis* nematodes at 20 μg mL^{-1} (Quaghebeur et al. 1994).

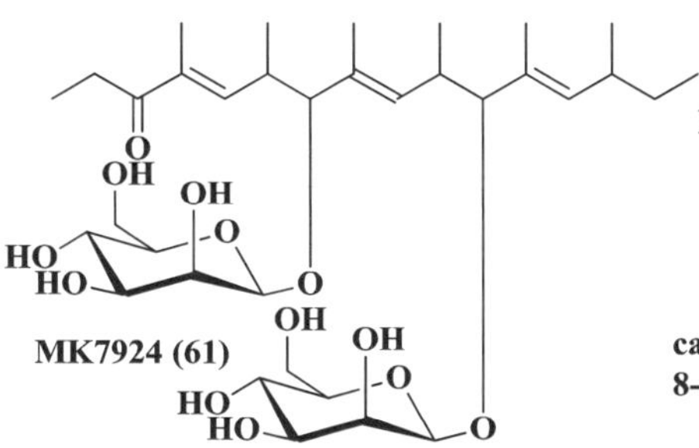

MK7924 (61)

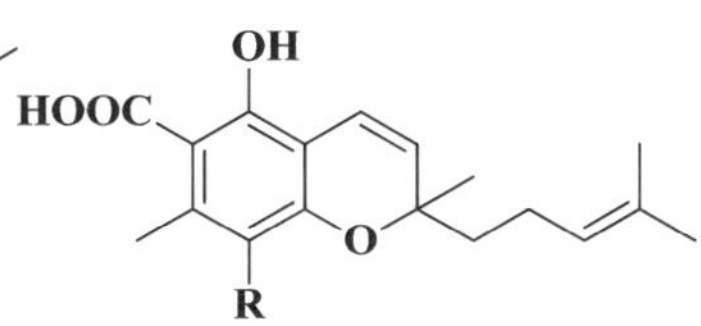

cannabiorcichromenic acid R=H (62)
8-chloro-cannabiorcichromenic acid R=Cl (63)

Nematicidal Metabolites from Daldinia Concentrica, Emericellopsis Poonensis and E. Synnematicola

Two naphthalenes 1-methoxy-8-hydroxynaphthalene (**64**) and 1,8-dimethoxynaphthalene (**65**) were isolated from *Daldinia concentrica* (Dasenbrock 1994) and both were active against *C. elegans* with LD_{50} values at 10 μg mL^{-1} and 25 μg mL^{-1} respectively. However, these two compounds were only weakly active against *Meloidogyne incognita* (Anke et al. 1995). The N-terminally acetylated lipophilic linear polypeptide antiamoebin I (**66**) had been obtained from fungal species *Emericellopsis poonensis* and *E. synnematicola*, which showed activity against helminthes (Thirumalachur 1968; Pandey et al. 1977). The structure of antiamoebin I (**66**) was determined by several spectral methods including X-ray crystallography (Brückner et al. 1980; Snook et al. 1998).

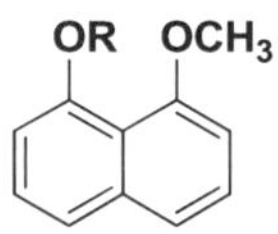

1-methoxy-8-hydroxynaphthalene R=H (64)
1,8-dimethoxynaphthalene R=CH_3 (65)

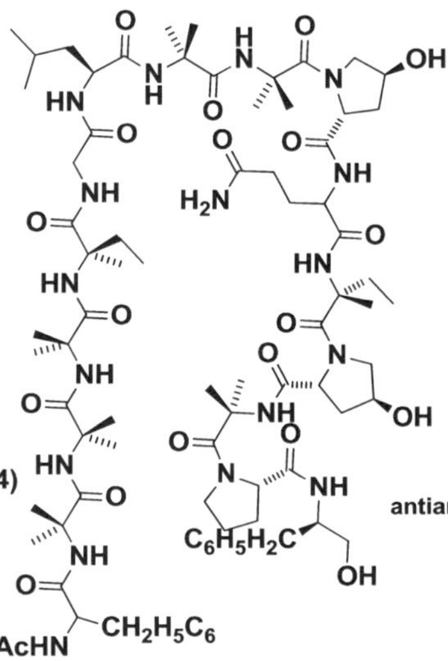

antiamoebin I (66

Nematicidal Metabolites from Fusarium Roseum, Trichothecium Roseum, Fusarium Sp. and Geotrichum Sp.

Trichothecolone (**67**) was obtained from *Fusarium roseum* and *Trichothecium roseum*, which had weakly activity against the nematode *Anguillula aceti* (Freeman et al. 1959; BačÍkovÁ et al. 1965; Konishi et al. 2003). Two cyclodepsipeptides, enniatin A (**68**) and enniatin B (**69**), were isolated from the culture broth of *Fusarium* spp. (Tomoda et al. 1992), and both were weakly active against *Meloidogyne incognita* (Mayer 1995). An endophytic fungus *Geotrichum* sp. AL4 was isolated from the leaf of *Azadirachta indica*. Two new metabolites, 1-((2*R*,4*S*,5*S*)-2-chloro-4-methyl-1,3-oxazinan-5-yl) ethanone (**70**) and 1-((2*R*,4*S*,5*R*)-2-chloro-4-methyl-1,3-oxazinan-5-yl)ethanone (**71**) as well as one known compound 2',4'-dihydroxyacetophenone (**72**) were isolated from this strain. The three compounds exhibited nematicidal activity against nematodes *Bursaphelenchus xylophilus* and *Panagrellus redivivus* (Li et al. 2007).

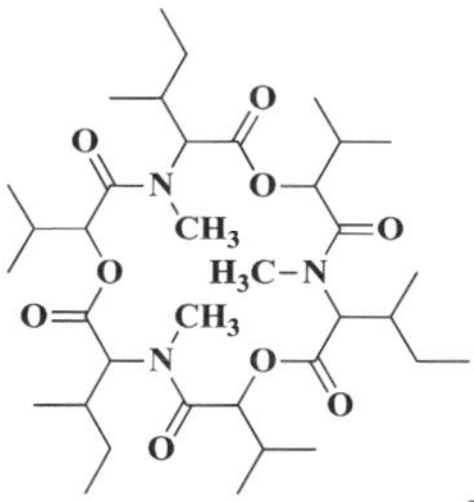
enniatin A (68)

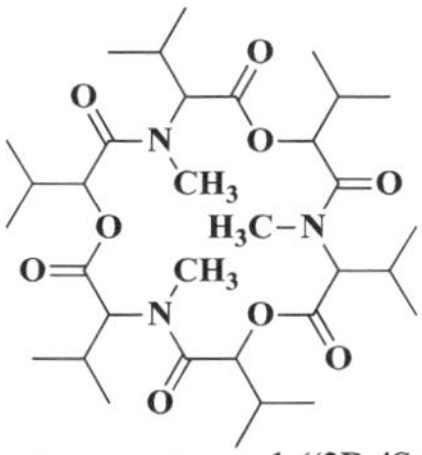
enniatin B (69)

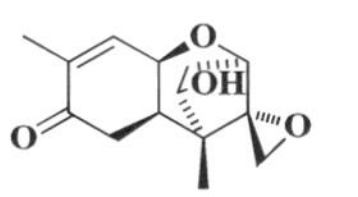
trichothecolone (67)

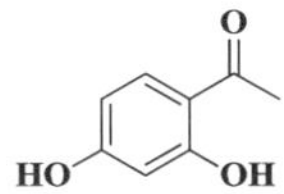
2',4'-dihydroxyacetophenone (72)

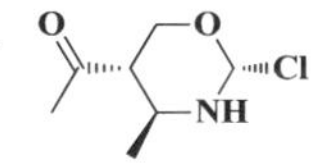
1-((2R,4S,5S)-2-chloro-4-methyl-1,3-oxazinan-5-yl)ethanone (70)

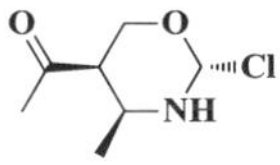
1-((2R,4S,5R)-2-chloro-4-methyl-1,3-oxazinan-5-yl)ethanone (71)

Nematicidal Metabolites from Gliocladium Roseum, G. Virens and Gymnoascus Reesii

A series of diketopiperazine compounds were isolated from the solid-substrate fermentation cultures of *Gliocladium roseum* (Dong et al. 2005, 2006). These compounds include gliocladine A (**73**), B (**74**), C (**75**), D (**76**), E (**77**), verticillin A (**78**), 11'-deoxyverticillin A (**79**), Sch52900 (**80**), Sch52901 (**81**) and glioclasine (**82**). The compounds showed nematicidal activities against *Caenorhabditis elegans* and *Panagrellus redivivus*. However, they showed little activity against *Bursaphelenchus xylophilus*. Compared to the other compounds in this group, glioclasine (**82**) showed the strongest activities against *Bursaphelenchus xylophilus*, *Caenorhabditis elegans* and *Panagrellus redivivus* with LD_{50} values at 15, 50 and 200 μg mL^{-1}, respectively (Dong et al. 2005). A sterol, viridian (**83**) was obtained from *Gliocladium virens* and some strains of *Trichoderma* (Blight et al. 1968), and it has been found to possess weak activity against *Anguillula aceti* (BačÍkovÁ et al. 1965). A nematicidal metabolite (3E,5E)-2,5-dihydroxy-2,7-dihydrooxepine-3-carboxylic anhydride (**84**) was isolated based on bioassay-guided fractionation from the extracts of the fungus *Gymnoascus reesii*, which showed activity against *M. incoginta* (Liu et al. 2011).

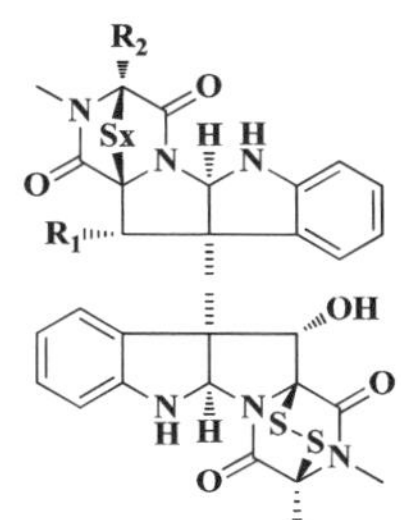
gliocladine A R1=OH, R2=CH3, x=3 (73)
gliocladine B R1=OH, R2=CH3, x=4 (74)

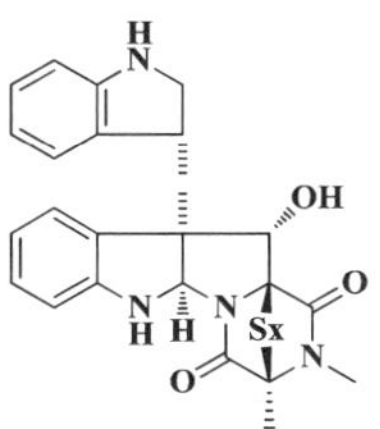
gliocladine C x=2 (75)
gliocladine D x=3 (76)
gliocladine E x=4 (77)

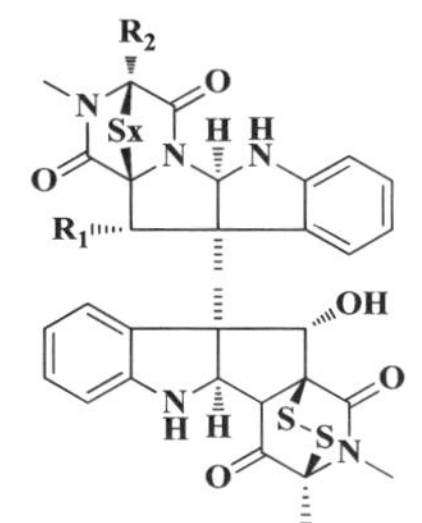
verticillin A R1=OH, R2=CH3, x=2 (78)
11'-deoxyverticillin A R1=H, R2=CH3, x=2 (79)
Sch52900 R1=OH, R2=CH2CH3, =2 (80)
Sch52901 R1=OH, R2=CH(OH)CH3, x=2 (81)

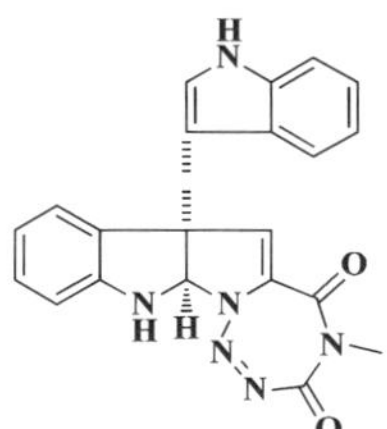
glioclasine (82)

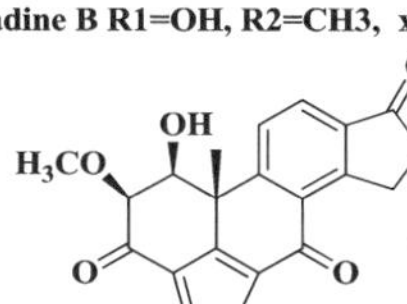
viridin (83)

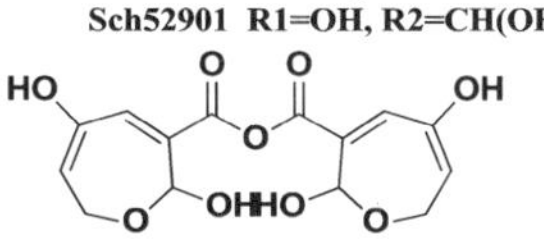
(3E,5E)-2,5-dihydroxy-2,7-dihydrooxepine-3-carboxylic anhydride (84)

Nematicidal Metabolites from Hirsutella Thompsonii Var. Synnematosa, Nigrospora Sphaerica, Verticillium Chlamydosporium and Hypomyces Sp.

A nematicidal compound phomalactone (**85**) was obtained by bioassay-directed fractionation from *Verticillium chlamydosporium* (Khambay et al. 2000). The mortality of *Meloidogyne incognita* reached 84% in 96 h when the concentration of phomalactone (**85**) was 500 mg L^{-1} (Khambay et al. 2000). This compound has been found in other fungi, e.g the entomopathogenic fungi *Hirsutella thompsonii* var. *synnematosa* (Krasnoff and Gupta 1994) and *Nigrospora sphaerica* (Kim et al. 2001). Two photosensitive compounds hypocrellin A (**86**) and elsinochrome A (**87**) were isolated from *Hypomyces* sp. (Dong et al. 2001). These two compounds were able to kill 50% of the nematode *Bursaphelenchus xylophilus* within 18 h at concentrations of 50 μg mL^{-1} for hypocrellin A (**86**), and 15 μg mL^{-1} for elsinochrome A (**87**) (Dong et al. 2001).

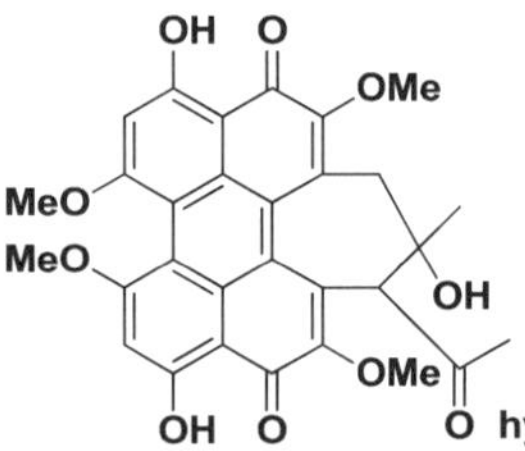

hypocrellin A (86)

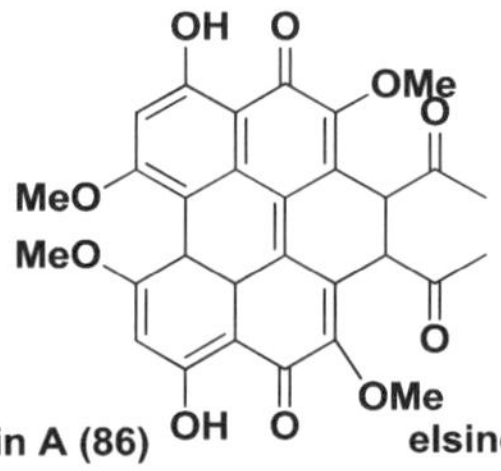

elsinochrome A (87)

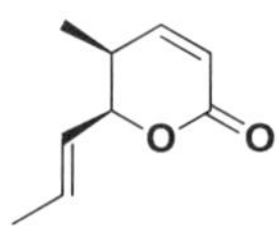

phomalactone (85)

Nematicidal Metabolites from Anamophic Fungi Strains D1084 and PF1022

Two novel depsipeptides bursaphelocides A (**88**) and B (**89**) were isolated from an unidentified anamorph strain, D1084. These two compounds were active against *B. xylophilus* at a dose of 100 μg per ball using the "cotton ball on the fungal mat method" (Kawazu et al. 1993). Another unidentified anamorph strain PF1022 produced a novel cyclodepsipeptide PF1022A (**90**) showed potent anthelmintic activity against *Ascaridia galli* (Sasaki et al. 1992). Importantly, no toxic effect was observed to the host animals. The efficacy of compound PF1022A (**90**) against anthelmintic-resistant nematodes in sheep and cattle was investigated and the result confirmed that PF1022A (**90**) was fully effective against these parasite nematode populations (Samson-Himmelstjerna et al. 2005).

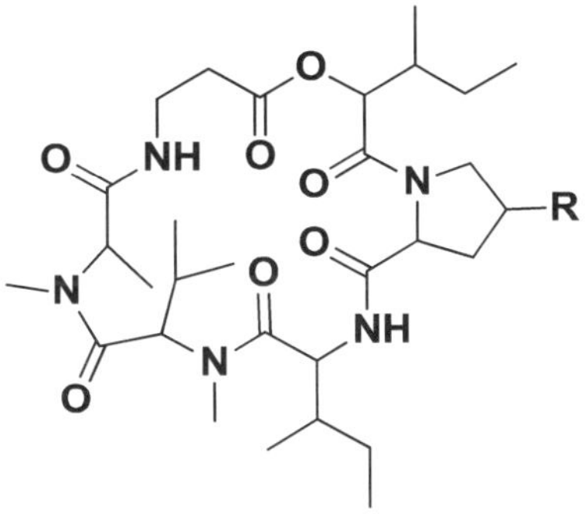

bursaphelocides A R=H (88)
bursaphelocides B R=CH_3 (89)

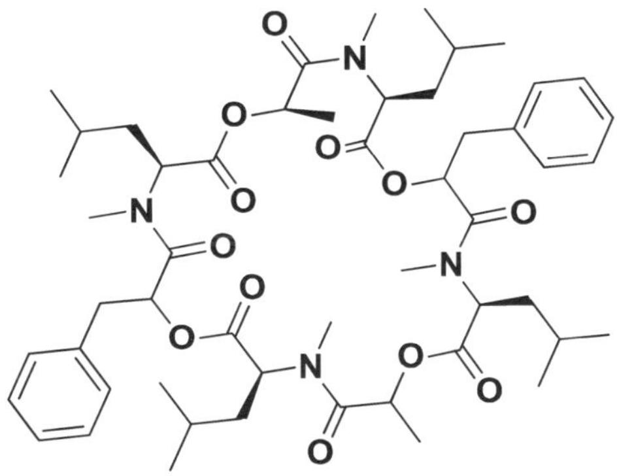

PF1022A (90)

Nematicidal Metabolites from Lachnum Papyraceum

Two new compounds lachnumfuran A (**91**) and lachnumlactone A (**92**) were obtained from *Lachnum papyraceum* (Shan et al. 1996). The two compounds had relatively weak activities against *Caenorhabditis elegans* with ND_{90} dosages at 100 and 50 μg mL^{-1} respectively (Shan et al. 1996). During the investigations of the influences of $CaBr_2$ on the biosynthesis of chlorinated secondary metabolites in *Lachnum papyraceum*, six isocoumarin derivatives, 6-hydroxymellein (**93**), 4-chloro-6-hydroxymellein (**94**), 4-bromo-6-hydroxymellein (**95**), 6-methoxymellein (**96**), 4-chloro-6-methoxymellein (**97**) and 4-chloro-6,7-dihydroxymellein (**98**) were obtained. Among them, compounds 4-chloro-6-hydroxymellein (**94**), 4-bromo-6-hydroxymellein (**95**), 6-methoxymellein (**96**) and 4-chloro-6,7-dihydroxymellein (**98**) were isolated for the first time from a natural source (Stadler et al. 1995a, b). These isocoumarin derivatives showed only weak nematicidal effects and the ND_{90} values of these compounds against *Caenorhabditis elegans* were all within the region of 100 μg mL^{-1} (Stadler et al. 1995a).

In additional, a series of mycorrhizins, mycorrhizin A (**99**), chloromycorrhizin A (**100**) and (1'-E)-dechloromycorrhizin A (**101**) were commonly found in normal fermentations of the fungus (Stadler and Anke 1993a). However, in fermentations in media containing a large amount of $CaBr_2$, additional mycorrhizins could be found. These included two brominated derivatives mycorrhizin B1 (**102**) and mycorrhizin B2 (**103**) as well as (1'Z)-dechloromycorrhizin A (**104**) (Stadler et al. 1995c, d, e; Shan et al. 1996). These mycorrhizins were all toxic towards *Caenorhabditis elegans* but were only weakly active against *Meloidogyne incognita*. Among these mycorrhizins, mycorrhizin A (**99**) showed the highest activity against *Caenorhabditis elegans* with an $LD_{50\ at}$ 1 μg mL^{-1}. Based on structural and functional comparisons, it was suggested that chlorine substitutions in the side chains could increase their biological activities, whereas chlorine substitutions within the ring systems seem to weaken their activities (Stadler et al. 1995a). The brominated mycorrhizins showed weaker activities than their chlorinated analogues. However, these differences were not statistically significant (Stadler et al. 1995c). Several minor metabolites, papyracons A (**105**), B (**106**), C (**107**) and D (**108**), 6-*O*-methylpapyraceum B (**109**) and 6-*O*-methylpapyraceum C (**110**) have also been isolated from *Lachnum papyraceum* and they are all weakly active against *Caenorhabditis elegans* (Stadler et al. 1995c, d, e; Shan et al. 1996). In addition, four isoepoxydon compounds were isolated from *Lachnum papyraceum* (Stadler and Anke 1993a, b; Stadler et al. 1995c, e). These compounds were lachnumon (**111**), lachnumon A (**112**), lachnumon B1 (**113**) and lachnumon B2 (**114**). Lachnumon (**111**) and lachnumon A (**112**) had similar activities against *Caenorhabditis elegans* with an $LD_{50\ at}$ 25 μg mL^{-1}. Their activities against *Meloidogyne incognita* were weak, with an LD_{50} exceeding 100 μg mL^{-1} for both (Stadler and Anke 1993a, b). The LD_{90}values of lachnumon B1 (**113**) and lachnumon B2 (**114**) against *Caenorhabditis elegans* were 25 μg mL^{-1} and 50 μg mL^{-1}, respectively. Their activities against *Meloidogyne incognita* were similar to those of compounds lachnumon (**111**) and lachnumon A (**112**) (Stadler et al. 1995c, e). 6-hydroxymellein (**93**) has also obtained from other taxa including *Discula* spp. (Venkatasubbaiah and Chilton 1991) and *Myxotrichum stipitatum* (Kimura et al. 2002). In addition, mycorrhizin A (**99**) and chloromycorrhizin A (**100**) have been isolated from a mycorrhizal fungus *Monotropa hypopitys* (Trofast and Wickberg 1977; Trofast 1978).

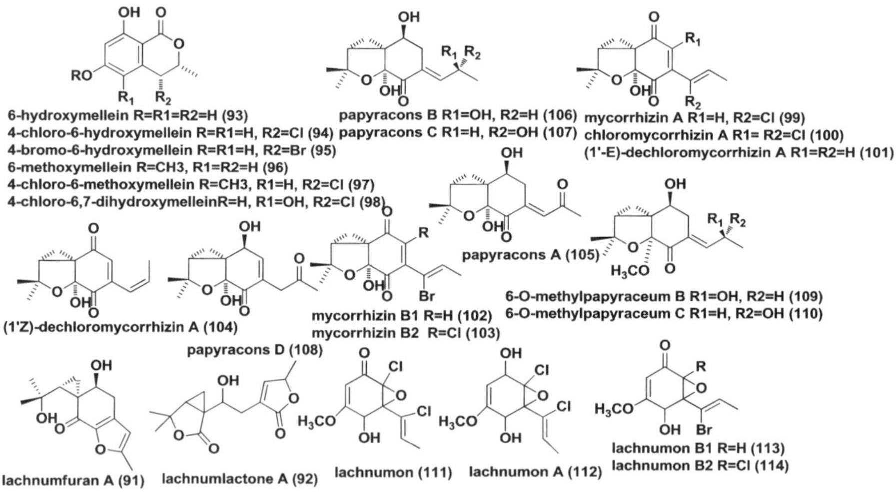

Nematicidal Metabolites from Melanconium Betulimum, Mycosphaerella Lethalis and Phomopsis Phaseoli

3-hydroxypropionic acid (3-HPA) (**115**) was isolated as the main nematicide from the submerged culture of fungus *Phomopsis phaseoli* originally found on a tropical tree. This compound has also been found from *Melanconium betuli-*

mum associated with *Betula pendula* and *B. pubescens* (Schwarz et al. 2004). The compound showed selective nematicidal activity against *Meloidogyne incognita* with an LD_{50} value of 12.5–15 µg mL^{-1}, and against *Caenorhabditis elegans* with an LD_{50} value about five times lower (Schwarz et al. 2004). A 9-lactide decane compound lethaloxin (**116**) isolated from *Mycosphaerella lethalis* (Arnone et al. 1993), which was proven capable of killing *C. elegans* with an $LD_{50\ at}$ 25 µg mL^{-1} (Stadler 1993).

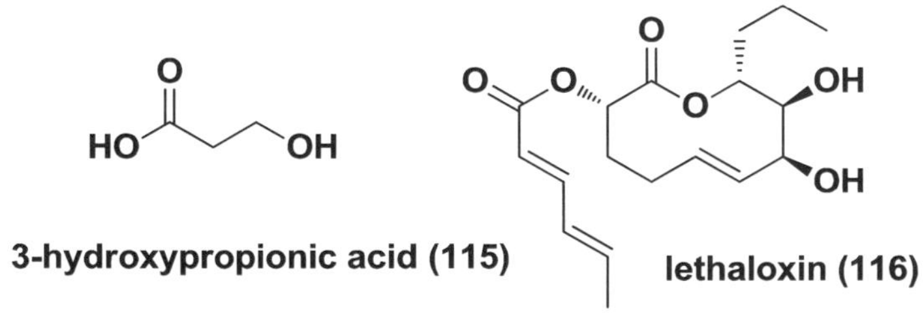

Nematicidal Metabolites from Nectria Radicola and Neobulgaria Pura

A macrolide radicicol (**117**) produced by *Nectria radicicola* was a highly cytotoxic antiprotozoal and antineoplastic agent (Mirrington et al. 1964). This compound has also been isolated from other fungal species such as *Monosporium bonorden*, *Penicillium luteo-aurantium* (Nozawa and Nakajima 1979) and *Chaetomium chiversii* (Kithsiri Wijeratne et al. 2006). Its two dialkoxy derivatives radicicol B (**118**) and radicicol C (**119**) also possessed nematicidal activities against an unidentified soil nematode with an LD_{50} value at 200 µg mL^{-1} (Stadler 1993). Compound 14-epi-cochlioquinone B (**120**) was isolated as a platelet aggregation inhibitor from the ascomycete *Neobulgaria pura* (Lorenzen et al. 1994). This compound had a strong nematicidal activity against *C. elegans* with LD_{50} value at 10 µg mL^{-1} (Anke et al. 1995). However, it was approximately 10 times less active against *Meloidogyne incognita* (Anke et al. 1995).

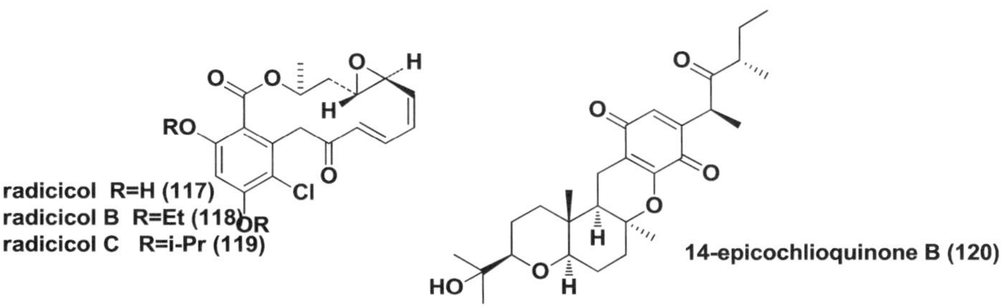

Nematicidal Metabolites from Oidiodendron Sp. and Ophioceras Dolichostomum

Two compounds including 4-hydroxyphenylacetic acid (4-HPA) (**121**) and oidiolactone D (**122**), were isolated from cultures of *Oidiodendron* sp.. The two compounds showed nematicidal activities against *Pratylenchus penetrans* and *Bursaphelenchus xylophilus* (Ohtani et al. 2011). Isoamericanoic acid A (**123**) and caffeic acid (**124**) were isolated from the freshwater taxon *Ophioceras dolichostomum* YMF1.00988. The LD_{50} values of the two compounds against *Bursaphelenchus xylophilus* were 133.7 and 46.8 μg mL^{-1} respectively (Dong et al. 2010).

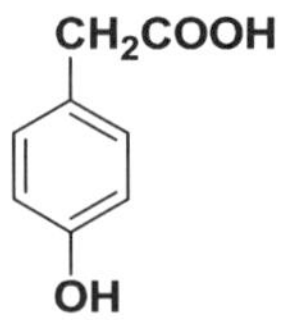

4-hydroxyphenylacetic acid (121)

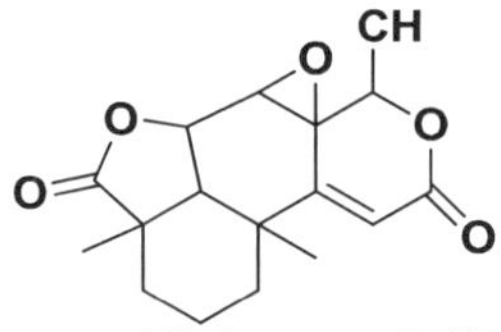

oidiolactone D (122)

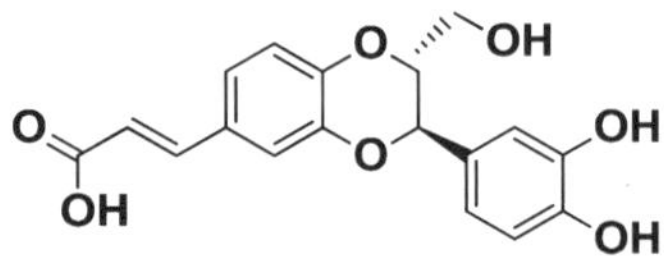

isoamericanoic acid A (123)

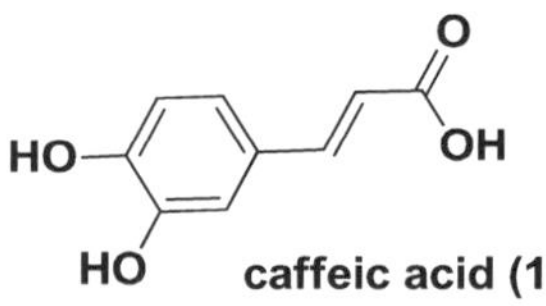

caffeic acid (124)

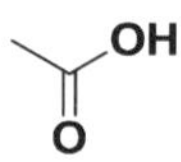

acetic acid (125)

Nematicidal Metabolites from Paecilomyces Lilacinus and Trichoderma Longibrachiatum

The common acetic acid (**125**) has been isolated from culture filtrates of *Paecilomyces lilacinus* and *Trichoderma longibrachiatum* (Djian et al. 1991; Park et al. 2004). Acetic acid (**125**) has been shown to have selective nematicidal activities against certain nematodes (Djian et al. 1991). *P. lilacinus* showed effective treatment on against root-knot nematode on tomato plants under greenhouse conditions (El-Din et al. 2012).

Nematicidal Metabolites from Paecilomyces Spp.

A new macrocyclic lactone derivative paeciloxazine (**126**) with the pyrrolobenzoxazine skeleton was isolated from the culture broth of *Paecilomyces* sp. BAUA3058 (Kanai et al. 2004). Biological assay showed that the compound was active against *Rhabditis pseudoelongata* at 50 μg mL^{-1}. Nematicidal compounds cerebroside A (**127**) and B (**128**) were isolated from another strain of *Paecilomyces* (Zhang et al. 2010). A new nematicidal compound 4-(4'-carboxy-2'-ethyl-hydroxypenty)-5,6-dihydro-6-methyl-cyclobuta[*b*]pyridine-3,6-dicarboxylic acid (**129**) was identified from *Paecilomyces* sp. YMF1.01761. The LD_{50} value of the compound within 48 h against *Panagrellus redivivus* was 50.86 mg L^{-1}, *Meloidogyne incognita* was 47.1 mg L^{-1}, and *Bursaphelenchus xylophilus* was 167.7 mg L^{-1} (Liu et al. 2009).

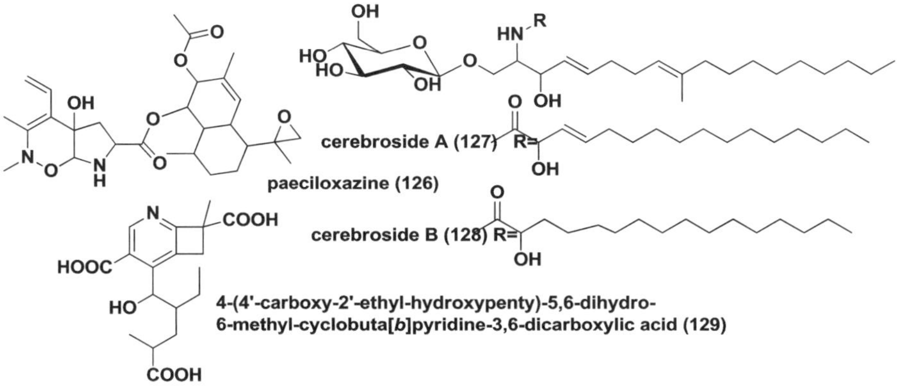

Nematicidal Metabolites from Paraniesslia Spp.

A new compound 3,5-dicarboxyaldehyde-4-hydroxy-acetophenone (**130**) was obtained from the freshwater fungus *Paraniesslia* sp. 83. This compound had a nematicidal activity against *Bursaphelenchus xylophilus* with an LD_{50} value at 200 ppm in 24 h (Dong 2005). Two sphingolipids including a new (2S,2'R,3R,3'E,4E,8E)-1-*O*-(β-D-glucopyranosyl)-3-hydroxyl-2-[N-2'-hydroxyl-3'-eicosadecenoyl]amino-9-methyl-4,8-octadecadiene (**131**) and a known (2S,2'R,3R,3'E,4E,8E)-1-*O*-(β-D-glucopyranosyl)-3-hydroxyl-2-[N-2'-hydroxyl-3'-oct adecenoyl]amino-9-methyl-4,8-octadecadiene (**132**) were isolated from another strain of *Paraniesslia* sp. YMF1.01400. Both compounds showed moderately nematicidal activities against *Bursaphelenchus xylophilus* (Dong et al. 2005).

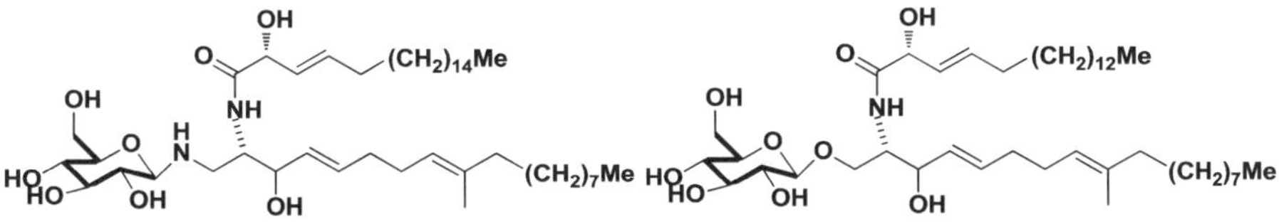

(2S,2'R,3R,3'E,4E,8E)-1-O-(b-D-glucopyranosyl)-3-hydroxyl-2-[N-2'-hydroxyl-3'-eico-sadecenoyl]amino-9-methyl-4,8-octadecadiene (131)

(2S,2'R,3R,3'E,4E,8E)-1-O-(b-D-glucopyranosyl)-3-hydroxyl-2-[N-2'-hydroxyl-3'-octadecenoyl]amino-9-methyl-4,8-octadecadiene (132)

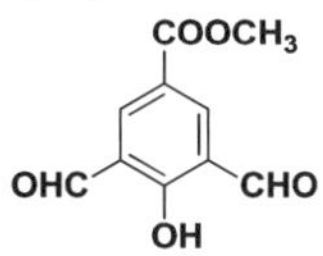

3,5-dicarboxyaldehyde-4-hydroxy-acetophenone (130)

Nematicidal Metabolites from Penicillium Bilaiae, P. Charlesii and P. Paraherquei

A new acetylenic nematicidal compound penipratynolene (**133**) was obtained from the culture filtrate of the fungus *Penicillium bilaiae* (Nakahara et al. 2004). The compound showed nematicidal activity against the root-lesion nematode *Pratylenchus penetrans*, capable of killing 77% of the nematode at a concentration of 300 mg L^{-1} (Alfaro et al. 2003). It was suggested that the alkyne carbons are likely to play an important role in the nematicidal activities of this group of compounds (Kimura et al. 1981; Mori et al. 1982). Two known compounds, 6-methoxy-carbonylpicolinic acid (**134**) and 2,6-pyridinedicarboxylic acid (**135**), were also obtained from the culture filtrate of the strain (Nakahara et al. 2004). By a bioassay at 300 mg L^{-1}, both the two compounds showed nematicidal activities with a mortality of 52% and 98% respectively against *Pratylenchus penetrans* (Nakahara et al. 2004). Nakahara et al. (2004) suggested that the carboxy groups in the compounds are likely to play important roles in their nematicidal activities. The oxindole alkaloid paraherquamide (**136**) was originally isolated from *Penicillium paraherquei* and its structure was determined by X-ray diffraction analysis (Yamazaki et al. 1981). Subsequently, paraherquamide (**136**) and its novel analogs paraherquamides B (**137**), C (**138**), D (**139**), E (**140**), F (**141**), G (**142**) were obtained from another species of *Penicillium charlesii* (ATCC 20841) (Ondeyka et al. 1990; Liesch and Wichmann 1990). All seven metabolites possessed nematicidal activities against *Caenorhabditis elegans* with LD_{50} values in the range of 2.5–160 μg mL^{-1}. Among them, paraherquamide (**136**) was the most potent with an LD_{50} value of 2.5 μg mL^{-1} (Ondeyka et al. 1990; Blanchflower et al. 1991).

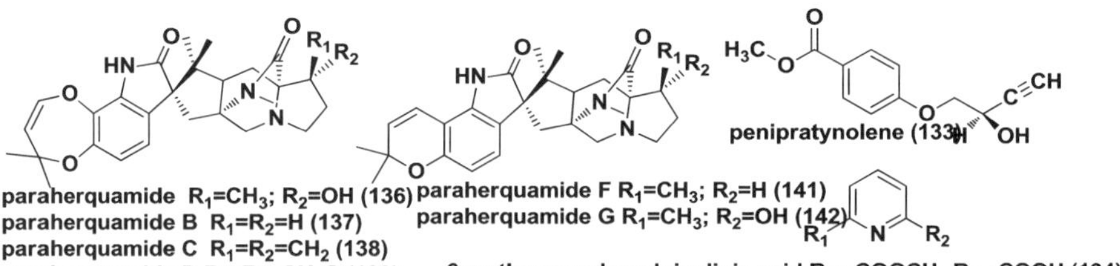

paraherquamide R_1=CH_3; R_2=OH (136)
paraherquamide B R_1=R_2=H (137)
paraherquamide C R_1=R_2=CH_2 (138)
paraherquamide D R_1=R_2=CH_2O (139)
paraherquamide E R_1=CH_3; R_2=H (140)
paraherquamide F R_1=CH_3; R_2=H (141)
paraherquamide G R_1=CH_3; R_2=OH (142)
6-methoxy-carbonylpicolinic acid R_1= $COOCH_3$ R_2= COOH (134)
2,6-pyridinedicarboxylic acid R_1= COOH R_2= COOH (135)

Nematicidal Metabolites from Penicillium Roqueforti and P. Simplicissimum

Three new alkaloids marcfortine A (**143**), B (**144**), and C (**145**) were obtained from the mycelium of *Penicillium roqueforti* (Polondky et al. 1980; Prangé et al. 1981). The chemical structures of marcfortine A (**143**) and C (**145**) were established by X-ray analysis. These three compounds possessed potent anthelmithic properties against plant-parasitic and animal-endoparasitic nematodes. A new alkaloid, peniprequinolone (**146**), together with the known compounds penigequinolones A (**147**), B (**148**) and 3-methoxy-4,6-dihydroxy-4-(4'-methoxyphenyl) quinolinone (**149**) were isolated from the liquid culture of *Penicillium* cf. *simplicissimum* (Kusano et al. 2000). The three known compounds were first isolated from other species of the genus *Penicillium* (Kimura et al. 1996; Hayashi et al. 1997). These compounds were active against the nematode *Pratylenchus penetrans* at the killing rates of 82.4 %, 69.2 % and 57.7 % respectively at the concentration of 1000 mg L^{-1}. These results indicated that either a phenolic hydroxyl group at C-5 or a tetrahydropyran ring in these compounds might be responsible for their nematicidal activities against *Pratylenchus penetrans* (Kusano et al. 2000).

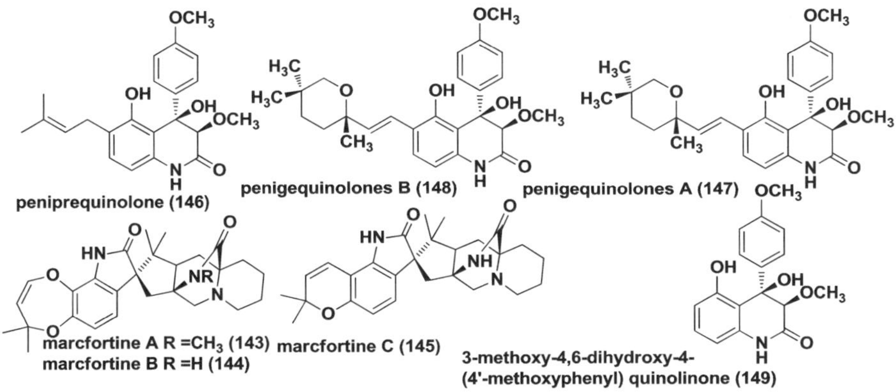

Nematicidal Metabolites from Penicillium Sp.

Paraherquamide (**136**) and its seven novel analogues VM55594 (**150**), VM54158 (**151**), VM54159 (**152**), VM55595 (**153**), VM55596 (**154**), VM55597 (**155**) and VM55599 (**156**) were isolated from a strain of *Penicillium*, IMI 332995 (Blanchflower et al. 1991, 1993). The nematicidal activities of compounds paraherquamide (**136**), VM55594 (**150**), VM54158 (**151**), VM54159 (**152**) were assayed against both *Haemunchus contortus* larvae and *Trichjostrongylus colubriformis* adults *in vitro*. Paraherquamide (**136**) and VM54159 (**152**) were more active than compounds VM55594 (**150**) and VM54158 (**151**), with MIC_{50} values of 31.2 and 25.6 μg mL^{-1} against *Haemunchus contortus* for paraherquamide (**136**) and VM54159 (**152**) respectively. In addition, paraherquamide (**136**) and VM54159 (**152**) could cause 99.5 % and 100 % reductions in faecal egg counts of the nematode *Trichjostrongylus colubriformis* at 4 mg kg^{-1}. The group of 14-de-hydroxy in paraherquamide (**136**) and VM54159 (**152**) was more potent than their corresponding the group of 14-hydroxy in analogues VM55594 (**150**) and VM54158 (**151**) (Blanchflower et al. 1991). Compound VM55596 (**154**) was the first *N*-oxide member in the paraherquamide family and it was found capable of eliminating 94 % faecal eggs of *Trichjostrongylus colubriformis* when dosed at 2 mg kg^{-1} (Blanchflower et al. 1993).

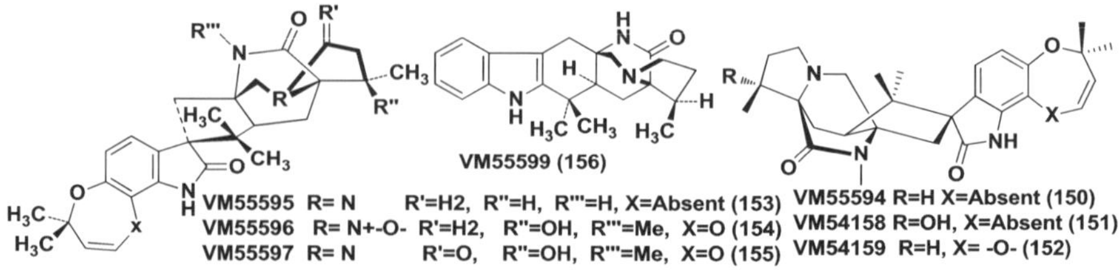

Nematicidal Metabolites from Pochonia Chlamydosporia

Two nematicidal aurovertin compounds aurovertins F (**157**) and D (**158**) were isolated from *Pochonia chlamydosporia*. The LD_{50} value of the two compounds against *Panagrellus redivivus* was 88.6 and 41.7 μg mL^{-1} respectively (Niu et al. 2010). Two new azaphilone metabolites pseudohalonectrin A (**159**) and B (**160**) were produced by the aquatic fungus *P. adversaria* YMF 1.01019 (Dong et al. 2006). These two compounds possessed nematicidal activities against the pine wood nematode *Bursaphelenchus xylophilus* (Dong et al. 2006).

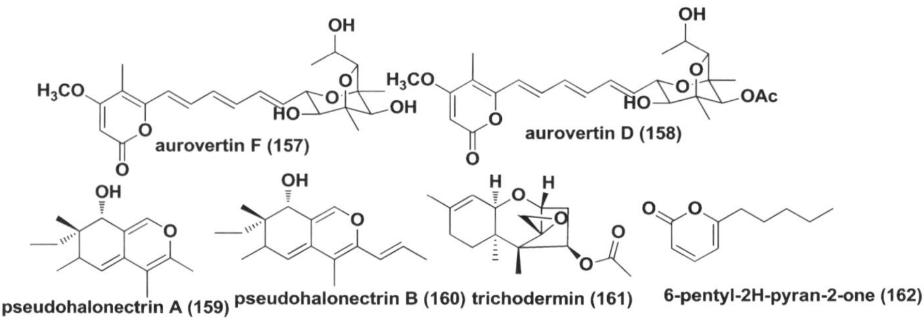

Nematicidal Metabolites from Trichoderma Spp.

A nematicidal sesquiterpene trichodermin (**161**) was isolated from ethyl acetate extract of *Trichoderma* sp. YMF1.02647. The compound could kill more than 95% both *Panagrellus redivivus* and *Caenorhabditis elegans* in 72 h at 0.4 g L^{-1} (Yang et al. 2010). Trichodermin (**161**) had been isolated from several species of *Trichoderma* including *T. viride*, *T. harzianum*, *T. longibrachiatum* and *T. reesei*, and other taxa such as *Stachybotrys cylindrospora* and *Memnoniella echinata* (Godtfredsen and Vangedal 1964; Watts et al. 1988; Nielsen et al. 1998; Reino et al. 2008). The volatile organic compound 6-pentyl-2H-pyran-2-one (**162**) was isolated from *Trichoderma* sp. YMF 1.00416. Nematicidal activity assays showed that the compound could kill > 85% of *Panagrellus redivivus*, *Caenorhabditis elegans*, and *Bursaphelenchus xylophilus* in 48 h at 200 mg L^{-1} in a 2 mL vial (Yang et al. 2012).

Nematicidal Metabolites from Talaromyces Thermophiles

A novel class of potent nematocidal thermolides A-D was isolated from a thermophilic fungus *T. thermophilus*. Thermolides A (**163**) and B (**164**) showed the strongest activities against *Meloidogyne incognita*, *Bursaphelenchus xylophilus* and *Panagrellus redivivus* with LC_{50} values ranging from 0.5–1.0 μg/mL, similar to those of avermectins. Thermolide C (**165**) displayed moderate activity, and weak inhibitory effect on the worms was observed for thermolide D (**166**) (Guo et al. 2012).

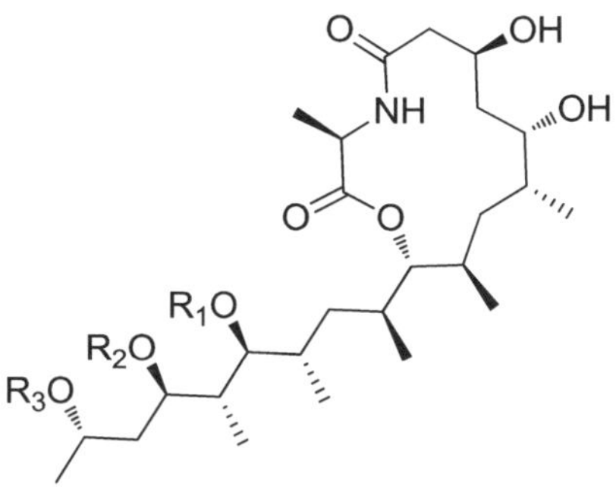

thermolide A R1=R3=H, R2=Ac (163)
thermolide B R1=R2=H, R3=Ac (164)
thermolide C R2=R3=H, R1=Ac (165)
thermolide D R1=R2=R3=H (166)

Nematicidal Metabolites from Unidentified Ascomycete

5-pentyl-2-furaldehyde (**167**) was isolated as the principal nematicide from an unidentified ascomycete belonging to the *Dermateaceae* (Anke et al. 1995). In addition, it has been found in other taxa such as a basidiomycete *Irpex lacteus* (Hayashi et al. 1981) and an unidentified fungal strain Kyu-W63 (Koitabashi et al. 2004). 5-pentyl-2-furaldehyde (**167**) was one of the few metabolites with nematicidal activity found in both ascomycetes and basidiomycetes. This compound was active against *Caenorhabditis elegans* with $LD_{50\ at}$ 75 μg mL^{-1}, against *Meloidogyne incognita* with $LD_{50\ at}$ 60 μg mL^{-1} and against *Aphelenchoides besseyi* with $LD_{90\ at}$ 200 μg mL^{-1} (Anke et al. 1995; Hayashi et al. 1981). Compounds ymf 1029 A (**168**), B (**169**), C (**170**), D (**171**), E (**172**), preussomerin C (**173**), D (**174**), (4*RS*)-4,8-dihydroxy-3,4-dihydronaphthalen-1(2*H*)-one (**175**) and 4,6,8-trihydroxy-3,4-dihydronaphthalen-1(2*H*)-one (**176**) were isolated from an unidentified freshwater fungus YMF 1.01029. These compounds had various nematicidal activities against *Bursaphelenchus xylophilus* (Dong et al. 2008). From the cultures of the ascomycete A111-95, four compounds 5-(2*E*)-2-buten-1-ylidene-3-(1*E*)-1-propen-1-yl-2(5*H*)-furanone (**177**), pregaliellalactone (**178**), and the mixture of 5(R)-(1E)-1,3-butadien-1-yl-3-(1E)-1-propen-1-yl-2(5H)-furanone (**179**) and 5(R)-(3-buten-1-yl) dihydro-3-vinyldihy-2(3H)-furanone (**180**) with nematicidal activity towards *Caenorhabditis elegans* and *Meloidogyne incognita* were obtained (Köpcke et al. 2002). Compound 5-(2*E*)-2-buten-1-ylidene-3-(1*E*)-1-propen-1-yl-2(5*H*)-furanone (**177**) was also obtained from the basidiomycete *Galiella rufa* (Hautzel and Anke 1990).

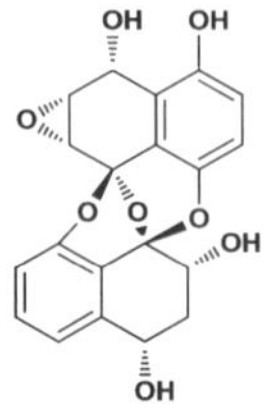
Ymf 1029A (168)

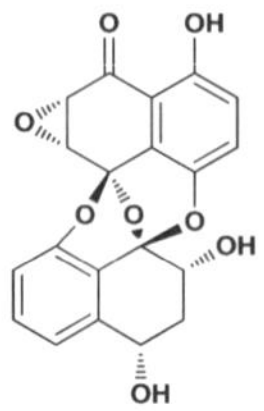
Ymf 1029B (169)

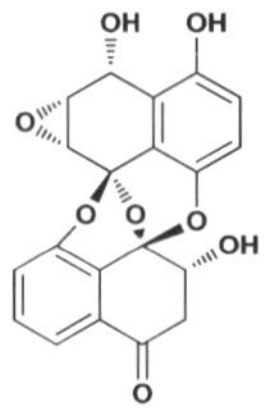
Ymf 1029C (170)

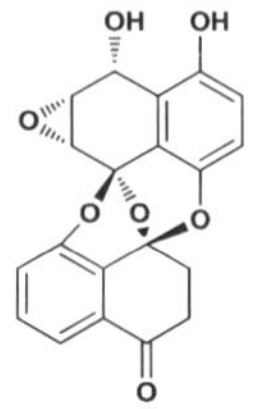
Ymf 1029D (171)

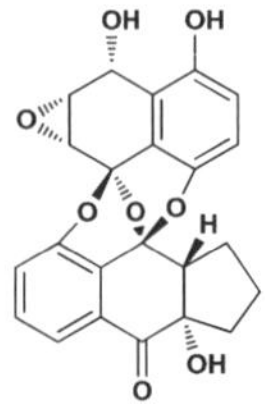
Ymf 1029E (172)

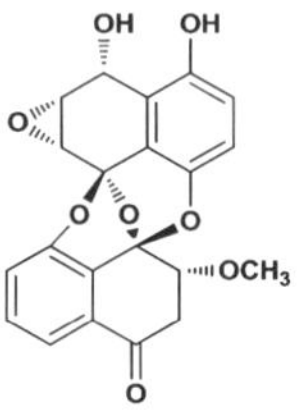
preussomerin C (173)

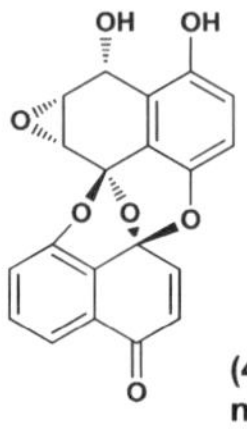
preussomerin D (174)

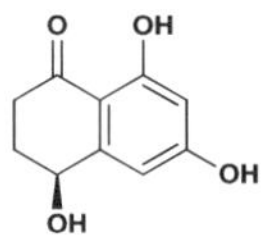
(4RS)- 4,8-dihydroxy-3,4-dihydro naphthalen-1(*2H*)-one (175)

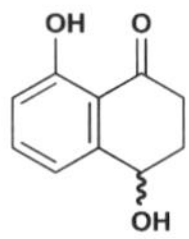
4,6,8-trihydroxy-3,4-dihydronaphthalen-1(*2H*)-one (176)

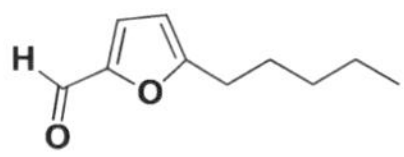
5-pentyl-2-furaldehyde (167)

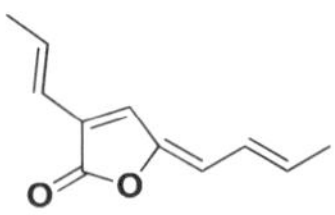
5-(*2E*)-2-buten-1-ylidene-3-(*1E*)-1-propen-1-yl-2(*5H*)-furanone (177)

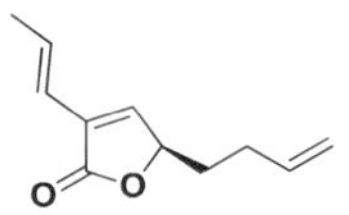
pregaliellalactone (178)

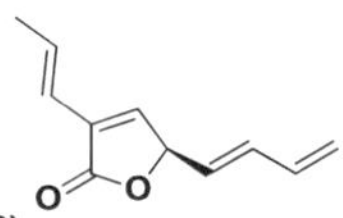
5(R)-(*1E*)-1,3-butadien-1-yl-3-(*1E*)-1-propen-1-yl-2(*5H*)-furanone (179)

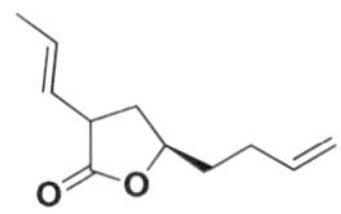
5(R)-(3-buten-1-yl)dihydro-3-vinyldihy-2(*3H*)-furanone (180)

Nematode-Toxic Basidiomycetes and Their Nematicidal Metabolites

Thorn and Barron (1984) reported that ten species of gilled fungi could attack and consume nematodes and considered that five species of *Pleurotus* could release a potent toxin which completely inactivated nematodes prior to penetration. Later, Barron and Thorn (1987) reported the details of *Pleurotus* used to attack its nematode victims and figured *Pleurotus ostreatus* produced tiny droplets of toxin from minute spathulate secetory areas and the toxin trans-2-decenedioic acid was isolated from *P. ostreatus* (Kwok et al. 1992) which inhibited 95 % *Panagrellus redivivus* at 300 ppm in 1 hour and was postulated to be identical with ostreatin. Subsequently, six nematicidal compounds were isolated from *Pleurotus pulmonarius*, and one of them was a new compound (Stadler et al. 1994c). Up to now, 23 species of *Pleurotus* have been reported to have nematicidal activity. Nematicidal acticity had been considered as one of characters of the genus *Pleurotus* (Hibbett and Thorn 1994). It is interesting that some edible mushrooms have such nematicidal activity. In addition, *Poria cocos*, a traditional Chinese medicine, had been also shown to possess nematicidal activity (Li et al. 2005).

Omphalotin A and its derivatives with potent and selective nematicidal activity were produced by the basidiomycete *Omphalotus olearius* (Mayer et al. 1997; Stener et al. 1997; Büchel et al. 1998). Omphalotin A and its derivatives had similar nematicidal activity with commercially available nematicide ivermectin.

Conversion of secondary metabolites as a response to injury in fruit bodies in basidiomycetes formed extracts with high nematicidal activity (Table 7.3) (Stadler and Sterner 1998). *Tricholoma terreum* and in all tested *Lepista* species, the nematicidal activity increased in response to injury, and *Tricholoma terreum* produced linoleic acid and S-coriolic acid while *Lepista* spp. produced free linoleic acid, along with unidentified metabolites (Stadler and Sterner 1998). *Melanoleuca cognate, M. melaleuca, Laccaria amethystine* and *Marasmius wynnei* produced fatty acids upon fruit body injury (Stadler and Sterner 1998). These findings reflect the presence of chemical defence systems in mushrooms, which is mediated by enzymatic conversions activated by physical injury.

Up to now, about 77 genera of basidiomycetes containing more than 160 species have been reported to possess nematicidal activity by producing active components. The nematode-toxic basidiomycete and their nematicidal compounds are listed in Table 7.3.

Nematicidal Metabolites from Calocybe Gambosa, Cheimonophyllum Candidissimum and Pycnoporus Sanguineus

The alkaloid phenoxazone (**181**) was isolated from the mycelial cultures of *Calocybe gambosa* and fruiting bodies of *Pycnoporus sanguineus* (Schlunegger et al. 1976; Gill 1994). This compound showed nematicidal activity against *Meloidogyne incognita* ($LD_{50:}$ 50 μg mL^{-1}) (Mayer 1995). Six new bisabolane sesquiterpenes cheimonophyllons A (**182**), B (**183**), C (**184**), D (**185**), E (**186**) and cheimonophyllal (**187**) were obtained from the submerged cultures of *Cheimonophyllum candidissimum* TA 8644. These compounds were active against *Caenorhabditis elegans* with LD_{50} between 10 and 100 μg mL^{-1} (Stadler et al. 1994a, b). In a further study, compound 1,2-dihydroxymintlactone (**188**) was isolated from the same fungus (Stadler et al. 1995). 1,2-Dihydroxymintlactone (**188**) was a new menithol monoterpene and possessed nematicidal activity. The LD_{50} of 1,2-dihydroxymintlactone (**188**) against *Caenorhabditis elegans* was 25 μg mL^{-1} This was the first compound in the *p*-menthane group reported from a basidiomycete (Stadler et al. 1995).

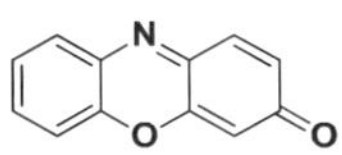

phenoxazone (181)

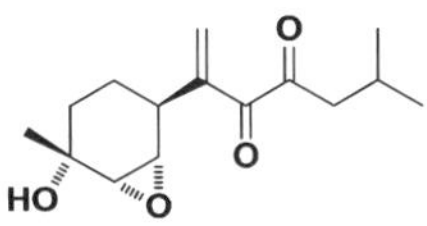

cheimonophyllons A (182)

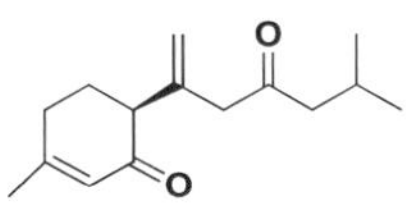

cheimonophyllons B (183)

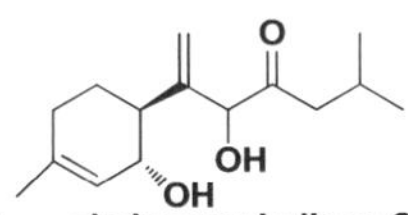

cheimonophyllons C (184)

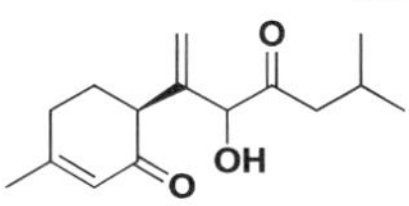

cheimonophyllons D (185)

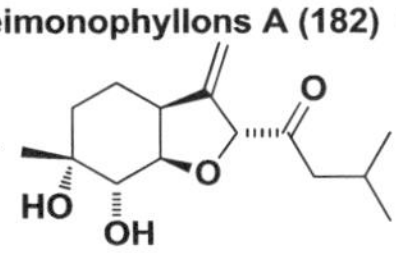

cheimonophyllons E (186)

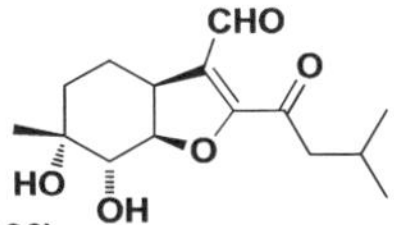

cheimonophyllal (187)

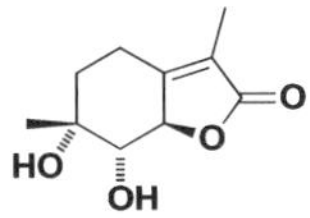

1,2-dihydroxymintlactone (188)

Table 7.3 Nematode-toxic basidiomycete and their nematicidal compounds

Species	Test nematodes	Nematicidal compounds	References
Agaricus arvensis	*M. incognita, H. glycines, Caenorhabditis* sp., *A. basseyi*	–	Chen et al. (2010)
Amanita excelsa	*M. incognita, H. glycines, Caenorhabditis* sp., *A. basseyi*	–	Chen et al. (2010)
A. fulva	*B. xylophilus*	–	Zhao (2004)
A. japonica	*B. xylophilus*	–	Zhao (2004)
A. rubescens	*B. xylophilus*	–	Zhao (2004)
A. virosa	*M. incognita, H. glycines, Caenorhabditis* sp., *A. basseyi*	–	Chen et al. (2010)
Amauroderma brunreopilus	*B. xylophilus*	–	Zhao (2004)
Auricularia auricular	*B. xylophilus*	–	Zhang and Zhao (2003)
Boletus albus	*B. xylophilus*	–	Zhao (2004)
B. fraternus	*B. xylophilus*	–	Zhao (2004)
Boletus sp.	*B. xylophilus*	–	Zhao (2004)
Calocera viscose (injured)	*C. elegans*	–	Stadler and Sterner (1998)
C. excipuliformis	*C. elegans*	–	Stadler and Sterner (1998)
Calocybe gambosa	*M. incognita*	Phenoxazone	Schlunegger et al. (1976); Mayer (1995)
C. tubaeformis (injured)	*C. elegans*	–	Stadler and Sterner (1998)
Cheimonophyllum candidissimum	*C. elegans*	Cheimonophyllal, Cheimonophyllons A, B, C, D and E, 1,2-dihydroxymintlactone	Stadler et al. (1994a, b); Stadler et al. (1995)
Chroogomphus rutilus	*C. elegans*	–	Stadler and Sterner (1998)
Clavulinopsis corniculata	*C. elegans*	–	Stadler and Sterner (1998)
Clitocybe sp.	*B. xylophilus*	–	Zhao (2004)
Clitocybe odora (injured)	*C. elegans*	–	Stadler and Sterner (1998)
Clitocybula oculus	*Aphelenchoides* sp.	–	William et al. (1998)
Collybia acervata	*M. incognita, H. glycines, Caenorhabditis* sp., *A. basseyi*	–	Chen et al. (2010)

Table 7.3 (continued)

Species	Test nematodes	Nematicidal compounds	References
C. confluence	*B. xylophilus*	–	Zhao (2004)
C. dryophila	*B. xylophilus*	–	Zhao (2004)
Coltricia sp.	*B. xylophilus*	–	Zhao (2004)
Coprinus clastophyllus	*P. redivivus*	–	Liu (2005)
C. comatus	*P. redivivus, M. arenaria* *M. incognita, H. glycines, Caenorhabditis* sp., *A. basseyi*	4,6-dimethoxyisobenzofuran-1(3*H*)-one, 5-methylfuran-3-carboxylic acid, 5-hydroxy-3,5-dimethylfuran-2(5*H*)-one, 4,6-dihydroxybenzofuran-3(2*H*)-one, 3-formyl-2,5-dihydroxybenzyl acetate, 5-hydroxy-3-(hydroxymethyl)-5-methylfuran-2(5H)-one, 4,6-dihydroxyisobenzofuran-1,3-dione	Luo et al. (2007); Chen et al. (2010)
C. disseminatus	*P. redivivus*	–	Liu (2005)
C. erythrocephalus	*P. redivivus*	–	Liu (2005)
C. micaceus	*M. incognita, H. glycines, Caenorhabditis* sp., *A. basseyi*	–	Chen et al. (2010)
C. plicatilis	*P. redivivus*	–	Liu (2005)
Coprinus. sp.	*P. redivivus*	–	Liu (2005)
C. verticillatus	*P. redivivus*	–	Liu (2005)
C. xanthothrix	*P. redivivus* *M. arenaria*	Xanthothone, 7,8,11-drimanetriol, 2-(1H-pyrrol-1-yl) ethanol	Liu et al. (2008)
Cyathus intermedius	*B. xylophilus*	–	Zhao (2004)
C. pallidus	*B. xylophilus*	–	Zhao (2004)
Cyathus sp.	*B. xylophilus*	–	Zhao (2004)
Daedalea biennis	*M. incognita, H. glycines, Caenorhabditis* sp., *A. basseyi*	–	Chen et al. (2010)
Dichomitus squalen	*B. xylophilus*	2*β*, 13-dihydroxyledol	Huang et al. (2004)
Fomitopsis pinicola	*M. incognita, H. glycines, Caenorhabditis* sp., *A. basseyi*	–	Chen et al. (2010)

Table 7.3 (continued)

Species	Test nematodes	Nematicidal compounds	References
Ganoderma sp.	*P. redivivus*	–	Li (2005)
Gastroboletus sp.	*B. xylophilus*	–	Zhao (2004)
Galiella rufa	*C. elegans* *M. incognita*	5-(2*E*)-2-buten-1-ylidene-3-(1*E*)-1-propen-1-yl-2(5*H*)-furanone	Hautzel and Anke (1990); Köpcke et al. (2002)
Gloeophyllum trabeum	*B. xylophilus*	–	Zhang and Zhao (2003)
Hericium coralloides	*C. elegans*	Linoleic acid, Oleic acid, Palmitic acid	Xiang and Feng (2001)
Hexagonia tenuis	*M. incognita, H. glycines, Caenorhabditis* sp., *A. basseyi*	–	Chen et al. (2010)
Hohenbuehelia grisea	*C. elegans*	–	Stadler and Sterner (1998)
Hydnum rufescens (injured)	*C. elegans*	–	Stadler and Sterner (1998)
Hygrophorus mesotephrus	*C. elegans*	–	Stadler and Sterner (1998)
H. albicans	*Aphelenchoides* sp.	–	Tzean and Liou (1993)
H. amoenum	*Aphelenchoides* sp.	–	Tzean and Liou (1993)
H. appendiculatum	*M. incognita, H. glycines, Caenorhabditis* sp., *A. basseyi*	–	Chen et al. (2010)
H. heterocystidium	*Aphelenchoides* sp.	–	Tzean and Liou (1993)
H. medioburiensis	*Aphelenchoides* sp.	–	Tzean and Liou (1993)
H. mutatum	*Aphelenchoides* sp.	–	Tzean and Liou (1993)
H. obtusiforme	*Aphelenchoides* sp.	–	Tzean and Liou (1993)
H. pallidum	*Aphelenchoides* sp.	–	Tzean and Liou (1993)
H. piceae	*C. elegans*	–	Stadler and Sterner (1998)
H. populneum	*Aphelenchoides* sp.	–	Tzean and Liou (1993)
H. pustulatus	*C. elegans*	–	Stadler and Sterner (1998)
H. radula	*Aphelenchoides* sp.	–	Tzean and Liou (1993)
H. setigerum	*Aphelenchoides* sp.	–	Tzean and Liou (1993)
H. typhicola	*Aphelenchoides* sp.	–	Tzean and Liou (1993)
Irpex lacteus	*C. elegans* *A. besseyi* *M. incognita*	5-pentyl-2-furaldehyd, 5-(4-pentenyl)-2-furaldehyd, Methyl 3-*p*-anisoloxypropionate	Hayashi et al. (1981)

Table 7.3 (continued)

Species	Test nematodes	Nematicidal compounds	References
Laccaria amethystine (injured)	*C. elegans*	–	Stadler and Sterner (1998)
Lachnella sp. 541	*M. incognita*	Marasmic acid	Kupka et al. (1983); Mayer (1995)
L. villosa	*M. incognita*	Marasmic acid	Sterner et al. (1985); Mayer (1995)
Lactarius deliciosus	*M. incognita, H. glycines, Caenorhabditis* sp., *A. basseyi*	–	Chen et al. (2010)
L. deterrimus	*C. elegans*	–	Stadler and Sterner (1998)
L. helvus	*C. elegans*	–	Stadler and Sterner (1998)
L. mitissimus	*C. elegans*	Lactarorufin A and B, Furantriol	Daniewski et al. (1990); Stadler and Sterner (1998)
L. porninsis (injured)	*C. elegans*	–	Stadler and Sterner (1998)
L. rufus (injured)	*C. elegans*	–	Stadler and Sterner (1998)
L. torminosus	*C. elegans*	–	Stadler and Sterner (1998)
L. trivialis	*C. elegans*	–	Stadler and Sterner (1998)
L. vellereus	*C. elegans* *M. incognita*	Isovelleral	Sterner et al. (1985); Mayer (1995); Hansson et al. (1995)
Lampteromyces japonicus	*P. redivivus*	–	Mo et al. (2000); Dong et al. (2000)
L. unuunescens	*P. redivivus*	–	Mo et al. (2000)
Leatiporus sulphureus	*B. xylophilus*	–	Zhao (2004)
Lentinus edodes	*B. xylophilus*	–	Zhang and Zhao (2003)
L. lepideus	*B. xylophilus*	–	Zhang and Zhao (2003)
L. similis	*M. incognita, H. glycines, Caenorhabditis* sp., *A. basseyi*	–	Chen et al. (2010)
Lenzites trabea	*B. xylophilus*	–	Zhang and Zhao (2003)
Lepista gilva (injured)	*C. elegans*	–	Stadler and Sterner (1998)
L. inversa	*C. elegans*	–	Stadler and Sterner (1998)
L. irina	*C. elegans*	–	Stadler and Sterner (1998)
L. nebularis (injured)	*C. elegans*	–	Stadler and Sterner (1998)
L. nuda (injured)	*C. elegans*	–	Stadler and Sterner (1998)

Table 7.3 (continued)

Species	Test nematodes	Nematicidal compounds	References
L. personata (injured)	*C. elegans*	–	Stadler and Sterner (1998)
Leucopaxillus albissimus var. *paradoxus* form *albiformis*	*N. braziliensis*	2-aminoquinoline	Pfister (1988)
Limacella illinita	*C. elegans*	Illinitone A	Gruhn et al. (2007)
Lycoperdon pyriforme	*C. elegans*	–	Stadler and Sterner (1998)
L. pusillus	*M. incognita, H. glycines, Caenorhabditis* sp., *A. basseyi*	–	Chen et al. (2010)
Marasmius alliaceus	*C. elegans*	–	Stadler and Sterner (1998)
M. conigenus	*M. incognita*	Marasmic acid	Kavanagh et al. (1949); Mayer (1995)
M. wynnei (injured)	*C. elegans*	–	Stadler and Sterner (1998)
Melanoleuca cognate (injured)	*C. elegans*	–	Stadler and Sterner (1998)
M. melaleuca (injured)	*C. elegans*	–	Stadler and Sterner (1998)
Meripilus giganteus (injured)	*C. elegans*	–	Stadler and Sterner (1998)
Mutinus caninus	*M. incognita, H. glycines, Caenorhabditis* sp., *A. basseyi*	–	Chen et al. (2010)
Mycena pura	*C. elegans*	–	Stadler and Sterner (1998)
M. sepia	*C. elegans*	–	Stadler and Sterner (1998)
Mycena sp.	*C. elegans*	Mycenon	Hautzel et al. (1990); Stadler (1993)
Nidularia sp.	*B. xylophilus*	–	Zhao (2004)
Omphalotus olearius	*M. incognita, C. elegans, H. glycines, Caenorhabditis* sp., *A. basseyi*	Omphalotin A, B, C and D	Mayer et al. (1997); Stener et al. (1997); Büchel et al. (1998); Anke et al. (1999)
Oudemansiella longipes (injured)	*C. elegans*	–	Stadler and Sterner (1998)
O. platyphylla (injured)	*C. elegans*	–	Stadler and Sterner (1998)
Peniophora laeta	*M. incognita*	Marasmic acid	Kupka et al. (1983); Mayer (1995)
Phanocollyleio sp.	*B. xylophilus*	–	Zhao (2004)
Pleurotus certicautus	*P. redivivus*	–	Li et al. (2001)
P. citrinopileatus	*P. redivivus*	–	Li et al. (2001)
P. colombinus	*P. redivivus*	–	Li et al. (2001)

Table 7.3 (continued)

Species	Test nematodes	Nematicidal compounds	References
P. cornucopiae	*P. redivivus*	–	Thorn and Barron (1984)
P. corticatus	*P. redivivus*	–	Chen et al. (2010)
P. cystidiosus	*P. redivivus*	–	Dong et al. (2010)
P. dryinus	*P. redivivus*	–	Li et al. (2001)
P. eryngii	*P. redivivus*	–	Li et al. (2001)
P. euosmus	*P. redivivus*	–	Li et al. (2001)
P. ferulae	*P. redivivus* *B. xylophilus*	Cheimonophyllon E, 5-hydroxymethyl-furancarbaldehyde	Li et al. (2001, 2007)
P. florrida	*P. redivivus*	–	Li et al. (2001)
P. levis	*P. redivivus*	–	Li et al. (2001)
P. memberancens	*P. redivivus*	–	Li et al. (2001)
P. ostreatus	*P. redivivus* *B. xylophilus*	*Trans*-2-decenedioic acid	Thorn and Barron (1984); Barron and Thorn (1987); Kwok et al. (1992); Stadler et al. (1994a)
P. pulmonarius	*C. elegans*	S-coriolic acid, Linoleic acid, *p*-anisaldehyde, *p*-anisyl alcohol, 1-(4-methoxyphenyl)-1,2-propanediol, 2-hydroxy-(4'-methoxy)-propiophenone	Stadler et al. (1994c)
P. sajorcaju	*P. redivivus* *A. composticola*	–	Sharma (1994); Li et al. (2001)
P. salmonseostraminus	*P. redivivus*	–	Li et al. (2001)
P. sapidus	*P. redivivus*	–	Li et al. (2001)
P. shodophyllus	*P. redivivus*	–	Li et al. (2001)
P. spodoleucus	*P. redivivus*	–	Li et al. (2001)
P. strigosus	*P. redivivus*	–	Thorn and Barron (1984)
P. subareolatus	*P. redivivus*	–	Thorn and Barron (1984)
P. tuberregium	*P. redivivus*	–	Hibbett and Thorn (1994)
Pluteus fulve	*B. xylophilus*	–	Zhao (2004)

Table 7.3 (continued)

Species	Test nematodes	Nematicidal compounds	References
Polyporus sulphureus	*M. incognita* *C. elegans* *B. xylophilus*	Beauvericin	Deol et al. (1978); Maye (1995); Shimada et al. (2010)
P. versicolor	*B. xylophilus*	–	Zhang and Zhao (2003)
Poria cocos	*P. redivivus* *M. arenaria*	2,4,6-triacetylenic octane diacid	Li et al. (2005)
Porphyrellus pseudoscaber	*C. elegans*	–	Stadler and Sterner (1998)
Psathyrella velutina	*C. elegans*	–	Stadler and Sterner (1998)
Pseudocolus fusiformis	*M. incognita, H. glycines, Caenorhabditis* sp., *A. basseyi*	–	Chen et al. (2010)
Pterula multifida	*C. elegans*	–	Stadler and Sterner (1998)
Pycnoporus cinnabarinus (injured)	*C. elegans*	–	Stadler and Sterner (1998)
P. sanguineus	*M. incognita*	Phenoxazone	Gill (1994); Mayer (1995)
Russula albonigra	*C. elegans*	–	Stadler and Sterner (1998)
R. alutacea	*M. incognita, H. glycines, Caenorhabditis* sp., *A. basseyi*	–	Chen et al. (2010)
Russula cuprea	*C. elegans* *M. incognita*	Isovelleral	Mayer (1995); Clerkyjzio and Sterner (1997)
R. decolorans	*C. elegans*	–	Stadler and Sterner (1998)
R. depalens	*B. xylophilus*	–	Kavanagh et al. (1949)
R. emetica (injured)	*C. elegans*	–	Stadler and Sterner (1998)
R. fellea (injured)	*C. elegans*	–	Stadler and Sterner (1998)
R. nigricans (injured)	*C. elegans*	–	Stadler and Sterner (1998)
R. ochroleuca	*M. incognita, H. glycines, Caenorhabditis* sp., *A. basseyi*	–	Chen et al. (2010)
R. sanguinea	*M. incognita, H. glycines, Caenorhabditis* sp., *A. basseyi*	–	Chen et al. (2010)

Table 7.3 (continued)

Species	Test nematodes	Nematicidal compounds	References
R. virescems	*B. xylophilus, M. incognita, H. glycines, Caenorhabditis* sp., *A. basseyi*	–	Zhao (2004); Chen et al. (2010)
Schizophyllum commune	*M. incognita, H. glycines, Caenorhabditis* sp., *A. basseyi*	–	Chen et al. (2010)
Sinotermitomyces carnosus	*B. xylophilus*	–	Zhao (2004)
Sparassis crispa	*C. elegans*	–	Stadler and Sterner (1998)
Spongipellis spumeus	*M. incognita, H. glycines, Caenorhabditis* sp., *A. basseyi*	–	Chen et al. (2010)
Stereum sp. 8954	*P. redivivus*	3,5-dihydroxy-4-(3-methyl-but-2-enyl)-benzene-1,2-dicarbaldehyde, Butyl 2,4-dihydroxy-6-methylbenzoate	Li et al. (2006)
Stereum sp. CCTCC AF 207024	*P. redivivus*	Stereumins A, B, C, D and E	Li et al. (2008)
Stropharia aeruginosa (injured)	*C. elegans*	–	Stadler and Sterner (1998)
Thelephora gambajun	*B. xylophilus*	–	Zhao (2004)
Trametes cinnabarina	*M. incognita, H. glycines, Caenorhabditis* sp., *A. basseyi*	–	Chen et al. (2010)
Trametes sp.	*P. redivivus*	–	Li (2005)
Tremella sp.	*B. xylophilus*	–	Zhao (2004)
Trichaptum sp.	*P. redivivus*	–	Li (2005)
Tricholoma columbetta	*C. elegans*	–	Stadler and Sterner (1998)
T. flavovirens	*C. elegans*	–	Stadler and Sterner (1998)
T. fulva	*B. xylophilus*	–	Zhao (2004)
T. terreum (injured)	*C. elegans*	–	Stadler and Sterner (1998)
T. ustale	*C. elegans*	–	Stadler and Sterner (1998)
Tricholomopsis sp.	*B. xylophilus*	–	Zhao (2004)
Tylopilus scabrusus	*B. xylophilus*	–	Zhao (2004)
Xerocomus chrysenteron (injured)	*C. elegans*	–	Stadler and Sterner (1998)
X. rubellus (injured)	*C. elegans*	–	Stadler and Sterner (1998)

No nematicidal compounds were reported from the fungus

Nematicidal Metabolites from Coprinus Comatus and C. Xanthothrix

Coprinus comatus had been proven to be active against several nematodes (Luo et al. 2007; Chen et al. 2010). A new furan 5-hydroxy-3-(hydroxymethyl)-5-methylfuran-2(5H)-one (**189**), two known furan compounds 5-methylfuran-3-carboxylic acid (**190**) and 5-hydroxy-3,5-dimethylfuran-2(5*H*)-one (**191**), as well as three benzofurans including a new 4,6-dihydroxyisobenzofuran-1,3-dione (**192**) and known 4,6-dimethoxyisobenzofuran-1(3*H*)-one (**193**), and 4,6-dihydroxybenzofuran-3(2*H*)-one (**194**), together with 3-formyl-2,5-dihydroxybenzyl acetate (**195**) were all obtained from *Coprinus comatus* (Luo et al. 2007). All compounds had nematicidal activities against *Panagrellus redivivus* and *Meloidogyne arenaria* at 400 ppm. The LD_{50} values of 5-methylfuran-3-carboxylic acid (**190**) and 5-hydroxy-3,5-dimethylfuran-2(5*H*)-one (**191**) were 100 ppm at 12 h (Luo et al. 2007). *Coprinus xanthothrix* produced three nematicidal compounds including a new compound xanthothone (**196**) and two known compounds 7,8,11-drimanetriol (**197**) and 2-(1H-pyrrol-1-yl) ethanol (**198**) (Liu et al. 2008). The LD_{50} of these compounds against *Panagrellus redivivus* and *Meloidogyne arenaria* was 125–250 ppm (Liu et al. 2008).

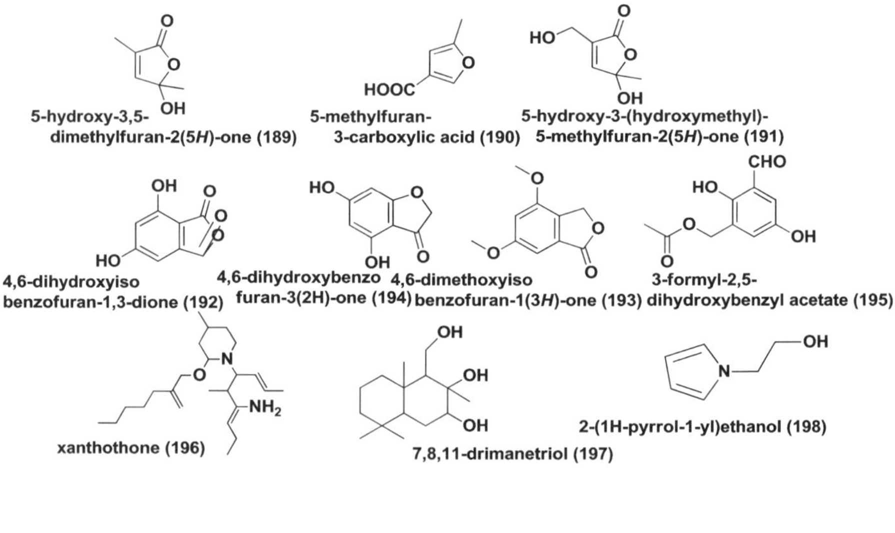

Nematicidal Metabolites from Dichomitus Squalens and Hericium Coralloides

A new aromadendrane, 2*β*,13-dihydroxyledol (**199**) was isolated from the solid mycelial cultures of *Dichomitus squalens* and this compound exhibited potent activity against *Bursaphelenchus xylophilus* with $LC_{50\ at}$ 35.6 μg mL^{-1} (Huang et al. 2004). A nematicidal fatty acid mixture containing linoleic acid (**7**), oleic acid (**200**), and palmitic acid (**201**) were obtained from the culture of *Hericium coralloides*. This mixture showed a nematicidal activity against *Caenorhabditis elegans* (Xiang and Feng 2001). These fatty acids have also obtained from several other taxa.

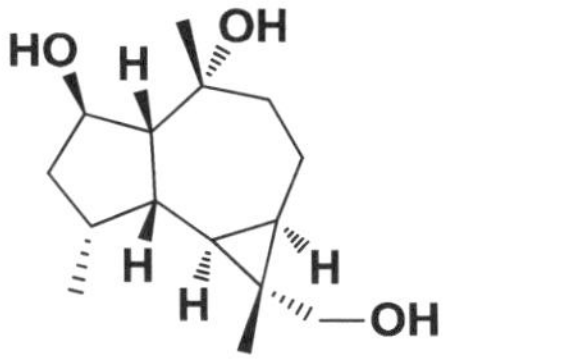

2β ,13-dihydroxyledol (199)

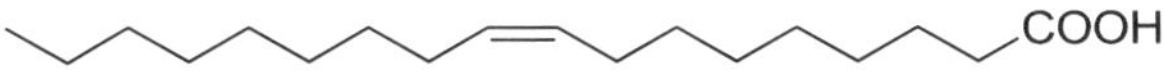

oleic acid (200)

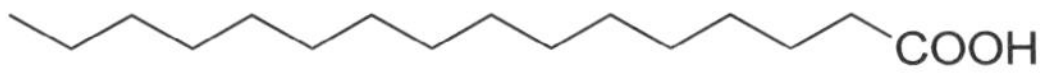

palmitic acid (201)

Nematicidal Metabolites from Irpex Lacteus

A new aromatic compound methyl 3-*p*-anisoloxypropionate (**202**) and a new furan compound 5-(4-pentenyl)-2-furaldehyde (**203**) were isolated from *Irpex lacteus*. The LD_{50} value of the two compounds against *A. besseyi* was 25 μg mL^{-1} and 50 μg mL^{-1} respectively (Hayashi et al. 1981). In addition, 5-pentyl-2-furaldehyde (**167**) was also isolated from *Irpex lacteus* (Hayashi et al. 1981). The compound was obtained from an unidentified ascomycete (Anke et al. 1995) and an unidentified fungal strain Kyu-W63 (Koitabashi et al. 2004).

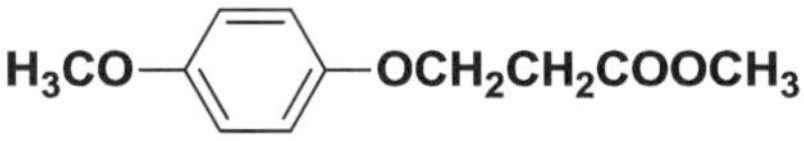

3-*p*-anisoloxypropionate (202)

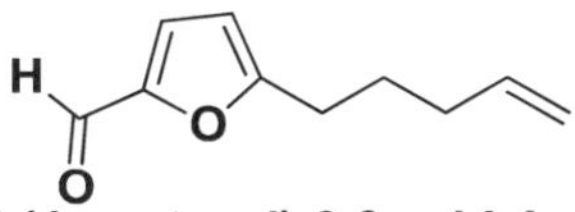

5-(4-pentenyl)-2-furaldehyde (203)

Nematicidal Metabolites from Lactarius Mitissimus, L. Vellereus and Russula Cuprea

Three furan sesquiterpenoids lactarorufin A (**204**), lactarorufin B (**205**) and furantriol (**206**) were isolated from *Lactarius mitissimus* and the all three compounds showed nematicidal activities against *Caenorhabditis elegans* with LD_{50} values at around 100 μg mL^{-1} (Daniewski et al. 1990; Stadler and Sterner 1998). Marasmane sesquiterpene isovelleral (**207**) could be found in injured fruiting bodies of the mushroom *Lactarius vellereus*. This compound was considered a key component of the chemical defense system against nematodes in this mushroom species (Sterner et al. 1985; Hansson et al. 1995). Isovelleral (**207** showed nematicidal activity against *Meloidogyne incognita* with $LD_{30\ at}$ 100 μg mL^{-1} and against *Caenorhabditis elegans* with $LD_{50\ at}$ 50 μg mL^{-1} (Mayer 1995). Isovelleral (**207**) has also been found in injured fruiting bodies of other mushrooms such as *Russula cuprea* (Clerkyjzio and Sterner 1997).

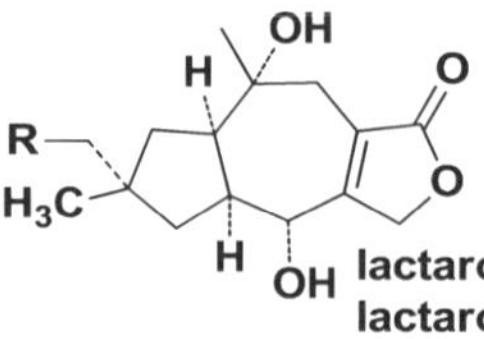

lactarorufin A R =H (204)
lactarorufin B R=OH (205)

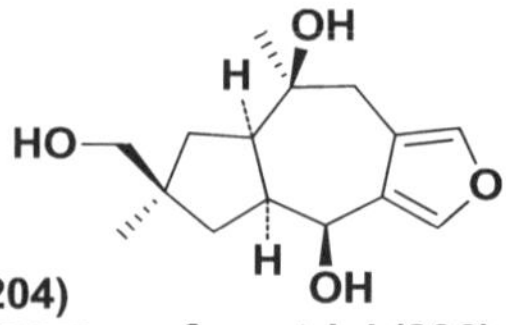

furantriol (206)

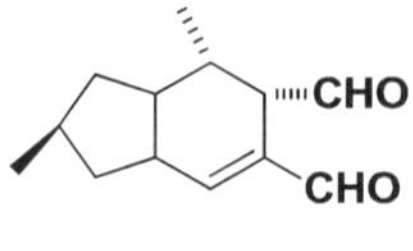

isovelleral (207)

Nematicidal Metabolites from Lachnella Villosa, Marasmius Conigenus and Peniophora Laeta

Sesquiterpene marasmic acid (**208**) was reported to have a weak activity against *M. incognita* by Mayer (1995). Marasmic acid (**208**) was originally isolated from *Marasmius conigenus* (Kavanagh et al. 1949). Subsequently, marasmic acid (**208**) was obtained from several other basidiomycetes including *Lachnella villosa*, *Lachnella* sp. and *Peniophora laeta* (Kupka et al. 1983; Sterner et al. 1985).

Nematicidal Metabolites from Leucopaxillus Albissimus Var. Paradoxus form Albiformis, Limacella Illinita and Mycena Sp.

A novel alkaloid 2-aminoquinoline (**209**) was isolated from the fruiting bodies of *Leucopaxillus albissimus* var. *paradoxus* form *albiformis* (Pfister 1988). At a concentration of 50 μg mL^{-1}, 2-aminoquinoline (**209**) caused 50% motility, 74% viability, and 52% cast formation reductions in the nematode *Nippostrongylus braziliensis* (Pfister 1988). A new compound illinitone A (**210**) was obtained from fermentations of *Limacella illinita*, which exhibited nematicidal activity on *C. elegans* with $IC_{50\ at}$ 25 μg mL^{-1} (Gruhn et al. 2007). Mycenon (**211**) is a chlorinated benzoquinone derivative isolated from the culture broth of a basidiomycete, *Mycena* sp. TA 87202 (Hautzel et al. 1990). It was shown to be active against *Caenorhabditis elegans,* with an $LD_{50\ at}$ 50 μg mL^{-1} (Stadler 1993).

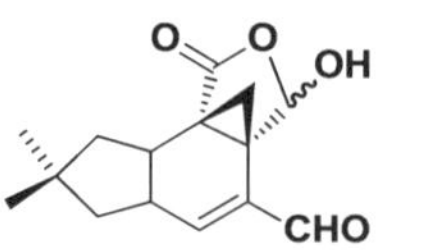

marasmic acid (208)

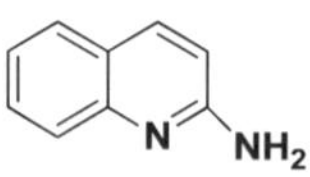

2-aminoquinoline 209

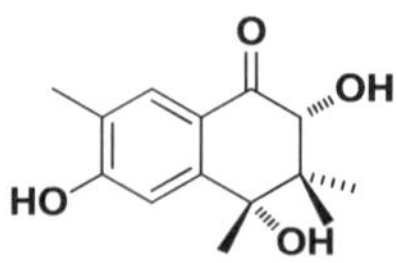

illinitone A 210

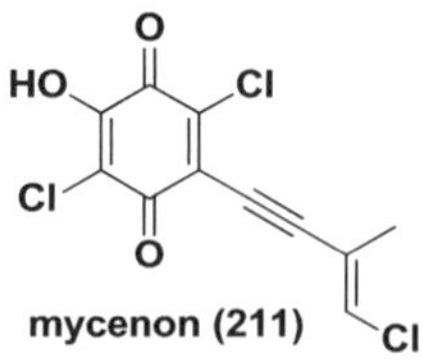

mycenon (211)

Nematicidal Metabolites from Omphalotus Olearius

Peptidal compounds omphalotin A (**212**) and its derivatives omphalotin B (**213**), C (**214**) and D (**215**) were obtained from *Omphalotus olearius*. All compounds possessed strong nematicidal activities against nematodes (Mayer et al. 1997; Stener et al. 1997; Büchel et al. 1998; Anke et al. 1999). Omphalotin A (**212**), a cyclic dodecapeptide, was highly toxic (LD_{90}: 0.76 μM) towards *Meloidogyne incognita*. However, it was approximately 50 times less active against *Caenorhabditis elegans* (LD_{90}: 38 μM) (Stener et al. 1997; Büchel et al. 1998). The corresponding LD_{90} values for the commercially available nematicide ivermectin were 4.6 μM and 0.46 μM respectively against *Meloidogyne incognita* and *Caenorhabditis elegans*. Omphalotin A (**212**) lacks any antimicrobial and phytotoxic activities, and contains only weak cytotoxic activity, making it a potentially useful nematicide. The three derivatives omphalotin B (**213**), C (**214**) and D (**215**) all possessed nematicidal activities similar to that of omphalotin A (Anke et al. 1999). Although the yield of these active compounds is low, the strong nematicidal activity showed it is possible to find new natural nematicidals from products of fungi.

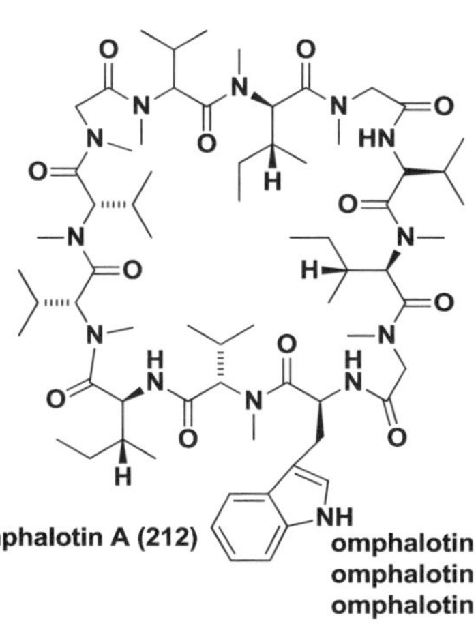

omphalotin A (212)

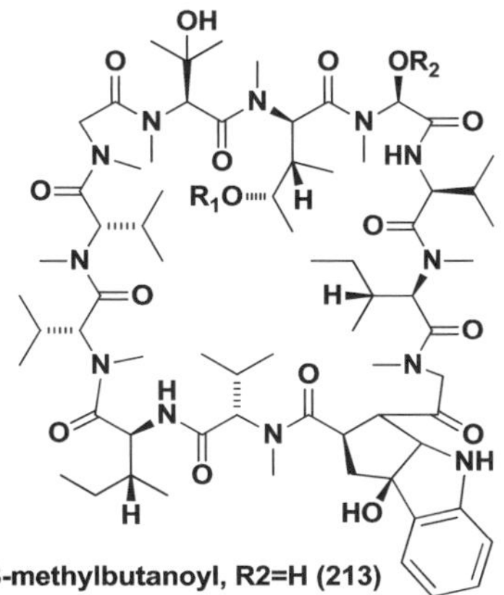

omphalotin B R1=3-hydroxy-3-methylbutanoyl, R2=H (213)
omphalotin C R1=3-hydroxy-3-methylbutanoyl, R2=acetyl (214)
omphalotin D R1=R2=acetyl (215)

Nematicidal Metabolites from Pleurotus Ferulae, P. Ostreatus and P. Pulmonarius

Pleurotus ferulae was shown to be active against *Bursaphelenchus xylophilus* and *Panagrellus redivivus* (Li et al. 2001, 2007) and two nematicidal compounds cheimonophyllon E (**216**) and 5-hydroxymethyl-furancarbaldehyde (**217**) were isolated from the taxon. *Pleurotus ostreatus* was also reported to be active against nematodes (Thorn and Barron 1984, 1987; Kwok et al. 1992; Stadler et al. 1994a). *Trans*-2-decenedioic acid (**218**) was isolated from *P. ostreatus* as the principal nematicide

(Kwok et al. 1992). The compound could immobilize 95% of the nematode *Panagrellus redivivus* at a concentration of 300 ppm (Kwok et al. 1992). Four aromatics *p*-anisaldehyde (**219**), *p*-anisyl alcohol (**220**), 1-(4-methoxyphenyl)-1,2-propanediol (**221**) and 2-hydroxy-(4'-methoxy)-propiophenone (**222**) were isolated from *Pleurotus pulmonarius* (Stadler et al. 1994a). The LD_{50} values of these compounds against *Caenorhabditis elegans* were all similar, at about 100 ppm (Stadler et al. 1994a). Extracts of the mushroom *Pleurotus pulmonarius* contained a nematicidal fatty acid *S*-coriolic acid (**223**). This fatty acid could kill the nematode *Caenorhabditis elegans* with an LD_{50} value at 10 ppm (Koitabashi et al. 2004). In addition, linoleic acid (**7**) was also found from the fungus (Stadler et al. 1994).

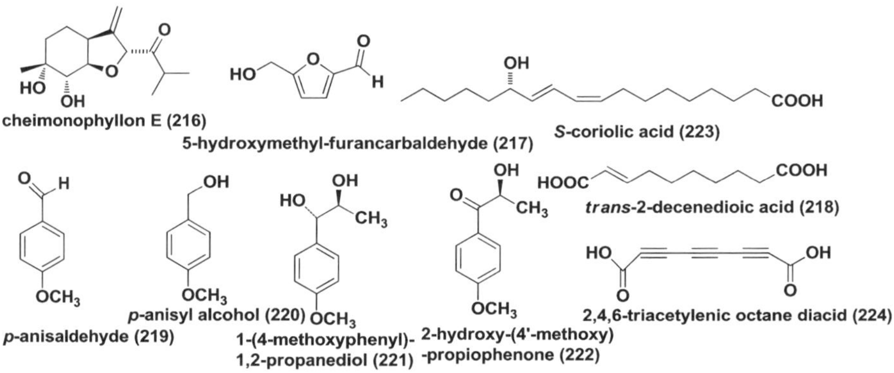

Nematicidal Metabolites from Polyporus Sulphureus and Poria Cocos

The cyclic depsipeptide beauvericin (**38**) was active against *Meloidogyne incognita*, *Caenorhabditis elegans* and *Bursaphelenchus xylophilus*. This peptide beauvericin (**39**) has also been isolated from several ascomycetes, but also was isolated from the basidiomycete *Polyporus sulphureus* (Deol et al. 1978). *Polyporus cocos* is a widely used traditional medicinal fungus, and produces a novel alkyne 2,4,6-triacetylenic octane diacid (**224**) found capable of killing 83.9% of *Meloidogyne arenaria* and 73.4% of *Panagrellus redivivus* at 500 ppm within 12 h (Li et al. 2005).

Nematicidal metabolites from *Stereum* spp.

Stereum sp. 8954 produced two new aromatics, 3,5-dihydroxy-4-(3-methyl-but-2-enyl)-benzene-1,2-dicarbaldehyde (**225**) and butyl 2,4-dihydroxy-6-methylbenzoate (**226**). 3,5-Dihydroxy-4-(3-methyl-but-2-enyl)-benzene-1,2-dicarbaldehyde (**225**) could kill about 90% of *Panagrellus redivivus* at 100 ppm in 12 h, while butyl 2,4-dihydroxy-6-methylbenzoate (**226**) was less active, capable of killing about 50% of the same nematode at 200 ppm in 24 h (Li et al. 2006). Five cadinane sesquiterpenoids, named stereumin A (**227**), B (**228**), C (**229**), D (**230**) and E (**231**) were isolated from the culture broth of the fungal strain *Stereum* sp. CCTCC AF

207024. The five compounds showed nematicidal activities against the nematode *P. redivivus* at 400 mg L^{-1}. Stereumin C (**229**) and stereumin D (**230**) killed 84.4% and 94.9% of *P. redivivus* respectively in 48 h (Li et al. 2008).

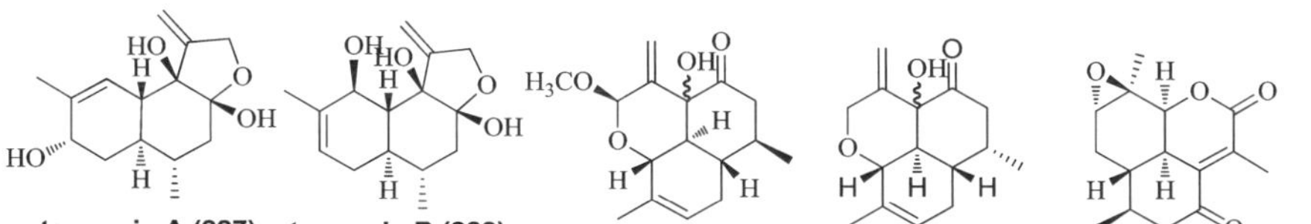

stereumin A (227) stereumin B (228) stereumin C (229) stereumin D (230) stereumin E (231)

3,5-dihydroxy-4-(3-methylbut-2-enyl)benzene-1,2-dicarbaldehyde (225)

butyl 2,4-dihydroxy-6-methylbenzoate (226)

Conclusions

More than 200 nematicidal compounds have been obtained from fungi, and their diversified structures mainly belong to alkaloids, quinones, isoepoxydons, pyrans, furans, peptides, macrolides, terpenoids, fatty acids, diketopiperazines, aphthalenes, simple aromatics and other kinds of compounds. Among these nematicidal compounds, about 60% are new natural isolates, which implies that searching for new compounds by screening with different models is an efficient method. Secondary metabolites in fungi are abundant, e.g., 24 compounds including 15 new isolates with nematicidal activities have been isolated from the cultures of *Lachnum papyraceum* and elucidated to be isoepoxydon, isocoumarin, mycorrhizin and furan compounds (Stadler and Anke 1993a, b; Stadler et al. 1995a, b, c, d, e; Shan et al. 1996), and six new sesquiterpenes and a novel monoterpene possessing nematicidal activity are fungal metabolites isolated from cultures of the basidiomycete *Cheimonophyllum candidissimum* (Stadler et al. 1995). Fungi are therefore a major source of biologically active natural products.

Many attempts have been made to find potent nematicidal substances. Ivermectin isolated from actinomycete is a commercially available nematicide up to now, but no major commercial product based on nematode-toxic fungi and the compounds isolated from fungi have been developed at present. Sharma (1994) reported that broth cultures of *Pleurotus sajor-caju* can immobilize the mushroom nematode in *Agaricus bisporus*.Xiang et al (2000) reported on the effects of *Pleurotus ostreatus* on the peanut root-knot nematode *Meloidogyne arenaria* in a greenhouse. The experiment showed that *Pleurotus ostreatus* could markedly lower *Meloidogyne arenaria* infection numbers and peanut root knot disease was also reduced by 87–94%. The key factor affecting control effectiveness was the application time of *Pleurotus ostreatus* in the soil (Xiang and Feng 2000). The potential of oyster mushrooms to attack and kill *Heterodera schachtii* was studied, and the result showed some mushrooms could significantly control the nematode (Palizi et al. 2009). Omphalotin A, B, C and D isolated from *Omphalotus olearius* had similar nematicidal activity to the commercially available nematicide ivermectin (Mayer et al. 1997; Stener et al. 1997; Büchel et al. 1998; Anke et al. 1999). It is necessary to search for nematode-toxic fungi and their active compounds to exploit novel nematicides.

References

Abraham, W. R., & Arfmann, H. A. (1992). *Penicillium camemberti* a new source of brefeldin A. *Planta Medica, 58,* 484.

Adams, K. B., Wu, M. T., & Salunkhe, D. K. (1976). Effects of gamma radiation on growth and patulin production of *Penicillium expansum* and *Penicillium patulim*. *Environmental and Experimental Botany, 16,* 189–193.

Alfaro, M. C., Urios, A., González, M. C., Moya, P., & Blanco, M. (2003). Screening for metabolites from *Penicillium novae-zeelandiae* displaying radical-scavenging activity and oxidative mutagenicity: Isolation of gentisyl alcohol. *Mutation Research, 539,* 187–194.

Anderson, M. G., Jarman, T. B., & Rickards, R. W. (1995). Structures and absolute configurations of Antibiotics of the oligosporon group from the nematode-trapping fungus *Arthrobotrys oligospora*. *Journal of antibiotics, 48*(5), 391–397.

Anke, H., Kolthoum, I., Zähner, H., & Laatsch, H. (1980a). Metabolic products of microorganisms. The anthraquinones of the *Aspergillus glaucus* group. I. Occurrence, isolation, identification and antimicrobial activity. *Archives of Microbiology, 126,* 223–230.

Anke, H., Kolthoum, I., & Laatsch, H. (1980b). Metabolic products of microorganisms. The anthraquinones of the *Aspergillus glaucus* group. II. Biological activity. *Archives of Microbiology, 126,* 231–236.

Anke, H., Stadler, M., Mayer, A., & Sterner, O. (1995). Secondary metabolites with nematicidal and antimicrobial activity from nematophagous fungi and Ascomycetes. *Canadian Journal of Botany, 73*(Suppl. 1), 932–939.

Anke, M., Michael, K., Birgit, H., Olov, S., & Heidrun, A. (1999). In-vitro and in-vivo nematicidal activities of the cyclic dodecapeptide omphalotin A. *Pesticide Science, 55,* 27–30.

Arai, K., Rawlings, B. J., Yoshizawa, Y., & Vederas, J. C. (1989). Biosynthesis of antibiobic A26771B by *Penicillium turbatum* and dehydrocurvularin by *Alternaria cinerariae*: Comparison of stereochemistry of polykide and fatty acid enoyl thiol ester reductases. *Journal of the American Chemical Society, 111,* 3391–3399.

Arnone, A., Assante, G., Montorsi, M., Nasini, G., & Ragg, E. (1993). Secondary mould metabolites. XLIII: Isolation and structure determination of lethaloxin, a fungal macrolide from *Mycosphaerella lethalis*. *Gazzetta Chimica Italiana, 123,* 71–73.

BačÍkovÁ, D., Betina, V., & Nemec, P. (1965). Antihelminthic activity of antibiotics. *Nature, 206,* 1371–1372.

Banks, R. M., Blanchflower, S. E., Everett, J. R., Manger, B. R., & Reading, C. (1997). Novel anthelmintic metabolites from an *Aspergillus* species; the Aspergillimides. *Journal of Antibiotics, 50,* 840–846.

Barron, G. L., & Thorn, R. G. (1987). Destruction of nematodes by species of *Pleurotus*. *Canadian Journal of Botany, 65,* 774–778.

Barrow, K. D., & Murphy, W. S. (1972). The structure of alboleersin and luteoleesin; the identity of luteoleersin with cochlioquinone A. *Journal of the Chemical Society Perkin Transactions, 1,* 2837–2839.

Beecham, A. F., Fridrichsons, J., & Mathieson, A. M. (1966). The structure and absolute configuration of gliotoxon and the absolute configuration of sporidesmin. *Tetrahedron Letters, 27,* 3131–3138.
Bernardini, M., Carilli, A., Pacioni, G., & Santurbano, B. (1975). Isolation of beauvericin from *Pacilomyces fumoso-roseus*. *Phytochemistry, 14,* 1865.
Blanchflower, S. E., Banks, R. M., Evertt, J. R., Manger, B. R., & Reading, C. (1991). New paraherquamide antibiotics with anthelmintic activity. *Journal of Antibiotics, 44,* 492–497.
Blanchflower, A. E., Banks, R. M., Everett, J. R., & Reading, C. (1993). Further novel metabolites of the paraherquamide family. *Journal of Antibiotics, 46,* 1355–1363.
Blight, M. M., Coppen, J. J. W., & Grove, J. F. (1968). The biogenesis, from mevalonic acid of the steroidal antifungal metabolite viridian. *Chemical Communications, 18,* 1117–1118.
Bodo, B., Rebuffat, S., Hajji, M. E., & Davoust, D. (1985). Structures of trichorzianine A IIIc, an antifungal peptide from *Trichoderma harzianum*. *Journal of American Chemical Society, 107,* 6011–6017.
Brown, A. E., Finlay, R., & Ward, J. S. (1987). Antifungal compounds produced by *Epicoccum purpurascens* against soil-borne plant pathogenic fungi. *Soil Biology & Biochemistry, 19,* 657–664.
Brückner, H., Nicholson, G. J., & Jung, G. (1980). Gas chromatographic determination of the configuration of isovaline in antiamoebin, samarosporin (emerimicin IV), stilbellin, suzukacillins and trichotoxins. *Chromatographia, 13,* 209–213.
Büchel, E., Martini, U., Mayer, A., Anke, H., & Sterner, O. (1998). Omphalotins B, C and D, nematicidal cyclopeptides from *Omphalotus olearius*. Absolute configuration of omphalotin A. *Tetrahedron, 54,* 5345–5352.
Burge, W. R., Buckley, L. J., Sullivan, J. D., McGrattan, C. J., & Ikawa, M. (1976). Isolation and biological activity of the pigments of the mold *Epicoccum nigrum*. *Journal of Agricultural and Food Chemistry, 24,* 555–559.
Chen, L. J., Chen, Y., Zhang, G. D., & Duan, Y. X. (2010). Nematicidal activity of extraction and fermentation filtrate of Basidiomycetes collected in liaoning province, China. Chinese. *Journal of Biological Control, 26,* 467–473.
Clerkyjzio, M., & Sterner, O. (1997). Conversion of velutinal esters in the fruit bodies of *Russula cuprea*. *Phytochemistry, 45,* 1569–1572.
Cutler, H. G., Crumley, F. G., Cox, R. H., Springer, J. P., Arrendale, R. F., Cole, R. J., & Cole, P. D. (1984). Ophiobolins G and H: New fungal metabolites from a novel source, *Aspergillus ustus*. *Journal of Agricultural and Food Chemistry, 32,* 768–772.
Daniewski, W. M., Gumulka, M., & Kibicki, P. (1990). Furantriol, a lactarane sesquiterpene from *Lactarius mitissimus*. *Phytochemistry, 29,* 527–529.
Dasenbrock, J. (1994). Isolierung und Strukturaufklärung neuer Wirkstoffe aus Höheren Pilzen. Ph.D. thesis, Department of chemistry, University of Bonn, Bonn, Germany.
Deol, B. S., Ridley, D. D., & Singh, P. (1978). Isolation of cyclodepsipeptides from plant pathogenic fungi. *Australian Journal of Chemistry, 31,* 1397–1399.
Djian, C., Pijarouvski, L., Ponchet, M., & Arpin, N. (1991). Acetic acid, a selective nematicidal metabolite from culture filtrate of *Paecilomyces lilacinus* (Thom) Samsan and *Trichoderma longibrachiatum* Rifai. *Nematologica, 37,* 101–102.
Dombrink-Kurtzman, M. A., & Blackburn, J. A. (2005). Evaluation of several culture media for production of patulin by *Penicillium* species. *International Journal of Food Microbiology, 98,* 241–248.
Dong, J. Y. (2005). Ph.D. thesis, Laboratory for conservation and utilization of bio-resource, Yunnan University, Kunming, China.
Dong, J. Y., Mo, M. H., Chen, J. H., Li, Q. Y., & Zhang, K. Q. (2000). The nematicidal stability of *Lampteromyces japonicus*. *Journal of Yunnan University, 22,* 365–368.
Dong, J. Y., Zhang, K. Q., Zhao, Z. X., Liu, W. Z., & Li, Q. Y. (2001). Nematicidal activity of perylenequinones photosensitive compounds. *Mycosystema, 20,* 515–519.

Dong, J. Y., Zhao, Z. X., Cai, L., Liu, S. Q., Zhang, H. R., Duan, M., & Zhang, K. Q. (2003). Nematicidal effect of freshwater fungal cultures against the pine-wood nematode, *Bursaphelenchus xylophilus*. *Fungal Diversity, 15,* 123–133.

Dong, J. Y., He, H. P., Shen, Y. M., & Zhang, K. Q. (2005a). Nematicidal Epipolysulfanyldioxopiperazines from *Gliocladium roseum*. *Journal of Natural Products, 68,* 1510–1513.

Dong, J. Y., Li, R., He, H. P., & Zhang, K. Q. (2005b). Nematicidal sphingolipids from the freshwater fungus *Paraniesslia* sp. YMF1.01400. *European Journal of Lipid Science and Technology, 107,* 779–785.

Dong, J. Y., Zhou, W., Li, L., Li, G. H., Liu, Y. J., & Zhang, K. Q. (2006a). A new Epidithiodioxopiperazine Metabolite Isolated from *Gliocladium roseum* YMF1.00133. *Chinese Chemical Letters, 17,* 922–924.

Dong, J. Y., Zhou, Y. P., Li, R., Zhou, W., Zhu, Y. H., Huang, R., & Zhang, K. Q. (2006b). Newnematicidal azaphilones from the aquatic fungus *Pseudohalonectria adversaria* YMF1.01019. *FEMS Microbiology Letters, 264,* 65–69.

Dong, J. Y., Zhu, Y. H., Song, H. C., Li, R., He, H. P., Liu, H. Y., Huang, R., Zhou, Y. P., Wang, L., & Zhang, K. Q. (2007). Nematicidal Resorcylides from the Aquatic Fungus *Caryospora callicarpa* YMF1.01026. *Journal of Chemical Ecology, 33,* 1115–1126.

Dong, J. Y., Song, H. C., Li, J. H., Tang, Y. S., Sun, R., Wang, L., Zhou, Y. P., Wang, L. M., Shen, K. Z., Wang, C. R., & Zhang, K. Q. (2008). Ymf 1029A E, Preussomerin Analogues from the Fresh-Water-Derived Fungus YMF 1.01029. *Journal of Natural Products, 71,* 952–956.

Dong, J. Y., Wang, L., Song, H. C., Wang, L. M., Shen, K. Z., Sun, R., Li, G. H., Li, L., & Zhang, K. Q. (2010). Ophiocerol, a novel macrocylic neolignan from the aquatic fungus *Ophioceras dolichostomum* YMF1.00988. *Natural Products Research, 24,* 1004–1012.

Dornberger, K., Ihn, W., Ritzau, M., Gräfe, U., Schlegel, B., Fleck, W. F., & Metzger, J. W. (1995). Chrysospermins, new peptaibol antibiotics from *Apiocrea chrysosperma* Ap101. *Journal of Antibiotics, 48,* 977–989.

Ekaterini, R., Lawrence, A. L., & Neussa, G. (2008). *Muscodor albus*, a potential biocontrol agent against plant-parasitic nematodes of economically important vegetable crops in Washington State, USA. *Biological Control, 45,* 380–385.

El-Din, H. K. M., Allam, A., & Tag, B. A. (2012). Nematicidal activity of some biopesticide agents and microorganisms against root-knot nematode on tomato plants under greenhouse conditions. *Journal of Plant Protection Research, 52,* 47–52.

Fardos, A. M. B. (2009). Efficacy of some *Trichoderma* species in the control of *Rotylenchulus reniformis* and *Meloidogyne javanica*. *Archives of Phytopathology and Plant Protection, 42,* 361–369.

Freeman, G. G., Gill, J. E., & Waring, W. S. (1959). The structure of trichothecin and its hydrolysis products. *Journal of the Chemical Society,* (0), 1105–1132.

Fukuda, T., Arai, M., Yamaguchi, Y., Masuma, R., Tomoda, H., & Ōmura, S. (2004). New beauvericins, potentiators of antifungal miconazole activity, produced by *Beauveria* sp. FKI-1366. *Journal of Antibiotics, 57,* 110–116.

Ghisalberti, E. L., & Rowland, C. V. (1997). 6-Chlorode-hydrocurvularin, a new metabolite from *Cochliobolus spicifer*. *Journal of Natural Products, 56,* 2175–2177.

Gill, M. (1994). Pigments of fungi (Macromycetes). *Natural Product Reports, 11,* 67–90.

Giuma, A. Y., & Cooke, R. C. (1971). Nematotoxin production by *Nematoctonus haptocladus* and *N. concurrens*. *Transactions of the British Mycological Society, 56,* 89–94.

Giuma, A. Y., Hackett, A. M., & Cooke, R. C. (1973). Thermostable nematotoxins produced by germinaling conidia of some endozoic fungi. *Transactions of the British Mycological Society, 60,* 49–56.

Godtfredsen, W. O., & Vangedal, S. (1964). Trichodermin, a new antibiotic related to trichothecin. *Proceedings of the Chemical Society, 6,* 188–189.

Grabley, S., Hammann, P., Thiericke, R., Wink, J., Philipps, S., & Zeeck, A. (1993). Secondary metabolites by chemical screening. 21 Clonostachydiol, A novel anthelmintic macrodiolide from the fungus *Clonostachys cylindrospora*. *Journal of Antibiotics, 46,* 343–345.

Gruhn, N., Schoettler, S., Sterner, O., & Anke, T. (2007). Biologically active metabolites from the basidiomycete *Limacella illinita* (Fr.) Murr. Zeitschrift fur Naturforschung—Section C. *Journal of Biosciences, 62,* 808–812.

Guo, J. P., Zhu, C. Y., Zhang, C. P., Chu, Y. S., Wang, Y. L., Zhang, J. X., Wu, D. K., Zhang, K. Q., & Niu, X. M. (2012). Thermolides, potent nematocidal PKS-NRPS hybrid metabolites from thermophilic fungus *Talaromyces thermophiles*. *Journal of the American Chemical Society, 134,* 20306–20309.

Hamill, R. L., Higgens, C. E., Boaz, N. E., & Gorman, M. (1969). The structure of beauvericin, a new depsipeptide antibiotic toxic to *Artemia salina*. *Tetrahedron Letters, 49,* 4255–4258.

Hansson, T., Sterner, O., & Strid, K. (1995). Chemotaxonomic evidence for a division of *Lactarius vellereus* and *L. bertillonii* as different species. *Phytochemistry, 39,* 363–365.

Harvey (2000). Strategies for discovering drugs from previously unexplored natural products. *Drug Discovery Today, 7,* 94–300.

Hautzel, R., & Anke, H. (1990). Screening of basidiomycetes and ascomycetes for plant growth regulating substances. Introduction of the gibberellic acid induced de-novo synthesis of hydrolytic enzymes in embryoless seeds of *Triticum aestivum* as test system. Zeitschrift fur Naturforschung—Section C.*Journal of Biosciences, 45,* 68–73.

Hautzel, R., Anke, H., & Sheldrick, W. S. (1990). Mycenon, a new metabolite from a *Mycena* species ta 87202 (Basidiomycetes) as an inhibitor of isocitrate lyase. *Journal of Antibiotics, 43,* 1240–1244.

Hayashi, M., Wada, K., & Munakata, K. (1981). New nematicidal metabolites from a fungus, *Irepex lacteus*. *Agricultural and Biological Chemistry, 45,* 1527–1529.

Hayashi, H., Nakatani, T., Inoue, Y., Nakayama, M., & Nozaki, H. (1997). New dihydroquinolinone toxic to *Artemia salina* produced by *Penicillium* sp. NTC-47. Bioscience Biotechnology and. *Biochemistry, 61,* 914–916.

Hayashi, A., Fujioka, S., Nukina, M., Kawano, T., Shimada, A., & Kimura, Y. (2007). Fumiquinones A and B, nematicidal quinones produced by *Aspergillus fumigatus*. Bioscience Biotechnology and. *Biochemistry, 71,* 1697–1702.

Haydock, P. P. J., Woods, S. R., Grove, I. G., & Hare, M. C. (2006). Chemical control of nematodes. In R. N. Perry & M. Moens (Eds.), *Plant Nematology* (pp. 392–410). Wallingford: CAB Interna-tional.

He, J., Wijeratne, E. M., Bashyal, B. P., Zhan, J., Seliga, C. J., Liu, M. X., Pierson, E. E., Pierson, L. S., VanEtten, H. D., & Gunatilaka, A. A. (2004). Cytotoxic and other metabolites of Aspergillus inhabiting the rhizosphere of sonoran desert plants. *Journal of Natural Products, 67,* 1985–1991.

Hibbett, D. S., & Thorn, R. G. (1994). Nematode-trapping in *Pleurotus tuberregium*. *Mycologia, 86,* 696–699.

Huang, Z. L., Dan, Y., Huang, Y. C., Lin, L. D., Li, T. H., Ye, W. H., & Wei, X. Y. (2004). Sesquiterpenes from the mycelial cultures of *Dichomitus squalens*. *Journal of Natural Products, 67,* 2121–2123.

Hyeon, S., Ozaki, A., Suzuki, A., & Tamura, S. (1976). Isolation of *αβ*-dehydrocurvularin and *β*-hydroxycurvularin from *Alternaria tomato* as sporulation-suppressing factors. *Agricultural and Biological Chemistry, 40,* 1663–1664.

Kanai, Y., Fujimaki, T., Kochi, S., Konno, H., Kanazawa, S., & Tokumasu, S. (2004). Paeciloxazine, a novel nematicidal antibiolic from *Paecilomyces* sp. *Journal of Antibiotics, 57,* 24–28.

Kang, S. W., & Kim, S. W. (2004). New antifungal activity of penicillic acid against *Phytophthora* species. *Biotechnology letters, 26,* 695–698.

Kavanagh, F., Hervey, A., & Robbins, W. J. (1949). Antibiotic substances from basidiomycetes. *Proceedings of the National Academy of Sciences, 35,* 343–349.

Kawazu, K., Murakami, T., Ono, Y., Kanzaki, H., Kobayashi, A., Mikawa, T., & Yoshikawa, N. (1993). Isolation and characterization of two novel nematicidal depsipeptides from an imperfect fungus, strain D1084. *Bioscience Biotechnology and Biochemistry, 57,* 98–101.

Kennedy, N., & Tampion, J. (1978). A nematotoxin from *Nematotonus robustus*. *Transantions of the British Mycological Society, 70,* 140–141.

Khambay, B. P. S., Bourne, J. M., Cameron, S., Kerry, B. R., & Zaki, M. J. (2000). A nematicidal metabolite from *Verticillium chlamydosporium*. *Pest Management Science, 56,* 1098–1099.
Khan, S. T., & Kgan, T. A. (1992). Effect of culture filtrates of soil fungi on the hatching and mortality of root knot nematode (*Meloidogyne incognita*). *Current Nematology, 3,* 53–60.
Khan, M. R., & Haque, Z. (2011). Soil application of *Pseudomonas fluorescens* and *Trichoderma harzianum* reduces root-knot nematode, *Meloidogyne incognita*, on tobacco. *Phytopathologia Mediterranea, 50,* 257–266.
Kim, H. L., & Kochevar, J. (1995). Isolation of brefeldin A. *General Pharmacology, 26,* 363–364.
Kim, J. C., Choi, G. J., Park, J. H., Kim, H. T., & Cho, K. Y. (2001). Activity against plant pathogenic fungi of phomalactone isolated from *Nigrospora sphaerica*. *Pest Management Science, 57,* 554–559.
Kimura, Y., Mori, M., Suzuki, A., & Kobayashi, A. (1981). Isolation and identification of two nematicidal substances from roots of *Erigeron philadelphicus* L. and nematicidal activities of their related compounds. *Agricultural and Biological Chemistry, 45,* 2915–2917.
Kimura, Y., Kusano, M., Koshino, H., Uzawa, J., Fujioka, S., & Tani, K. (1996a). Penigequinolones A and B, pollen-growth inhibitors produced by *Penicillium* sp. No. 410. *Tetrahedron Letters, 37,* 4961–4964.
Kimura, Y., Nakahara, S., & Fujioka, S. (1996b). Aspyrone, a nematicidal compound isolated from the fungus, *Aspergillus melleus*. Bioscience Biotechnology and. *Biochemistry, 60,* 1375–1376.
Kimura, Y., Shimada, A., Kusano, M., Yoshii, K., Morita, A., Nishibe, M., Fujioka, S., & Kawano, T. (2002). Myxostiolide, myxostiol, and clavatoic acid, plant growth regulators from the fungus *Myxotrichum stipitatum*. *Journal of Natural Products, 65,* 621–623.
Kimura, Y., Tani, S., Hayashi, A., Ohtani, K., Fujioka, S., Kawano, T., & Shimada, A. (2007). Nematicidal activity of 5-hydroxymethyl-2-furoic acid against plant-parasitic nematodes. *Zeitschrift Fur Naturforschung C, 62,* 234–238.
Kind, R., Zeeck, A., Grabley, S., Thiericke, R., & Zerlin, M. (1996). Secondery metabolites by chemical screening. 30. Helmidiol, a new m acrodiolide from *Alternaria alternaya*. *Journal of Natural Products, 59,* 539–540.
Kithsiri Wijeratne, E. M., Paranagama Priyani, A., & Leslie Gunatilaka, A. A. (2006). Five new isocoumarins from Sonoran desert plant-associated fungal strains *Paraphaeosphaeria quadriseptata* and *Chaetomium chiversii*. *Tetrahedron, 62,* 8439–8446.
Kobayashi, A., Hino, T., Yata, S., Itoh, T. J., Sato, H., & Kawazu, K. (1988). Unique spindle poisons. Curvularin and its derivatives, isolated from *Penicillium* species. *Agricultural and Biological Chemistry, 52,* 3119–3123.
Koitabashi, M., Kajitani, Y., & Hirashima, K. (2004). Antifungal substances produced by fungal strain Kyu-W63 from wheat leaf and its taxonomic position. *Journal of General Plant Pathology, 70,* 124–130.
Konishi, K., Iida, A., Kaneko, M., Tomioka, k, Tokuda, H., Nishino, H., & Kumeda, Y. (2003). Cancer preventive potential of trichothecenes from *Trichothecium roseum*. *Bioorganic & Medicinal Chemistry, 11,* 2511–2518.
Köpcke, B., Johansson, M., Sterner, O., & Anke, H. (2002). Biologically active secondary metabolites from the ascomycete A111-95. 1. Production, isolation and biological activities. *Journal of Antibiotics, 55,* 36–40.
Krasnoff, S. B., & Gupta, S. (1994). Identification of the antibiotic phomalactone from the entomopathogenic fungus *Hirsutella thompsonii* var. *synnematosa*. *Journal of Chemical Ecology, 20,* 293–302.
Kumazawa, S., Kanda, M., Utagawa, M., Chiba, N., Ohtani, H., & Mikawa, T. (2003). MK7924, a novel metabolite with nematocidal activity from *Coronophora gregaria*. *Journal of Antibiotics, 56,* 652–654.
Kupka, J., Anke, T., Mizumoto, K., Giannetti, B. M., & Steglich, W. (1983). Antibiotics from Basidiomycetes. XVII The effect of marasmic acid on nucleic acid metabolism. *Journal of Antibiotics, 36,* 155–160.

Kusano, M., Koshino, H., Uzawa, J., Fujioka, S., Kawano, T., & Kimura, Y. (2000). Nematicidal alkaloids and related compounds produced by the fungus *Penicillium* cf. *Simplicissimum*. Bioscience Biotechnology and. *Biochemistry, 64,* 2559–2568.

Kusano, M., Nakagami, K., Fujioka, S., Kawano, T., Shimada, A., & Kimura, Y. (2003). *β,γ*; -dehydrocurvularin and related compounds as nematicides of *Pratylenchus penetrans* from the Fungus *Aspergillus sp*. Bioscience Biotechnology and. *Biochemistry, 67,* 1413–1416.

Kwok, O. C. H., Plattner, R., Weisleder, D., & Wicklow, D. T. (1992). A namatical toxin from *Pleurotus ostreatus* NRRL 3526. *Journal of Chemical Ecology, 18,* 127–136.

Lai, S., Shizuri, Y., Yamamura, S., Kawai, K., Terada, Y., & Furukawa, H. (1989). Novel curvularin-type metabolites of a hybrid strain ME 0005 derived from *Penicillium citreo-viride* B. IFO and 4692. *Tetrahedron Letters, 30,* 2241–2244.

Lai, S., Shizuri, Y., Yamamura, S., Kawai, K., Terada, Y., & Furukawa, H. (1990). New metabolites of two hybrid strains ME 0004 and 0005 derived from *Penicillium citreo-viride* B. IFO 6200 and 4692. *Chemistry Letters, 19,* 589–592.

Li, Y. Z. (2005). Researches on the screening of nematophagous Basidiomycetes and their effects on nematodes. Master thesis. Laboratory of Plant Pathology, College of Agriculture, Guizhou University, Guiyang.

Li, G. H., Dong, J. Y., Mo, M. H., & Zhang, K. Q. (2001). Nematicidal activity of nematophagous *Pleurotus* and allied fungi to *Panagrellus redivivus*. Chinese. *Journal of Biological Control, 17,* 26–29.

Li, G. H., Shen, Y. M., & Zhang, K. Q. (2005). Nematicidal activity and chemical component of *Poria cocos*. *The Journal of Microbiology, 43,* 17–20.

Li, G. H., Li, L., Duan, M., & Zhang, K. Q. (2006). The Chemical Constituents of the Fungus *Stereum* sp. *Chemistry and Biodiversity, 3,* 210–216.

Li, G. H., Wang, X. B., Zheng, L. J., Li, L., Huang, R., & Zhang, K. Q. (2007a). Nematicidal metabolites from the fungus *Pleurotus ferulae* Lenzi. *Annals of Microbiology, 57,* 527–529.

Li, G. H., Yu, Z. F., Li, X., Wang, X. B., Zheng, L. J., & Zhang, K. Q. (2007b). Nematicidal metabolites produced by the endophytic fungus *Geotrichum* sp. AL4. *Chemistry and Biodiversity, 4,* 1520–1524.

Li, G. H., Duan, M., Yu, Z. F., Li, L., Dong, J. Y., Wang, X. B., Guo, J. W., Huang, R., Wang, M., & Zhang, K. Q. (2008). Stereumin A-E, sesquiterpenoids from the fungus *Stereum* sp. CCTCC AF 207024. *Phytochemistry, 69,* 1439–1445.

Liesch, J. M., & Wichmann, C. F. (1990). Novel antinematodal and antiparasitic agents from *Penicillium Charlesii*. II. Structure determination of paraherquamides B, C, D, E, F, and G. *Journal of Antibiotics, 43,* 1380–1386.

Liu, Y. J. (2005). Ph.D. thesis, Laboratory for conservation and utilization of bio-resource, Yunnan University, Kunming, China.

Liu, Y. J., Liu, Y., & Zhang, K. Q. (2008). Xanthothone E, a new nematicidal N-compound from *Coprinus xanthothrix*. *Chemistry of Natural Compounds, 44,* 161–162.

Liu, Y. J., Zhai, C. Y., Liu, Y., & Zhang, K. Q. (2009). Nematicidal activity of *Paecilomyces* spp. and isolation of a novel active compound. *The Journal of Microbiology, 47,* 248–252.

Liu, J. H., Wang, L., Qiu, J. Y., Jiang, L., Yan, J. Y., Liu, T., Liu, W. C., & Duan, Y. X. (2011). Nematicidal activity of *gymnoascus reesii* against *meloidogyne incognita*. *African Journal of Microbiology Research, 5,* 2715–2719.

Lopez-Diaz, T. M., & Flannigan, B. (1997). Production of patulin and cytochalasin E by *Aspergillus clavatus* during malting of barley and wheat. *International Journal of Food Microbiology, 35,* 129–136.

Lorenzen, K., Anke, T., Anders, U., Hindermayr, H., & Hansske, F. (1994). 14-epidihydrocochlioquinone B, and 14-epocochliqquinone B, antibiotics from fermentation of the ascomycete Neobulgaria pura: Structure elucidation and effects on platelet aggregation. *Zertschrift Fur Naturforschung C, 49C,* 313–320.

Luo, H., Liu, Y. J., Fang, L., Li, X., Tang, N. H., & Zhang, K. Q. (2007). *Coprinus comatus* Damages Nematode Cuticles Mechanically with Spiny Balls and Produces Potent Toxins. *Applied and Environmental Microbiology, 73,* 3916–3923.

Martínez-Luis, S., González, M. C., Ulloa, M., & Mata, R. (2005). Phytotoxins from the fungus *Malbranchea aurantiaca*. *Phytochemistry, 66,* 1012–1016.

Mayer, A. (1995). Ph.D. thesis, University of Kaiserslautern, Kaiserslauten.

Mayer, A., Anke, H., & Sterner, O. (1997). Omphalotin, a new cycle peptide with potent nematicidal activity from *Omphalotus olearius*. I. Fermentation and Biological activity. *Natural Product Letters, 10,* 25–32.

Meira, B. E., Edna, S., & Yitzhak, S. (2006). Nematicidal activity of *Chrysanthemum coronarium*. *European Journal of Plant Pathology, 114,* 427–433.

Meng, Q., Shi, X., Meng, F., Feng, X., & Sun, J. (2012). Isolation of an *Acremonium* sp. from a screening of 52 seawater fungal isolates and preliminary characterization of its growth conditions and nematicidal activity. *Biotechnology Letters, 34,* 1847–1850.

Metzger, J., Schlegel, B., Fleck, W. P., Dornberger, K., Ihn, W., Schade, W., & Gräfe, U. J. (1994). Canadian patent Appl. No. 2,115,966 (18.02.94).

Meyer, S. L. F., Huettel, R. N., Liu, X. Z., Humber, R. A., Juba, A., & Nitao, J. K. (2004). Activity of fungal culture filtrates against soybean cyst nematode and root-knot nematode egg hatch and juvenile motility. *Nematology, 6,* 23–32.

Mirrington, R. N., Ritchie, E., Shoppee, C. W., Taylor, W. C., & Sternhell, C. (1964). The constitution of radicicol. *Tetrahedron Letters, 7,* 365–370.

Mo, M. H., Li, G. H., Dong, J. Y., & Zhang, K. Q. (2000). *Lampteromyces japonicus*- A new nematophagous fungi. *Mycosystema, 19,* 529–533.

Mori, M., Hyeon, S., Kimura, Y., & Suzuki, A. (1982). The nematicidal activity of acetylene compounds. *Agricultural and Biological Chemistry, 46,* 309–311.

Moulé, Y., & Hatey, F. (1977). Mechanism of the in vitro inhibition of transcription by patulin, a mycotoxin from *Byssochlamys nivea*. *FEBS Letters, 74,* 121–125.

Munro, H. D., Musgrave, O. C., & Templeton, R. (1967). Curvularin. Part V. The compound $C_{16}H_{18}O_5$, *αβ*-dehydrocurvularin. *Journal of the Chemical Society C,* 947–948.

Nakahara, S., Kusano, M., Fujioka, S., Shimada, A., & Kimura, Y. (2004). Penipratynolene, a novel nematicide from *Penicillium bilaiae* Chalabuda. Bioscience Biotechnology and. *Biochemistry, 68,* 257–259.

Nielsen, K. F., Hansen, M. O., Larsen, T. O., & Thrane, U. (1998). Production of trichothecene mycotoxins on water damaged gypsum boards in Danish buildings. *International Biodeterioration & Biodegradation, 42,* 1–7.

Nitao, J. K. B., Meyer, S. L. F., Oliver, J. E., Schmidt, W. F., & Chitwood, D. J. (2002). Isolation of flavipin, a fungus compound antagonistic to plant-parasitic nematodes. *Nematology, 4,* 55–63.

Niu, X. M., Wang, Y. L., Chu, Y. S., Xue, H. X., Li, N., Wei, L. X., Mo, M. H., & Zhang, K. Q. (2010). Nematodetoxic aurovertin-type metabolites from a root-knot nematode parasitic fungus *Pochonia chlamydosporia*. *Journal of Agricultural and Food Chemistry, 58,* 828–834.

Nozawa, K., & Nakajima, S. (1979). Isolation of radicicol from *Penicillium luteo-aurantium*, and meleagrin, a new metabolite, from *Penicillium meleagrinum*. *Journal of Natural Products, 42,* 374–377.

Ohtani, K., Fujioka, S., Kawano, T., Shimada, A., & Kimura, Y. (2011). Nematicidal activities of 4-hydroxyphenylacetic acid and oidiolactone D produced by the fungus *Oidiodendron* sp. Zeitschrift fur Naturforschung—Section C. *Journal of Biosciences, 66,* 31–34.

Ōmura, S., Miyadera, H., Ui, H., Shiomi, K., Yamaguchi, Y., Masuma, R., Nagamitsu, T., Takano, D., Sunazuka, T., Harder, A., Kölbl, H., Namikoshi, M., Miyoshi, H., Sakamoto, K., & Kita, K. (2001). An anthelmintic compound, nafuredin, shows selective inhabition of complex I in helminth mitochondria. *Proceedings of the National Academy of Sciences USA, 98,* 60–62.

Ondeyka, J. G., Goegelman, R. T., Schaeffer, J. M., Kelemen, L., & Zitano, L. (1990). Novel antinematodal and antiparasitic agents from *Penicillium Charlesii*. I. Fermentation, isolation and biological activity. *Journal of Antibiotics, 53,* 1375–1379.

Palizi, P., Goltapeh, E. M., Pourjam, E., & Safaie, N. (2009). Potential of oyster mushrooms for the biocontrol of sugar beet nematode (*Heterodera schachtii*). *Journal of Plant Protection Research, 49,* 27–33.

Pandey, R. C., Meng, H., Cook, J. C., & Rinehart, K. L. (1977). Structure of antiamoebin I from high resolution field desorption and gas chromatographic mass spectrometry studies. *Journal of American Chemical Society, 99,* 5203–5205.

Park, J. O., Hargreaves, J. R., Cole, A. L. J., Ghisalberti, E. L., Gams, W., & Sivasithamparam, K. (2001). Cuticular disruption and mortality of *Caenorhabditis elegans* exposed to culture filtrate of *Byssochlamys nivea* westling. *Nematology, 3,* 355–363.

Park, J. O., Hargreaves, J. R., McConville, E. J., Stirling, G. R., Ghisalberti, E. L., & Sivasithamparam, K. (2004). Production of leucinostatins and nematicidal activity of Australian isolates of *Paecilomyces lilacinus* (Thom) Samson. *Letters in Applied Microbiology, 38,* 271–276.

Pfister, J. R. (1988). Isolation and bioactivity of 2-aminoquinoline from *Leucopaxillus albissimus. Journal of Natural Products, 51,* 969–970.

Polondky, J., Merrien, M. N., Prangé, T., Pascard, C., & Moreau, S. (1980). Isolation and structure (X-ray analysis) of marcfortine A, a new alkaloid from *Penicillium roqueforti. Journal of the Chemical Society—Chemical Communications, 13,* 601–602.

Prangé, T., Billion, M. A., Vuilhorgne, M., Pascard, C., Polonsky, J., & Moreau, S. (1981). Structures of marcfortine B and C (X-ray analysis), alkaloids from *Penicillium Roqueforti. Tetrahedron Letters, 22,* 1977–1980.

Quaghebeur, K., Coosemans, J., Toppet, S., & Compernolle, F. (1994). Cannabiorci- and 8-chlorocannabiorcichromenic acid as fungal antagonists from *Cylindrocarpon Olidum. Phytochemistry, 37,* 159–161.

Radwan, M. A., Farrag, S. A. A., Abu-Elamayem, M. M., & Ahmed, N. S. (2012). Biological control of the root-knot nematode, Meloidogyne incognita on tomato using bioproducts of microbial origin (2012). *Applied Soil Ecology, 56,* 58–62.

Rao, A. V. R., Murthy, V. S., & Sharma, G. V. M. (1995). The first synthesis and determination of absolute stereochemistry of clonostachydiol-Part II. *Tetrahedron Letters, 36,* 143–146.

Regaieg, H., Ciancio, A., Raouani, N. H., Grasso, G., & Rosso, L. (2010). Effects of culture filtrates from the nematophagous fungus *Verticillium leptobactrum* on viability of the root-knot nematode *Meloidogyne incognita.* World. *Journal of Microbiology and Biotechnology, 26,* 2285–2289.

Reimerdes, E. H., Engel, G., & Behnert, J. (1975). Investigation on the production of mycotoxins and quantitative evaluation. I The production of penicillic acid by *Penicillium cyclopium. Journal of Chromatography, 110,* 361–368.

Reino, J. L., Guerrero, R. F., Hernández-Galán, R., & Collado, I. G. (2008). Secondary metabolites from species of the biocontrol agent *Trichoderma. Phytochemistry Reviews, 7,* 89–123.

Robertson, A. P., Clark, C. L., Burans, T. A., Thompson, D. P., & Geary, T. G. (2002). Paraherquamide and 2-deoxy-paraherquamide distinguish cholinergic receptor subtypes in Ascaris Muscke. *The Journal of Pharmacology and Experimental Therapeufics, 302*(3), 853–860.

Reyes-Estebanez, M., Herrera-Parra, E., Cristóbal-Alejo, J., Heredia-Abarca, G., Canto-Canché, B., Medina-Baizabal, I., & Gamboa-Angulo, M. (2011). Antimicrobial and nematicidal screening of anamorphic fungi isolated from plant debris of tropical areas in Mexico. *African Journal of Microbiology Research, 5,* 1083–1089.

Robeson, D. J., & Strobel, G. A. (1981). αβ-dehydrocurvularin and curvularin from Alternaria cinerariae. *Zeitschrift fur Naturforschung C, 36c,* 1081–1083.

Robeson, D. J., & Strobel, G. A. (1985). The identification of a major phytotoxic component from *Alternaria* macrospore as *αβ*-dehydrocurvularin. *Journal of Natural Products, 48,* 139–141.

Ruanpanun, P., Tangchitsomkid, N., Hyde, K. D., & Lumyong, S. (2010). Actinomycetes and fungi isolated from plant-parasitic nematode infested soils: Screening of the effective biocontrol potential, indole-3-acetic acid and siderophore production. *World Journal of Microbiology and Biotechnology, 26,* 1569–1578.

Samson-Himmelstjerna, G. V., Harder, A., Sangster, N. C., & Coles, G. C. (2005). Efficacy of two cyclooctadepsipeptides, PF1022A and emodepside, against anthelmintic-resistant nematodes in sheep and cattle. *Parasitology, 130,* 343–347.

Sankaranarayanan, C., Hussaini, S. S., Kumar, P. S., & Prasad, R. D. (1997). Nematicidal effect of fungal filtrates against root-knot nematodes. *Journal of Biological Control, 11,* 37–41.

Sasaki, T., Takagi, M., Yaguchi, T., Miyadoh, S., Okada, T., & Koyama, M. (1992). A new anthelmintic cyclodepsipeptide, PF1022A. *Journal of Antibiotics, 45,* 692–697.
Schaeffer, J. M., Frazier, E. G., Bergstrom, A. R., Williamson, J. M., Liesch, J. M., & Goetz, M. A. (1990). Cochlioquinone A, a nematocidal agent which competes for specific [3H] ivermectin binding sites. *Journal of Antibiotics, 43,* 1179–1182.
Schlunegger, U. P., Kuchen, A., & Clemencon, H. (1976). Mycelium products in higher fungi. I. phenoxazine derivatives in Calocybe gambosa. *Helvetica Chimica Acta, 59,* 1383–1388.
Schwarz, M., Köpcke, B., Weber, R. W. S., Sterner, O., & Anke, H. (2004). 3-Hydroxypropionic acid as a nematicidal principle in endophytic fungi. *Phytochemistry, 65,* 2239–2245.
Shan, R., Stadler, M., Sterner, O., & Anke, H. (1996). New metabolites with nematicidal and antimicrobial activities from the ascomycete *Lachnum papyraceum* (Karst.) Karst. VIII. Isolation, structure determination and biological activities of minor metabolites structurally related to mycorrhizin A. *Journal of Antibiotics, 49,* 447–452.
Sharma, V. P. (1994). Potential of *Pleurotus sajor-caju* for biocontrol of *Aphelenchoides compost-icola* in *Agaricus bisporus* cultivation. *Mushroom Research, 3,* 15–20.
Shimada, A., Fujioka, S., Koshino, H., & Kimura, Y. (2010) Nematicidal activity of beauvericin produced by the fungus Fusarium bulbicola. *Zeitschrift fur Naturforschung C, 65,* 207–210.
Shiomi, K., Ui, H., Suzuki, H., Hatano, H., Nagamitsu, T., Takano, D., Miyadera, H., Yamashita, T., Kita, K., Miyoshi, H., Harder, A., Tomoda, H., & Ōmura, S. (2005). A γ-lactone form nafuredin, nafuredin-γ; also inhabits helminth complex I. *Journal of Antibiotics, 58,* 50–55.
Singh, S. B., Smith, J. L., Sabnis, G. S., Dombrowski, A. W., Schaeffer, J. M., Goetz, M. A., & Bills, G. F. (1991). Structure and conformation of opiophiobolin K and 6- epiophiobolin K from Aspergillus ustus as a nematocidal Agent. *Tetrahedron, 47,* 6931–6938.
Singleton, V. L., Bohonos, N., & Ullstrup, A. J. (1958). Decumbin, a new compound from a species of Penicillium. *Nature, 181,* 1072–1073.
Snook, C. F., Woolley, G. A., Oliva, G., Pattabhi, V., Wood, S. F., Blundell, T. L., & Wallace, B. A. (1998). The structure and function of ntiamoebin I, a proline-rich membrane-active polypeptide. *Structure, 6,* 783–792.
Stadler, M. (1993). PhD thesis, University of Kaiserslautern, Kaiserslautem, Germany.
Stadler, M., & Anke, H. (1993a). Lachnumon and lachnumol A, new metabolites with nematicidal and antimicrobial activities from the ascomycete *Lachnum papyraceum* (Karst.) Karst. I. Producing organism, fermentation, isolation and biological activities. *Journal of Antibiotics, 46,* 961–967.
Stadler, M., & Anke, H. (1993b). Lachnumon and lachnumol A, new metabolites with nematicidal and antimicrobial activities from the ascomycete *Lachnum papyraceum* (Karst.) Karst. II. Structural elucidation. *Journal of Antibiotics, 46,* 968–971.
Stadler, M., & Sterner, O. (1998). Production of bioactive secondary metabolities in the fruit bodies of macrofungi as a response to injury. *Phytochemistry, 49,* 1013–1019.
Stadler, M., Sterner, O., & Anke, H. (1993). New biologically active compounds from the nematode-trappingfungus Arthrobotrys oligospora Fresen. *Zeitschrift fur Naturforschung C, 48c,* 843–850.
Stadler, M., Anke, H., & Sterner, O. (1994a). Six new antimicrobial and nematicidal bisadolanes from the Basidiomycete *Cheimonophyllum candidissimum. Tetrahedron, 50,* 12649–12654.
Stadler, M., Anke, H., & Sterner, O. (1994b). New nematicidal and antimicrobial compounds from the Basidiomycete *Cheimonophyllum candidissimum* (Berk & Curt) Sing, I. Producing organism, fermentation, isolation, and biological activaties. *Journal of Antibiotics, 47,* 1284–1289.
Stadler, M., Mayer, A., Anke, H., & Sterner, O. (1994). Fatty acids and other compounds with nematicidal activity from cultures of basidiomycetes. *Planta Medica, 60,* 128–132.
Stadler, M., Anke, H., & Sterner, O. (1995a). Metabolites with nematicidal and antimicrobial activities from the ascomycete *Lachnum papyraceum* (Karst.) Karst. III. Production of novel isocoumarin derlvatives, isolation and biological activities. *Journal of Antibiotics, 48,* 261–266.
Stadler, M., Anke, H., & Sterner, O. (1995b). New metabolites with nematicidal and antimicrobial activities from the ascomycete Lachnum papyraceum (Karst.) Karst. IV. Structural elucidation of novel isocoumarin dervatives. *Journal of Antibiotics, 48,* 267–273.

Stadler, M., Anke, H., & Sterner, O. (1995c). Metabolites with nematicidal and antimicrobial activities from the ascomycete Lachnum papyraceum (Karst.) Karst. V. Production, isolation and biological activities of bromine-contaning mycorrhizin and lachnumon derivatives and four additional new bioactive metabolites. *Journal of Antibiotics, 48,* 149–153.

Stadler, M., Anke, H., Shan, R., & Sterner, O. (1995d). New metabolites with nematicidal and antimicrobial activities from the ascomycete Lachnum papyraceum (Karst.) Karst. VI. Structure determination of non-halogenated metabolites structurally related to mycorrhizin A. *Journal of Antibiotics, 48,* 154–157.

Stadler, M., Anke, H., Shan, R., & Sterner, O. (1995e). New metabolites with nematicidal and antimicrobial activities from the Ascomycete Lachnum papyraceum (Karst.) Karst. VII. Structure determination of brominated Lachnumom and mycorrhizin A derivatives. *Journal of Antibiotics, 48,* 158–161.

Stadler, M., Anke, H., Dekermendjian, K., Reiss, R., Sterner, O., & Witt, R. (1995f). Novel bioactive azaphilones from fruit bodies and mycelial cultures of the Ascomycete Bulgaria inquinans. *Natural Products Letters, 7,* 7–14.

Stadler, M., Fouron, J. Y., Sterner, O., & Anke, H. (1995). 1,2-Dihydroxymintlactone, a new nematicidal monoterpene isolated from the Basidiomycete Cheimonophyllum candidissimum (Berk & Curt) Sing. *Zeitschrift fur Naturforschung C, 50c*, 473–475.

Sterner, O., Bergman, R., Kihlberg, J., & Wickberg, B. (1985). The sesquiterpenes of *Lactarius Vellereus* and their role in a proposed chemical defense system. *Journal of Natural Products, 48,* 279–288.

Stener, O., Etzel, W., & Mayer, A. (1997). Omphalotin, a new cycle peptide with potent nematicidal activity from *Omphalotus olearius*. II. Structure elucidation. *Natural Products Letters, 10,* 33–38.

Sun, J. H. (1997). The control of growth and reproduce to *Bursaphelenchus xylophilus* by fungal cultures. *Acta Scientiarum Naturalium University Nankaiensis, 30,* 82–85.

Sun, J., Wang, H., Lu, F., Du, L., & Wang, G. (2008). The efficacy of nematicidal strain *Syncephalastrum racemosum*. *Annals of Microbiology, 58,* 369–373.

Suzuki, Y., Tanaka, H., Aoki, H., & Tamura, T. (1970). Ascotoxin (decumbin), a metabolite of Ascochyta imperfecta. *Agricultural and Biological Chemistry, 34,* 395–413.

Tarbell, D. S., Carman, R. M., Chapman, D. D., Huffman, K. R., & McCorkinadale, N. J. (1960). The structure of Fumagillin. *Journal of the American Chemical Society, 82,* 1005–1007.

Thirumalachur, M. J. (1968). Antiamoebin, a new antiprotozoalantihelmintic antibiotic. Part I. Production and biological studies. *Hindustan Antibiotics Bulletin, 10,* 287–289.

Thorn, R. G., & Barron, G. L. (1984). Carnivorous mushrooms. *Science, 224,* 76–78.

Tomoda, H., Nishida, H., Huang, X. H., Masuma, R., Kim, Y. K., & Ōmura, S. (1992). New cyclodepsipeptides, enniatins D, E and F produced by *Fusarium* sp. FO-1305. *Journal of Antibiotics, 45,* 1207–1215.

Trofast, J., & Wickberg, B. (1977). Mycorrhizin A and chloromycorrhizin A, two antibiotics from mycorrhizal fungus of *Monotrapa hypopitys* L. *Tetrahedron, 33,* 875–879.

Trofast, J. (1978). Chloromycorrhizinol A, a furochroman from an isolate of the roots of *Monotropa hypopitys*. *Phytochemistry, 17,* 1359–1361.

Tsipouras, A., Adefarati, A. A., Tkacz, J. S., Frazier, E. G., Rohrer, S. P., Birzin, E., Rosegay, A., Zink, D. L., Goetz, M. A., Singh, S. B., & Schaeffer, J. M. (1996). Ophiobolin M and analogues, noncompetitive inhabitors of ivermectin binding with nematocidal activity. *Bioorganic & Medicinal Chemistry, 4,* 531–536.

Tzean, S. S., & Liou, J. Y. (1993). Nematophagous resupinate basidiomycetous fungi. *Ecology and Epidemiology, 83,* 1015–1020.

Ui, H., Shiomi, K., Yamaguchi, Y., Masuma, R., Nagamitus, T., Takano, D., Sunazuka, T., Namikoshi, M., & Ōmura, S. (2001). Nafuredin, a novel inhibitor of NADH-fumarate reductase, produced by *Aspergillus niger* FT-0554. *Journal of Antibiotics, 54,* 234–238.

Van Dessel, P., Coyne, D., Dubois, T., De Waele, D., & Franco, J. (2011). In vitro nematicidal effect of endophytic *Fusarium oxysporum* against *Radopholus similis*, *Pratylenchus goodeyi* and *Helicotylenchus multicinctus*. *Nematropica, 41,* 154–160.

Venkatasubbaiah, P., & Chilton, W. S. (1991). Toxins produced by the dogwood anthracnose fungus *Discula* sp. *Journal of Natural Products, 54,* 1293–1297.
Verdejo-Lucas, S., Viera, A., Stchigel, A. M., & Sorribas, F. J. (2009). Screening culture filtrates of fungi for activity against *Tylenchulus semipenetrans*. *Spanish Journal of Agricultural Research, 7,* 896–904.
Vurro, M., Evidente, A., Andolfi, A., Zonno, M. C., Giordano, F., & Motta, A. (1998). Brefeldin A and α,β-dehydrocurvularin, two phytotoxins from *Alternaria zinniae*, a biocontrol agent of *Xanthium occidentale*. *Plant Science, 138,* 67–79.
Wagner, C., Anke, H., & Sterner, O. (1998). Rubrobramide, a cytotoxic and phytotoxic metabolite from *Cladobotryum rubrobrunnescens*. *Journal of Natural Products, 61,* 501–502.
Wang, J. F., Huang, Y. J., Fang, M. J., Zhang, Y. J., Zheng, Z. H., Zhao, Y. F., & Su, W. J. (2002). Brefeldin A, a cytotoxin produced by *Paecilomyces* sp. and *Aspergillus clavatus* isolated from *Taxus mairei* and *Torreya grandis*. *FEMS Immunology and Medical Microbiology, 34,* 51–57.
Watts, R., Dahiya, J., Chandhary, K., & Tauro, P. (1988). Isolation and characterization of a new antifungal metabolite of *Trichoderma reesei*. *Plant and Soil, 107,* 81–84.
Wei, L. X., Zhang, H. X., Tan, J. L., Chu, Y. S., Li, N., Xue, H. X., Wang, Y. L., Niu, X. M., Zhang, Y., & Zhang, K. Q. (2011). Arthrobotrisins A-C, oligosporons from the nematode-trapping fungus *Arthrobotrys oligospora*. *Journal of Natural Products, 74,* 1526–1530.
William, A. A., Rudong, S., Latchezar, S. T., & Leonard, J. H. (1998). Sesquiterpenes from the nematicidal fungus *Clitocybula oculus*. *Phytochemistry, 49,* 589–592.
Willis, A. J., & Thomas, M. B. (1998). Biocontrol-risky but necessary? *Trends in Ecology & Evolution, 13,* 325–329.
Wirth, J. C., Gilmore, T. E., & Noval, J. J. (1956). Penicillic acid, the antibiotic responsible for the activity of culture filtrate of a strain of *Penicillium martensii*. *Archives of Biochemistry and Biophysics, 63,* 452–453.
Xiang, H. Q., & Feng, Z. X. (2001). The nematicidal toxicity of the fruits bodies of *Pleurotus ostreatus*. *Journal of Shenyang Agricultural University, 32,* 173–175.
Xu, L., Wang, J., Zhao, J., Li, P., Shan, T., Wang, J., Li, X., & Zhou, L. (2010). Beauvericin from the endophytic fungus, *Fusarium redolens*, isolated from *Dioscorea zingiberensis* and its antibacterial activity. *Natural Product Communications, 5,* 811–814.
Yamazaki, M., Okuyama, E., Kobayashi, M., & Inoue, H. (1981). The structure of paraherquamide, a toxic metabolite from *Penicillium Paraherquei*. *Tetrahedron Letters, 22,* 135–136.
Yan, X. N., Sikora, R. A., & Zheng, J. W. (2010). Activities of seven fungal endophytes from cucumber plants to the second stage juveniles of meloidogyne incognita. *Chinese Journal of Biological Control, 26,* 181–185.
Yang, Z. S. (2008). Screening of nematicidal activities of *Trichoderma* spp. and the isolation and identification of active compounds. Master thesis, laboratory for conservation and utilization of bio-resource, Yunnan University, Kunming, China.
Yang, Z. S., Li, G. H., Zhao, P. J., Zheng, X., Luo, S. L., Li, L., Niu, X. M., & Zhang, K. Q. (2010). Nematicidal activity of *Trichoderma* spp. and isolationof an active compound. *World Journal of Microbiology and Biotechnology, 26,* 2297–2302.
Yang, Z. S., Yu, Z. F., Lei, L. P., Xia, Z. Y., Shao, L., Zhang, K. Q., & Li, G. H. (2012). Nematicidal effect of volatiles produced by *Trichoderma* sp. *Journal of Asia-Pacific Entomology, 15,* 647–650.
Yao, T., Liang, Z. Q., Mo, M. H., & Wang, C. Y. (2006). Nematicidal activity of *Paecilomyces* isolates from Yunnan province. *Chinese Journal of Biological Control, 22,* 226–229.
Yuan, Y. P., Xiang, M. M., Xi, P. G., & Jiang, Z. D. (2010). Screening of Nematicidal isolates from the Endophytic Fungi of Quisqualis indica and optimization of culture conditions for the isolate. *Chinese Journal of Biological Control, 26,* 474–479.
Zhang, Y. (2004). Master thesis, TianJin University of Science and Technology, TianJin, China.
Zhang, J. P., & Zhao, B. G. (2003). Influences of fungi isolated from dead pine trees ans other wood decaying fungi on the population growth of *Bursaphelenchus xylophilus*. *Journal of Fujian College of Forestry, 23,* 245–248.

Zhang, Y. G., Yuan, W. P., Xia, X. Q., Liu, X., Meng, X. M., Wang, X. J., Zhang, J. S., & Liu, C. H. (2010). Isolation and identification of the nematicidal secondary metabolites from one strain of entomogenous fungi. *Chinese Journal of Pesticide Science, 12,* 225–228.

Zhao, Z. X. (2004). Master thesis, laboratory for conservation and utilization of bio-resource, Yunnan University, Kunming, China.

Zhou, Y. P., Shen, K. Z., Dong, J. Y., Wang, L. M., Wang, C. R., Wang, L., & Zhang, K. Q. (2009). Nematicidal metabolites of the aquatic fungus *Coelomycetes* sp. YMF1.01029. *Chinese Journal of Antibiotics, 34,* 74–78.

Zhu, Y. H., Dong, J. Y., Wang, L., Li, L., He, H. P., Liu, H. Y., & Zhang, K. Q. (2008). Screening and isolation of antinematodal metabolites against *Bursaphelenchus xylophilus* produced by fungi. *Annals of Microbiology, 58,* 375–380.

Chapter 8
Future Study

Lianming Liang, Chenggang Zou and Ke-Qin Zhang

Abstract Previous studies based on traditional techniques have revealed much about nematophagous fungi. Modern techniques however, can now help to elucidate the molecular mechanisms underlying infection of nematodes by the nematophagous fungi, and help us to understand virulent factors, the role of proteases, chitinases and small chemical molecules, and the regulation of trap formation. Further studies using modern molecular techniques are needed to better understand these important mechanisms and the overall functioning of nematode trapping fungi.

Keywords Molecular mechanisms • Virulence factors • Serine protease • Trap formation

While previous studies have broadened our understanding of the evolution, ecology, and infection of nematophagous fungi, there are several mechanisms in the biology of nematode-trapping fungi that need further study. Investigation of these issues in the future will lead to a better understanding of the infection of nematodes by the nematophagous fungi and lead to new generation of fungal nematicides.

The molecular mechanisms of nematophagous fungi against nematodes are limited to partial understanding of only a very few genes, such as subtilisin-like serine proteases and chitinases, involved in pathogenic process. However, the functions and regulation of these genes during infection of nematodes are not very clear. For example, the available evidence indicates that subtilisin-like serine proteases play an important role in the infection of nematodes. It has been shown that the expression of these serine proteases is stimulated by the addition of nematode cuticle to the medium (Ahman et al. 1996). However, some genes encoding subtilisin-like serine proteases are present in nematophagous fungi. For instance, there are 20 subtilisin-like serine protease genes in the genome of the nematophagous fungus *Arthrobotrys oligospora*. Our results revealed that only two serine protease (designated P12 and P186) gene expressions were markedly induced by nematode cuticles in *A. oligospora*. In contrast, the expression of other protease genes was not influenced by nematode cuticles. These observations are similar to the results of Fekete et al.

K.-Q. Zhang (✉) · L. Liang · C. Zou
Laboratory for Conservation and Utilization of Bio-Resources, and Key Laboratory of Microbial Diversity in Southwest China, Ministry of Education, Yunnan University, 650091 Kunming, Yunnan, China
e-mail: kqzhang1@ynu.edu.cn

K.-Q. Zhang, K. D. Hyde (eds.), *Nematode-Trapping Fungi,* Fungal Diversity Research Series 23, DOI 10.1007/978-94-017-8730-7_8,

(2008a). These authors found that apart from two serine proteases (spr1 and spr2), no other serine protease-encoding genes were identified among the infection-regulated genes in the nematophagous fungus *Monacrosporium haptotylum* (Fekete et al. 2008a). Disruption of the most highly up regulated gene (P186) significantly reduced the degradation of nematode cuticle by *A. oligospora* and pathogenicity of *A. oligospora* to the nematode. The functions of these protease genes remain unclear. The evolution analysis revealed that these genes encoding serine proteases originated from an ancestral gene. The origin of these genes encoding serine proteases resulted from gene duplication. However, the analysis does not answer the question why the *A. oligospora* genome contains high levels of serine proteases. The analysis of signal peptide prediction indicates that each of remaining18 serine proteases in *A. oligospora* has a signal peptide in its N-terminus, suggesting that most of them are secretary proteins. This raises a possibility of functional redundancy in degradation of the nematode cuticle. The Blast of protein sequences reveal that one of these genes (designated P4) has a higher similarity to serine proteases in fungi involved in root location, implicating a mycorrhizal role. Furthermore, one of these protease genes (designated P42) has a higher similarity to serine proteases ATG1 in yeast and fungi, suggesting that it probably plays a role in autophagy. Thus, these genes probably have distinct functions which are different from those involved in degradation of nematode cuticles. Chitinases in nematophagous fungi have been shown to inhibit the development of the root-knot nematodes by degrading chitinous components of their eggs (Gan et al. 2007; Huang et al. 2004). A previous study identified three chitinases (cr-ech58, cr-ech42 and cr-ech37) from the nematophagous fungus *Clonostachys rosea* (Mamarabadi et al. 2008). Only cr-ech42 and cr-ech37 expression is induced by colloidal chitin. Since cr-ech58 is not influenced by chitin, it plays less important role for the antagonistic function and instead perhaps has other roles in morphogenesis and in the breakdown of cell walls in *C. rosea* during growth. However, there are 16 chitinase genes in the genome of the nematophagous fungus *A. oligospora*. Thus, the situation for chitinases is similar to that for serine proteases. The function of most serine proteases and chitinases remains unknown.

With the exception of serine proteases and chitinases, the virulence factors for pathogenicity of nematophagous fungi to nematodes need to be elucidated. Tunlid and his colleagues have studied the transcriptional response in the parasitic fungus *M. haptotylum* during adhesion, penetration and digestion with *Caenorhabditis elegans* using cDNA microarrays (Fekete et al. 2008b). These authors identified two up regulated genes during infection. However, more experiments are needed to clarify which fungal genes encompass the virulence factors during infection. After completion of the genome sequence of more nematophagous fungi, other virulent factors will be revealed by combining with the application of emerging and novel genomic (e.g. fungal gene deletion and silencing) and proteomic strategies. Identification of the virulence genes will provide further insights into the molecular basis of infection process.

The molecular mechanism underlying the formation of trap organs in the nematode-trapping fungi induced by nematodes needs to be elucidated. Prey on nematodes by the nematophagous fungi is considered as an adaptive strategy in nutrient-poor habitats, since direct assimilation of nitrogen sources from nematode seems to be an advantage (Schmidt et al. 2008). Under normal conditions, the nematode-trapping fungi grow as saprotrophs in soil. After the induction of nematodes, they enter the parasitic stage by developing specific morphological structures called traps, including adhesive three-dimensional nets, constricting rings, branches and knobs (Dijksterhuis et al. 1994). A previous study tried to investigate the genes involved in knob formation by comparison of gene expression in trap cells and vegetative hyphae of the parasitic fungus *M. haptotylum* during infection by *C. elegans* using cDNA microarrays (Ahren et al. 2005). Although many genes differentially expressed in knobs versus mycelium of *M. haptotylum* were identified, it remains unclear whether these genes were involved in knob formation due to the lack of genetic testing. The recent completion of *A. oligospora* genome sequence has fuelled the potential for further mechanism studies. We are studying the differential expression of genes in the presence and absence of nematode cuticles using two-dimension and RT-PCR techniques. Based on these results, we propose a model of trap formation in *A. oligospora* after nematode induction. The majority of nematode-trapping fungi are closely related (Ahren et al. 1998). Thus, it could be speculated that the infection mechanism appears to be rather similar. However, it is still unknown whether the proposed model in *A. oligospora* can be used to predict trap formation in the other nematophagous fungi. After completion of the genome sequence of other nematode-trapping fungi, clarifying the common pathways involved in the trap formation may contribute to our understanding of the mechanism underlying the transition between saprotrophic and predatory lifestyles in fungi.

Nematophagous fungi produce chemical substances that lure nematodes to the site of trap formation in predatory species or to the site of spores in the case of endoparasites. It is a key factor for effectively parasitizing. The accumulation of both plant-parasitic and non-parasitic nematodes around plant roots and fungal and bacterial colonies is well known. Several factors, such as temperature, electrical potential, carbon dioxide and various organic and inorganic substances, are thought to be involved (Croll 1970; Green 1971). The attraction of nematodes to fungi has been investigated primarily to show the host-finding mechanisms of fungus-feeding nematodes (Klink et al. 1970; Townshend 1964). Researchers have further demonstrated the attraction of nematodes to culture filtrates and living mycelia of several nematophagous fungi. It was suggested that certain exudative and volatile compounds, which were produced by living fungal mycelia, including most nematophagous fungi, might be responsible for the attraction (Balan et al. 1976; Field and Webster 1977; Jansson 1982; Jansson and Nordbring-Hertz 1979, 1980; Monoson et al. 1973). Recently, the attraction of pinewood nematode was studied. It was confirmed that the living mycelia of *Esteya vermicola* were attractive not only to pinewood nematodes *in vivo*, but also to that in the discs of infected pine seedling, dead blocks of infected pine tree, which may provide information for the application of *E. vermicola* as biological control agent of pinewood nematode (Wang et al. 2009). Future study of nematode attraction should focus on the identification of attracting substances secreted by nematophagous fungi, or as well as hosts of plant-parasitic nematode, i.e., root tissues. A good example is the study of infection mechanism of a nematode-parasitic bacteria (Niu et al. 2010). The regulation of the secretion pathways should also be studied when the genomic data for nematophagous fungi is available (Yang et al. 2011a).

The molecular mechanism of biogenesis of the compounds with nematicidal activities secreted by nematophagous fungi should be further studied. Until now, more than 200 compounds secreted by about 270 species fungi of 140 genera have been purified and characterized and are able to kill nematodes effectively (see Chapter 7 for details). These substances could be commercially developed, if they have high activity but are environmentally friendly. Many more compounds are likely to be discovered, especially from nematode trapping fungi in special habitats, such as extreme environments or those with endophytic lifestyles.

Studying the biogenesis of these compounds may give clues for developing high production strains using gene engineering strategies. The first genome of nematophagous fungi (*A. oligospora*) has been sequenced (Yang et al. 2011a). This genome contains five putative polyketide synthetase (PKS) gene clusters and seven nonribosomal peptide synthetase (NRPS) genes. These genes may be responsible for the synthesis of the nematicidal compounds. Study should become easier when the genetic background is clear.

The first generation of fungal nematicides including Royal 300, Royal 350, Biocon (Bioact^{R}WG), Ditera, Deny and ZK-7 have been used in several countries (Mo and Mou 2009). They were obtained by screening thousands of candidate strains. The main disadvantages of these fungal nematicides are low and unsteady efficacy. To solve these problems, the next generation of fungal nematicides should be modified by gene engineering. Firstly, fungal strains expressing external toxic proteins (proteases, chitinases) were found to have high activity (Ahman et al. 2002; Yang et al. 2011b), and other toxic genes like BT and scorpion neurotoxin, may also be used to elevate nematicidal activity (Marroquin et al. 2000; Wang and St Leger 2007). Secondly, for survival in the host body against the host immune system, some protective coat proteins may be helpful (Wang and St Leger 2006). Thirdly, the nematode-trapping fungi with various trapping-organs have great potential as biocontrol agents, but the low level of sporulation restricts their utility because of the low abundance in the environment. Recently, some genes suppressing sporulation have been characterized, and thus enhanced sporulation may be acheived by knocking out these genes (Sun et al. 2011). Although at present transgenic strains are restricted for use by governmental policy, the utility of the next generation of fungal nematicides have obvious advantages over the previous generation and chemical nematicides. As the study of fungal infection mechanisms against nematodes continues, the next generation of fungal nematicides with high and steady efficacy are not far away.

References

Ahman, J., Ek, B., Rask, L., & Tunlid, A. (1996). Sequence analysis and regulation of a gene encoding a cuticle-degrading serine protease from the nematophagous fungus *Arthrobotrys oligospora*. *Microbiology (Reading, England), 14,* 1605–1616.

Ahman, J., Johansson, T., Olsson, M., Punt, P. J., van den Hondel, C. A., & Tunlid, A. (2002). Improving the pathogenicity of a nematode-trapping fungus by genetic engineering of a subtilisin with nematotoxic activity. *Applied and Environmental Microbiology, 68,* 3408–3415.

Ahren, D., Ursing, B. M., & Tunlid, A. (1998). Phylogeny of nematode-trapping fungi based on 18S rDNA sequences. *FEMS Microbiology Letters, 158,* 179–184.

Ahren, D., Tholander, M., Fekete, C., Rajashekar, B., Friman, E., Johansson, T., & Tunlid, A. (2005). Comparison of gene expression in trap cells and vegetative hyphae of the nematophagous fungus *Monacrosporium haptotylum. Microbiology (Reading, England), 151,* 789–803.

Balan, J., Krizkova, L., Nemec, P., & Kolozsvary, A. (1976). A qualitative method for detection of nematode attracting substances and proof of production of three different attractants by the fungus *Monacrosporium rutgeriensis. Nematologica, 22,* 306–311.

Croll, N. A. (1970). *The behaviour of nematodes: Their activity, senses and responses.* London: Edward Arnold Ltd.

Dijksterhuis, J., Veenhuis, M., Harder, W., & Nordbring-Hertz, B. (1994). Nematophagous fungi: Physiological aspects and structure-function relationships. *Advances in Microbial Physiology, 36,* 111–143.

Fekete, C., Tholander, M., Rajashekar, B., Ahrén, D., Friman, E., Johansson, T., & Tunlid, A. (2008a). Paralysis of nematodes: Shifts in the transcriptome of the nematode-trapping fungus *Monacrosporium haptotylum* during infection of Caenorhabditis elegans. *Environmental Microbiology, 10,* 364–375.

Fekete, C., Tholander, M., Rajashekar, B., Ahrén, D., Friman, E., Johansson, T., & Tunlid, A. (2008b). Paralysis of nematodes: Shifts in the transcriptome of the nematode-trapping fungus *Monacrosporium haptotylum* during infection of Caenorhabditis elegans. *Environmental Microbiology, 10,* 364–375.

Field, J. I., & Webster, J. (1977). Traps of predacious fungi attract nematodes. *Transactions of the British Mycological Society, 68*, 467–469.

Gan, Z., Yang, J., Tao, N., Liang, L., Mi, Q., Li, J., & Zhang, K. Q. (2007). Cloning of the gene Lecanicillium psalliotae chitinase Lpchi1 and identification of its potential role in the biocontrol of root-knot nematode *Meloidogyne incognita. Applied Microbiology and Biotechnology, 76,* 1309–1317.

Green, C. D. (1971). Mating and host finding behaviour of plant nematodes. In B. M. Zuckermawn, W. F. Mai, & R. A. Rohd (Eds.), *Plant parasitic nematodes*, (Vol II pp. 247–266). New York: Academic Press.

Huang, X., Zhao, N., & Zhang, K. Q. (2004). Extracellular enzymes serving as virulence factors in nematophagous fungi involved in infection of the host. *Research in Microbiology, 155,* 811–816.

Jansson, H. B. (1982). Attraction of nematodes to endoparasitic nematophagous fungi. *Transactions of the British mycological Society, 79,* 25–29.

Jansson, H. B., & Nordbring-Hertz, B. (1979). Attraction of nematodes to living mycelium of nematophagous fungi. *Journal of General Microbiology, 112,* 89–93.

Jansson, H. B., & Nordbring-Hertz, B. (1980). Interactions between nematophagous fungi and plant-parasitic nematodes: Attraction, induction of trap formation and capture. *Nematologica, 26,* 383–389.

Klink, J. W., Dropkin, V., & Mitchell, J. (1970). Studies on the host-finding mechanisms of Neotylenchus linfordi. *Journal of Nematology, 2,* 106–117.

Mamarabadi, M., Jensen, B., & Lübeck, M. (2008). Three endochitinase-encoding genes identified in the biocontrol fungus Clonostachys rosea are differentially expressed. *Current Genetics, 54,* 57–70.

Marroquin, L. D., Elyassnia, D., Griffitts, J. S., Feitelson, J. S., & Aroian, R. V. (2000). *Bacillus thuringiensis* (Bt) toxin susceptibility and isolation of resistance mutants in the nematode *Caenorhabditis elegans. Genetics, 155,* 1693–1699.

Mo, M., & Mou, G. (2009). Control of plant parasitic nematode by using Nematophagous fungi. *Journal of Dali University, 8,* 71–76.

Monoson, H. L., Galsky, A. G., Griffin, J. A., & McGrath, E. J. (1973). Evidence for and partial characterization of a nematode attraction substance. *Mycologia, 65*(1), 78–86.

Niu, Q., Huang, X., Zhang, L., Xu, J., Yang, D., Wei, K., Zou, C. (2010). A Trojan horse mechanism of bacterial pathogenesis against nematodes. *Proceedings of the National Academy of Sciences of United States of America, 107,* 16631–16636.

Schmidt, L. M., Mouton, L., Nong, G., Ebert, D., & Preston, J. F. (2008). Genetic and immunological comparison of the cladoceran parasite *Pasteuria ramosa* with the nematode parasite *Pasteuria penetrans. Applied and Environmental Microbiology, 74,* 259.

Sun, X., Zhang, H., Zhang, Z., Wang, Y., & Li, S. (2011). Involvement of a helix-loop-helix transcription factor CHC-1 in CO_2-mediated conidiation suppression in *Neurospora crassa. Fungal Genetics and Biology, 48,* 1077–1086.

Townshend, J. (1964). Fungus hosts of *Aphelenchus avenae* Bastian, 1865 and *Bursaphelenchus fungivorus* Franklin & Hooper, 1962 and their attractiveness to these nematode species. *Canadian Journal of Microbiology, 10,* 727–737.

Wang, C., St Leger R. J. (2006). A collagenous protective coat enables *Metarhizium anisopliae* to evade insect immune responses. *Proceedings of the National Academy of Sciences of United States of America, 103,* 6647–6652.

Wang, C., St Leger R. J. (2007). A scorpion neurotoxin increases the potency of a fungal insecticide. *Nature Biotechnology, 25,* 1455–1456.

Wang, C. Y., Wang, Z., La Lee, M., Li, Z., Zhang, D. L., Liu, L., Sung, C. K. (2009). Attraction of pinewood nematode, *Bursaphelenchus xylophilus*, to the endoparasitic fungus *Esteya vermicola. African Journal of Microbiology Research, 3,* 637–640.

Yang, J., Wang, L., Ji, X., Feng, Y., Li, X., Zou, C., Wu, J. (2011a). Genomic and proteomic analyses of the fungus *Arthrobotrys oligospora* provide insights into nematode-trap formation. *PLoS Pathogens, 7,* e1002179.

Yang, J., Zhao, X., Liang, L., Xia, Z., Lei, L., Niu, X., Zhang, K. Q. (2011b). Overexpression of a cuticle-degrading protease Ver112 increases the nematicidal activity of *Paecilomyces lilacinus*. *Applied Microbiology and Biotechnology, 75,* 21–31.

Index

Note: Page numbers followed by "*f*", "*t*" indicate figures and tables respectively

K.-Q. Zhang, K. D. Hyde (eds.), *Nematode-Trapping Fungi,* Fungal Diversity Research Series 23, DOI 10.1007/978-94-017-8730-7, © Mushroom Research Foundation 2014

Printed by Printforce, the Netherlands